GOD, SCIENCE, AND THE PROBLEM OF FAITH

How Modern Science Debunks the Tenets of Religion and Why People Still Believe

Corey D. Fox, Ph.D.

SPRING CEDARS®

First edition, 2026

Book cover and design by Spring Cedars

ISBN 978-1-963117-92-9 (paperback)
ISBN 978-1-963117-93-6 (hardcover)
ISBN 978-1-963117-94-3 (ebook)

Published by Spring Cedars
Denver, Colorado
www.springcedars.com

CONTENTS

PREFACE

This book is about gods, science, and irrational beliefs. Questions surrounding belief and atheism have been around for a few thousand years. Throughout history, there have been countless books on the matter of gods, and many on atheism have appeared in the last few decades. Intense debates about such things continue in academic circles, professional journals, popular media, and living rooms.

I have always been preoccupied with the question of why people believe in God, despite there being no actual evidence for it. That is what this book is about—why the belief is so strongly held and why it is irrational. I have been inspired, of course, by the works of Richard Dawkins, Christopher Hitchens, Daniel Dennett, and Sam Harris. Nevertheless, as a psychologist, I came to believe that behavioral science has something unique to offer about these issues.

I must also cite the work of the Polish mathematician and astronomer Nicolaus Copernicus, who demonstrated that the Earth orbits the Sun and not the reverse—that we don't occupy any special place in this Universe. Perhaps we're not special in any other way. In my first college philosophy course, we were assigned to write a paper on "What does it mean to be alive?" I realized that one would first have to look at the question of "What is life?" This led me to an examination of viruses and the question of whether these tiny things are alive or not. The scientific debate on this continues. But perhaps the boundary between living and non-living things is not as stark as most people believe. Perhaps "life" is simply one type of physical and chemical phenomena in our universe, which bears on the issue of how special or unique we humans are. That perspective, which has become known as the Copernican Principle, casts serious doubt on the notion of the existence of God. This book delves into that issue in a number of ways.

There is a long history to the connections among religion, human cruelty, and war. This affected me even more personally as I provided psychological help for people and various organizations in New York after the religiously motivated tragedy of 9/11. I saw firsthand how faith in a god can lead to incredible destruction, suffering, and death. My decades of clinical experience have also led me to this point. As a health psychologist, I have treated patients with serious diseases, chronic intractable pain, and severe disabilities, and I've seen how faith and religion in general offer people only temporary relief and false hope.

What I've done here is to address a number of seemingly independent ideas that are relevant to the question of whether any supernatural entity could possibly exist. These are not new subjects, of course; but I have tried to bring together the major arguments that bear on this question. But I've also introduced new information—from neuroscience and cognitive and behavioral science—to drill down on why people still believe in gods, despite there being no evidence for such faith.

This book deals with a number of subjects which are somewhat alien to most readers—physics, cosmology, biology, anthropology, and behavioral and neuroscience. This is not a text on those disciplines, merely an examination of them in relation to the problem of gods and why people continue to believe. Some of this will get a bit technical, but will be understandable to most readers. For the interested reader, I've provided references to the most useful sources on all these subjects.

I am indebted to a few people for their help and encouragement in writing this book. My family has been incredibly supportive in this long journey. I am indebted to my long-time friend and colleague, Dr. Richard Geisler, who has helped to inspire many of my ideas. I'm especially grateful for the late professor Carl Pitts, who taught me that the behavioral approach is the best-evidenced and the most humane way of understanding human behavior and coping with our problems.

1

SOME VERY BIG QUESTIONS

The pivot round which the religious life, as we have traced it, revolves is the interest of the individual in his private personal destiny. Religion, in short, is a monumental chapter in the history of human egotism. —William James[1]*

The whole problem with the world is that fools and fanatics are always so certain of themselves, but wiser people so full of doubts. —Bertrand Russell

Some people are sure God exists. Many affirm there is no God. Others have doubts. Some are ambivalent. Some people just go about their lives and don't care about the question. This book is about addressing these issues, about religion and gods, and about some big questions—some very big questions. Is there a god? Why do people believe in gods? Where did everything come from? Why am I here? There is also the notion that, in the grand scheme of things, are we human beings somehow special and unique, and has some God so endowed us? These are the biggest questions humans have dealt with throughout history. But they all come down to the problem of whether any supernatural being actually exists.

My answer: This book is about the irrationality of belief in God and why many people continue to believe anyway. Humans do show a variety of irrational beliefs, but belief in the fantasy of a god is a particularly pervasive and dangerous one, as history well attests. We'll examine in detail why these beliefs are specifically irrational and why religious behavior is so pervasive in our culture. We'll see that no god is necessary to explain our world, our universe, or ourselves. And critically, we'll see that using "common sense" to understand some difficult concepts about our universe and the idea of God is just not possible.

This is an incredibly ambitious endeavor—to say the least. Nothing will be absolutely proven here. My purpose is simply to bring together and integrate various perspectives and lines of evidence to suggest that our reality does not necessitate the existence of a god, that we human beings are not as special as we think, and that the idea of an omnipotent god is literally nonsensical. To do this, we'll look at the relevant issues, from

---

*Superscript numbers refer to Notes. Numbers in parentheses refer to References.

cosmology to morality, from philosophy to quantum mechanics, from evolutionary biology to neuroscience and free will, from anthropology to human behavioral science, and a few others. Looking at such a broad swath of fields would seem off the point; but a question we're concerned with here is why human beings actually believe in gods.

These topics have been addressed by numerous scientists and countless philosophers throughout history; and, for good reason, you will see a number of them cited and quoted here. We do advance our understanding principally on the shoulders of others. However, along with integrating various fields of knowledge, believe it or not, there are some new information and perspectives I introduce here; and I believe that these will shed additional light on the god problem and humans' continuing belief.

Why Do You Believe?

Why do people believe in gods? In this book, I will describe what is essentially *the evolution of the human conception of a god and how it is maintained*. That is, we will review human, and prehuman, evolution and the factors that led to such superstitious behavior, their associated rituals, and humans' invention of deities to explain events in their environments. We'll see how the idea of a supernatural being came into existence purely through human biological and cultural evolution and that therefore we must consider such an idea as merely human behavior. I show here that what is proposed and worshiped as God is really no more than a human idea or fantasy, whose continuation is purely a function of known natural laws, including those governing human behavior.

A further provocative issue is, "Do you decide to believe or not to believe?" Perhaps that is *the* question. Faith in a supreme being is often touted as a "choice" to believe in a god or other entity. We'll see that you do not have such a choice. The question is generally defined as the problem of "free will." I'll argue in Chapter 6 that such a choice or "free will" is, in fact, an illusion. I'll debunk the idea of free will, as this is commonly construed, and show that your faith or lack of faith has already been determined. Relatedly, belief or faith, we'll see, is not some ethereal sense. It is human behavior. We'll see that human behavior, including religious behavior, has nothing at all to do with any god or supernatural power. I'll demonstrate that we "believe" simply and only because we are conditioned to do so.

What Do We Mean by "God?"

In order to discuss this topic, it might be reasonable to define what is meant by the term "god." Unfortunately, this is impossible. It depends on who you ask, what religion they follow, what sect or variety is endorsed, and people's personal conceptions. We'll get into this in some detail in later chapters.

Nevertheless, if we are to have any reasonable discussions about this, we need to have some common frame of reference—some ability to actually communicate about this. Communication among humans about such matters leaves a great deal to be desired. Yet, we need to have some common language that allows us to communicate in a way that we might understand each other. When we refer only to subjective experiences, we are lost. Everyone's experience, by definition, is unique and inaccessible to others. We need to be able to share language and other communication which have referents that we can agree to, so I'll briefly attempt to address this matter.

When we talk about a god or gods, there is an important distinction, and this will be important throughout the book. There is the idea that there was a supernatural force that "created" the universe; and then did nothing after that. This is called *deism.* No specific or personal character is generally assigned to this "force"; and, when it's discussed, the term is just left without further clarification. Then there is the notion of what's referred to as a "personal god," some sort of "being" who not only creates the universe but directs its operation and takes an active interest in the affairs of human beings. This is generally referred to as *theism.* I will focus my comments principally on the theistic notion of a god who was the putative creator of the universe and all reality, including us—principally the God of the Judeo-Christian tradition. We'll also have to consider, of course, some of the other gods, with similar or various other powers, which have appeared throughout history.

Language

Throughout the text, I use the term 'god' or 'gods' to denote what we're generally talking about. I have also often not capitalized the word 'god' as a generic noun, since we are indeed talking just about an idea, even though the concept of a god is specifically undefinable (or at least it is an idea with various definitions). In fact, since 'god' is often an idea, not a person or personage, there is no need for a proper noun for this used generically. I make exceptions and capitalize the term where the reference is to a specific character, such as Osiris or Zeus or Allah or the God of Judeo-Christian mythology. We capitalize the names of all sorts of characters, including those from myth or other literature, even though they do not exist in reality. I capitalize the Biblical "God" much as we do Rumpelstiltskin and Cinderella or Odysseus or Aladdin or Santa Claus. Characters of fantasy are all proper nouns. As we develop the arguments, when capitalizing God, remember that such doesn't imply an actual existence or reality of such a being. We can't avoid the awkwardness of this, but remember that it's all merely a grammatical convention.

I also use articles, adjectives, or names throughout to signify that what we are talking about are concepts or ideas—not one individual "being." While the theme of this book is that certainly no such being exists, just to keep the discussion flowing, I do need to use pronouns, such as "he" or "she," as ordinary. Modern parlance sometimes uses he/she, which always seems awkward. Of course, scriptures notwithstanding, the use of a personal pronoun to refer to an omnipotent and omniscient being is rather ludicrous. So I do this with tongue planted firmly in cheek. If we remove gender from this, it becomes "It" (You'll notice the absurdity of the idea when "It" might need to be capitalized.). It does seem nonsensical to provide a designation of gender to a putative supernatural being. It only reflects the imaginary idea that humans were created in His image, while the reverse is actually the case (which I'll get into later), along with the obvious bias toward the putative superiority of the male gender. We should rightly just use the word "It"; but for the sake of convenience and language, I'll variously use "He" or "She" or sometimes "It," understanding that the pronoun does not reflect any suggestion that such a "being" actually exists. I will also randomly vary the use of genders when talking about mundane ideas, e.g., "Her prayers weren't answered," or "He went to church," but only for the sake of simplicity and convenience. No sexual or gender issues are implied by such usage.

Obviously, the idea of a god or gods varies greatly from one culture to another, from one period in history to another, and from one person to another. But for now, we need to be clear that there is no singular god I'm referring to when I use that term generically. And you'll notice that I often refer to the concept with the phrase "a god." There is a need to be careful and use this nonspecific phrase, as people tend to hear the term "god" as equivalent to *their* god. I also use the term 'God' metaphorically, on occasion, to refer to some ideas or actions that "He" may have taken, such as noted in Bible passages or opinions espoused by theologians or other religious acolytes. "God commanded the Israelites," or "God hears my prayers," for example. But in no way should you take such comments to suggest that I'm endorsing any reality of such a being. Again, these are merely conventions of our language.

God as Substance

There are also conceptions of "god" referring to some sort of universal substance, spirit, or essence which pervades the universe and everything in it. This type of philosophy might refer to everything from animism to Buddhism to pantheism to various spiritual or other historical or contemporary beliefs. Then there is the conception that the term "god" refers essentially to the universe itself, as espoused by the seventeenth-century philosopher Baruch Spinoza, or the closely related view of Albert Einstein that god might be a term applied the beauty or awe inspired by the universe and the natural laws which describe it. I am not generally talking about these two later conceptions, though they will be referred to for clarification at times.

God as a Meme

The idea of a supernatural being has been around for thousands of years, in fact, in some form, for hundreds of thousands of years. Over the last few millennia, the idea of God, or a god, has evolved as a *meme*; that is, an idea which survives throughout a culture because of its value to that culture. The term "meme" was coined by evolutionary biologist Richard Dawkins in 1976 in the book *The Selfish Gene* (111); and it now has permeated our modern culture to signify anything from gestures to ideas to cultural traditions to internet emojis. Memes, as technically conceived, are "units of cultural inheritance." (113) Whether memes represent superstitions or not is not relevant at this point. When they are important to a culture, or any part of it, they survive, much as a gene variant survives in a pool of organisms because of its survival value to those organisms. A meme is a word, an idea, a picture, an object. It is a symbol which is loosely subject to a type of Darwinian evolution, if you will—that which survives is the one best suited to the environment in which it "lives." Memes which are functional in specific environments—cartoons, putting your hand over your heart—tend to persist. Those that lose functionality—the butter churn, the abacus or even now a slide rule, "know your onions" (meaning to be very knowledgeable about something)—*tend* to become extinct. What I'll address in various ways are the reasons that the God meme has persisted.

God of the Gaps

We humans have an inherent need—actually an incredibly strong genetic tendency—to explain things that we observe. That was a critical skill that allowed us to survive and evolve—in short, to become human. But there are many things we can't explain, almost by definition an infinity of them. And science cannot explain everything we care about. Neither can our insights gained from ordinary observations and reasoning. When we can't explain things, we often just invent explanations for them.

For several millennia, humans have been inventing explanations for things they don't understand. We have developed brains which allow us to imagine things, a skill which also allows us to invent explanations that aren't actually factual. Such fantasies have developed as a putative explanation for mysterious events. Filling in knowledge gaps about mysterious happenings has been commonplace throughout history, and they've often generated various god memes. Early humans fabricated spirits moving things around our environments. And later, these spirits evolved into angry gods punishing humans for unwanted or sinful behavior, and much later to monotheistic gods responsible for our world and everything in it.

When Darwin published *On the Origin of Species* in 1859, there was widespread discomfort with the idea. After all, God was supposed to have "created" everything and every living creature. And even though evolution may have explained a great deal, it certainly did not explain everything about how all the Earth's living creatures came to be. These gaps in our knowledge still lead to some uneasiness. When such gaps appear for scientists or other curious humans, they generate a desire to learn more. For religious apologists, they generate notable discomfort. The gaps generate fantasies and superstitions and, in the modern Judeo-Christian tradition, of course, God.

It's a bit unclear who first used the phrase "God of the Gaps" or some variant of this to describe this phenomenon. It was invoked, in one way or another, by many throughout history. It may have originated in the ancient Greek tradition of *deus ex machina*, in which a Greek god would intervene directly in some human affairs ("God from the machine," a metaphor for a god who would suddenly intervene to resolve a plot problem in a play). In modern times, it has been more formally ascribed to the 19th-century Scottish evangelist Henry Drummond. Drummond was also a biologist and clearly understood Darwin's thesis, but he also argued for the idea that evolution can be God's method of creation (still a view of some theologians today). However, he also warned that the use of God to explain the various "gaps" in our understanding of the world might eventually render God moot.

> When things are known, that is to say, we conceive them as natural, on Man's level; when they are unknown, we call them divine — as if our ignorance of a thing were the stamp of its divinity. If God is only to be left to the gaps in our knowledge, where shall we be when these gaps are filled up? And if they are never to be filled up, is God only to be found in the disorders of the world? Those who yield to the temptation to reserve a point here and there for special divine interposition are apt to forget that this virtually excludes God from the rest of the process. (140)

We still have that argument today. The concept "god of the gaps" is a strategy still invoked by Christians to overcome the idea of Darwinian evolution by natural selection. Where evolution has not explained, or cannot yet explain, something in the living world, then God is there to fill in the gaps. It is a common polemic technique employed by Creationists, but of course it explains nothing. And the "gaps" do not end with specific evolutionary details. Theists often cite seemingly ultimate gaps—advanced life on this planet is too complex to have been the result of biological evolution, so there must be a Creator who intelligently designed it. We don't know *why* we're here, so some god must have put us here, they assert; and humans are a totally unique form of life, the only explanation for this being God. They also assert that the world (actually, now the universe) was created, so there must have been a Creator.

The notion of God (or various other conceptions of an omnipotent "being") is a human fabrication. It comes about because of our human proclivity to construct explanations for things we do not understand and our limited human capacity (which will always be present) to conceive of the notion that there are things that cannot be explained with science. Science, by definition, can never explain all events over all time. It is limited in this regard, and the pursuit of knowledge, through science, is an ever-ongoing process. The "gaps" in our knowledge remain, and our knowledge will always be incomplete. Nevertheless, when seen in this empirical context, it is obvious that the "god of the gaps" is ever-shrinking, which is the threat that Drummond was concerned about.

The "God of the gaps" notion is an important superstition for our purposes, and I'll refer to it multiple times in the coming chapters. Religious apologists do not need to fill in the gaps with anything substantial. After all, God works in "mysterious" ways, and we cannot possibly understand His mind or His motivations—two terms usually reserved for human beings. That is enough "evidence" for them. There are other putative "escape hatches" for theologians; but the seeming philosophical legerdemain that they use to promote "gap" ideas are patently absurd.

We're Not So Special

We tend to believe that it is *we* humans who are special, and that we are here for a purpose, a unique purpose, and a purpose that can only be satisfied by a view of reality based on some god. The arguments boil down to the idea that we humans are somehow unique or distinct, and we are thus created by a personage outside the laws of nature. This is one of the main disputes I'll address in this book. So it goes—human beings think they are somehow special, as forms of life and even as unique creatures in this universe and that, therefore, there is a God who rendered us that way. While I enthusiastically endorse that we humans are wonderful creatures, I will formally debunk the idea of our specialness or uniqueness in a number of ways.

I will focus on both humans and other animals in this book. Yes, we'll often consider humans as animals, which they are. That may seem a bit cold or dehumanizing, and in a sense, it is. But we *are* animals, attempts by religious apologists to avoid this notwithstanding. But even when we acknowledge this, we also tend to think of ourselves as unique in the animal kingdom, comparing ourselves only to currently extant species. This is notably shortsighted. We've only been around for a couple of hundred millennia, a tiny fraction of time in the evolution of life on this planet and a trivial duration of the time since our immediate ancestors existed. When we look around, we do see ourselves as unique. But that is a narrow perspective. Human beings are indeed amazing creatures. But we do exist on a continuum of species across the billions of years our Earth has been here. We've also learned that human beings are not totally unique in the living world but have evolved from more primitive forms, beginning with simple single-celled creatures. We are not truly unique but represent only the latest form of a species along a branching continuum of primates.

The Copernican Principle

Our view of "specialness" also extends to the world around us. The Earth must be the center of the universe, what used to be only our own solar system. For most of human history, this is the view that prevailed. The Greek scholars Aristotle and Ptolemy thought all heavenly bodies revolved around the earth. It was the Polish mathematician and astronomer Nicolaus Copernicus who is famous for developing the heliocentric theory of the solar system in the sixteenth century. The Earth does orbit the sun and not vice versa, as most of us are aware, but it was Copernicus who's generally credited with the theory. Of course, Copernicus' view was a patently heretical one at the time, and his work wasn't published until he was virtually on his deathbed. His revolutionary idea about our planet, that we do not have a *privileged* place in this universe, has been extended to suggest that we are not that special in many other ways as well. We've come to know this as the *Copernican Principle*, which I'll address in various ways in many of the following chapters.

More Time and Space

We are a very parochial species. For most of us, the world, for example, is limited to our personal experience. When we refer to such ideas as "my world," we usually mean our family, our house, our group of coworkers, etc. In a larger sense, we may use that term as essentially synonymous with planet Earth. But to some ancient humans, the world was the geography their tribes could actually view or consider from their experience. Much later, it was a small area of Africa and the Middle East. Over time, humans there would have been unaware, for example, that there were people and civilizations thousands of miles away on the other side of the globe. Most of us are now aware that our planet is just a small rock orbiting an average star in an arm of an enormous spiral galaxy, which probably contains probably hundreds of billions of other stars, and that our Milky Way galaxy is but one of hundreds of billions of other galaxies in our known universe (perhaps trillions of them); and we'll see that there may, in fact, be other universes, perhaps an infinite number of them. We do not yet know what else is "out there," including other "life." Current cosmological theories open the likelihood that there are multiple universes, including ones in which other humanities exist—a blow to our specialness or unique status if there ever was one. As fantastical as all that seems, I'll get into the details of this subject in Chapter 3. Our place in the vastness of whatever reality exists is not at all special. The Copernican Principle suggests that when you take this long view, the "specialness" we infer about ourselves is only a perception. We are not at the "center" of anything, particularly when you consider space, time, or the nature and possibilities of life.

We're far from "special" as living creatures as well. It took over 13 billion years for us to come into existence in this universe, including nearly 4.5 billion years just on this planet. Someday life will cease to exist on this planet; and even if we, quite inconceivably, seed other planets, their stars will die, and human life will be extinguished there. Our so-called privileged place in the universe will again disappear. And in the distant future, the expansion of the universe will make life impossible anywhere. And some day our universe will also cease to exist as such, all of which I'll detail in Chapter 3.

We should also consider the possibility that "life" itself has more than one definition and that there may well be other planets or other realities where some sort of life exists. And now there's the possibility that progress in artificial intelligence may generate consciousness and further reduce our uniqueness. The notion that we are special and unique clearly fades into the background of history and scientific discovery.

A Little Perspective

In sum, the Copernican Principle suggests that there is nothing truly unique about us or where we live but that we are part of a continuum of space and time within which we exist. We are essentially meaningless to a vast universe. We may be "special," though only to ourselves, and not really unique, except at one point in time and space. It's incredibly parochial to consider that we are somehow "special."

We also believe we are here for some purpose. Theists hold that it is to serve God. But since we are not the center of the universe, physically, temporally, or conceptually, it is hard to accept that. The Copernican Principle suggests that understanding ourselves demands we put our perspective of uniqueness on something of a back burner. Science historian Michael Shermer puts this in perspective:

> The Copernican principle . . . holds that we are not special. We prefer religious and anthropic explanations that the universe was created and fine-tuned for us because they put humans right back in the center of the cosmos anthropocentrically—it is all about *us*. But 500 years of scientific discoveries have revealed that it isn't about us. (470) (I'll delve further into the so-called anthropic principle in Chapter 3.)

Conclusion

Throughout history there have been many—and I mean *many*—books written about gods, the origin of gods, cultural and biological theories about gods, the irrationality of belief in gods, and the destructive nature of religion. I'll be alluding to findings and opinions from these and others as we go through the relevant subject areas. There is a good deal of redundancy to the arguments and the social/cultural issues in these types of books. It is unavoidable. But there is a good deal of new material I'll address here, and we will take a detailed journey through the arguments about the existence of gods and the reasons for the perilous, widespread, irrational faith in them.

Much of this book might impress the reader as cold or dehumanizing. That is certainly not my intention. Quite the contrary. You might conclude that we are "merely" collections of atoms obeying the laws of physics, biology, and behavior. You might feel that, without God and "free will" life is meaningless. And you might infer that without God there is no moral compass to guide us. None of these conclusions is valid, as I'll address in detail. Most of us continue to behave and cooperate in ways that advance our knowledge and promote our wellbeing and the well-being of others, whether we "believe" in God or not.

There is little reason to palliate the concerns of those who profess the social and psychological benefits of organized theistic religion. This view and rationalizations for theistic religion have been laid out by many, which I'll address in more detail later. In any case, they are not valid arguments for God; and they're grossly outweighed by organized religion's long-time destructive force on humanity. Religious behavior has been responsible for incredible hatred, suffering, and widespread death since the dawn of religiosity. If nothing else, this compels us to continue to look at the god problem in whatever ways we can. That is what I have done here.

Certainly, not everyone will endorse these ideas, even if he or she reads carefully. It is also clearly the case that anyone who doesn't want to be convinced will be less likely to read these words. Many hold their beliefs very tightly and are reluctant to acknowledge anything which tends to contradict them, as Bertrand Russell so eloquently noted at the top of this chapter.

Let me close here with the obvious. There is a certain arrogance to writing a book like this, an understatement if there ever was one. The questions we're talking about really have no certain answers. But I think that the new information, arguments and discussion laid out here will bring into serious doubt the need for a god, and even more clearly why people believe in such things. Enjoy.

2
WHAT IS SCIENCE

All our science, measured against reality, is primitive and childlike—and yet it is the most precious thing we have. —Albert Einstein

The good thing about science is that it's true whether or not you believe in it. —Neil deGrasse Tyson

Facts do not cease to exist because they are ignored. —Aldous Huxley

There are two alternatives to an appreciation of the existence of our reality—God or science. They are not compatible ways of thinking; and we'll delve more into this conflict below and in Chapter 8. For now, we will focus on what science is and how it is the most appropriate way to understand our world, without god fantasies. Further, I need to talk a bit more about science itself, as we'll employ a number of scientific concepts in the pages that follow—including the notion of human religiousness.

We gain knowledge about the world by observing it. We employ many perspectives in doing so, and we "report" it in just as many ways. We write, we paint, we make music, dance, run, act, meditate, etc. These are all valid and indeed wonderful ways of knowing and expressing the nature of the world and ourselves—all of which contribute to our humanness. But they are not science, which is the focus of what's to come in the following chapters.

Science is often relegated to a small, insignificant corner when it comes to many areas of human behavior and our culture—education, public policy, morality, even sometimes health and medicine. Science in such areas is often ignored or suppressed in favor of certain traditions or preconceptions, often derived from religious superstitions. What we need to know now is that science gives us a better way to understand our world than the fantastical notions of religion, and we should therefore appreciate how it operates—an understanding that is in shortening supply these days. Since "science denial" has become more widespread in some parts of our culture and is often used to support the fantasies of religion, we need to take a closer look—to see what science actually *is*.

There are a lot of things we don't know. We don't know why there's so much tragedy in the world; we don't know the causes and potential cures of many diseases; we don't know if there is life elsewhere in our universe. The list is endless. When we can't figure something out, the tired old excuse for religious apologists is that "god works in mysterious ways;" but this rings so hollow as to be laughable. It is just "god of the gaps" extended to anything for which an explanation doesn't come immediately to mind. What it really means is that any particular event which can't be explained is mysterious, though this really means mysterious *to the observer*. Mysteries are there to be solved, and science continues to solve them. Things that were mysterious a few hundred or thousand years ago are no longer mysteries—from what lightning is to why we have eyes. Sometimes it takes a long time for science to fill in the "gaps;" but many of the gaps are rapidly disappearing.

What is commonly apparent to us, our perceptions, may not be consistent with what science reveals. Returning to analogies about the Earth, we once thought, based on our experience and perception, that the Earth was flat. That is how we observed it. It may have been Pythagoras who was responsible for the initial idea of the Earth as a sphere as far back as around 500 BCE, although it was Aristotle who provided the scientific reasoning probably in the fourth century BCE (CE or "common era" and BCE "before the common era" are the current terms of art for our secular purposes). It was only the circumnavigation of the globe that convinced some of us that the world was round. That was not a perception, per se; it was a scientific inference. The dramatic photograph of the Earth taken by Apollo astronauts on December 7, 1972, showed the clear and beautiful image of the spherical Earth. That we have confidence in the scientific instrument called a camera led most to believe the finding. Of course, quite remarkably, there were—and still are—some people who still believe the Earth is flat (There is actually still an International Flat Earth Society, which continues to propagate such beliefs). To reiterate from the last chapter, does this planet orbit the sun? Of course, these days most people believe that the ancient Christian views of the Earth being the center of all existence is unfounded, as such has been refuted by the theories of Copernicus, Galileo, Newton and others.

In Chapter 4 we'll see how Darwin, and others, have successfully debunked our conception of a god directly creating humans. Nothing new here, but it's worth taking another look at some details not often considered. Science always provides us with more valid accounts of nature, through advances in technology and mathematics, to challenge our limited human perceptions of reality. The ideas of human uniqueness, the distinct possibility that life, as we may conceive of it (or in other forms not yet conceived of), is not unique to this planet, and that our solar system and even our galaxy are not particularly unique should convince us that human perception is sometimes a very poor way to appreciate reality. *Common sense*, so to speak, which grows out of the Copernican Principle, often gives a distorted picture of our world. So let's consider how science is a better tool, what science actually is, and in subsequent chapters what it has to say about us and the superstition of God.

History

The physical nature of the world around us has been the subject of human scientific endeavor for quite a while. In ancient Greece, those who "studied" the world were called natural philosophers. They came up with some interesting "findings," things we've largely refined but were incredibly thought-provoking at the time. Even before Socrates, the Greek philosopher Empedocles, who lived in the fifth century BCE, thought that the four basic substances in the world were earth, air, fire, and water, which, in different combinations, made up everything we see. Later, Aristotle adopted Empedocles' idea that these were the "elements" that made up all things, with the exception of a fifth element, called the "quintessence," which was the pure element making up the heavens. But what were all the "elements" themselves made of? Even earlier, the Greek philosopher Democritus suggested that all matter was made of what he called *atomos* (unbreakable or uncuttable), tiny invisible, indivisible particles that, when combined in different ways, were the basis for all matter. Democritus may not have been the first to suggest such an idea, but we do generally acknowledge him as the father of this concept. It is a shame that this idea never caught on well at the time and was revived only in the early 19th century by the British chemist John Dalton.

Of course, scientific thought took a back seat to the Catholic Church during the Middle Ages. Religious dogma tended to suppress enquiry and free thought, as it always has. All through these centuries, the Church dominated philosophical thinking, suppressing any natural philosophy. But at the dawn of the Renaissance, human curiosity began to feel its oats again, and the scientific thought which had permeated the ancient Greek civilization began to reemerge, first in the Islamic world and later in the Renaissance in Europe. It was philosopher, Franciscan, and scientist Roger Bacon who is generally considered the father of what we now call the "scientific method," which he promulgated in the 1260s, most notably for the use of systematic observation and experiment to generate knowledge, as opposed to philosophical musings. Famously, scientific methods were refined and applied to the problems of the heavenly bodies by Copernicus and Galileo in the 16th and 17th centuries. The methodologies they used were further formalized in the 16th–17th century by philosopher Sir Francis Bacon, who is often credited with pioneering the specific ideas of data collection and the experimental method which underlie modern scientific methods. And these were further refined by Isaac Newton in his *Principia* (*The Mathematical Principles of Natural Philosophy*), first published in 1687. Science was off and running, at least walking, but nevertheless, the Church continued to suppress scientific thought wherever it could, as I'll touch on later.

Common Sense

In the pages that follow much of what I'll discuss will sometimes defy our *common sense*. We think of "common sense" as ordinary practical knowledge; and it is, in essence, our shared perceptions of the world. But common sense can be deceiving. Our sense of the world comes about because we're human and perceive the world in certain ways. It is oriented to the things around us that we can see, hear, feel, touch, and smell, giving us the abilities that enable us to solve everyday problems. That sense is responsible for our evolution and how we've survived, but it is limited and sometimes well off the mark.

History teaches us that our own experience can gradually become trumped by scientific discovery, and we'll see that common sense is inadequate and sometimes contradictory to what is actually true. Common sense gives us access to only a very limited understanding of the known universe. It is based on only a small slice of the electromagnetic spectrum, for example, what we can see. It considers things in a certain range of size and weight. It cannot psychologically conceive of vast distances or incredibly small ones—from billions of light years down to billionths of an inch. It's generally helpless when it comes to things that are incredibly complex, like nuclear physics or the workings of the human brain. Science helps us understand the rest of the universe where common sense is inadequate.

History is replete with scientific facts that have come to be appreciated as such, even though they were initially inconsistent with common sense. Revisiting the Earth problem, we once thought the world was flat; that view was derived from our senses and our shared knowledge. The natural philosophy of the ancient Greeks suggested that the Earth was round, although the notion that it is flat continued as the commonsense view, since that was consistent with our limited perceptions. When you look out the window, the Earth *is* flat. We commonly thought the sun revolves around the Earth, a view that persisted until Copernicus came up with the heliocentric idea and Galileo provided the evidence, only a few hundred years ago. However, when you look up to the sky, you do see the sun appearing and disappearing as it seems to revolve around the Earth. That is common sense, and obviously it can be very deceiving.

In the twentieth century, the discovery of quantum mechanics has further challenged our commonsense thinking. Quantum mechanics, which we'll touch on in subsequent chapters, is well-established science; and it offers descriptions of reality which the human mind is presently incapable of grasping. How can something be in two places at once; how can information travel faster than the speed of light, in contradiction to Einstein's theories; how can something be both a particle and a wave at the same time; how can something come in and out of existence? Our common sense does not allow for understanding how such things are possible—even to the point that many think such things are *not* possible.

Common human experience and logic cannot replace what science tells us. We are forced to accept the inescapable fact that human common sense and reasoning are limited. We depend on science to enlighten us in many areas. Of course, there is often resistance to this, something which unfortunately is now growing—at our own peril. We need to broaden our views of reality, including with science, in order to appreciate what is actually true. This applies to who we are, what we believe, and to the notion of some supernatural authority.

Science and Knowledge

The critical issue for science—and the important problem for science deniers—is the question: How do we know what we know? Let's take a look. There are many ways of "knowing," and philosophy has a long tradition of thinking and talking about such things, something known as epistemology. To state it more generally, how do we come to know anything? Humans affirm all sorts of ways that their knowledge comes about, or "ways of knowing." In addition to science as a way of knowing, there are such things as introspection, reason, meditation, intuition, innate understanding, and even faith. Let's look briefly at this issue, as it's relevant to how humans have conceived of God.

Though it goes back thousands of years, the philosophy of *introspection* was most formally proposed by the 17th-century French philosopher and mathematician Rene Descartes—that knowledge comes about merely by examining one's own thoughts and feelings. If you want to know what is actually true, look inside at your own thoughts and emotions. Religious faith often works just that way. This was extended by the Irish philosopher George Berkeley in his philosophy of "immaterialism," the idea that reality *is* one's mind. Of course, that perspective leads to what we call solipsism, the idea that the only thing whose existence we can be sure of is one's own mind. For our purposes here, I'll assume that there does exist a reality outside of one's mind and relegate "immaterialism" to the dustbin of history (more on this later).

Rationalism

During the period called the Enlightenment, spanning approximately the 17th and 18th centuries, rationalism became popular among philosophers. It was felt that reason or logic was the appropriate source of all knowledge or truth. It is the philosophy promoted by such thinkers as Gottfried Leibniz, Emanuel Kant, and others (whom we'll encounter later). The source of truth itself could be boiled down to what we call deductive logic. You can reason out the answer to any problem through deductive logic. It's most notably illustrated by the classic Socrates syllogism:

All men are rational.
Socrates is a man.
Therefore, Socrates is rational.

As you see, all deductive logic is a tautology, merely affirming the antecedent (the first premise). That is, we could not say that "all men are rational" unless we had already examined Socrates in that regard. The concept of deductive reasoning is often associated with the idea of "logical truth." However, as you see, such "truth" is only a reflection of the assumptions used (and of course, the notion that "all" men are rational is a bit strained). Deductive logic has been often used in so-called "proofs" for God's existence, though it too suffers from that logical fallacy, as we'll discuss in Chapter 8. Yes, our brains both process experience and engage in logical reasoning, which increases our knowledge. I readily admit and embrace this. But inferring "truth" by reason alone—without sufficient basis in reality—is always a fool's errand.

Empiricism

Science stands in contrast to some perspectives of rationalism. Science is based on the idea of objectivity. That is, we believe that there exists a world outside of our own minds, and we know about it through observation. This is what we call *empiricism*. The validity of what we know comes from our systematic observations of the world.

Although such a philosophical approach goes back to Aristotle and others, it was the so-called Enlightenment empiricists who most famously espoused this view, most notably the English philosopher John Locke and David Hume in Scotland. What we know comes from what we observe about the world. This became the approach of modern science—we gain knowledge from observing the world and organizing our perceptions into some coherent understanding of how things work.

In the view of objectivity or empiricism, our knowledge appears because we perceive the world via our five senses (and our sensing instruments).[1] We are not born with any substantive knowledge, essentially a *tabula rasa*, as Locke put it. We have only a reasonable number of responses available that keep us alive until our own learning can take over. If we were not bombarded by chemical, thermal, mechanical, tactile, and electromagnetic energy from our environments, we would never actually know very much. We may not be born fully *tabula rasa,* but it's pretty close. We do assume there is an external world and that we learn about it through our senses. This is empiricism, the basis for all science.

Reality

I will not consider here the various philosophical conceptions of the ultimate nature of reality, what's sometimes called metaphysics.

Metaphysics is not physics, though I'll necessarily deal with the subject of philosophy in Chapter 8. However, for our present purposes, again, we must assume that there does exist a reality external to our own brains.

Science is based on empiricism—what we know based on observation (including observations made by our tools or devices) and our shared observations and inferences. It therefore explores and explains reality—at least some of it, at this point. There are some notable critics, of course, one example being the aforementioned George Berkeley. Berkeley had originally promoted the idea of empiricism, and he was generally known as an empiricist. However, he believed that if one takes empiricism to its logical end, that the world is specifically and only what we perceive, such a view would eventually lead to atheism. There would be no need for a god. Therefore, he considered that no objective reality actually exists other than one's perception. Human perception *was* reality. And if no human was doing the perceiving, it was God doing it. Perception was everything. Nothing was truly objective. Modern science would await, but Berkeley's argument was most famously rebutted by the noted English writer Dr. Samuel Johnson. In his biography of Johnson (*Life of Johnson,* 1791), James Boswell reports that Johnson, in negating Berkeley's argument and affirming that the world actually exists, kicked a large stone and said, "I refute it thus." Empiricism won the day. The story has been recounted many times but, although a bit metaphorical, it is one of the more poignant, compelling arguments against rationalism or introspection as ways of knowing truths about the world. Anything else just boils down to the sophistry of solipsism, an idea that gets us nowhere. We can proceed to consider that reality actually exists beyond our internal experience and that empiricism and the science that uses it is the best way to increasingly understand it.

What is Science?

It is important to consider what we mean by science—what science actually is. It seems even more important these days, as the knowledge science generates is exploding; and at the same time, it is often ignored, belittled, or denounced. Unfortunately, there are too many people who just don't know what science *is*. Many think of science as merely the development of technologies, the things encountered every day which are hard to understand—computers, smartphones, engineering marvels, and so forth. Or maybe it's very educated people sitting in a laboratory mixing up chemicals to create materials or medicines, or looking through microscopes at cells or through telescopes at the stars. But it is somewhat different and much more than these types of activities. And it's much more than doing experiments, though that's a critical piece. It is a way of knowing about the world, a way totally different from any rationalism or introspection. At base, science is really a way of observing, systematically thinking, and solving problems.

More generally, science is that area of human behavior characterized by empirically describing the universe, or some parts of it, including ourselves. We speak about the *wonder* of the universe for a very good reason. We are simultaneously in awe of what we see and are always unsure and questioning about it—from the telescope images of the earliest galaxies to how our brains remember things to how an electron moves. This is the perspective of science. By definition, we will never know the full answers, but that does not keep us from being amazed, continuing to question, and searching for better understandings.

Science is Natural

We think of science as a modern invention, at least a way of thinking which has come into being over the last few thousand years, particularly among the ancient Greeks. But let's take a step even further back. Scientific thinking evolved naturally in human beings (and in some primitive form in other creatures). It is what is responsible for our evolution, enabling us to solve problems that advanced our species. We became capable of making specific observations about the world and then making predictions about what would happen. And humans are a curious lot. We want to know why things happen. As I've said, that's largely responsible for our success as a species. We observe and make inferences about what causes the things we see.

Early humans engaged in this as a way to solve problems of survival. Many other animals—predominantly mammals—do this. It is natural to attend to "what causes what." Individuals naturally consider their surroundings and develop a prediction of what's going to happen. We started out wanting to know what caused water to run and how to slay animals. We wanted to know the relationship between seeds and some plants we could eat. We gained knowledge about the world; and as our brains developed more capacities, we began to solve more complex problems. This type of thinking was obviously present well before the advent of modern human civilization, and even before the evolution of *Homo sapiens*. Our prehistoric ancestors showed this ability as do other primates, mammals, birds, and certain other organisms. Learning from one's environment and using that learning to enhance survival was associated with increasing brain size and cognitive abilities, which has led us to where we are. Such learning enabled the activity of predicting the future based on past and current conditions. That is *the* basis of science.

Consider how a child develops. It is a condensed analogy of how we humans came to a scientific way of thinking. In the beginning, children can't do much, but they quickly become very curious. They explore and test things out. That's the way they experience details about the world. There is continual experimentation. Some things "work" and others don't. This results in what we call *learning*. There's a clear tendency to get a sense of the world and discover new things, including through play, which enhances the ability to test things out. They eventually start to ask "why" things are the way they are. Answers are not always readily available, but children are "hard-wired" to seek out answers. Eventually, children grow up and begin to ask more detailed or pragmatic questions. Why does it get dark? Where does food come from? Where did the herd go, and why? And of course, there are many more that even early humans pondered. Such are the roots of scientific ways of thinking. (I'll have more to say about children in subsequent chapters.)

Definition

Science is the systematic enterprise of gathering knowledge about the world and organizing and condensing that knowledge into testable hypotheses, theories, and laws of nature that predict events. There are many formal definitions of "science." Let's start with the U.S. edition of the Oxford English Dictionary [OED], which defines science as "The systematic study of the structure and behavior of the physical and natural world through observation, experimentation, and the testing of theories against the evidence obtained." For our purposes here, one of the more useful and understandable definitions was used by the judge in the famous 2005 court opinion, Kitzmiller v. Dover Area School District. (273) This was the legal decision which struck down the teaching of intelligent design in public schools. The judge partially cited the definition from the National Academy of Sciences:

> Science is a particular way of knowing about the world. In science, explanations are restricted to those that can be inferred from the confirmable data — the results obtained through observations and experiments that can be substantiated by other scientists. Anything that can be observed or measured is amenable to scientific investigation. Explanations that cannot be based upon empirical evidence are not part of science. (361)

Science is a set of behaviors in which humans engage to generate new knowledge and understanding of our world. Its essential mission is to predict events based on past events and develop the theories which follow from them. There are two important concepts to clarify in this regard as we discuss the various issues I'll address. These are the questions of "why" and "how"? We often use the terms interchangeably; but there is a subtle, though important, distinction. To be clear, asking "why" generally implies that there is a "reason," and we are left to then ask what that reason is. It also implies some sort of purpose. Science breaks down at that point. The question science actually addresses is "how," i.e., the *manner* in which something comes about. It is the search for the objective conditions that are associated with any particular phenomenon. Asking "how" enables one to conduct observations and make inferences based on empirical data. That is science. The interchangeable use of the terms is very common, and you may see that occasionally happen here—language is often very limiting. If it does, remember that we're actually talking about is "how" certain things come to be.

Consensus

Science works by empirical observation and deductive reasoning. However, any given experiment can lead to spurious results—and it often does. Empirical observation is itself flawed since we all perceive things a bit differently. We all have different perceptions of reality, including often to exactly the same stimuli. Our measurements also can vary. Our measuring devices have gotten incredibly better—from space telescopes and gravitational wave observatories down to electron microscopes and even the identification of fundamental aspects of nature discovered in particle accelerators. However, any two experiments, often for unknown reasons, can come up with different results. Bias can also be a factor. What is true empirically, therefore, demands repeated observations and experiments. And it usually takes a long time, with repeated experiments done by scientists in order to have some confidence in the results.[2] Most critically, it requires consensus among various scientists who've conducted the relevant experiments, along with those who independently examine the results. Many people are seduced into believing the results of one study, and such are often hyped by popular media. The appearance of "discoveries" and "breakthroughs" is indeed rare and too often labeled that way. Nearly all don't bear out as such. It takes consensus over time. That is how science works. Understanding the world empirically takes time, painstaking study, and agreement among observers.

Conclusion

In science, being an "authority" does not convey "truth" about anything. The empirical truth about the world must originate from the observations we make and the theories that follow to explain them. In contrast to science, religion proceeds on authority, not on empirical truth. When one asks, "How does science deal with the God question," the answer is clear. It does not. Science is not in the business of explaining gods. As the National Academy of Sciences put it, "Science is a way of knowing about the natural world. It is limited to explaining the natural world through natural causes. Science can say nothing about the supernatural. Whether God exists or not is a question about which science is neutral." (361) But science does have a lot to say about religious behavior and superstitions, and we'll get into that issue in later chapters.

Let's conclude our endorsement of science with the following caveat. Let me again acknowledge that there are clearly other human activities which increase our knowledge, principally the arts and humanities in all their wondrous forms. They do increase knowledge and enrich the lives of human beings. However, they are not science. The salient difference here is that those fields affect the knowledge and lives of people *individually*. Everyone "learns" something different from art, music, poetry, etc. Science, on the other hand, increases human knowledge in ways that lead to a shared view of reality, the consensus I alluded to. There are those who'll disagree with this analysis, but so be it. Some have suggested that science might be enhanced by its integration with other fields. But the effort to integrate science with other fields of knowledge obfuscates the term. Some have quaintly offered that many fields, including the humanities, are engaged in "the same goal of improving our understanding of the true nature of things." (469) This is a provocative notion but has nothing to do with what science actually is. It is true that, although practitioners of the arts and humanities use empirical methods, they do so only to inform their own behavior. An artist or novelist leans on his own empirical assessment of the world and his reasoning in producing his work; but any consensus about it is not necessary for it to be regarded as artful. While there are clearly impressions of beauty in scientific discoveries, as Einstein himself endorsed, the converse is not true—there is no objective, empirical standard, for example, for what is good art, or even what art *is*. Often, there is splendid creativity in conducting science, though we seldom refer to such as art. Science is science, and art is art. What we are confining our discussion to here is science.

The Laws of Nature

Let's start with a term that makes most people shudder—*determinism*. Science is based on the idea of determinism, i.e., that all events in the universe unfold in a predictable way, i.e., that any event can be predicted by the previous state of the universe. It was the "natural philosophy" of the ancient Greeks and has guided scientific knowledge and inquiry ever since. Determinism has been variously defined. For our purposes, we'll use the definition proposed by American philosopher Richard Taylor in 1967, "For everything that ever happens there are conditions such that, given them, nothing else could happen." (514) Specifically, any event or occurrence is determined by those that precede it. The future is always *determined* by the past. This is the perspective we'll take in examining various issues in this book.

Science is in the business of discovering the principles which describe how this occurs. These are often referred to as the *laws of nature*. We should be clear about this term. Those Laws are not actual laws. They do not exist in and of themselves outside of the events they describe, and they are not "codes of conduct" handed down by some "lawgiver" to control the universe. The Laws are abstractions—convenient summaries—of our observations of physical reality and the language of mathematics we use to describe them.[3] The laws describe nature and allow us to communicate efficiently, but they do not exist as "reality" themselves. When we "discover" various laws of nature, we are not finding some new entities, though, as you'll correctly argue, we continue to find new particles, events, etc. We're simply developing a way to accurately describe *how* physical events or phenomena take place. To further clarify, when I sometimes suggest that phenomena "obey" the laws of nature, I'm not saying that they are controlled by some unseen, overriding entity. I'm merely describing, somewhat metaphorically, the patterns they embody.

The idea of determinism, as I'll discuss further in the next chapter, often implies that events are determined in some fixed way. However, there are some current scientific conceptions of the universe, such as quantum theory, which sometimes employ probabilistic notions of reality (see Chapter 3). Whether such a conception represents the "deepest" level of understanding of reality or not is still an open question. Einstein, of course, famously thought that it did not; and most physicists today do not think that quantum theory is a threat to determinism. The notion that some aspects of reality may be *currently* described in probabilistic ways does not negate the idea that events are consistent with the laws of nature, generally construed, though some of them remain currently hidden to us. We must proceed with the notion that all events *are* determined, including our thoughts and behavior, which I'll get into in Chapters 6 and 10.

Determinism is also often confused with the idea of "pre-determined." While this may be linguistically convenient, the notion of "pre-determined" suggests that there was an ultimate referent or cause; perhaps that some entity is responsible for setting off some plan. Of course, there is no evidence for this. We use the term "determinism" merely to convey that the future is predictable from previous events. There is no reason to entertain the notion of "pre-determined." The universe exists as it is and changes over time; and, as we'll see later, there is no reason to believe it hasn't always existed. Determinism is also not "predestination," a theological concept we'll revisit in Chapter 6.

Some philosophers, and even some scientists, have promoted the idea of "indeterminism," that some particular events in the universe do not conform to the known or knowable laws of nature. If there are events in the universe which are not "determined" consistent with the laws of nature, then there are only four possibilities for them. Either they are random (and we know there is no "true randomness" in the universe, which we'll get into further); or there are laws yet to be described, which has always been the case; or there is some divine intervention; or there is some sort of "free will" on the part of humans independent of the laws of nature. In particular, the notion of quantum mechanics being probabilistic to our measuring devices is not an argument for indeterminism, as we'll see. Finally, that there is, as yet, no coherent unified theory of our universe is not an argument for indeterminism. Either there are potential explanations for all events, or else a god decides which events are determined and which are not. There can be no coherent theory for that!

Calculational Impasse

Some say that science can't know everything. I agree, as do all scientists. And by definition, there are no final answers in science. But this is a vacuous criticism. As astronomer Edwin Hubble noted, "The scientist explains the world by successive approximations," and such approximations never end with any final "truth" (more on Hubble's famous discovery in Chapter 3). Scientific knowledge is always tentative, and it becomes more refined over time. To "know" everything about any given subject—or somehow believe you know everything—is not science. It is a religion.

We assume that the laws of nature, in their expression throughout the history of our universe, explain—or have the *potential* to explain—events as we may observe them. Although we have our laws of nature, and they can be pretty exquisite; for our universe as we presently know it, they are notably incomplete. Our level of understanding is just that. And it will always be that way. It is a fascinating and challenging endeavor, and it is forever beyond our full appreciation. As physicist Max Planck once observed, "Science . . . means unresting endeavor and continually progressing development towards an aim which the poetic intuition may apprehend, but which intellect can never fully grasp."

This dynamic exists in all of science—in physics, chemistry, biology, and certainly in human psychology. We cannot precisely predict a tornado, when a leaf will fall off a tree, when you will burp, when or why a species will go extinct, or when you'll be too tired to continue reading this. This is because at any specific point in time, the data would be too massive, too rapidly changing, too inaccessible to our ability to observe, and too complex to actually calculate. We are also limited by the tools we have and the time in which we exist. This problem has been dubbed the *calculational impasse*, introduced by physicist Brian Greene. (191) It was actually (and somewhat sarcastically) presaged by Pierre-Simon Laplace, the Renaissance French mathematician, physicist, astronomer, and polymath. He called it his "divine calculator" (The phrase is but a metaphor. Laplace, by most accounts, was actually an atheist.) That is, if there was an entity who could know all the material in the universe, how every particle and force in the universe was acting, he could perfectly describe the history of the universe and its future in its most granular detail. Of course, that's impossible, by definition. Criticisms of science, especially when it comes to the behavior of human beings (Chapter 10), would be well advised to remember that.

We are also limited by the fact that science is inescapably human behavior. To assume that we are capable of both measurement and reasoning adequate to fully describe the dynamics of our universe may not be justified. Science is and will always be thus limited. I'll be talking a lot about the *calculational impasse* in the coming chapters, but, even given that, we do have theories that, to one degree or another, help to explain our reality. Though imperfect, science does explain an incredible amount of our natural world. As Carl Sagan notes, "Science is far from a perfect instrument of knowledge. It is just the best we have." The calculational impasse—or any other way of describing our scientific limitations—is rarely considered by religious apologists. Using a "God of the gaps" for yet unsolved "mysteries" requires a lot less of their effort.

Theory

Science works by formulating theories, which we'll see in many subjects that follow. But what is a theory? We regularly hear that there are scientific theories about how the world works. People often see these as "guesses," or think that scientists are particularly uncertain about what they espouse. This is not the case. A scientific theory is a well-tested explanation of events we see—from subatomic particles to the causes of many diseases to climate change to human behavior. Scientific theories are not guesses. They are based on data accumulated with appropriate "scientific method"—that is, observation, formulating hypotheses, subjecting them to experimentation, repeated experiments by various scientists, and consensus on results and their significance. That is the key. Experiments *isolate* the reasons that any particular phenomena exist. That is the essential basis of science. As Sherlock Homes once said, "When all other possible explanations are eliminated, whatever remains, however improbable, must be the truth." Most importantly, any theory advanced must be *falsifiable*. That is, you must be able to test a theory in order to prove that it is *not* a valid way of explaining the world. "Theories" which do not meet that test are not scientific theories. You cannot falsify the notions of reincarnation, or ghosts, or intelligent design (which I'll address in Chapter 4). So-called theories which don't meet such experimental criteria are not science. As physicist and Nobel Laureate Richard Feynman bluntly put it, "If it disagrees with experiment, it is wrong." At bottom, scientific theories are ways to reliably predict future events. Keep that in mind as we address matters of religion, morality, and human behavior in subsequent chapters.

Levels of Theory

In the following chapters, I discuss various "kinds" of science bearing on human behavior, including the absurdities of religion. The first thing to know about those discussions is that we are ruling out things—or more specifically "ideas"—that are inconsistent with the laws of nature and theories that explain them. That is, we are approaching science without the influence of any supernatural force. The only forces that can be considered are the ones described by our most basic physics—the various fundamental particles and forces in our universe. Without belaboring details here, these presently include the strong and weak nuclear forces, electromagnetism, and gravity (Of course, this does not rule out the possibility that other forces, particles, etc., may be discovered in the future. (More on this later.) But fundamental physics doesn't give us convenient descriptions—or more rightly usable predictions—of what happens on larger scales of reality—such as cells, trees, or humans. For that, we depend on such theories as offered by chemistry, biology, neuroscience, anthropology, behavioral science, and others. While we consider all these as appropriate levels of theory to understand our world and ourselves, we must realize that they all flow from the basic theories of physics. We will discuss human behavior and religion, for example, on the basis of neuroscience and behavioral science while appreciating that these flow from biological processes, which in turn arise out of chemistry, which in turn flow from basic physical processes of atomic entities and fundamental forces. We need not delve into all these levels of theory. That would be unrealistic. It is sufficient here to assert that our discussion—regardless of level of inference—is based on the fundamentals of science and that the laws of nature are the best descriptors, and predictors, of human thinking and behavior which require various principles to explicate.

In practical terms, we cannot conveniently predict common events in our world by describing the fundamental forces which come to play in any situation—growth of a tree, evolution of the eye, or human religiosity. We do not know how to bridge the chasm between the most fundamental physics and more ordinary observable phenomena. We generally do not describe a table with details about protons; we don't describe the evolution of language with details of neuron physiology—even though we know that all these things are relevant. Principles used in chemistry or anthropology, for example, make conversation easier and more meaningful, but they do not envisage theories of some type of new reality that cannot be explained—theoretically—by the laws of physics. Since basic physics cannot directly predict those kinds of events, particularly individual human behavior, social phenomena, etc., various approaches have been offered, which I'll refer to in later chapters. Physicist Sean Carroll offers the concept of "poetic naturalism." (78) (With a nod to Max Planck, who first used "poetic" in this way.) This is described basically as natural law plus the idea that there are many ways of simply talking about reality. Physicists Steven Hawking and Leonard Mlodinow use the concept of "effective theories" (214), which they consider as ways of talking about phenomena that cannot be described by the equations of physics. That's probably a poor term for such, as the "effective" ability of many of the "theories" they allude to is sorely lacking. Physicist Stephen J. Blundell uses the concept of "emergent narratives" to argue for models of reality with levels of inference beyond basic physics. (52) All these merely suggest that the limit of our knowledge is a considerable burden, à la Greene's "calculational impasse," and we need to talk about events in some meaningful way. We do readily accept theories of different levels of inference when it comes to such things as chemistry, biology, and geology, for example, even though we acknowledge that these fields are merely higher levels of inference about phenomena which might be conceivably described with more specific fundamental physics—though that would be totally impractical.

Things Are Always Changing

Now let's take a popular opinion or misconception often stated as "Science can't prove that." We see this offered in many areas, particularly when it comes to things like human individual or social behavior. This is actually correct. We cannot formally prove things in science. Scientific theories are not "proofs" of anything; they are not *absolute truths*.[4] They are, however, the best empirical explanations of how things "work," if you will; that is, the best way to understand and predict events that happen. That is why theories are always being revised. As new and better data come to light through experiment, a theory gets discarded or updated. Ptolemy's geocentric theory gave way to Copernicus' better explanation. Newton's theory of gravity was an incredibly good one—until Einstein came along and made it better. Darwin's theory of evolution made pretty good predictions, but the discoveries of modern paleontology and genetics have made the theory even more precise. Any theory "merely" provides a better explanation for the world than those which came before it.

To reemphasize, there are no ultimate answers in science, merely the agreement, based on observation, among us humans on various issues—from quantum mechanics to cosmology to human behavior—about what is real about the world. Science isn't perfect (by definition), but invoking God to fill gaps just complicates matters. Scientific theories are the most economical ways of understanding the world, and the notion of God is always superfluous. Invoking God also violates what we call the principle of economy. The concept of scientific economy has come to be known as Ockham's Razor—the simplest adequate explanation is probably the best one—after Sir William of Ockham, the 14th-century English Franciscan friar and theologian. He offered, "Plurality should not be posited without necessity." He also had the intrepidity to confront the Pope and was tried for heresy for the notion that one does not need to invoke God to understand the world around us. The principle was later and probably most famously illustrated by Laplace, who espoused the idea of a determined universe. When Napoleon Bonaparte asked him what role God played in the construction of the universe, Laplace answered, "I had no need for that hypothesis."

Why Do Some People Reject Science?

Science is under attack these days, and by a significant part of the American population. This is a dangerous trend. In recent years, especially in the United States, there has been a degrading of respect for science, an actual antipathy in some quarters, and an evisceration of science education.

We need to pause here and consider why there is such widespread doubt and hostility toward science. Science has led to technological improvements that enabled us to become the dominant species on earth. It is truly remarkable to see how many so readily abandon it.

Science has created all sorts of things that we take for granted, yet many continue to doubt it. From the wheel to ancient aqueducts to modern technologies—the phonograph, the automobile, the internet, MRI machines, communication and GPS satellites, lifesaving medications—the list is endless. Many people reject the idea of science even though they take medicine, drive a car, and use a microwave oven. Many wouldn't be caught dead without a smartphone in their pockets. Many also accept scientific explanations for the behavior of most living creatures, but reject it where humans are concerned. When it comes to our own bodies, to biological evolution, to human behavior, the doubt continues; and the belief among science deniers about how such things work, they assert, must remain the province of God.

Cultural Memes

People reject science as an explanation for events in the world for many reasons. It seems boring. It seems too difficult. Formal education doesn't give it enough time (an increasing problem). Most critically, it's a threat to theistic religious beliefs. Let's begin with some of the related memes which permeate our culture. There are strange ideas and attitudes which develop among people, and they can take on a life of their own. When such memes develop and spread, they become cultural phenomena (and sometimes "conspiracy theories," which I'll address further in Chapter 11). And science which may conflict with them may get ignored or rejected. For example, alien UFOs do exist, and this means they're spying on us. The Loch Ness monster, Bigfoot, and other incredible creatures do exist. The moon landing was faked.

The antivaccination movement is a prime example. One of its tenets is the fallacious idea that vaccines don't work and actually cause disease. This was exemplified by the hysteria decades ago that certain vaccines (MMR[5]) cause autism, something which has been empirically and unequivocally rejected. The physician who published the study allegedly "proving" the false idea, in fact, was shown to have been a fraud, and the "study" was retracted by the medical journal *The Lancet*.[6] Nevertheless, the "legend" developed; many people continued to believe the fantasy; and a serious antivaccine movement about this has developed. The meme of "science suspicion" grew. While this issue was a significant stimulus for the contemporary antivaccination movement, we must acknowledge that an antivaccination culture goes back at least to Edward Jenner's vaccine for smallpox in 1796. Many people were unwilling to accept this effective vaccine, and many needlessly died. Benjamin Franklin, in fact, openly regretted that he had not inoculated his son, who died of the disease at age 4 (there was reportedly hesitation because the child had a severe cold). Science does seem to get ignored when conflicting, potent cultural memes evolve.

Information

Active denial of other well-established scientific facts is also disappointingly high. Many people just don't believe them. We are all biased in what we believe, the facts notwithstanding. Science is based on facts. And we have opinions about the world, whether they're based on facts or not. As Mark Twain once quipped sarcastically, "I am not one of those who in expressing opinions confine themselves to facts." Sometimes it takes some patience and focus to appreciate them. And those skills and attitudes are in short supply in many quarters.

But also, when it comes to science, those facts can be difficult to understand. It often requires a bit of education to adequately appreciate them. Sadly, Americans are woefully undereducated in this regard. Much of our population is poorly educated, at least about science and its value. For example, as of a decade ago, some surveys found that about a quarter of Americans still believed that the sun revolves around the Earth. (370, 382) (Current reliable data on this are not available.) That's a truly sad commentary about our trust in science.

Sometimes people are unwilling to put forth the effort to find out more about things that are a bit confusing. Actual empirical facts are often unimportant, even annoying, to people when they are not relevant to their everyday experience. They accept that science exists and that its technology makes the car go and makes a computer do its thing. But when it comes to issues that are not part of personal experience, making efforts to appreciate, and even accept, scientific facts, are sometimes just not worth the effort.

One Swallow Doesn't Make a Summer

This is an important aphorism to remember. We hear about scientific "discoveries" all the time. When an important scientific study comes to light—and modern media are quick to publish such things—great interest can be generated. Then something may change. And people "freak." Eggs are bad for you; now OK. GMOs are dangerous; now not evidenced. Cell phones cause cancer; now not a problem. Things change in science, by definition, and many people don't appreciate or focus on that. Thus, an attitude often emerges—"Science really doesn't know."

The issue of foods has been especially confusing for people. Which foods are good for us, and which one should we avoid? Scientific skepticism grows. Many people were confused by changing advice during the COVID-19 pandemic. We even once thought smoking was innocuous. But this all misses the point, of course. Scientific knowledge is always changing, based on new information. That's what science is. As I've said, science works by replication and consensus. One study or set of facts rarely means much, but popular press about it can set off a cultural meme which can be hard to break. And when contrary information comes to light, the reputation of science itself falters. Information is always changing. We continue to hear information from scientists about diet, infections and vaccines, and general matters like climate change. The information is never exactly the same, and many people don't appreciate that. They think science produces unchanging ideas; and when they do change, people lose trust. This is unfortunate but comes from a poor understanding of what science actually is. Again, people think that science, if it is valid, must provide a final, definite answer. Of course, in science, as I've said, that is impossible. There is no final empirical answer or explanation to the world around us, only the best knowledge we have at the time. That's a difficult concept for many to accept.

Religious Conflict

Most importantly, the conflict between religion and science is often responsible for the growing disrespect for science. Usually, it is because science conflicts with religious fantasies. It is difficult to cognitively hold both empirical facts and religious ideas at the same time. Thus, people can become willfully ignorant about science. One irrefutable truth remains: knowledge threatens religious faith. There is a long history of this clash. In the fourth century CE, the theologian and philosopher St. Augustine warned, "There is another form of temptation, even more fraught with danger. This is the disease of curiosity. It is this which drives us to try and discover the secrets of nature, those secrets which are beyond our understanding, which can avail us nothing and which man should not wish to learn."

That attitude remains prevalent. We know that many people are unwilling to consider facts about the natural world, for various reasons, especially because of religious indoctrination, which begins during childhood (I'll discuss this more in Chapter 10).[7] Science operates on facts. Religion operates on authority. Belief and trust in authority are the fundamental dynamic of theistic religion; and this curtails curiosity, critical thinking, and empirical truth, as St. Augustine warned. As Einstein once said, "Authority gone to one's head is the greatest enemy of the truth." (149)[8]

There is also a certain amount of fear among religious folks. A threat to their belief systems is certainly on the line. Fear is a powerful motivator, and religious beliefs tend to be held mightily. Science is greatly feared by the faithful. Perhaps Thomas Jefferson said it best. In an 1820 letter to Portuguese philosopher and diplomat José Corrêa da Serra, Jefferson noted, "The priests of the different religious sects dread the advance of science as witches do the approach of day-light." (248)

People of "faith" are acutely uncomfortable with this conflict. The "need to know" and the discomfort of not knowing is replaced with an artificially constructed idea—God—to fill in the gaps. Religious apologists prefer biblical and various apocryphal teachings to science. St. Augustine's insight thus confirmed. Let's take what's become a prime example—evolution. There are still those who deny the incontrovertible fact of biological evolution on this planet, and this is perhaps the primary area of scientific knowledge that they tend to reject. There is no question that Charles Darwin's theory created a psychological and social storm, though it has dissipated somewhat over the decades, with the unfortunate exception of a significant minority of the population who are threatened by it, especially in America. In various surveys, only about half of Americans believe that we humans evolved from earlier species of animals. (346) This is up only about 14% from various surveys over the last three decades, and the percentages and trends have been pretty stable over the last few years. (95, 326) Currently, about 40% of the American population actually endorse the idea of Creationism—that God created humans in their present form (61); and for those acknowledging the reality of evolution, only about a third view it as a natural process, while about a quarter aver that God has directed it. (96) America is almost an outlier when it comes to these issues. In 2005, out of 34 developed nations, only Turkey had a lower acceptance rate for affirming the idea of biological evolution (27%). (345)[9]

Evolution is also still a problem in science education. Only a small minority of high school teachers feel comfortable teaching it in any detail. (406) Biologist Jerry Coyne observes the remarkable fact that, "Nearly two-thirds of Americans feel that if evolution is taught in the science classroom, creationism should be as well. Only 12 percent—one in eight people—think that evolution should taught without mentioning a creationist alternative." (100) Nevertheless, state legislatures in America continue to advance bills that promote the teaching of Creationism in public schools. It was actually not until 1987 that the U.S. Supreme Court "seemed" to settle this issue, finding that if biological evolution was taught in schools, then "creation science" need not be. (146) Even then, two justices dissented, Antonin Scalia and William Rehnquist. The decision actually did not prohibit the teaching of "creation science" (Note the misuse of the term "science"); and our political bodies continue to push "creation science" and "intelligent design" into children's classrooms. As of this writing, at least four states have enacted legislation to weaken the Supreme Court's decision. Such laws enable teachers to discuss other "theories" about life, whether or not "creation" terms are mentioned. Of course, the word "theory" is never defined, often enabling the education of children with what's being called "stealth creationism." (342) This is a dismal picture of our culture's embrace of science and the continuing efforts to tear it down.

Religionists also reject Earth science, often rather dramatically. They profess a so-called Young-Earth history: that the Earth is about 6,000 years old. I think one of the most bizarre examples of this I've seen was Richard Dawkins' criticism of British Prime Minister Tony Blair's defense of a religious school's academics. Dawkins notes that there was a school inspection which "gave a rave report to a school whose head of science teaches that the entire universe began after the domestication of the dog." (113) Actually, this would be *well* after the domestication of the dog. It is likely that dog domestication happened multiple times, but most recent archeological evidence ranges from about 19,000 to 32,000 years ago in different parts of the world. (195) It is indeed remarkable (almost comical) how some people deny science to this degree and accept what they know in their "gut" to be pretty ludicrous.

Cognitive Processes and Attitudes

The ability to discern fact from fiction is a matter of both attitude and information. In the face of growing scientific knowledge (and in many areas, this is increasing at incredible rates), it does take some willful ignorance to maintain one's religious faith. That is, it takes more and more rationalization to ascribe events to God's will and eschew what science has to say. Personal attitudes also create a psychological barrier to accepting or even acknowledging scientific truth. There are often matters of personal identity which influence the process, for example, strongly identifying with certain political movements, phobias around science (e.g., rejecting medical procedures out of fear), subscription to conspiratorial ideas, having certain vested financial interests, or identifying with a particular social group that has certain cultural views. All these can contribute to some rejection of science. (225)

On a more granular basis, the Nobel Prize-winning psychologist Daniel Kahneman lays out how we are often victims of our own emotional and primitive decision-making processes, which disfavor processing more information to come to conclusions. (259) (Daniel Kahneman's a name we'll periodically encounter in this book.) It usually takes a good deal more effort to process scientific facts than to merely, and somewhat reflexively, believe in God's will. In a more recent work, neuroscientist Tali Sharot notes how we tend to seek and value information depending on our emotional state, fears, levels of stress, how much purchase one has over the information, possible rewarding aspects of the information, and many other cognitive and emotional factors. (463) Belief in God and rejection of science is a prime example of such. Theistic religion takes hold as stress, fear, possible reward and punishment, and despondent emotional states come into play, as we'll further discuss in Chapter 10. In such contexts, science is often rejected in favor of less effortful religious ideas.

Another critical factor in rejecting science in favor of religion has come to be known as "confirmation bias." That is, we tend to believe what's consistent with our current views. It's become a modern, well-known term, but the matter was most cogently summarized by Bertrand

Russell a century ago.

> If a man is offered a fact which goes against his instincts, he will scrutinize it closely, and unless the evidence is overwhelming, he will refuse to believe it. If, on the other hand, he is offered something which affords a reason for acting in accordance to his instincts, he will accept it even on the slightest evidence. The origin of myths is explained in this way. (434)

This is technically referred to as "cognitive dissonance," a theory developed by psychologist Leon Festinger. (161) We tend to seek consistency in our beliefs, favoring ideas that agree with held positions and rejecting those which are at odds with them. Tension arises when disparate or conflicting ideas arise. The tension is reduced when one of those ideas is chosen or favored and the other one rejected or minimized. Essentially, we are reinforced by reducing tension between disparate states, which explains, in part, Russell's sagacious view and why it is difficult to change minds. Conflicts between religion and science thus become resolved for any individual in this way. It is less and less possible to influence belief in a religious person with contradictory scientific facts, since cognitive dissonance works against that. Of course, the exposure to contrary ideas is also a threat when you're a bit unsure of the things you profess or where you can't rationalize them. That is why reliance on authority is the hallmark of theistic religion. One need not worry about conflicting ideas. God tells me what's right, or, in His absence, the Bible, priest, or other clergy does.

Finally, when the cause of events is unknown, we have a great reluctance to just say, "I don't know." That's sometimes uncomfortable; it's just inconsistent with our tendencies. We are more inclined to offer a constructed interpretation, idea, or fantasy. A scientific approach, however, may offer that "it is unknown at this point, and we'll keep studying." That there is discomfort with not knowing is natural, but dealing with such "anxiety" or uncomfortableness does not logically lead to an inference of God. One doesn't have to exert much effort to invoke a deity to explain events. You could say that this is the "lazy explanation," and it can be used for almost anything, but the notion that unexplained events are "God's will" is merely a convenient way to end conversations, especially about science.

Consequences

Disrespect for science can have unfortunate consequences. It can limit job opportunities, impair your ability to exploit useful information, and negatively impact your health. Sometimes the consequences can be disastrous. Returning to the example of vaccines, many people just don't "believe" in them, despite the fact that they're some of the most remarkable and well-evidenced technological inventions in human history.

In fact, belief is so strong among some in the antivaccination movement that they would rather die than get vaccinated. Let's take one example. During the recent COVID-19 pandemic, many doubted the efficacy of the vaccines, including people who were at high risk of dying from the virus. Many of those people did die. Many evangelicals refused vaccinations—on putative religious grounds (though such grounds are unclear to me). They continued to espouse that God would either save them; or that, if He let them die, that was His plan. His so-called "will" was more important than the science that might have saved them. Many antivaxxers of various religious persuasions died from that virus, some actually claiming that the virus wasn't killing them even as they were dying in the hospital. These rejections of science led to countless deaths. As the result of such ignorance—as well as the arrogance of some politically biased authorities—over a million Americans have succumbed to the virus (so far) in the U.S, and millions more around the world.

Conclusion

There is a notion that you believe in science, or you believe in God, or perhaps both (We'll discuss more later how these are incompatible ideas). However, the notion of "belief" in science is a bit misleading. Science exists independent of anyone's belief. It is a fundamental human behavior. Science will persist in any case, as it has for thousands of years. Science sometimes conflicts with our everyday experience, as we've seen, and it takes some effort to look at the facts. Religion often impairs that.

Does science want religion to go away? It matters little what individual scientists desire, and a small few, quite remarkably, are actually believers (which I'll address in Chapter 10). But it's clear that the incompatibility of religion and science is a problem for modern human culture. History is crystal clear on that. History is also clear on the notion that religion cannot be suppressed. It has also been always unwise to attempt to limit speech and "free" expression; and both behavioral science and history tell us that suppressive methods just don't work in the long run. Rather, the relative weakening of organized religion, and consequent respect for science, can only be accomplished in positive ways, as history has shown. This includes economic development, enhanced education (especially for women), and the shaping of critical thinking and respect for science, especially in the classroom, all of which are associated with reduced religiosity. More detail on this in Chapter 11.

We all are scientists at heart; at least we start out that way. It's just that as we grow older, we're less and less open to new ideas or new information. As childhood gives way to adolescence and adulthood, the curiosity and sense of exploration we're born with tends to fade away. We become more closed to ideas that are different from our own. We avoid information that conflicts with our views of reality and focus mainly on information that reinforces our thinking. We start out very curious, and some humans often maintain this well into later childhood. We ask such questions as, "Why are the leaves green?" or "What's over the horizon?" or even "Why is the sky blue?" Note: If you're now trying to answer the latter question in your mind, you still have some really good curiosity left (even reading this book verifies that). If you're still curious, note that the sky question piqued the curiosity of such thinkers as Aristotle, Plato, Albert Einstein, Leonardo da Vinci, Rene Descartes, even Goethe. (399) The difference between them and many of us may be that they kept asking such questions. The people who spend their days asking such questions we call scientists.

Most of us tend to associate with people who agree with us. We read or otherwise consume fewer things that are likely to challenge our world views, and we consume media that mainly agree with us. And when we encounter contradictory information, we're a bit uncomfortable and retreat into our very familiar conceptions of reality. We may lose some of our sense of wonder and merely maintain our mental (and behavioral) positions, which are well-hardened and comfortable. Of course, age does this to us all, but it's a matter of degree. But we can see this issue along a continuum. Religion, well-practiced, is at one end of the continuum, the epitome of rigidity and closure to new ideas, and science, well done, is at the other end, the epitome of curiosity and mental openness.

There is an aversion to looking at scientific details by many who just consider them dull and boring. But the world of science can be dramatic and beautiful. As I've said, one can argue, quite well I think, that the arts and the humanities, in all their wondrous forms (visual art, music, dance, literature, etc.) should also be considered as "ways of knowing." I couldn't agree more. In point of fact, I'm continually impressed at how keenly artists respect the sciences and scientists appreciate the arts. One can't help recalling Einstein here. On the one hand, he often found playing the violin to be helpful in dealing with difficult problems. And on the other hand, he often saw the universe as wondrous and beautiful. For example, he viewed the music of Mozart to be "so pure it seems to have been ever-present in the universe." (235) The difference is that in science—as opposed to other "ways of knowing"—is that facts and ideas so generated are "disprovable" whereas in other "methods" they are not. We do need both science and the arts and humanities. We do not "need" theistic religion.

One may not necessarily be convinced by the arguments in this book. What is relevant at this point is "merely" a respect for science. If such occurs, there is less superstitious reliance on a God to "explain" events (which are actually not explanations, of course). So let us proceed to understand how religious behavior has come about, how the world we know is explainable without reference to some supernatural force, and why the notion of God continues as an irrelevant, yet dangerous, superstition.

3
THE COSMOS AND US

There are more things in heaven and earth, Horatio,
Than are dreamt of in your philosophy. —Hamlet to Horatio (*Hamlet*, Act 1)

If you want to make an apple pie from scratch, you must first invent the universe. —Carl Sagan

As Hamlet intimated, the universe is a pretty complicated thing. But it *is* understandable—to a certain extent. Here I'll demonstrate that in understanding our cosmos (including ourselves) and how it came about, there is no need for a god. In this chapter, we'll consider our universe, everything we "see" or can access in some way. One of the principal arguments for the existence of a god is that there must have been a Creator for all that. This is known as the "first cause" argument, which we'll also look into further below and in Chapter 8. The purpose now is to debunk the fantasy that a god was necessary for our existence—or the existence of anything we know as reality. The other argument is that we, in this vast universe, are somehow special. I'll address this here, too.

Let's begin with the notion that we're fairly ignorant about our world. We don't know enough about some things we think are very important, e.g., how did the universe actually come into existence; is there life elsewhere in our universe; how will it all end? These and many other important questions remain somewhat ambiguous—from biology, physics, anthropology, and other disciplines. But science has made incredible strides in explaining phenomena in our world, and in this chapter, I'll review some of them that relate to our universe. We don't know everything, especially the full answers to the above questions; and, as I've said, there can never be "complete" scientific knowledge—by definition. That can be uncomfortable to some, but in the end, you either trust what science can reveal or resort to the fantasy that some supernatural being is responsible for our world. Let's see why.

One of the crucial tenets of theistic religion is how the "world," now we might consider the universe, came into being. In this chapter, we'll examine some critical questions related to this problem. How did everything get started? Was a god necessary to create the universe? Why are we here? In fact, one of the principal arguments for God is that It *is* the Creator, the constructor and designer of all things. Of course, this is not a novel idea invented by Christians, which I'll address more in Chapter 7. The whole idea of theistic religion is that there *must* be a god who created everything, including the supposedly unique and special creatures called human beings. These are merely assumptions, of course, surrounded by a few thousand years of superstitious dogma. In this chapter, we'll look at the fallacy of these assumptions and render the god idea unnecessary.

We'll talk about physics and cosmology in this chapter, and some of this discussion will seem absurd. The human mind cannot conceive of many things which are consistent with accepted physics. Can something like an electron be in two places at once? In fact, it could be anywhere. Can a change in one event cause a change in another event a long distance away without communicating with it? (i.e., faster than the fastest thing that can happen, the speed of light). Can things come into and out of actual existence? Can something come from nothing? The human mind is severely limited (perhaps completely) when it comes to psychologically envisioning the world as it really exists at its most fundamental levels. That's understandable. I simply suggest that we should be guided in our understanding by science as the best supplement to what we actually think and perceive.

This is not a chapter on physics or cosmology, per se, far from it; but we will outline some of the current thinking about these topics for a very specific purpose: Where did we and everything else come from?[1] In this chapter, I hope to show that this is a provocative but somewhat meaningless question and that *a god is not necessary for our existence*. I've attempted to outline some of the current scientific thinking about the universe, or multiple universes, as we'll see, and the problem of whether a god is necessary for such. This may sound like philosophical mumbo jumbo, but the issues will become clearer as we get into the meat of the problem.

Horizon Problem

We humans have always suffered from what has been called the horizon problem. For example, our horizon used to be merely as far as we could see with our eyes, which made the Earth flat, of course. People saw the horizon if they looked out at the ocean, and they could see no further. Our horizon was limited. The same kind of problem is where the Earth fits in the solar system, as I've reviewed. The horizon was changing. Galileo's powerful telescope allowed him to "see" that the Earth and other planets really do revolve around the sun, confirming Copernicus' theory. The horizon began expanding further. In the 17th century, our horizon into the cosmos was limited to the stars and planets we could easily observe. They were all in our own galaxy, and it was not until the 20th century that Edwin Hubble demonstrated that some of the lights in the sky were actually in other galaxies.

Now our horizon is our universe, though we can't actually see all of it. We can "see" vast distances into the cosmos. However, our horizon is limited by the light that has had a chance to reach us. It may be that this is not "all there is" since we do not know what else may be too far away. Also, some light that we do observe, of course, is from sources that no longer exist, e.g., extinct stars, at least in relation to our own "time" here on Earth.[2]

Our horizons are also limited by time. The universe is incredibly large and old, by current estimates being about 13.8 billion years old.[3] It is also incredibly large, with an estimated visible diameter of about 95 billion light years. The difference here is that despite the universe's age, the universe itself has been expanding, which accounts for its huge size. We don't know if it's actually bigger even than that, and it continues to expand. (The new James Webb Telescope has apparently imaged galaxies at least 33 billion light years away.)

Our horizons can be limited by the small things we can "see." It wasn't until the seventeenth century that the Dutch scientist Antonie van Leeuwenhoek used a compound microscope to discover a whole world of little creatures that we couldn't see and didn't know existed. More powerful techniques have allowed us to extend small horizons down to molecules, atoms, and even to record some subatomic particles. Our horizons have always been limited by our scientific measuring instruments and methods. We once could "see" only the matter that reflected light. Now we understand that this is only about 5% of the entirety of matter and energy in our universe—the rest being the dark matter and energy we're currently searching for. The more our technology develops, the better understanding we're able to achieve about ourselves and the universe around us. There's still a lot we don't know, and we don't even know how much of that there is. Our ignorance, however, is a horizon problem; but it is not an excuse for invoking a "god of the gaps," as we'll see. (We also have a horizon problem with respect to "complexity," e.g., the human brain, which we'll deal with in subsequent chapters.)

Early Historical Conceptions of the Universe

Let's review a bit. What is the universe, anyway, and where did everything come from? Humans have always thought about such questions, with or without gods being involved. We only have records dating back to the times that things seem to have been written down. Many of the ancient Greeks believed that the universe was created out of a formless void or "chaos" (the word itself comes from the Greek term for nothingness or void). They saw the cosmos as arising out of that chaos. Aristotle thought all the matter in the universe was fashioned by the gods out of this state of disorder, from the chaos that may have existed eternally, without form. That is, the universe must have always existed, according to Aristotle. He also thought the universe was finite in extent; and he believed that it was eternal—it would continue to exist forever.

Then there's the problem of what the universe is made of. We saw how the ancient Greeks considered the four basic elements arising in the new order, the cosmos—earth, air, fire, and water. They also conceived of some of the dynamics of these "elements." We've come a long way since then. For example, today we know that what we call matter and what we call energy are interchangeable. But the ancient Greek philosopher Heraclitus of Ephesus, who lived around 500 BCE, even presaged a bit of this knowledge when he noted (217),

> *All things change to fire,*
> *and fire exhausted*
> *falls back into things.*

We all remember $E=mc^2$. Heraclitus clearly saw, in metaphorical terms at least, the notion of Einstein's Special Theory of Relativity—that the universe is merely a dynamic morphing between energy and matter.

The ancient Greeks also considered humans the center of the cosmos. Humans and our world were the center of all creation. Of course, this is now of historical interest only, as the center has disappeared at the hands of scientists over the centuries. Aristotle also believed everything rotated in spheres around the Earth. The Greeks observed some of the lights in the sky, naming some of these "planetes," or wanderers, since they appeared to move across the night sky, as opposed to the other "stars." It did appear that all these "planetes" were moving around the Earth.

Fast forward to the Ptolemaic view of our reality in the second century CE—that the sun revolved around the Earth, which persisted for hundreds of years. It later gave way—much later of course—to the better and the more parsimonious theory of Copernicus. That Earth revolved around the Sun, the heliocentric theory of the solar system, was indeed revolutionary. In fact, it was a major heretical view at the time, and Copernicus' work wasn't published until he was virtually on his deathbed. Interestingly, he actually included in the work a dedication to Pope Paul III. (Copernicus was actually a devout clergyman. In fact, many of the people historically we think of as scientists were actually clergy or trained in religious traditions, from ancient times to more modern, e.g., Georges Lemaitre, who first proposed the Big Bang theory, and even Darwin.) Then, it was the Italian astronomer Galileo Galilei who quite famously made things even worse for the Church. Through his meticulous observations, Galileo was able to confirm Copernicus' heliocentric theory, and his findings were published in 1632. (As is well known, he was charged with suspicion of heresy and sentenced to house imprisonment by the Pope for the rest of his life. More on Galileo in Chapter 8). And it was the German mathematician and astronomer Johannes Kepler (1571–1630 CE) who further reinforced the Copernican theory, working out the specific laws of motion of the planets. Nevertheless, his view of the heliocentric solar system obviously also stood in contrast to the tenor of the times.[4] This occurred during a crossroads period in history—when science was struggling to exist as we were escaping medieval times. To wit, as he was developing the laws of planetary motion, Kepler's mother was arrested on charges of witchcraft. Kepler actually spent many years developing her legal defense, and she was eventually acquitted, though she died shortly after a long period of imprisonment.

Nevertheless, the Earth-centric view of the universe, the clear view of the Church, continued to persist among the populace, and any books referring to the heliocentric theory were banned. Today, even most theologians acknowledge that Copernicus was right, even as they promote the idea that we are the center of all so-called creation. In actuality, our Earth, our Solar System, our Milky Way galaxy—none of them—are at the center or any special place in the universe. Our location is not the center of anything. The Big Bang (below), which led to the universe we know and love, did not occur in one place. It occurred "everywhere." Though it seems paradoxical, modern cosmology tells us that no point in space has any more right to the label of "center" than any other. But it was actually the seventeenth—century French philosopher and mathematician Blaise Pascal who presaged this current conception of the universe. "The universe is a circle whose center is everywhere and whose circumference is nowhere." In sum, we are not the center of the universe that religion has always told us. And as the Copernican Principle suggests, we are not special in any other way, as we'll see later.

Quantum Theory

For now, let's take a tangent to an area of physics which does relate to some fundamental aspects of reality, what has become known as quantum mechanics. My purpose here is partly to illustrate—and rather dramatically—that you can't use your common sense or reasoning to appreciate some aspects of our world. This is a difficult subject, well beyond our scope here, but I'll attempt a general description, since it is highly relevant to a number of issues I'll discuss later—specifically the ideas of multiple universes, consciousness, and "free will."

I noted in Chapter 2 that Democritus had postulated that matter was composed of tiny particles that we couldn't even see, what we now know as atoms. But now we know there are even smaller things—subatomic particles and other particles which are fundamental to our reality—electrons, quarks, neutrinos, bosons, and many, many others (and still others that have been legitimately theorized but haven't been "found" yet). However, they do not behave in ways governed by the same laws of nature we know and love. It is a whole different world described by quantum mechanics. Objects in our everyday experience are governed by Einstein's theory of general relativity, but the theory doesn't work for the components of an atom and the other fundamental aspects of nature. And the quantum mechanics I'll discuss doesn't work for larger objects. This issue has been a dilemma for physicists for over a century.

Let's start with one of the fundamental constituents of our universe—light. Isaac Newton had considered that light was composed of corpuscles. But it was the 17th-century Dutch astronomer Christiaan Huygens who thought light could be a wave (although he thought it was something vibrating through an "ether" that suffused all of space, an idea that was later discounted.[5]) But it was Einstein who contributed greatly to quantum mechanics in 1905 with his work on what's called the photoelectric effect. Although light can be a wave, he demonstrated that light can also be thought of as particles or quanta, what we now call photons. So we have light as both a particle *and* a wave, an example of a fundamental dilemma of quantum mechanics.

Quantum mechanics reveals some very weird things about our world. Can something really be a particle or a wave, or both, or neither? Can something come in and out of existence? Does it make sense that something can be in two places at once? How can information travel faster than the speed of light—in fact, instantaneously across great distances without any time elapsing? Is the world actually determined, or is it a matter of probability?

Quantum theory is one of the most well-evidenced theories in the history of science. It is the most widely accepted current explanation of what goes on in our universe—at the most fundamental levels. But before delving into this area, I should be remiss if I didn't point out two very important considerations. First, quantum mechanics certainly cannot be described here in much detail and is certainly not understandable by us ordinary mortals. In fact, it was physicist Richard Feynman, who actually won the Nobel Prize for his work on quantum mechanics, who noted, "I think I can safely say that nobody understands quantum mechanics." (162) And second, it does clearly illustrate that you can't rely on any notion of common sense to understand how the most fundamental aspects of reality work. Feynman, in fact, observed that quantum mechanics "describes nature as absurd from the point of view of common sense." (164)[6]

The Measurement Problem

What are the most fundamental entities in nature? (Leaving out the controversial String Theory for the moment, which I'll discuss below.) Are they particles, like electrons or photons or quarks? Or are they waves, like the electromagnetic waves that we commonly know as light or radio waves or X-rays? Well, quantum mechanics tells us that they are both—or neither, a pretty confusing situation. Or perhaps there is a more fundamental level of reality we don't know about, as Einstein argued.

How *can* something be a particle *or* a wave? There is still no consensus "answer" to that question; but let's look at how the question actually arose. As it turns out, whether some phenomenon is a particle or a wave depends on how you look at it. That is extremely counterintuitive, but it is extremely well-evidenced. This is the so-called *quantum measurement problem*. Without going into much of the laboratory work on this, it may be helpful to look briefly at a classic experiment that illustrates the problem. It's called the *double-slit experiment*. The experiment has been repeated many times in various ways, but let's look at the basic paradigm. Bear with me.[7] In a typical experiment, a stream of particles, e.g., electrons (one of nature's fundamental particles), is fired at a metal plate which has a vertical slit in it.[8] A wall on the other side of the plate records an image, an imprint of what the electrons created. The image is basically a vertical pattern of dots, like one would get from ordinary pellets hitting a wall after going through the aperture. Now consider the same experiment when there are two vertical slits in the plate, placed some distance apart. When the stream of electrons is fired at the plate, what do the electrons do? To assess this, let's examine the image created at the recording wall. Amazingly, it's a whole array of vertical lines. This suggests that each electron goes simultaneously through both slits as a *wave*; and those waves actually *interfere with each other*, creating an interference pattern on the wall, such as any waves would make. You can see the same pattern when looking at ocean waves lapping the shore, where the waves interfere with each other as they approach the beach. This is exactly what happens with the electrons—they act as waves. When there is one slit, each electron acts like a particle. When there are two slits, each electron acts like a wave, showing a clear interference pattern.

Now, to complicate matters further, we can put "cameras" (actually some sort of sophisticated detection devices) next to each of the slits in the metal plate to "observe" the electrons as they go through. Now, with two slits, the pattern on the back wall looks like it did originally, but with two vertical lines on the wall—the electrons again act as particles. What? When the electrons are "observed," they seem to be particles. When they are not observed, they seem to be waves. This is the classic quantum measurement problem. It all seems absurd, but that's the well-established finding of quantum experiments. A fundamental particle, such as an electron, *might* be a wave. But we're left with the notion that observation, e.g., by a "camera," apparently changes the nature of reality—our reality; and it "becomes" a particle to us. When it is not observed, it impresses us as a wave. How can this be? It certainly doesn't make any common sense, as Feynman suggested.

In point of fact, it turns out that all of what we know as reality can be thought of as either *particles or waves*. Physicist Greene notes that, "*All* matter has wavelike character." (191) Particles are things we think of as matter, existing in some defined form and having mass. Waves are things that transmit some kind of energy from one place to another, but they have no mass of their own. When you see a wave in the ocean, you're seeing a change in how the water moves, but the wave doesn't have any mass itself —only the water does. When a sound wave is propagated, it has energy, but it has no mass of its own; the wave is merely certain movements of particles in the air. Electromagnetic waves, for example, also do not have any mass of their own; they are types of energy. To simplify a bit, electromagnetic waves go from very long wavelengths like radio waves or microwaves to infrared radiation to our visible light to ultraviolet radiation to things like gamma rays and x-rays. All these can be described as waves, but they can also be described as photons, "particles" of light, or quanta, as Einstein called them. Photons and electromagnetic waves are essentially two ways of looking at the same phenomenon. And all aspects of fundamental reality work the same way, from electrons to quarks to bosons (the particles that carry forces), to even atoms, molecules, and perhaps larger entities. Everything seems to have a wavelike character, as Greene suggests.

Copenhagen Interpretation

How do you explain this? There are a few theories which have been offered. We'll look at only a few of the issues that are relevant for our further discussions in Chapters 5 and 6. In the early twentieth century, physicists Niels Bohr and Werner Heisenberg proposed what has now become known as the classic Copenhagen interpretation of quantum mechanics. In this model, a fundamental particle, for example an electron, exists in some sort of haze or *superposition*, being neither a particle *nor* a wave. Its actual character is unknown. And it could be anywhere, suggesting some sort of randomness; we don't actually *know* where it is. However, when "it" is *observed*, its character as a particle and in a particular location becomes known. That is, observation reduces the probability of alternatives to some definite result, for example, a measured particle. The phenomenon is described with a "wave function," that is, a mathematical description of the possible alternatives for the phenomenon (well beyond our focus here). When the phenomenon is observed, the wave function is said to "collapse" from a probability of alternatives to a definite state. This is a gross simplification of the theory. But it does illustrate, as some physicists suggest, that perhaps reality at its most fundamental level is probabilistic, and it emerges as definite only as a function of observation or measurement in some way.

Physicist Erwin Schrödinger had a slightly different take on this. He developed an equation that shows how the nature of some quantum state evolves over time, but he did not feel that the "wave function" necessarily collapses when observation occurs. He questioned the notion that observation "forces" a particular quantum state to be known. To illustrate the problem, in 1935 he famously proposed the now well-known thought experiment of a cat in a box which was randomly subject to possible poisoning by a device, something in which a decaying radioactive substance may or may not trigger the poison. We don't know whether the cat has been poisoned or not, and as a result we should think of the cat as both dead and alive at the same time. The problematic idea is that the cat's nature cannot be known until the box is opened and the cat is observed. Observation creating reality? It's often cited that Schrödinger may have merely been sarcastic with his cat thought experiment, though he was clearly illustrating the absurdity or at least the limits of the Copenhagen interpretation.

The Copenhagen interpretation has actually been the dominant theory in quantum mechanics, but it has been variously questioned over the years. Astrophysicist Adam Becker has recently offered that too much credence has been given to this interpretation. (40). Is superposition a real thing, or is it just a term we use as a placeholder for the unknown? In addition, we can question whether reality can actually be changed merely by observation. Einstein, for one, doubted this, holding out for a "deeper" reality. Going even beyond Schrödinger, he once asked, "If a person, such as a mouse, looks at the universe, does that change the state of the universe in any way?" (475) Einstein clearly knew that an observer is an observer, despite the sarcastic equivocation of mice and people. Despite this perspective, the phenomenon found in the double-slit experiment has been evidenced over and over again, with ever-more sophisticated experiments. A few other theories have been offered, though the concept of "observation," whatever that means, is still not finally settled.

Particle-Wave Duality

One would be right to ask the question of whether there is something else going on to account for this seemingly probabilistic nature of reality. Some physicists have suggested that the electron has properties of *both* a particle and a wave, what's been called "particle-wave duality." Physicist David Bohm offered that a particle may exist "as such" but also has a wave function as an additional property of its reality, which may be hidden from view in some circumstances. (192) He noted that there are "hidden variables" which may account for the uncertainty that exists prior to measurement. The electrons have definite positions and velocities, but these properties are hidden. Particles and waves are actually separate elements of one "reality." We just don't know them under ordinary circumstances.

This particular theory aside, current thinking is that, in fact, all matter does have both particle and wavelike character. For example, interference effects have even been observed not only for fundamental particles such as photons or electrons but for molecules, even large molecules, e.g., molecules containing as many as 60 carbon atoms. (14) It seems that all fundamental entities can be characterized as having both particle and wave properties. It's just that those we're familiar with in everyday life have incredibly large probabilities of being seen, with more definite position and movement. This is true of molecules, cats, and everything else, including brains (which we'll come back to in Chapter 6 on Free Will).

Many Worlds

One of the most provocative explanations for quantum phenomena is the *Many Worlds* interpretation, which was developed by physicist Hugh Everett in the 1950s. It is becoming more and more popular among physicists as an interpretation of quantum mechanics. In this theory, the wave function never collapses, and superposition does not exist as such. What happens is that all values of the wave function *do* exist, but in some number of "parallel universes."

What? The explanation is that when any quantum systems or fundamental particles interact, our universe splits or branches into other universes. We don't yet know what that means empirically. "Worlds" here, as we see that word, does not necessarily connote separate planets or galaxies or other physical entities as we might conceive of them—though that's a bit uncertain. It merely suggests that realities exist in ways that are defined mathematically to be different than what we can actually observe at any one moment. This all seems weird and defies all common sense. One of the best descriptions of this comes from physicist Sean Carroll. He describes the quantum multiverse, as it's termed, in clear language. Building on Hugh Everett's Many Worlds theory, he notes that in any quantum event, such as measurement of an electron, there are "*multiple possible observers* . . . [in] *multiple worlds*, each of which contains a single person with a very definite idea about where the electron was seen." (79) Certainty of any event does occur; we just might not be able to observe it from where *we* are at a given time. The theory is gaining some traction among physicists and cosmologists. It also addresses Einstein's notion of there being some deeper understanding of the world and seems to explain the "measurement problem" that was posed by the Copenhagen interpretation over a century ago. I'll discuss this multiverse concept further below.

Entanglement

One of the weirdest, most interesting, and most confusing findings of quantum mechanics is that information can appear to be transmitted instantaneously and over long distances, i.e., with *no time* elapsing. It is one of the most poignant illustrations of the fact that you cannot rely on your common sense to understand the world.

This strange feature of reality seems to violate a fundamental physical principle—called *locality*—which says that some particle or force must act directly on another one to produce a change in the latter. One thing must be in "contact" with another in order to influence it, whether it be matter or energy. It must be local. A baseball can't be thrown unless some object, like an arm, imparts the momentum. Water doesn't boil unless energy from the stove directly interacts with the pot which interacts with the water. Your skin can burn if the ultraviolet energy from the sun directly interacts with it. But a strange feature of quantum mechanics is that fundamental particles can violate this feature, something known as *quantum entanglement*. That is, a pair of photons, for example, can behave as though they were "joined at the hip," even though they may be separated by great distances, even many miles. In a typical experiment, a pair of photons is specially prepared, and then one of them is taken some distance away. Then a change is stimulated in one of the pair (for example, changing the "spin" property). Voilà—a related change occurs (reverse of the spin) at the exact same time in its partner located somewhere else. There is no contact between them, and it happens instantaneously. That is, *no time* elapses for this to occur. It is said that the particles are somehow "entangled," something which violates the fundamental principle of locality. It is one of the great mysteries in physics, something Einstein called "spooky action at a distance." Is there some communication we don't know about? Could something be intervening between them, causing them to somehow communicate?

In 1964, Irish physicist John Bell demonstrated that such could not be the case, essentially that there can be no "hidden variables" to account for this phenomenon. His work suggested that there can be nothing that communicates between the pair of photons. Locality in quantum mechanics was ruled out. He showed that Einstein's "spooky action at a distance" was, in fact, really spooky. He found that there can't be any intervening variables in how one particle could "cause" an entangled partner to behave in a certain way. Our world, at its most fundamental level, became "non-local."[9] That is, at the most basic level, there is no local "cause and effect" directly between certain fundamental particles. We say that they are merely correlated. They are correlated in experiments, even over several miles, and, by extension, they may be correlated across many light years. Any two photons can, theoretically, be entangled over vast distances in the universe. And it may be that other particles can be entangled. For example, entanglement has been demonstrated in the "top quark," the heaviest of the known fundamental particles. (12) If the entanglement of fundamental particles wasn't weird enough, there is now some evidence that even molecules can be entangled. (223)

So far, it does seem that at its most basic level the universe is nonlocal. Of course, no one knows where the universe of correlations comes from. We could assume that they come from the very initial state of the universe, perhaps the singularity that existed at the Big Bang (i.e., the very early state of affairs of the universe where the principles of general relativity don't seem to apply.). This would make all subsequent events in the universe both fully determined and correlated. This is sometimes referred to as superdeterminism, which is not altogether popular with physicists. Perhaps there is a "wave function" for the entire universe, which quantum mechanics is telling us, that might account for the universe of correlations. But there's something equally important and provocative here. It suggests that our ordinary notions of "cause and effect"—our commonsense reasoning—can't be generalized to all the events in our universe, which we do not yet understand, and perhaps to our universe itself.

Decoherence

Nevertheless, in our observation of everyday events, locality just seems to rule the day. Regardless of the "spookiness" of quantum entanglement, or regardless of any probabilistic nature for fundamental particles, we see (or in some cases can theoretically see) that common events are fully predictable, at least at our levels of observation.

But let's look at the wave aspect that is theorized for any phenomenon. In any type of wave, there are crests and troughs, like the ones you see in the ocean. Waves have a smooth and orderly sequence of crests and troughs, what is referred to as coherence. Electromagnetic waves are the same, with smooth patterns of crests and troughs. The only differences among various electromagnetic waves are their frequencies (for example, visible light waves, X-rays, etc., have different frequencies) or the amplitude of the wave, i.e., its strength (for example, the brightness of a light). But when waves of any kind, including electromagnetic waves, encounter objects in their environment—such as your skin, or an eye, or a molecule of some sort, or a cat—the orderly pattern or smoothness of the wave is disturbed. The wave is said to be "*decohered*," and a definite position and movement can be understood. Interaction with the environment (including measuring devices) decoheres the wave function, and ordinary probabilities emerge, as we saw with tags or "cameras" interacting with electrons.

Decoherence is essentially the reduction of uncertainties such that quantum effects are not observed for larger objects, for example, in everyday life. We do know the position and motion of baseballs or cars. You can look at it this way, as science writer Charles Seife eloquently put it, "The larger something is, and the less isolated it is from its environment, the quicker it decoheres and loses its superposition." (458) Thus, uncertainties are reduced, and the nature of what we're looking at becomes clearer. This is the case even when we look back to fundamental particles such as the electron. When it interacts with its environment, the vague probability notion disappears, and its reality becomes specifically known. The "properties or character" of the particle, if you will, becomes known—or can theoretically become known, given adequate technologies —in some definite way. This is highly relevant for the chemistry of our brains, as I'll discuss later, which forms the basis for all our human behavior—and yes, including religiosity.

Conclusion

Nevertheless, at its most fundamental level, we're still left with the question of whether the universe is probabilistic, or does it follow known or potentially known empirical laws such that it is fully determined? Albert Einstein was never convinced that, at its most fundamental level, the universe is probabilistic, and many physicists still share that view. Quite famously, he is thought to have said, "God does not play dice with the universe."[10] His actual language, "Quantum mechanics is certainly imposing. But an inner voice tells me that it is not yet the real thing. The theory says a lot, but it does not really bring us any closer to the secrets of the Old One. I, at any rate, am convinced that He does not play dice." (235)

(As we'll discuss later, of course, Einstein did not believe in a personal god or a supernatural presence in our lives. Use of the terms by him here are but metaphors.)

Thus, he was denying any randomness or probabilistic nature of reality. He believed that the universe was fully determined. But physicist Bohr, one of the pioneers of quantum mechanics, with only a hint of sarcasm, told Einstein, "Stop telling God what to do." And Stephen Hawking went a bit further, noting, "God not only plays dice with the universe, but sometimes throws them where they can't be seen." But we understand none of these gentlemen were talking about some personal, omniscient being. They were metaphorically referring to the notion that we certainly don't know everything. No god is necessary for an argument about quantum mechanics. The scientific search for solutions goes on.

Metaphors aside, current thinking is that the probabilistic nature of quantum phenomena is not contrary to the notion of a specific determinism to the universe. *Randomness derives from the measurement problem*—we just don't know what character something has at any given moment. But events are still determined. The bottom line is that appearance of quantum uncertainties is not an argument against determinism for our universe, as physicist Carroll notes:

> The quantum state of a system, left alone, evolves in a perfectly deterministic fashion, free even of the rare but annoying examples of non-determinism that we can find in classical mechanics. But when we *observe* a system, it seems to behave randomly, rather than deterministically. The wave function 'collapses,' and we can state with very high precision the relative probability of observing different outcomes, but never know precisely which one it will be. (78)

When we can predict events from observations, determinism holds. Physicist Brian Greene says, "The probability that any particular event will occur at some chosen time in the future is fully determined by knowledge of the wave function at any prior time." (191) Quantum mechanics is not contradictory to the notion that our world is, in fact, determined. It is not random. And Einstein may have been right.

If the Copenhagen interpretation doesn't hold, it is possible that the Many Worlds theory of quantum mechanics does suggest that reality is still deterministic, and likely, there may well be a deeper reality than what we observe. Current thinking is that, in fact, there is no "true randomness" in the universe. (358) (See also *Many Worlds* below.) Our universe operates as a deterministic system, even though the quantum measurement problem endures. Events evolve according to the known laws of physics, even though the weirdness of quantum phenomena persists. We must accept the fact that events in our world, and our lives, are subject to the laws of a determined universe. Despite various interpretations of quantum mechanics, the future is still specifically determined by the past, and it is not necessary for any god to intervene in this system.

Origin of the Universe According to Current Physics

Did God bring our world, our universe, into existence? There is no evidence—and no rationality—to support this fantasy, but we should give some space to how our universe actually came into being. The origin of our universe is not a finally settled matter. I readily admit this. However, there is good theory about this, based on known facts. If you don't accept the science in this regard, you're left with only divine creation, a patently unsatisfying explanation. This may be the ultimate invocation of "God of the gaps" because you don't know where the universe came from (the illogic of which we'll discuss more in Chapter 8). This may not make you comfortable, but then again, that's irrelevant to the argument. It's no different than the myriad of invocations of God throughout history, from the motion of the planets to what causes disease to how bees can communicate with each other. One by one, science fills in the "gaps" and relegates God to a growing dustbin of history. The origin of our universe is no exception. So let's look at the actual science on this.

Big Bang

In 1929, Edwin Hubble observed that there exist galaxies beyond our own Milky Way. He also found that our universe was expanding, with all the other galaxies moving away from our own. Einstein had previously believed that the universe was static, only later conceding to the data that it was expanding. In fact, scientists have confirmed Hubble's initial impression that the expansion of the universe is actually accelerating; that is, the rate of expansion is itself increasing. If you "wind the clock back" on this phenomenon, it's clear that in the past the universe was smaller—a great deal smaller—than it is today. And, to grossly oversimplify, at one time it was incredibly small and then suddenly it "exploded," leading to the expanding universe we have today. Most people know about this Big Bang theory of the universe, which I'll discuss a bit more. But it actually wasn't Hubble, but the Belgian Catholic priest Georges Lemaître who was first to propose that the universe began as a single point, a "primeval atom" or "cosmological egg." In 1931, he used Hubble's data to show that the universe probably started from a very small point. Interestingly, Pope Pius XII saw Lemaître's theory as consistent with the Christian philosophy of Creation. God created everything, so He must have merely started it off with a Big Bang, another gap argument, of course, and Pius started promulgating this idea. But in 1951, Lemaître actually dissuaded the Pope from continuing to push the idea. Of course, many Christians still believe in that perspective today.

So how did the Big Bang happen? No god made it happen. The Big

Bang arose purely from an incredibly hot, "infinitely" dense area of quantum energy fluctuations. The energy which morphed into the universe we know was "there" "before" the Big Bang actually "happened." That was all that existed at that "time." (I use "there," "before," and "time" as though these are meaningful ideas in this context. They may not be, which I'll discuss further).

And it happened a long time ago, a very long time ago. The current consensus among cosmologists is that our universe began with a Big Bang some 13.8 billion years ago. This is a bit of a metaphor. Of course, there was no "bang." Bangs and other noises are the result of our ears converting sound waves from the air into a perception, but there was no air, of course, at the time the universe was born. The universe began as an incredibly hot, infinitely small point, though there is no understanding about just what that means.[12] If you go back in time, the classical laws of physics as we know them break down at the point of the Big Bang. This is known as a singularity.[13] Nevertheless, science tells us that there was then an "explosion" of some sort followed by a very sudden inflating of the universe. Since then, the universe has continued to expand to its present form, and it appears that it will continue expanding. The future is uncertain (see below), but it's clear that everything we know—galaxies, planets, humans, everything—was born of that single point of incredible energy.

And why did the Big Bang happen, and why did it happen 13.8 billion years ago? No one really knows.[14] Why did it happen with the force and other characteristics that it did? Why do our laws of physics, as we presently know them, break down at the time of the Big Bang? (I use "time" here somewhat metaphorically since time as we know it did not exist "before" the Big Bang. See Section on *Time* below for more discussion of this). No one knows the answer to these questions, but I'll discuss some of the alternatives.

Contrary to what most people think, the Big Bang was not the start of existence for our universe. It was *merely* a change, an energy burst that "grew" incredibly rapidly into the universe we know and love. The energy that was present "before" the explosion and the total mass-energy which we know today is exactly the same as it was then (Again, I use the word "before" advisedly. There was no "time," as we know it, until shortly after the explosion.) The Big Bang was *not a creation*; it was merely a change in the nature of the energy that was present at the "time." We know that matter can become energy and vice versa, again remembering Einstein's famous equation, $E=mc^2$. The quantum energy fluctuations at the point of the Big Bang gave rise to all the matter and energy in our universe (and maybe others).[15] We don't know why it happened. That is a matter for continuing investigation. We're unlikely to find a satisfactory answer in the near future, but it did happen. To reiterate, the Big Bang was not "creation"; and we must clearly distinguish its meaning here from what religious apologists refer to in using the term, i.e., the position that there was "nothing" until God created it. The "god of the gaps" might again be invoked here to "explain" the Big Bang (what many theists may see as an ultimate God of the gaps), as the Pope wanted to conceive of it. But it was not God's creation out of "nothingness."

Nothingness

Let's devote some space to this concept. Some of the ancient Greeks thought the universe had always existed, even in a state of chaos. However, at some point the notion of a "creation" began to take on more traction. Did some being "create" the universe? And did She/He/It do it "out of nothing?" And what does that mean? It was St. Augustine, the fourth-century theologian, who had offered that God *created* the world *out of nothing*. St. Thomas Aquinas, the thirteenth-century Dominican priest, also thought so, and his "reasoning" about this has been the theists' argument ever since (I'll get into the problems with this in Chapters 8). This may be the crux of the argument, and it's an old one. Was there "nothing" before the universe arose—before anything at all came into being? Is a supernatural being responsible for "creation" itself?[16]

Though this question has been an issue pondered since early civilizations; it was the German mathematician and philosopher Gottfried Leibniz who most famously posed the problem as an intriguing question, "Why is there something rather than nothing?" (301) His answer was God. The matter is also captured, with some sarcasm, by physicist Steven Hawking, who asked the question, "Why does the universe go to all the bother of existing?" (210) It seems like a nonsensical question, but it is an important one to consider. Philosophers and physicists (and their predecessors, the natural philosophers) have wrestled with Leibniz's question for hundreds of years, even before he formally proposed it. Science gives no answer at present. But the imposition of God, the ultimate god of the gaps, is also no answer. To analyze Leibniz's question further, perhaps we need to say what we mean by "nothing." We can say that we know what we mean by "something." We also *seem* to know what we mean by "nothing." But drilling down on this makes it almost a meaningless question: We cannot envision "nothing"—except, of course, as the actual absence of something. We cannot really conceive of the existence of "nothing" since that idea is actually based on our own limited experience and the constraints of language.[17] Is "empty" space nothing? When we refer to "space," we're actually referring to "something." It's *space*. And, by the way, it's not empty. Consider what happens when you physically remove all "something" from what we call space. Are you left with "empty" space? That's what most humans would consider "nothing." In actuality, space has its own energy. Even if you could remove all matter, there would still be "something." Einstein has also shown us that space itself *is* "something"—technically *spacetime*.[18] You might consider that a god or gods created the universe out of the vast "empty" void, as many early philosophers and religions proposed. But the void—the void itself—still existed, which is "something." Leibniz' question obviously has no answer, and cannot have an answer, unless you assume a god, a continually unsatisfying one.

In point of fact, according to some modern cosmological theories, the universe may indeed have arisen *spontaneously* "out of nothing." As theoretical physicist Lawrence Krauss notes, "The structures we can see, like stars and galaxies, were all created by quantum fluctuations from nothing" (284), though we can reasonably rule out that quantum fluctuations are not "nothing," as most would define that word. Nevertheless, quantum mechanics affirms that particles *do* come in and out of existence. If that is the case, it's clear that a universe of particles "out of nothing" is also possible. "Out of nothing" makes no sense, of course, in terms of human "common" sense or understanding, but the idea has gained some traction. Physicist Steven Hawking notes, "According to M-theory (which I'll discuss further below) ours is not the only universe. Instead, M-theory predicts that a great many universes were created out of nothing. Their creation does not require the intervention of some supernatural being or god." (2014)[19] This may seem a bizarre conception, but it is accepted by most cosmologists. In addition, I'll discuss the idea of the multiverse below, but Greene puts the "nothingness" argument in perspective:

> In the Ultimate Multiverse, a universe consisting of nothing *does* exist. As far as we can tell, nothingness is a perfectly logical possibility and so must be included in a multiverse that embraces all universes. . . . The answer to Leibniz, then, is that in the Ultimate Multiverse there is no imbalance between something and nothing that calls out for explanation. Universes of both types are part of this multiverse. (193)

If universes of "nothing" are possible, Leibniz has his answer, but we humans still have a problem with the idea. And that is because we invoke our common sense to attempt an understanding. To reiterate, language fails us, as the idea of "nothing" in this regard is impossible to operationally define. There is no affirmative definition of nothing; it comes down only to the absence of something. Second, we humans are able to pose questions which in themselves do not have any inherent meaning and which are unanswerable in any meaningful way. We often do this, and we know and admit it. As I noted in Chapter 2, humans have a natural tendency to seek answers to questions that have enabled improved functioning and survival. And thus, we have developed the clear propensity to ask questions and seek answers to all sorts of things, both real and imagined. We can formulate questions that nearly all humans would agree cannot be answered and are admittedly absurd. Imagination and grammar can be strange bedfellows. Some examples: What color is the number seven? When will the Tooth Fairy come? Are trees more moral than grasses? What does the letter Z taste like? What is the best number? What temperature is that poem? Is heaven near Bakersfield? How many aliens live in your basement? (There may be varying opinions on the latter two). Just because you can grammatically formulate a question doesn't mean that it isn't absurd—by definition. Some are just logically absurd or are valid only by assumption. Why is 2+2 four? It just is. And some are self-contradictory. How far is infinity?

Just because one can pose a question doesn't mean that it is reasonable to do so or that it can be answered. The failure to answer cosmological questions meaningfully does not give logic to fill the gap with God or anything else. Further, that something could arise from nothing, as some cosmologists have proposed, may leave you with a sense of emptiness or incompleteness. But that subjective experience is not a rationale for a god or supernatural power; it is merely a psychological problem. In sum, the classic "nothingness" argument is not a rationale for a god.

Was There a Beginning?

If you believe in a creation, you must assume, of course, that there was "nothing" (whatever that may mean) at some point in "time" and that what we know as the universe actually *began* at that point. However, in fact, we might ask, "Was there even a beginning?" Theologians assume that there must have been a beginning. The only reason for God to exist, based on any reasonable logic, is the so-called "creation" of our world—in today's understanding, creation of our universe or any reality *in toto*. This is the bottom line in religious arguments for the existence of God, an ultimate "God of the gaps," if you will. Even if religious apologists give up the idea of filling in supposed evolutionary gaps in the history of living organisms, an increasingly and uncomfortably failing endeavor (see Chapter 4), they always resort to the idea that there must be a god who got everything started. This is no longer a viable argument, as we'll see.

The notion of a creation or beginning has been argued since humans began fantasizing about supernatural forces. Plato and Aristotle mused about it, but it was probably most famously put forth by St. Thomas Aquinas. In general terms, he proposed to prove the existence of God—essentially by noting that everything has a cause, that the universe exists, and that God created it, as He is the only "uncaused" thing that can exist. This is the classic "first cause" or "cosmological argument," which has now been around for many centuries (We'll get into the logic and philosophy of the so-called First Cause or cosmological and other arguments for the existence of God in Chapter 8).

But what if there was no beginning and therefore no need for Creation—by a god or anything else? We humans like the idea that everything has a beginning (We're less comfortable with the notion that everything has an end, but I'll talk about that later.). It's fundamental to how we think and behave on a day-to-day basis. However, it may not be warranted in cosmological terms. As far back as the ancient Greeks, there was the notion of an eternal universe, that it had always existed and would persist forever. The problem with Aquinas' First Cause argument was cogently summarized by Bertrand Russell in 1927:

> There is no reason why the world could not have come into being without a cause; nor, on the other hand, is there any reason why it should not have always existed. There is no reason to suppose that the world had a beginning at all. The idea that things must have a beginning is really due to the poverty of our imagination. Therefore, perhaps, I need not waste any more time upon the argument about the First Cause. (435)

Theologians, of course, disagree. Most Christian theists assume a god which is "infinite in time," whatever that means. However, many cultures throughout history have gods which are not. Obviously, there is no meaningful answer here. If God—any god—was the creator and *not* infinite in time, then we're left with the uncomfortable question of what or who created God, and so on and so forth. It's reminiscent of the ancient Hindu lore about creation of the world—that the world (the Earth, all that was known at the time) was held up by a giant elephant (or elephants), which was held up by a giant turtle, and by yet another turtle, etc.—what has come to be known by the phrase "turtles all the way down." (The phrase is often used, in general, to illustrate the problem of an "infinite regress," often attributed to the nineteenth-century preacher Joseph Berg.)

But it is equally logical to assume that the universe (or the multiverse, below) has always existed, as Russell cogently opined. As theoretical physicist Lawrence Krauss puts it, we can ask, "Who created the creator? After all, what is the difference between arguing in favor of an eternally existing creator versus an eternally existing universe without one?" (284) Philosopher Peter Williams specifically argues against Krauss' view by suggesting that the universe must have an explanation and that explanation is God. (547) This, however, is merely a modern retelling of Aquinas' language. Assuming an omnipotent and eternal god is no different than assuming an eternal universe, and since it is eternal going forward, there is no reason to doubt that it is eternal going "backward in time." It is infinite.[20] It is the most economical explanation of our reality. Interposing a god in this context is illogical and at best redundant. It is only our psychological need that is associated with this superstition. If the universe is eternal, going forward and backward, then the notion of some creation is meaningless.

Additional cosmological theories have been promulgated about this eternality. Stephen Hawking developed the formal theory that the universe actually need have no beginning or end. This is based, in part, on the concept of "imaginary time." This is not what we think of as time, or more specifically spacetime, which emerged into reality out of the Big Bang. In this scheme, the universe would have no boundaries in time or space. Consider the analogy to the surface of the Earth. If you go north from any point, you will no longer be able to go north after you reach the North Pole. There is nothing north of the North Pole. If you proceed, you're actually going south. There are no linear boundaries to the Earth. It is a closed surface. Imaginary time would operate similarly. It might proceed, but it has no boundaries. Thus, conceiving of a beginning or end is meaningless. (211) Therefore, it would make no sense to use "time" to talk about an "origin" or creation for the universe. This obviates the need for some type of god altogether. Carroll astutely summarizes the point as, "There's nothing wrong with the universe existing without any external help. . . . in unmodified Newtonian physics the universe has no beginning . . . . Quantum mechanics allows us to consider universes that are fundamentally timeless, but in which time emerges at a coarse-grained level of description. . . . there's nothing we know about how such laws work that suggest the universe needs any help to exist." (78)

The theists' alternative, of course, is that God *preceded* the universe. Of course, this assumes a common notion of "time," which I've just addressed and which seems absurd here. However, if we accept their questionable assumption about time, we could then ask the ludicrous question, "How long did God wait to create the universe?" If He has existed eternally and the universe is *only* 13.8 billion years old (pretty small compared to an eternity), why the delay? We'll not give much attention here to St. Augustine's answer about this matter, i.e., his offering that, "God created hell for people who ask such questions." Ignoring the absurdity of the question, the matter was most aptly captured by Albert Einstein when he somewhat sarcastically offered, "What I want to know is whether God had any choice in the creation of the universe." In that case, you need to explain God's apathy or boredom, at least as we paraphrase such things in human terms, which led to a delayed creation. All this is cosmological and theological nonsense, of course. But it does again render meaningless the question of Creation.

In sum, the most reasonable explanation for the existence of our universe, or reality in general, is that reality is eternal and universes appear and die in this reality all the "time" (see *Multiverse* below). In that sense, *existence* itself is infinite, so to speak, which also renders Leibniz' question irrelevant. Thus, the ultimate "god of the gaps" argument when it comes to "creation" of the universe fails. Must there be a god to fill that "cosmic gap?" This has become a bankrupt argument. God is fully redundant or unnecessary in this regard. Again, the thesis here is that *it is no more logical to assume a divine creation or eternal being than to assume an eternal universe.* This may again be uncomfortable for some, but that's where we are.

Our current state of understanding does make God unnecessary. But you may certainly feel personally compelled to continue to ask the question about there being an initial cause, as Aquinas did and theologians continue to do. The difficult step psychologically is to realize that your asking the question is more about human nature and behavior than it is about posing an answerable philosophical question. The notion of "beginning" is *Homo sapiens'* reasoning. As I've intimated, we need to invoke various mundane "causes"—that some things come before others—in order to function (I'll discuss "cause and effect" more below), a skill that has evolved over the millennia and enabled survival. However, you can't project that idea onto the universe, a possible multiverse of any sort, or any conception of some ultimate ontology.

The universe does not need to have a beginning just because you want it to. Nevertheless, in all this, you are still left with the subjective notion of "but there *must* have been a cause." You may continue to insist that there must be a final, linear answer to the creation question. This becomes fatuous at some point, with the only escape being the invention or fabrication of a supernatural being. But again, this is your "common sense" subjective notion, which doesn't support any notion of cosmological reality. The bottom line of the "beginning" argument is foiled in two ways. Scientifically, there need not be a beginning, and it is our human need to ascribe causes that merely leads to an assumption of a beginning and the fabrication of a god.

End of the Universe

Our universe *will* end, at least in its current form. Just how and when is a matter of some debate, which I'll get into. But we should consider that if we human beings are so special, if we are God's children, if we are put here to worship Him, then why would He end it all? We're not talking here about some Biblical apocalypse. We know our Earth will be uninhabitable, gone in just a few years—perhaps just a few hundred million. Even in the event that we master human-feasible interstellar travel (a very unlikely scenario), humans are doomed, as the universe itself is destined for destruction in its present form. It is inevitable. Theists have no explanation for this, the one about an afterlife being irrelevant here. We're left only with the bizarre question of why God would destroy His own creation.

There are a few interesting theories about how this will happen. Let's consider just a couple of the more scientifically relevant ones.

The Big Chill

One of the most accepted theories among physicists is that the universe will end in a "heat death," sometimes dubbed the Big Chill or Big Freeze. The universe will expand past the point where any gravitational attraction can hold things together—galaxies, planets, people, everything. This may be due to the postulated, so-called "dark energy" which pervades the universe, but we don't really know. Scientists now think that dark energy will force the indefinite expansion of the universe and the "breakup" of all its constituents.[21] All matter in the universe may end up as energy. Of course, per the First Law of thermodynamics, it cannot be destroyed; it must persist. Then what? Nobody knows. But one thing is clear—"creation" will be over, a pretty cynical act for any supposed God.

Big Crunch and Bounce

But what if the universe expands to a certain point and then stops? Perhaps attractive gravity takes over and the universe starts to contract. The reverse of what has happened occurs, and eventually the universe contracts back to a single point. You can imagine what might happen. It might just "sit there," a Big Crunch, or perhaps it would undergo another expansion. This is the Big Bounce theory, which is gaining some scientific popularity. (355) The universe as we know it will destroy itself, and we cannot rule out that it, or something like it, might start over.

Maybe it's already happened. It may well be that our universe will collapse in the future into an infinitely small point and inflate again, leading to another iteration of "us"—or something like the universe we know, or perhaps different, and it may have already happened, a large or infinite number of times. This is the infinitely "bouncing" universe, one that grows, stops, contracts, and expands again and again.[22] This is a less accepted theory than one of eternal expansion or "heat death," or other ending, but it does have its strong supporters among cosmologists. (505) In this model, the universe—at a certain point—recedes back to its previous state, and then expands again. In this scenario, there is no actual beginning or end to the universe, just a never-ending cycle. We merely find ourselves in one period or iteration of it.[23] This theory would potentially also be consistent with the eternality of our universe. If our universe is a cycle of expansions and contractions to an initial quantum state, then it may be that it's been going on forever and will presumably continue infinitely. Thus, the notion of creation becomes irrelevant. Why would God create something only to see it expire and start over again? Perhaps God keeps trying until he "gets it right." But that's pretty inconsistent with the notion of an omnipotent deity.

Anything Else?

In any case—in any scenario for the end of the universe—it *will* end, *as we know it*, even though it will likely take trillions of years. But it will not end in an absolute physical sense. The energy that exists will continue to exist. It is eternal. And it *was* eternal at the point of the Big Bang, whether there was one bang or an infinite number of them. No creator God would be necessary. I should point out that there are other theories about the end of the universe not addressed here. For example, there is what is dubbed the "big rip." This is when all matter, and spacetime itself, is ripped apart, leaving only energy. Then there's the Higgs field, which is what imparts mass to other fundamental particles. If the field collapses (no one knows when or if that will happen), some or all the material universe we know would no longer exist as matter. Any of these scenarios may happen, though unlikely in our lifetime. The point is that the universe, as we know it, *will* end in its current form in some way. We humans are insignificant in this process and have no specialness here.

Multiverse

If God created our world, did He create others as well? Are there other universes? If so, we're not very special after all. We need to give this subject some consideration since it is important, something highly relevant to the idea of God and belief, and quite interestingly, it is relevant to other related problems, such as—unbelievably—the faith-related notion of free will, which we'll cover in Chapter 6.

The term multiverse, or parallel universes, has become popular in recent years. It's not just a subject for sci-fi novels, comics, or movies anymore. It has become a subject of interest for serious scientists. Perhaps we are not that unique, even to the point of having the only universe. That's a question which has been on the minds of scientists—and science fiction writers—for a long time (I find it interesting how often science fiction becomes realized by actual science. One need only look at space travel, wireless communication, and artificial intelligence, which were once only fanciful science fiction ideas.) However, in view of the current thinking on this subject, there's a certain kind of hubris or arrogance to assume that our universe is the only reality. We must admit that this is a controversial subject, but there are compelling reasons to take this seriously, and we'll take a look at what is known and theorized about this.

We're a little hamstrung here by language. We think of the term universe as meaning "everything," and of course if there are other realities other than our own, the term loses a bit of meaning. I use the term "universe" here to mean only a certain, particular reality which may be different in some respect(s) from other realities. This will become clearer in the pages that follow.

Historical

The concept of a "multiverse" is not an entirely new one. As far back as ancient Greece, Democritus reasoned that since everything was composed of atoms, including all of Earth, and since atoms must also exist in the heavens, he proposed that they could form themselves into other worlds, perhaps an infinite number of them. Plato also proposed that the three-dimensional reality we experience may not be the entire story. He envisioned our ancestors observing their shadows on a cave wall enabled by the fire around which they were sitting; their perception of the two-dimensional shadows may have been seen as actually some other reality, the "true" reality.[24] There have been many multiverse fantasies though history. There was the 19th-century French socialist and revolutionary Louis-Auguste Blanqui, who was repeatedly imprisoned for his actions (often related to advocacy for a free press and his criticism of the government). However, from his prison cell, he developed a "theory" that the universe is infinite and that what we actually see repeats over and over again in many places, including the elements and atoms, even all the way up to the existence of twin Earths. In fact, we'll see that presages some other, modern multiverse theories.

In modern times, there is a long history of the idea that there exist other realities or parallel universes, especially in science fiction, from many films and television shows to numerous novels, comic books, and other media. For me, one of the most provocative ones comes from H. G. Wells' 1923 novel *Men Like Gods.* (535) Wells developed the idea that humans could travel to parallel universes. The protagonist in the story claimed he was suffering from a condition called neurasthenia (This describes a medical condition associated with fatigue, sleep problems, and other symptoms. It is a term no longer used.); and he felt the need "to get away for a while." While driving his car, he found himself transported to an alternate world called Utopia, with many similarities to Earth but with some important differences—there were no longer any religions or politics, and advanced science was flourishing. His time away was all-too-short, but he returned to our world with a renewed intent to make it more utopian. This is science fiction, of course, but is it conceivable in some theories of multiverse? And wouldn't it be nice to see our world become a bit more like Wells' vision?

General

The question we considered, "What started our universe?", may not be comprehensive enough. In fact, we may be but one of a number—and an incredibly large number—of universes which have come into and out of existence. To assume we're somehow special, even as a universe, may be unjustified. There may be billions of others, perhaps an infinite number of them. This may seem like wild speculation, and some think it is. But some cosmologists have developed multiverse theories which would be consistent with our knowledge of general relativity and quantum mechanics and are "potentially" testable, though no one has figured out exactly how just yet. I'll address that a bit below.

Current cosmological theories do propose that we are not alone: There are universes other than our own. I'll describe a few of the theories. Some universes may be functionally separate from ours. Some may be characterized by different dimensions. Some may exist only for fractions of a second or fail to achieve any "form" at all. Some may have existed but have gone extinct. In this context, we might use the idea of multiple "realities" to describe the various theories of multiverse. In any case, as we'll see, the notion that our universe, as we know it, as the only one, might not be correct.

We should understand that these are only unconfirmed theories at this point. It should be pointed out that there are no empirical data to support these, but they are provocative theories based on existing science which can't be easily dismissed. But remember that *all* explanations of reality are theories, whether they be of gravity, evolution, electricity, or why humans do what they do. We judge a theory based on evidence and how well it predicts future observations. Some theories are better than others. And we should realize that a theory is never fully complete. Again, that would be religion. Evolution by natural selection, for example, is one of the most well-evidenced theories in the history of science, yet we continue to refine it with more data, so the theory becomes even more precise. The theory of gravity is also one of the best-evidenced and broadly accepted ones in science. That was true before Einstein came along and revolutionized it, though it did grow out of the theory which Newton had developed over two centuries earlier. Physicists continue to struggle, however, to refine gravity to fit more recent experiments and theories, and we still don't know the relationship between gravity and quantum mechanics, a critical chasm, perhaps the most critical one, in the broad breadth of modern science. So let's not dismiss the idea of parallel universes out of hand. Various theories of multiple universes are still somewhat controversial, and experimental work continues to try to provide better models. Given that caveat, there is a growing number of cosmologists who do support theories of multiple or parallel universes. Let's examine what we know about the possible frameworks for multiple universes that might constitute some ultimate reality.

Finite or Infinite

Let's start with a dilemma, a big puzzle, a very big one. Our universe is either finite or infinite. It *does* seem to have to be one or the other. However, it may be that this is too simplistic. It's not that clear; perhaps it can be both. The paradox has profound implications for the notion of multiple universes or multiple realities.

Our *cosmic horizon*, the current perceptible size of our universe, is believed to be about 95 billion light years, i.e., its diameter. That's pretty big, but it may be bigger still. Or it may go on in terms of spatial extent "forever." We just don't know. Or it may have a boundary, or it may not (most cosmologists predict that it does not).[25] Nevertheless, at the present time, cosmologists tend to support the idea that our universe is *infinite* in scope. There is no specific proof of this. However, at the "time" of the Big Bang, the universe was pretty small, extremely small, and, if it had a "size" ("size" being a term we associate with a finite number), that size is probably not easily definable.[26] Yet, many cosmologists theorize that the universe is infinite in scope, and it was also so at the time of the Big Bang. Is this a contradiction? Let's see. As Greene offers, "If the universe is spatially infinite, there was already an infinite spatial expanse at the moment of the Big Bang." The analysis goes something like this. If you roll the clock farther back in time, the denser the universe becomes, and ". . . its contents are squeezed even closer together, making the density of matter even larger. But its overall extent remains *infinite*." (193) That's a difficult concept to grasp. As an analogy, he argues that if you theoretically/mathematically "shrink" something—a table, a room, a baseball, any reality—it becomes smaller, let's say by half. And if you progressively do that over and over again to an infinite universe, decreasing its size by half each time, it may *appear* smaller, but it's still infinite in extent. Winding the clock backwards to the Big Bang, the universe becomes very small, to an indefinably small extent, as some "external observer" might assess, but it is still infinite in extent as we, inside it, might perceive it, being part of this universe (This might make sense in an abstract sense. However, in mathematical terms, you cannot halve or otherwise create a fraction of infinity. Infinity is an idea, not a number). Of course, this is an impossible idea for the human mind to grasp. How can something be both finite *and* infinite—infinite from the inside and finite from the outside? This is a paradox not likely to be soon resolved. In the meantime, we might accept this notion, as the matter is a pivotal issue for the various multiverse theories we'll discuss. For example, if it is finite from the outside, isn't it more than reasonable to ask—what else exists?

Laws of Nature

To examine the idea of multiple universes, let's again visit the idea of the "laws of nature." Now we must ask why those laws exist as they do. In our universe, we acknowledge that there are certain laws which describe our reality—the law of gravity, evolution by natural selection, etc. (Remember, from Chapter 2, that these Laws don't have their own tangible existence. They are mathematical descriptions of reality that we use to communicate about the world.) At a more underlying level, we recognize various "fundamental constants" of nature which are associated with those laws. So far, we've identified about 26 fundamental constants—Newton's gravitational constant, the speed of light, Einstein's cosmological constant, the masses of fundamental particles, etc.[27]

One must ask, as many have, why the fundamental constants of our universe have the values that they do. Are they random or arbitrary or inevitable? Or are they set by some external influence? We know that if they were much different in our universe, life as we know it would not exist.[28] If the extant values of the fundamental constants were different, for example, if gravity were a bit weaker, no stars would have formed, and we wouldn't be here (See also the *Anthropic Principle* below). But if other universes exist, the fundamental constants might be different, perhaps very different from ours. Perhaps there would also be other constants, and even the laws of nature themselves could vary, enabling all sorts of different universes. Some universes may barely exist before becoming extinct. Some may arise and expand forever without anything interesting happening. Some might be composed entirely of dark matter or antimatter. Some may have life similar to our own; some may have very different types of life. The possibilities are seemingly endless. If ours is not the only "universe," we likely are merely a lucky accident in a possibly infinite panoply of them.

Theories of Multiverse

Most (though not all) theories about a multiverse support the notion that the Big Bang *did* occur. That is, the quantum fluctuations present at the "time" gave rise to a different form than it had at time zero, setting off the universe we know and love. But it may not be the only one. As physicist Greene observes, we likely . . . "find ourselves inhabiting one of an infinite collection of parallel worlds." (193) There are a number of theories about such parallel universes. Let's examine just a few that are relevant for our subsequent discussions. I'll move somewhat quickly through them, but for the interested reader, there are a number of excellent lay resources to examine for these and others. (193, 517, 526) I'll note again that there is not universal agreement on these, but they are certainly provocative, and scientific curiosity continues to build.

Cosmic Inflation

When the Big Bang occurred, the universe rapidly expanded; and then shortly (very shortly) thereafter, it inflated at incredible speed, faster than the speed of light (Recall that the speed of light only refers to the velocity of photons *through* space, not space itself.). This is known as *cosmic inflation*, the most generally accepted theory of how the structure of our universe got started.

The theory also suggests that the Big Bang gave rise not only to our infinite universe, but that other universes may have also arisen. It suggests the possibility that there were so-called bubble universes that arose and are expanding infinitely themselves. Cosmic inflation also suggests that there have been and will be many big bangs, enabling an infinite multiverse—an infinite number of metaphorical bubble universes.

We've noted that our universe is infinite in extent. However, inflationary cosmology predicts that each of the bubble universes *can* have an infinite extent, at least as observed from the inside. This seems impossible, of course. Again, Greene notes that each bubble universe can have the same finite-infinite dynamic as our own. He states that "Each of the bubble universes appears to have *finite* spatial extent when examined from the outside, but *infinite* spatial extent when examined from the inside." (193) This again is a difficult concept to grasp. We certainly cannot envision an observer examining our universe from the outside. It is beyond our human ken. Nevertheless, for our purposes, we are left with the notion that our universe, as we perceive it, is infinite in extent; and this is true for other surviving bubble universes which emerge from any Big Bang.

The basic physics may be different in each bubble. They may or may not have characteristics similar to ours. Some may be "empty." Some may "collapse" before they really get started. Some may expand forever with nothing interesting happening. Some may develop with matter or antimatter. Some may evolve matter which leads to life, though some may be of our kind, and some may be different. The imagination can run wild here, for example, life composed of antimatter. But we may never know since we currently envision no way to access them; they're physically separate from us.

Cosmic inflation also predicts that all possible events can occur, and any event can happen infinitely many times. Let's see how this works. In our universe, if it is infinite, then any region of space has large enough areas or "patches" in which combinations of events (matter and energy) necessarily repeat. Since there are only a finite number of particle arrangements possible in each patch, and given a large enough patch of space and no limit to time, any combination of particles *can and does occur*. This means that any particular combination of particles can be repeated, and given an infinite universe, they necessarily are. Water molecules should exist elsewhere, as should phosphorous, carbon, methane, and granite. In an infinite universe, all entities thus exist in an infinite number of patches. This implies that there are other places where stars form, other planet Earths, other organisms such as we have on Earth, and other humans, including "you." This seems ludicrous on its face, but the logic is clear. We are *not* unique in this scenario. This is difficult to accept, but the reasoning is clear: everything that exists occurs multiple times at some points somewhere. The notion of infinity demands it.

Now let's consider the other bubbles or baby universes inflation produces (other than ours), perhaps an infinity of them. The process of repetition occurs in each of them. As Greene notes, "Parallel universes arising from inflation generate quilted cousins." (193) That is, there are other pocket universes, each with its own patch of infinitely repeating combinations of particles. Since there is an infinity of bubble universes, and each may lead to an infinity of patches, events and matter which occur somewhere can and do repeat again and again. Again, the possible arrangement of particles in each gives rise to any possible combination and recognizable reality. Greene goes on, "The best available cosmological data leads us to think of ourselves as occupying one of a vast inflationary system of parallel universes, each of which harbors its own vast collection of quilted parallel universes. Cutting-edge research yields a cosmos in which there are not only parallel universes but parallel parallel universes. It suggests that reality is not only expansive but abundantly expansive." (193)

We humans may exist not only multiple times in our own universe, for example, but an infinity of times in other universes. Human uniqueness certainly takes another blow.

String Theory

Now let's consider a different type of multiverse theory, one that looks at reality quite differently. This is an extraordinarily complex idea, which we'll not adequately describe here. (One of the better lay ones is contained in Greene's *The Elegant Universe*.) (191) Classically, we have always thought of the fundamental particles of nature as "things," things that have mass and occupy particular points in the universe. The electron, quark, neutrino, etc. (Note—the photon, of course, might be a "point" or quanta, but it has no mass.) There's a huge list of these "point particles," contributing to what is known as the Standard Model of particle physics. But that may be an incomplete picture (We've already seen that particles can be waves and vice versa.). There is another theory that suggests that fundamentally all these "particles" are not particles at all. This is *String Theory*, a still very controversial idea. String Theory has been proposed—in great detail—to explain the most basic "units" of nature. All the fundamental particles are essentially minute one-dimensional "strings." As Hawking notes, "According to string theory, particles are not points, but patterns of vibration that have length but no height or width—like infinitely thin pieces of string." (214)

These strings come in various shapes and sizes, can be short or long, open or closed-looped, and have other characteristics, and they vibrate in different ways, all of which produces our impressions of all the various fundamental aspects of nature. They can lead to our observation of what we think of as quarks or electrons or the weak force or any other particle in the Standard Model, but the reality of each is a one-dimensional string. They cannot be "seen" at this time. They are incredibly small, beyond our ability to measure them. Our technology is far from being able to experimentally assess them in any way, and the theory is still only hypothetical at this time. It has elaborate mathematical underpinnings and is only a theory that "makes sense" to a lot of physicists.

There are actually five superstring theories; and these are unified into a more general theory called M-theory. Although still controversial, some physicists think that M-theory might eventually solve the most elusive problem in physics—that is, to develop a unified picture of the universe which accounts for the principles of both quantum mechanics and general relativity. As I've said, quantum mechanics doesn't get along with Einstein's general relativity. Relativity explains the behavior of larger objects in our world (baseballs, buildings, the moon and the Earth, other planets, etc.), but it comes up short when it comes to subatomic particles and other fundamental forces of nature (e.g., quarks, electrons, etc.). And it's just the reverse for quantum mechanics, which doesn't explain the behavior of larger objects. Coming up with a theory that accounts for the behavior of both fundamental particles and of larger entities has been the elusive quest of physics for over a century. Einstein never conceived of a way to do this—what physicists now refer to as the elusive "theory of everything." The efforts continue, but so far have come up empty. Now, by describing all aspects of reality, M-theory may provide an answer.

If it does, we should consider that M-theory does predict the existence of universes or realities other than the one we're familiar with. Although controversial, M-theory does suggest the existence of realities different from our everyday experience. It suggests that there are realities we can't even begin to comprehend. For example, we think of reality, our universe, as existing in four dimensions—three spatial dimensions and one of time. However, that may not be the whole picture. M-theory predicts that there could also be other dimensions that we cannot so far access—actually 11 dimensions (ten spatial ones plus time), not just the four we're familiar with. We just cannot "see" or experience them because they're curled up within the others. And they're pretty small. It is estimated that the extra dimensions are not very different from what's called the Planck scale (10^{-33} cm). And they can be "curled up" in various ways, which possibly creates seemingly innumerable kinds of realities. In each reality, there can be different laws of nature, different organizations of matter, and of course, the possibility of different kinds of life. Hawking summarizes the model:

> The laws of M-theory therefore allow for *different universes* with different apparent laws, depending on how the internal space is curled . . . . Also, M-theory can contain not just vibrating strings but also point particles, two-dimensional membranes, three-dimensional blobs, and other objects that are more difficult to picture and occupy even more dimensions of space, up to nine. . . . depending on how the eternal space is curled . . . M-theory has solutions that allow for many different internal spaces, perhaps as many as 10^{500}, which means it allows for 10^{500} different types of universes, each with its own laws. (That's 10 followed by 499 zeros.) (214)

M-theory is supported by only a notable minority of physicists at this time, and it is something that has taken a lot of criticism and rebuke—especially for not being testable with extant technology or for not making any reasonable predictions. This is serious criticism. We don't yet have an empirical test for the validity of String Theory. Most physicists doubt that any such test could be conducted, and they cannot even conceive of one. The absence of adequate measurement technology has been a problem, often solved, in the history of science. But we shall see.

Still, if M-theory cannot unify the fundamental forces or fails to find extra dimensions, we wouldn't fill in those gaps with God. Einstein could not have envisioned a multiverse based on M-theory, but he didn't resort to a god for the universe. He used the concept of "god" in his wonderment about the universe. But he wasn't referring to some supernatural being. He was referring to the laws of nature. Stephen Hawking had a similar view. "If you like, you can call the laws of science 'God,' but it wouldn't be a personal God that you would meet and put questions to. Although, if there were such a God, I would like to ask however did he think of anything as complicated as M-theory in eleven dimensions." (213) But one thing is clear from the theory: "According to M-theory, ours is not the only universe. Instead, M-theory predicts that a great many universes were created out of nothing. Their creation does not require the intervention of some supernatural being or god." (214) The Copernican Principle reminds us of how insignificant we are in this grand scheme of things—an instance of human life in one *type* of universe out of 10^{500} types of them (though we presently have no way of knowing if such exists in any of the others).

Many Worlds

Another multiverse theory is based on the Many Worlds hypothesis of quantum mechanics introduced above (this is an important idea for what we'll talk about in Chapter 6). Since physicist Hugh Everett's initial development of the theory in the 1950s, it has gained more and more support from physicists. This will seem truly bizarre. (More lay descriptions of the theory are available.) (358, 378)

Let's start with the question that quantum theory raises. Is reality inherently probabilistic? The Many Worlds theory says no. But is it probabilistic only to our observations? If nature is deterministic, as physicists agree, then reality almost demands some type of a multiverse. What we "see" in our universe is determined by our biased observation, i.e., our location in this universe. All values of the wave function *do* exist, but they exist in an infinite number of "parallel universes." In this scenario, anything that could happen does happen in some given reality, not necessarily ours. We are just not aware of the "other" universes because we actually live *in* this one. They are functionally decoupled from us. For example, in some experiment in one universe a scientist observes an electron going one way, in another a scientist observes it going another way. The wave function splits or branches into different realities, what could be termed "parallel universes." Multiply this by any or all events, human decisions, etc., and you see that the splitting can be described as a multiple of realities. This is hard to envision with common sense, remembering that *our* common sense is a function of the reality or universe that *we* happen to inhabit.

In the Many Worlds interpretation of quantum mechanics, there are an infinite number of realities that can arise. Rather than some physical space as we know it from our own mundane observations, parallel worlds exist in a "probability space." That is, every possible outcome of all events can and does occur in determined fashion, and they all occur at the same time. Science writer George Musser cogently summarizes this perspective:

> This way of thinking about determinism also makes sense of an interpretation of quantum theory that was developed in the years after Einstein's death in 1955: the many-worlds interpretation. Advocates argue that quantum mechanics describes a collection of parallel universes—a multiverse—that behaves deterministically in the large but looks indeterministic to us because we are able to see only a single universe. For instance, an atom might emit a photon to the left or to the right; quantum theory leaves the outcome open. According to the many-worlds interpretation, that is because the same situation arises in a zillion [metaphorically, zillion is not a number] parallel universes; in some, the photon goes deterministically left, and in others, it goes right. Not being able to tell which of those universes we reside in, we cannot predict what will happen, so the situation from inside looks inexplicable. 'There is no true randomness in the cosmos, but things can appear random in the eye of the beholder,' says cosmologist Max Tegmark of the Massachusetts Institute of Technology, a proponent of this view. 'The randomness reflects your inability to self-locate'(358.)

This resolves the problem of determinism and quantum uncertainties, supporting the former and doing away with the latter. Any elementary "particle" is described as a probability wave, describing the probability of "finding" it "anywhere." Thus, anything that is possible does happen in some reality—in some "universe," which fully retains the idea of determinism. "The future follows uniquely from the present," as physicist Carroll describes it. (78) There is actually only the "wave function of the universe." When an event occurs, different parts of the wave function or "branches" come about, and they do this consistent with deterministic reality. All of reality "is just the quantum wave function . . . [everything else] is a convenient way of talking." (78) For example, if you make a decision in this universe, e.g., to turn right, there is another reality in which "you" turn left. All possible outcomes exist; we are only aware of the ones we experience ourselves *here*. Science writer Charles Seife summarizes this perspective.

> If our universe is part of an enormous *multiverse* that constantly proliferates and sprouts new universes, the rules of quantum mechanics begin to make a bit more sense. One form of this *many worlds* hypothesis states that every time a quantum object makes a 'choice' - alive or dead, spin up or down, left slit or right slit - our universe splits in two. If the many worlds scenario

> is true, there might be lots and lots of different universes out there with different constants. Some collapse in a millisecond. Some have almost no matter in them. We just happen to inhabit one that is suitable for life. (458)

Experimental Confirmation and Theory

There are many cosmologists who reject the various multiverse theories, understandably, because they cannot be tested experimentally with current technologies. Evidence for universes or realities other than our own still lacks empirical evidence. Thus, there is intense debate among cosmologists whether the idea of the multiverse, in one form or another, even qualifies as science. That is, is it reasonable to propose something without any empirical evidence? Or are good scientific theories also developed by conceiving of explanations for phenomena and then testing them against observations? Likely, it is both.

Of course, that does not rule out the possibility that novel ideas or methods for testing these ideas may emerge in the future, as farfetched as that may seem. History is full of examples of ideas that later proved valid, but only when empirical methods and technologies came along appropriate to the problem. Darwin developed the theory of biological evolution by painstaking observations over many years. He observed that the origin and survival of species, and the evolution into other species, could be dictated by the environments in which they existed. Nevertheless, doubts remained. However, over the past century and a half, the theory has been supported by sophisticated paleontology and dating procedures, as well as by advanced genetics, a seemingly infinite number of observations and experimentally confirmed over and over again in everything from mammals down to single-celled organisms (I'll highlight some of this work in Chapter 4). It is one of the most well-evidenced theories in all of science, which grew purely out of basic empirical observations.

The advent of new tools and procedures is often what enables scientific progress. Modern astronomy was enabled by Galileo's use of a telescope. The "germ theory" of disease was enabled only after invention of the microscope and the subsequent work of Pasteur, Joseph Lister, and Robert Koch. American biologist James Watson and English physicist Francis Crick are generally credited with the discovery of the structure of DNA in the 1950s. However, the work was based, in part, on the technique of X-ray crystallography which was developed by British chemist Rosalind Franklin. Technology was a key. The notion that our universe arose from a Big Bang, as Georges Lemaître proposed, was accepted only after we had the technology to support it, enabling Edwin Hubble to find that the universe was rapidly expanding. The existence of subatomic particles had been proposed for quite a long time, but they have now been confirmed experimentally. One of the fundamental particles of our universe was proposed only theoretically many years ago. In 1964,

Peter Higgs (and others) hypothesized that a particle should exist, based on some pretty esoteric reasoning, which gives every other one its mass. It took a long time, but in 2012, existence of the aptly named Higgs particle was finally confirmed at the Large Hadron Collider in Switzerland.

On the other hand, merely keen insight has led to other, equally correct, theories. For example, Einstein engaged in years of what are called "thought experiments." He imagined how a person on a moving train would perceive a couple of lightning strikes ahead of and behind him —as opposed to the perception of those by someone standing on the platform. And, with some additional effort, out came the Special Theory of Relativity. We all know what gravity is, or we think we do. Actually, no one has ever *observed* gravity, per se; we only see events which are consistent with the theory. Albert Einstein didn't observe gravity either. He imagined how a person in a windowless elevator would feel floating through space or, on the other hand, being accelerated through space. And, with painstaking mathematics over several years, out came the theory of General Relativity, which is more commonly called gravity.[29] When Einstein proposed General Relativity, it was rejected out of hand by many scientists. The theory was actually not empirically confirmed until many years after he developed it. It was validated by the description of light bending around the sun by the telescopic observations of Sir Arthur Eddington in 1919. The finding was something so important to our understanding of the world that it made the front page of the *New York Times*. No one had "bought" this theory for quite a while, but it was an example of scientific thought growing out of keen insight and imagination, evidence being gathered only later. It has now stood the test of time, having been confirmed though countless empirical observations to the present day. The theory also predicted other phenomena unknown at the time, e.g., the existence of black holes and gravitational waves, both of which have now also been confirmed. Great insight *can* generate good scientific theories.

There are a number of theories of "parallel worlds" or multiverses, developed by keen insight and mathematical reasoning. But do they really exist? Scientists are busy trying to find experimental confirmation of such things. A number of strategies are being pursued, though none of them have yet panned out.

Let's take just one example. Inflationary cosmology predicts that the Big Bang produced many (perhaps infinite) bubble universes. Shortly thereafter (or at some other time) these could exist "nearby" to us. Although they would be functionally separate from our own universe (light would not travel between them), some cosmologists have theorized that such universes could "bump" into each other or somehow otherwise interact. If that happens, or has ever happened, such an event might leave its mark on our universe in some way. If that happens, it might be detected. How? When we look out into the "sky," we're looking out into the entire universe. That "looking" has been observed with different kinds of telescopes over many years. What's been found is that the universe is suffused with "light" left over from the Big Bang. Actually, this light is in the microwave range, and the phenomenon is called the *cosmic microwave background radiation* or CMBR. The vibrations of the CMBR have contributed a great deal to what we know about the universe. The CMBR is everywhere. It has an incredibly uniform pattern all over the universe, no matter where you point your telescope. The variation in this pattern, as described by temperature, is incredibly small. If there is some detectable change in this somewhere in our universe (more than the very minor variations we "see" now), that imprint "could" be evidence of an interaction with some other entity, another universe. In fact, it "may" have already happened. There is some suggestion that a known particularly identified "cold spot" in the CMBR could represent such a "collision" in our early universe. Other explanations for the phenomenon have so far come up short, but the inference does remain controversial. (27) More sophisticated observations and experiments may enable us to gather better evidence.

An interesting historical parallel. The 17th-century poet John Milton had even noted that if we could "hear the music of the spheres," time would reverse itself. In *De Sphaerarum Concentu* ('The Music of the Spheres'), Milton, who was also a musician, essentially suggested that the universe works according to the idea of music, which had been eternal. Even before that, the Greek mathematician Pythagoras noted that the heavenly spheres or planets created some sort of music. Milton was also referring to the planetary spheres, but he was essentially right. "Time" *is* reversed when we look at the CMBR, which are the vibrations that existed over 13 billion years ago—as we look back in time. Perhaps continuing to examine it will lead to some indication of parallel universes.

The concept of multiverse, in numerous forms, continues to gain traction among cosmologists, as does doubt, but the search for evidence and new ways of thinking about the problem go on. As cosmologist Lawrence Krauss notes, we and the universe are but an "accident." There are likely many other realities, and currently evolving technologies may "perhaps even shed light on the possible existence of other universes." (285)

Anything That Can Happen Does

The bottom line is that all the multiverse theories (including others not alluded to here) are consistent with the notion that *anything that can happen does happen*. Our planet is not unique. Perhaps neither is our known universe . . . or you. That is an incredibly difficult concept to accept.

This is nonsense from a commonsense view, but it is perfectly logical from the view of an infinite universe. Given our infinite universe—and an infinity of infinite universes—and given that any combination of particles or wave branching can and does occur, anything that can happen *does*. I should make clear that, within this universe, such must occur within the known laws of physics and mathematics we have. There are some things that could not happen. Two plus two will never equal five. Humans as we know them will never be able to fly, though we can't rule out that other hominin species may evolve wings somewhere. There's the old metaphor that, given a typewriter and enough time, a monkey will compose a Shakespearean play. Monkeys and typewriters can certainly occur. However, a monkey Hamlet cannot happen. This is not possible given the assumed character of typewriters and monkeys, as we define them, who, by current definition, do not have the attention or lifespan to sit at a typewriter that long. There could be variants, of course (see "Planet of the Apes"), and in other "patches," or in other universes with different laws of physics, the "anything can happen" paradigm might hold for such things. Perhaps monkeys in some other universes are longer-lived, literate, and more creative than the ones we have here.

And let's not discount the possibility of various religions and superstitious worship. There are thousands of religions and various gods that have been worshiped here on our Earth. We do not know that there aren't advanced civilizations of living things that have a well-developed polytheism: some that may worship trees (yes, we have that too), and some that long ago settled the matter of a godless universe. They may currently exist; they may have existed and died out, or they may arise in the future. In some civilizations, there is only blind faith and superstitions. In some—like our own—there is a mix of both science and superstitions among our species. There may be worlds like H.G. Well's *Utopia*, without any religion, and where humans fully respect science and discount the idea of a Creator. Of course, in some, humans have not even come into being, and in some, human civilizations have died out or been destroyed (notably inconsistent with a loving God). In some, other types of creatures have superstitious beliefs, and in others, some creatures believe in science and are atheists. And in most, life itself (as we presently know it) is not even possible. We live in a universe in which the organization of matter is as we (mostly) know it, and that is only a matter of the fact that we are here (see also the *Anthropic Principle* below). It does not make our "beliefs"—or for that matter, any set of ideas or behaviors—any more valid than those in any other universe or any other "patch" of our own.

You might ask yourself: Why would a god go to the trouble of creating multiple universes—some with life, some like life as we know it, some with physical matter and no life, some with exact duplicates of ours (including exactly you), some that collapse immediately, and some with nothing at all? It certainly renders the issue of any human "specialness" meaningless. This is possibly the ultimate expression of the Copernican Principle. Science has not yet provided empirical support for any of the multiverse theories. But we shouldn't attempt another theistic gap explanation for why we're here or even why our universe is here. That's just another philosophical dead end.

Time

In order to address the notion of Aquinas' "First Cause" logic about God, the cosmological argument, we'll need to examine the idea of "time." (This is related to cause and effect, which I'll also discuss below.) It goes something like this. If God "caused" the universe to come into being, then He must have existed *before* it. In fact, St. Augustine had already supposedly "resolved" the temporal problem by saying that it was God who invented time. Not so fast.

What is It?

What do we mean by the idea of "time?" Our current notion of time comes down from our ancestral wonderment about the world, specifically the order in which things occur. Some things happen before other things. That individual cognition, if you will, is what allowed complex organisms to survive. And it is what led to our idea of time. The idea of "time" is so ingrained in our experience that we find it difficult—if not impossible—to shed it. We think of everything that happens as occurring over "time."

Other animals also have some conceptions of time, if you will. They "know" that some things come before others. Otherwise, they couldn't function. Time to all living organisms is essentially the sequence of events. When the prey is near, I attack. When the infant chimpanzee cries, I feed it. The sun will come out tomorrow. The falling leaves suggest cold is on the way. Any change in the environment is "time." Events seem to "precede" each other in our everyday lives as we perceive them. But time, as a distinct entity, may not even exist, and I'll examine this. Nevertheless, in this chapter, in order to merely communicate, I am forced to use more mundane ideas of time with words such as before, after, and the like. Think of this as a grammatical convenience or merely a way to make the issues easier to appreciate. This is an area where language again fails us. But let's proceed.

Some physicists hold the notion that "time" itself may not exist. Of course, we know that there is no *absolute* time. Einstein put that to rest. It is always relative to motion and mass. This is what Einstein conceived in the theory of general relativity and the idea of spacetime. *There is no absolute time.* The faster we go, the slower time passes. For example, your watch will lose time when you're on a jet plane going on a long trip, though the change is too small for you to notice. We've all heard the fantasy of traveling around the solar system on a spaceship at very high velocities; and when you return to Earth, its history has moved faster than your traveling clock would indicate. *Your* time has slowed down. And gravity slows down time as well. The greater the gravitational field, the slower time passes. It's been shown that clocks on earth at sea level run slower than those placed at altitude, like on top of a mountain, though again too small a difference to be measured by ordinary clocks. The penultimate of this principle is when you get close to a black hole. The gravity there is so strong that your clocks start to move extremely slowly. And it is hypothesized that time actually "stops" at the center of a black hole, something we cannot envision with common sense.

In fact, time itself may actually be an illusion. We don't actually notice "time" itself, something that's been recognized for a long "time." Aristotle, in fact, noted that time is merely the relation between events, as he said, "There is no time apart from change." We "measure" it with clocks, but what is it that's being measured? Most practically, what we perceive as "time" depends only on where we are and our sense of movement. Our conception is actually based on ordinary movement or change. Specifically, it is how we perceive the speed of something relative to the distance it travels. When a baseball is thrown, or a car speeds by us, or we see a snail crawling along, we infer time only because of how it seems to be moving. Our *perception*, what we call time, is only the ratio of the distance an observed object travels to its velocity.

The 18th-century Scottish philosopher David Hume acutely summarized the notion, arguing that there could be no such thing as absolute time. He opined, "As it is from the disposition of visible and tangible objects we receive the idea of space, so from the succession of ideas and impressions we form the idea of time, nor is it possible for time alone ever to make its appearance." (230) He rightly conceived that time is an inference based on the movements we observe. His insight is also what is felt to have been one of the influences on Einstein in his development of his theory of relativity. (235)

There Is No "Before"

Let's take a closer look at St. Augustine's notion that "time did not exist until God created it." That is, it did not exist *before* that point. But "until" and "before" are temporal constructs; they assume the existence of time. Of course, this is merely another gap fantasy. The universe began with the Big Bang, and it is that which enabled "time" as we may "know" it. In Big Bang theory, all the dimensions of space and time appeared only at that "point." In actuality, as Hawking notes, time was initially "just" a spatial dimension. (214)[30] There was no "before" for Augustine's God to invent something.

To be clear, let me restate the current conception of "time." It is again reasonable to state that time has no boundaries, and thus conceiving of a beginning or end is meaningless. (211) As Hawking notes, "Maybe the universe has no boundary in space and time." (213) In fact, it was New Zealand astronomer and cosmologist Beatrice Tinsley who may have first offered that asking what came before the Big Bang is like asking what is north of the North Pole, as I suggested above. It makes no sense. The analogy is again apt—if you keep walking through the North Pole, you're going south. Time has no boundary—no beginning or end, though we'd be remiss in leaving out Aristotle here, who offered a similar notion two and a half millennia earlier. It is difficult to envision a no-boundary condition for time. But try this. Just try to actually imagine an end to time, or a beginning, or even that there is no time. It is impossible as a cognition. Thus, some being "inventing" time becomes even more ludicrous.

Does It Flow?

We also think of time as flowing, almost like a river, with its energy moving from one place to another. Asked another way, do some events actually follow others? It is our mundane perception that creates such an inference—from our perception of movement. On larger scales, in order to explicate this, we need to consider one of the fundamental laws of physics, the concept of *entropy*. Entropy is a measure of disorganization or disorder. Though it's grammatically awkward, the higher the entropy of a system, the more disordered it is. When we slowly pour some cream into a cup of coffee, the mixture is initially in a state of low entropy. That is, there is some notable separation of the cream and coffee, a relatively low state of disorder. You can discriminate some of the cream from the coffee. However, as it dissolves, the system becomes more disordered; you can't tell the cream from the coffee. It evolves to a state of higher entropy. From this, we infer "time" because there is a change in the entropy of the system. The same principle holds for our universe. It was initially in a state of low entropy and has become more disordered, more disorganized. It would seem not, but you must rule out the parts of the universe where some organization is caused by the forces of gravity and electromagnetism, for example, which create the organized objects we see—molecules, stars, planets, and people. And entropy can never decrease. At the largest levels, the universe is constantly in a state of increasing disorder or entropy. (This is the Second Law of Thermodynamics, which will be referred to again later).

We believe that time flows in one direction—forward, and never backward. Science fiction notwithstanding, there is no reason to believe that. It is only our perception. This is what's often referred to as the "arrow of time." Everything must happen "over time." But such an "arrow" cannot be found in the basic laws of physics. There's no law of physics that predicts the direction of time. We only infer this based on the principle of entropy. It is what gives us the *illusion* of time. As physicist Carroll notes, the "Property of entropy is responsible for *all* the differences between past and future that we know about. Memory, aging, cause and effect—all can be traced to the second law of thermodynamics and in particular to the fact that entropy used to be low in the past." (78) What we mean by "past" is merely a state of lower entropy and the "future" a state of higher entropy. Therefore, the question of time and what came "before" becomes meaningless in this large context. Entropy is why we think time, if "it" exists, moves in a particular direction. Thus, Augustine's and others' notions of such ideas as "before" and "until" are irrelevant.

Conclusion

Time is merely one aspect, a mathematical one, of spacetime, something that arose shortly "after" the Big Bang. And, in our common perception, it is no more than our perception of motion. If the universe stopped all its motion—the Earth rotating, the rain falling, tissues decomposing, molecules moving around, subatomic particles changing, etc.—"time" would come to an end, i.e., there would be no time. Without motion, there would be no perception of what we call "time."[31] St. Augustine's conception that God created time is merely another assumption or religious fantasy. It is just a theistic "temporal" gap, so to speak. There was actually no "time" in which a creator could exist. Augustine's proffer is patently illogical in that it merely assumes a God and assumes His "invention" of time, once again rather ultimate gap fantasies.

Cause and Effect

What "caused" the universe to come about? There is no answer to that question. Did God "cause" the universe to come into being, as Aquinas suggested? To examine the cosmological argument a bit more, let's consider what we actually mean by "cause and effect." And as we'll see, the notion of "cause"—the commonsense view—as a reason for there being a god actually makes no sense.

Human Experience

As with time, "cause and effect" is how we generally see things in our *everyday experience*. We and other organisms have used this to the benefit of our own survival. We *have to* know what leads to what. We perceive the world as linear; one thing comes before another and causes it to happen. If we do one thing, some other thing happens. It is only our everyday experience that leads us to an impression of cause and effect. Our arm causes the baseball to be thrown. Touching a key causes a character to appear. A sad movie causes a tear to appear. The autumn cold causes leaves to change color and fall off their trees. Everything we experience reinforces the idea of a linear cause and effect. We see causes as preceding effects. If A comes first, it causes B, and not the reverse. The baseball doesn't cause our arm to hurl. The appearance of letter characters does not cause us to press the keys. Tears do not cause images in the movie.

It is subjectively difficult to shed such commonsense ideas of cause and effect. You can no more easily do that than suppress your sense of time in scheduling a meeting or observing the growth of a plant. It is natural, and how evolution got us here. But this generalization is plainly and purely an aspect of human or other animal behavior and perception and does not bear on any logical reasoning about cosmology and the infinity or eternal nature of the universe.

All Events Are Merely Correlated

It may be that certain events do not even "cause" others. It may be just coincidence. David Hume, back again, astutely observed that our impression of "cause and effect" merely "arises entirely from experience, when we find that any particular objects are constantly conjoined with each other." (229) He sagaciously noted that one does not actually "observe causes" in the real world, only associations between events (228), what we now call correlation. Hume argued that when we seek causal relations, we can never discover the real power, the "metaphysical glue," as it were, that binds events together. All we are able to see are *regularities*—the "constant conjunction" of certain sorts of observation. He concluded from this that any talk of causal powers is illegitimate. Which is not to say that he was ignorant of the central importance of causal reasoning; indeed, he said that it was only by means of such inferences that we can "go beyond the evidence of our memory and senses." Causal reasoning was somehow both indispensable and illegitimate. We appear to have a dilemma. We use causal reasoning in everyday life while understanding that particular events do not have specific causes.

When a leaf turns color and falls off a tree in the fall, we "assume" it's the weather. (Of course, twenty thousand years ago it was the spirit within the tree). What *does* cause leaves to fall off trees in the fall? Ordinary observation suggests that it's cold weather; but that's not a complete answer. Yes, it's colder in the fall. But it's largely because the tree is shutting down water transport, which is triggered by the change in daylight, something that helps a tree survive the winter (and ask yourself how the tree knows the winter is coming). Your common sense reveals only a correlation with the weather. You have to look deeper. At one point the tree was only a seed, and leaves falling off trees did not occur before there were deciduous plants. You might have to go back to early Earth to appreciate some of the relationship, based on evolution. It is only a function of how far back you look that you might discover a "root cause," so to speak (pun only mildly intended).

In the macro world, there is a rather fundamental principle of science that says, "Correlation is not causation." By that we mean that if two things occur together over time and one precedes the other, the former is not necessarily the cause of the latter. This makes intuitive sense, and it is why scientists do experiments—to lend credence to the idea that one event may specifically *lead* to some other one happening. But consider the following "true" statements:

- The consumption of ice cream predicts the number of murders and shark attacks.

- Female-named hurricanes have killed twice as many people as male-named hurricanes.
- The marriage rate in New York State predicts the number of murders by blunt objects.
- The number of lawyers in North Carolina predicts suicides by hanging, strangulation, and suffocation.
- Money spent on movie theater admissions in the U.S. predicts levels of precipitation in Pennsylvania.
- The number of lawyers in Puerto Rico predicts the number of people who died falling out of bed in the U.S. (The increase over time for that one has a whopping correlation of .97, an almost perfect relationship.)

We would not generally consider these "causal" statements. Causation in this sense does not mean that certain events are *responsible* for the appearance of subsequent events. It merely suggests something about possible intervening variables. But they all share something fundamental—there may be some "root cause" for each of these relationships. We just don't quickly know what they are. It's interesting to consider the possible root origin of these. Some are harder to figure out than others. It's pretty easy to diagnose the first couple; the others are a bit more difficult. Nevertheless, they are "correlations," legitimate mathematical relationships. However, they are merely associated with each other mathematically and do not reveal what we normally think of as causation.

For example, the correlation "Ice cream consumption leads to increased murder rate" is mathematically valid. However, ice cream, in most circumstances, does not cause a murder. The root cause, if you will, can go back much further than we might consider. Let's assume a few obvious things. Hot weather brings people out more; tempers can rise; other crime increases, which raises more violent possibilities, and so forth. Before ice cream shops and cities, the correlations were probably a lot smaller. The further you wind the clock back, the more the correlations make sense. Before ice cream, the correlation disappears entirely.

The Universe Just Evolves

All we know is that "things" or events interact with each other. They do not "cause" each other. We only infer a cause and effect because of the "arrow of time," which is merely our perception. We see some things happening before other things and we infer that the first causes the second. But remember, in the largest sense, using "time" to make such inferences is not justified.

What we see as causes and what we see as effects are determined by the underlying laws of nature. It is only because we perceive time the way we do that we infer a direct cause-and-effect relationship between seemingly discrete events. You believe that when you throw a ball, its flight was determined by *you*. But both your actions and the action of the ball are due to the underlying laws of the universe which have determined its evolution—you did not spontaneously create a "cause." We're left with the possibly inescapable conclusion that all events in the universe are merely correlated with each other. Quantum theory tells us that the universe evolves in a particular way. What we actually observe as causes and effects are both natural events which are part of this process. One particular observed event does not "cause" the one we observe next. They are both part of the same process of the "unfolding" of the universe. One can trace back the evolution of falling leaves or ice cream related violence back as far as you want—theoretically, and the concept of particular causes fades out, as Carroll suggests.

> There is no reference to 'causes' in the fundamental laws of physics; there isn't an arrow of time either. The laws treat the past and future on the same equal footing. Events simply are arranged in a certain order, with no special responsibility attributed to one over any of the others. We can't pick out one moment, or a particular aspect of one moment, and identify it as 'the cause.' Different moments in time in the history of the universe follow each other, according to some pattern, but no one moment causes any other. (78)

No God Causation

St. Thomas Aquinas' argument (the so-called cosmological argument we'll revisit in Chapter 8) falls on its face. Aquinas was a human, after all, as are all theists and others who propose that God "caused" the world to come into existence. However, that is human, commonsense reasoning. Not only does it imply a time parameter, the commonsense notion of cause and effect is irrelevant to the existence and behavior of events in our universe or the universe itself. We merely *want* there to be a cause for the universe since there are "perceived" causes for things we experience in everyday life, i.e., based on our limited human experience with linear "cause and effect." Aquinas got it wrong. The bottom line is that your intuitions about time and "cause and effect" really can't be used to inform some theological argument. When you invoke God for that cause, you're really just falling back on your commonsense reasoning, which doesn't apply to this problem. You're merely engaged in a fantasy based on your own personal experience. Physicist Paul Davies summarizes the temporal and causal argument well.

> Perhaps the most serious objection, however, to the causal version of the cosmological argument is the fact that cause and effect are concepts that are firmly embedded in the notion of time. Yet, as we have seen, modern cosmology suggests that the appearance of the universe involved the appearance of time itself. It is usually accepted that cause

> always precedes effect in time: the target shatters after the gun is fired, for example. In that case it is clearly meaningless to talk about God creating the universe in the usual causal sense, if that act of creation involves the creation of time itself. *If there was no 'before' there can be no cause (in the usual sense) of the big bang, either natural or supernatural.* [emphasis added] (108)

The notion that there must be or must have been a god to "get things started" is becoming harder and harder to defend. As many cosmologists have pointed out, no cause was necessary for our universe to come into existence. Carroll amplifies, "We don't yet know the final laws of physics, but there's nothing we know about how such laws work that suggests the universe needs any help to exist." (78) Bottom line—you cannot use your human conception of cause and effect to infer that the universe had a cause, and of course, go on to say that some personal god must have been the agent. We only resort to concepts of origin or "creation" to resolve our subjective discomfort with the idea of infinity or eternality or to affirm a superstitious belief. It is the ultimate and fallacious gap argument.

Anthropic Principle

An interesting question arises. If there's no known cause for the universe to exist, and if we're not so special, how is it that we can understand all these things about the universe and its possible place in the possibly larger cosmos? For some, this is almost tantamount to "Why are we here?" A provocative approach to that question has come to be known as the *Anthropic Principle*.

The "anthropic principle" is variously defined. In general, it is the principle that the universe has evolved in such a way as to create beings that can understand it. As such, it is seen as being "fine-tuned" for life and for the evolution of life into human beings who can consider the nature and laws of our universe. More specifically, the fundamental constants of nature must exist in a way that has enabled the evolution of life, including human beings. If they did not, life would not exist, and we would not be here to observe and explain the physical universe. It seems to be a contrarian view to the Copernican Principle; and it is not altogether a popular one among physicists.

Strong and Weak

Many thinkers about this problem divide it into *strong* and *weak* anthropic principles. A "weak anthropic principle" holds merely that only in a universe capable of eventually supporting life will there be living beings capable of observing any such fine-tuning and other aspects of it. Again, some physicists formulate the principle to suggest that "things are the way they are in the universe because if they weren't, we wouldn't be here to notice." (191) This seems more than obvious. Those laws of nature (more specifically, all the phenomena that they describe), in effect, would exist anyway. Obviously, the "fine-tuning" of the fundamental constants is obviously the key to why we are actually here. Of course, there was a time when they existed and were not observed. And there will be a time when they will not be observed and interpreted by human beings—or perhaps other life forms. I'd suggest that the weak Anthropic Principle gets you no closer to a reasonable answer about the existence of our universe or its properties or why we are here.

The *strong anthropic principle*, on the other hand, offers an "explanation." It's just not a very good one. It suggests that the universe is compelled, in some sense, for conscious life to eventually emerge, and the force behind it is something supernatural. It is argued, in other words, that life (at least as we know it) could not have come about randomly. The fundamental constants were "fine-tuned" in a way that allowed life to develop, and eventually for us humans to exist. Of course, "fine-tuning" does imply that it was actually done by some agency. For religious apologists, that agent is God. As physicist Michio Kaku puts it, "The strong form of the Anthropic Principle [states] that God or some designer had to create a universe 'just right' to make life possible." (261) Of course, the invocation of a "living" fine-tuner or supernatural god gets you no closer to, and in fact farther away from, an understanding of the cosmos and the existence of humans. To be annoyingly repetitive, it's just another gap argument.

Multiple Universes

Some proponents of the Anthropic Principle reason that it explains why the universe has the age and the fundamental physical constants necessary to accommodate conscious life. As a result, they believe it is unremarkable that the universe's fundamental constants happen to fall within the narrow range thought to be compatible with life. If you see that as a tautology, you're probably right.

However, we should ask: why *do* the fundamental constants have the values that they do—Einstein's cosmological constant, mass of the electron, the strong force, and all the others? Is there any rhyme or reason for this situation? If there are other universes, per theories we've outlined (and others), those universes may not share the values of the fundamental constants we experience here. If there are other universes, the laws of physics may be different in each. In some, gravity is too weak to enable stars . . . and people. In some, a universe may exist only for a brief second. There may be other fundamental forces we can't envision. There may be universes in which almost everything is composed of antimatter, or maybe dark matter. One possibility is certainly that the fundamental constants can randomly vary, and we happen to live in a particular universe in which they have the values that they do. If, in any of the multiverse scenarios, these constants can have many possible "degrees of freedom," there is no reason to infer that there's anything special about the ones we have.[32] The "fine-tuning" may not be so fine after all, and the Anthropic Principle loses all meaning.

Life

If there are parallel universes, some with the same laws of physics and many with different ones, and this dynamic proceeds infinitely, then we are likely an "accident" of a much larger cosmos, one that has given rise to human beings. We might call it cosmic natural selection, at least metaphorically. Life has evolved here because the conditions have been favorable for that to happen. In other universes, that may—or may not—happen. With a pretty big number of possible universes (10^{500} types that M-theory predicts, or an infinity of parallel universes that many theories predict), the notion that life may exist elsewhere can't be definitively ruled out. In some universes, there can be people like us; in some, very different advanced beings; in some, only single-celled creatures. And, given an infinite multiverse, there are, by definition, people like you and me contemplating the questions we've raised.

We should also consider the issue that life as we know it can only exist within the narrow conditions we know about. There are growing theories suggesting that even slightly modifying some of the fundamental values that nature has used to create life in our universe, e.g., slightly different masses of some quarks or absence of the weak force, would allow the evolution of organic life not very much dissimilar to that in our own universe—though such life would have somewhat different chemistry and it would have to get its energy from somewhere other than the stars we're familiar with. (250) Life as we know it, or life eerily similar to what we know, could exist in other universes with values just a bit different.

Scientific Behavior

The other anthropic principle notion is that we are a *product* of this universe and can therefore "understand" it. If we again assume the concept of multiverse, we might look at some other contexts. That we are here should be thought of as but a statistical oddity—not that we are somehow special enough to understand what has happened in our reality. There may be other universes with "beings" that came about in those, and those beings can appreciate how their universes happened. In some, we can conceive of beings who never made it to a stage of awareness or scientific thinking. There may be universes where no humans exist, but where the laws of nature *are* potentially understandable. There are some in which very intelligent humans exist, but they explain their existence erroneously (as we've done historically). And there are universes with understandable laws but no life. A universe less compatible with some sort of life will go unbeheld, but it will still operate according to whatever laws may exist there. The Anthropic Principle offers no help here. Finally, if multiple universes are possible, we must ask if "*we* are asking these questions" in other realities. Again, some multiverse theories, e.g., the Many Worlds hypothesis, suggest that "we" *do* ask such questions in other realities. And "we" may come up with different answers. Perhaps beings in such universes have their own type of anthropic principle.

Conclusion

How can there be a principle which actually doesn't explain anything? At its most basic level an anthropic principle, however conceived, is either a tautology or an argument from design. However, we can see that this is but another facet of the Copernican Principle—we believe that there is something unique about our universe and that it involves us. But if we accept that some type of multiverse does exist, we are merely an accident of cosmic history. The fundamental constants we know must exist, irrespective of our presence and observation. The notion of "fine-tuning" is often merely a metaphor suggesting that someone or something did the tuning—deliberately or unwittingly. Many physicists will admit that it is just a metaphor, but the continued use of the term by many, particularly theologians, seems to have propagated the idea that the universe was, in fact, "fine-tuned" for us. And it was done by God.

Yes, we ask the question of why we are here to observe the universe. But we should also be curious as to why we actually ask such things (I'll get into human behavior more in Chapter 10). It is because your history—and the history of humankind, including the history of the universe—has shaped you to do so, including asking hard questions. Again, our evolutionary behavioral history points clearly to the overwhelming tendency of humans to learn, seek answers, and solve problems. It has given us an incredible survival advantage (at least temporarily). Its generalization to various problems is what we call science, but leaping over this to reframe the problem as the Anthropic Principle seems to obscure the question. We ask hard questions simply because we have evolved to do so.

There are also interesting extensions of the anthropic argument which seem to make it moot. Let us assume, for example, that we create some sort of artificial intelligence, or more likely that it evolves over the long term from present AI to some rather unimaginable intelligence, such that it could contemplate the questions of big science, the origin of the universe, and its own existence in it. Given an infinity of time (or its existence in other universes), this is a plausible assumption. We could speculate that such AI would come to the correct conclusion: that it arose because of human intervention, with further evolution of its own. It might well believe that there was a cybernetic god who created it along with the rest of the universe, a universe which is just a computer simulation (I'll address the issue of conscious AI in Chapter 5). In fact, there are serious thinkers who hold that our reality might actually *be* a computer simulation (11); and some AI in the universe of our future might contemplate the question. But of course, human existence would be seen as merely an intermediate step in the program. On the other hand, it might have no knowledge that such a "step" occurred, possibly long after we're gone. In either case, an inevitable "AI-thropic" Principle would suggest that the universe was fine-tuned for "it."

God Not Necessary

Is it necessary to postulate a god as responsible for our universe? As I've laid out, the answer is clearly "no." We can see now that there's no reason to invoke a personal god for the existence of our reality. Let me summarize some main points:

If not God, what explains our universe? Ockham's Razor, as we introduced in the last chapter, is that the simplest adequate explanation is probably the best one. This is clearly incompatible with God since the mechanism of "creation" requires incredible assumptions of complexity, the unresolvable problem of defining God, and the whole problem of "cause and effect" of specific events. To reiterate: *An eternal god is just an assumption, no more valid than an eternal universe or multiverse.* To posit a supernatural creation becomes meaningless. The consensus is that the universe—or multiverse (per many of the above)—is eternal—there doesn't seem to be any conception of an end, or a beginning. Whether there is or has been a cyclic pattern to our own universe, whether it will suffer a big chill, whether it finds some other fate, or whether it is itself just one of an inconceivable number of universes, is immaterial. The eternality of our universe goes both forward and backward. Thus, the whole notion of Creation falls on its face. With the infinity and eternality of the universe, there is no need for a god to "get things started."

Take another quandary. If God is omnipotent and infinite, as religious apologists assert, and if He actually existed, why would "He" hang around for some unknown period before creating a particular universe? One wonders why a presumably omnipotent god with particular interests in the affairs of humans would wait 13.8 billion years to satisfy them within our particular universe. And if humans are so special to God, why did it take over four billion years for Him to prepare Earth itself for human life? If the laws of physics hold, He also plans to destroy it at some point. That makes no sense, a conclusion not rationally supported by the bizarre "god works in mysterious ways" argument. Such a God would either not be the omnipotent one we hear about from theists, or else He has a bizarre sense of humor, finding trial and error entertaining; or maybe something else created Him, which leads us back to the same question of what created God—turtles all the way down.

Further, the complexity of the universe is baffling. If we are so special to God, it's a bit puzzling, to say the least, that there exists such a massive universe around us (and perhaps other universes). What is all the rest of this for? There is no reason for our incredibly large universe to exist just for the purpose of us recent humans being here. But it does exist, and it exists, according to some physicists, on an infinite scale. We are by no means special in that context. And the notion that God created the "world" for us is just a wholly risible idea. That may make you feel uncomfortable, mainly because our common sense and ordinary experience limit our understanding of the universe. But the universe owes you no comfort. It's merely a psychological problem, and invention of a god to make you feel better really doesn't solve that.

4
WHO ARE WE AND HOW DID WE GET HERE

Man is embedded in nature. —Lewis Thomas (*The Lives of a Cell*)

Evolution is cleverer than you. —Francis Crick (Co-discoverer of DNA)

One of the theists' principal arguments for the existence of God is that humans were created by Him and that biological evolution is just wrong about that. This is an old, nonsensical story, but we need to take a harder look at it here. I'll offer that, in understanding life on our planet, there is no need for a personal god. And we're not all that special in the grand scheme of things biological. We are but a point on a continuum of organic processes that have been occurring on this planet for over 4 billion years.

Religious apologists usually (perhaps always) offer the idea that there must be a god because humans are "so special" and unique that we must have been "created" and therefore there must be a Creator. Of course, this is absurd logic. But our job here is to show that, in fact, (1) we're not that special, not really "unique" creatures, even in our thinking and behavior, (2) we're only somewhat different from other creatures if you ignore all of history, (3) humans did evolve from previous animal forms, despite the theists' denial of evolution, and (4) our evolution has led to the natural development of supernatural superstitions and other religious ideas.

In order to address these matters, we must take another look at the fact of biological evolution, something totally anathema to fundamentalist Christians. Organic evolution is one of the most well-supported theories in all of science—as much so as gravity, the earth orbiting the sun, or how oxygen keeps us alive. We'll consider our place in the tree of life, which must also consider how life began on this planet.

We humans have basically the same body and brain as our ancestors did about three hundred millennia ago. Our culture has admittedly become a lot more complicated, but we are basically the same people as we were then, with the same urges, the same capabilities, the same desires, and the same native ability to solve problems. We have much more information today, more complex tools, and more sophisticated ways of communicating. Nevertheless, we are essentially the same people we were a quarter million years ago. That is a long journey, but in the absence of a god depositing us on the planet, let's consider how we got here—and how we became human.

More basically, the Creationist argues that only God could create life. Of course, this is just a blind assumption. It falls to science to explain how life arose on Earth. There is, as yet, no clear answer to that, although research continues to make progress. But that doesn't suggest the need for a god. This is one more example of Creationists' use of Drummond's "God of the Gaps," and admittedly a pretty big one. Of course, incomplete answers are not a reason for such fantasy. I'll talk a bit about what life is and how it actually may have arisen.

Finally, we'll see how the human species arose and became a social animal and how that enabled the development of religious superstitions, including gods. To wit, the superstition about gods comes directly out of how we evolved. It arose as a natural process, not as a communication with some supernatural force (We'll see how this evolved into organized religions, per se, in Chapter 7.).

Origins of Life

Let's begin with life itself. Take a moment to eschew the notion of any god magically creating life on this planet we call Earth. So how *did* life originate here? When we take a scientific approach to this problem, we're still a bit in the dark. But significant progress has been made in understanding how this could have happened. Biological life may have arisen here in a couple of different ways. Nobody knows for sure, but we can consider a couple of promising theories addressing this. One is that life arose here from the combination of less complex molecules evolving into "living" matter. Another is that primitive life, or at least the biochemical precursors of life, originated elsewhere in the universe and were transported here in some way.

Chemical Evolution

Life arose out of inorganic matter in some way, what we might call chemical evolution. Apparently, it was not that difficult for life to spontaneously arise, taking only less than half a billion years after the Earth was well-formed. The "delay" may have been only due to a more hospitable environment coming about, e.g., not enough water and certain other conditions being initially present in the early years. Given certain geological and atmospheric conditions, the advent of life as we know it could well have developed naturally on the planet. The theories about life arising from non-living chemicals is a process referred to as *abiogenesis*, and there are some ideas about how such abiogenesis could have happened. In any case, the initial evolution of "life" as we know it must have followed the laws of chemistry and physics as we know them.

To grossly oversimplify, there is now some evidence that RNA, the precursor of life as we know it, conceivably could have formed "on its own" in the early primordial soup on earth. (413) There continues to be debate on this, though further research is clarifying some of the contested issues. (501) Amino acids, which are critical pieces of RNA-type molecules, are abundant on Earth, and they easily combine with other molecules to enable more complex structures. The tendency of such molecules to replicate continues to be a subject of research, but the evolution of terrestrial life may certainly have begun in this way. There is also what's called the "metabolism-first hypothesis" for life's origins. This suggests that various kinds of chemical reactions occurred, and they absorbed energy from the environment, which led to the evolution of more complex molecules. Eventually, those formed particular structures, which sustained themselves by using the available energy. Only later did they begin to replicate using some sort of RNA or DNA molecule. It is possible that these processes may have arisen around warm, alkaline hydrothermal vents deep in the ocean. Such life may have begun there, with high temperatures and pressures being the sources of energy needed for the chemical reactions; and as such reactions became more and more complex, they led to actual functioning entities.

Opponents of life "evolving" from non-life argue that if it was so easy for this to occur (life beginning perhaps only a few hundred million years after formation of the planet), why did it occur only once (presumably), with the implication that it must have been "created?" This is a fallacious argument, of course. In fact, we don't know if it occurred more than once. We have no idea of whether biochemical processes subserving primitive RNA or similar molecules arose more than once, but conditions may not have been favorable for previous ones to persist, or they may have persisted but didn't achieve some critical steps toward an actual self-replicating entity. If it occurred more than once, it is likely that alternate forms became extinct rather quickly. At this point, we have no way of knowing. All we know is that current biological life has a consistent basis across organisms, an RNA-like molecule, which is now DNA in all current organisms.

Evidence

What about the evidence? Obviously, there are no fossilized records of the initial chemistry to examine. However, one of the ways that earthly abiogenesis is being investigated is in the laboratory. If life can be engineered in the lab—a prospect awfully threatening to the theist—we'll perhaps know a lot more about how it may have arisen on the early Earth. Can a living "thing" actually be "created" by us? It hasn't happened yet, but let's look at some of the activity here.

Back in the 1950s, two American chemists, Stanley Miller and Harold Urey, performed the now-classic experiments showing how the building blocks of life could have spontaneously originated on Earth. (One of the original papers on this is indeed very interesting. (348)) They set up an experimental apparatus which approximately simulated the conditions on the early Earth. When it was electrically charged, organic molecules, including amino acids, were found to have been "created" out of purely inorganic ones. In fact, over 20 different amino acids appeared, including many that are necessary for life as we know it. In retrospect, this was only a crude approximation of how the molecules necessary for life appeared, but it was a start. Even though there were some problems with their experiment, e.g., they might not have perfectly simulated the conditions on Earth at the time life arose, since then it has been refined and essentially replicated and extended in various other laboratory experiments. (334)

Many of the ingredients necessary for primitive life have been spontaneously generated from simple ingredients using similar methods, including amino acids, and indeed certain sugars, membrane-forming lipids, and nucleobases (which are critical pieces in RNA and DNA molecules, including the bases for encoding genetic information). However, even though such are important as the basic building blocks of life, they have not yet reached the level of "living" tissue. There are compelling recent experiments (41) and excellent reviews (334) which have extended Miller and Urey's findings. I'll not review details of these experiments here. However, to summarize, there has been considerable success in the laboratory production of these complex prebiotic molecules, including the sugar and other components of RNA, as well as analyses of how they may have operated to conjure up life from prebiotic conditions. In addition, it has actually been observed that certain synthetic RNA enzymes will engage in *spontaneous replication* by themselves, over and over again, a truly provocative development, as this is a possible basis for similar genetic mechanisms in living tissue. (314) Though predictions are impossible, construction of a self-replicating entity in the lab is certainly conceivable.

Everyone knows that our genome has been sequenced, as has the genome of some other organisms, which only adds to our knowledge. And, in recent years, there have been many successful clonings of animals. This is not the "creation" of life, but such research brings us closer to understanding it. And though it may be many years away, we're coming ever closer to understanding how to create a living cell in the laboratory. There are, as yet, no laboratory-created cells, capable of nutrition, replication, and other functions we might consider "life", but Reverend Drummond's gap continues to narrow here. The "state of the art" for actually building an artificial cell continues to advance rapidly. (411)

While "life" as we now know it—according to definitions we use on this planet (see below)—has not yet been experimentally replicated in the laboratory, we have come closer to doing so, and it's only a matter of time before rudimentary self-replicating molecules or cells are artificially created—and without the aid of divine intervention. If and when it happens, it would be truly remarkable. Of course, the anticipated argument of some theists that this would demonstrate only that God allowed humans to do such might ring pretty hollow.

Panspermia

If life didn't evolve chemically here, perhaps Earth may have been "seeded" with primitive life, which may have hitched a ride to Earth on some extraterrestrial rock. This is known as panspermia. Numerous organic compounds have been identified from asteroids or meteorites that have hit Earth (and now from more samples obtained from asteroid space return missions); and these include amino acids, sugars, nucleobases, and literally thousands of other organic molecules. It would be shortsighted to ignore that such findings might suggest life arising elsewhere, or at least some of its critical components being transported here. One cannot totally rule out that there may have been multiple seedings, either of "life" or some of its very complex components, with the result that what finally "took" became the type of life we're familiar with. After all, Earth had been hit with comets and asteroids for at least a few hundred million years before our earliest life forms appeared. Whether life originated here or was deposited on the planet, we must accept that life as we know it did "evolve" from nonliving chemistry. Of course, the gap will take more than a serious blow if we find life itself, or specific chemistry suggestive of it, on some other planet.

What is Life?

All this brings forth the question, "What *is* life, anyway?" When you get down to it, life as we know it boils down to a set of chemical reactions. That's "all" it is. There are chemical reactions in the "non-living" world too. It's just that we have many more of them, and they are much more complex. When you look at history, as one origin theory goes, there was a time when certain chemical reactions could be considered "life" as opposed to non-life. There were complex organic molecules. One theory (and there are others) suggests that at some point they began to replicate, e.g., the possibility of self-replicating molecules like RNA. The transition from non-replication to replication may have been very quick or taken a very long time. No one knows. Were the first self-replicating molecules "alive?" You have to strain yourself a bit to consider a self-replicating molecule alive, but self-replication is a key feature of biological life. If there had not been self-replication, there would have been no evolution to the kind of life we're familiar with. This puts a different perspective on what we consider life. It does inform a view of life to the point that it must be seen as a continuum of natural processes, not something qualitatively different from the non-living world.

Yes, life is unique on this planet, but its evolution from non-living chemical reactions is fairly trivial in the context of our universe. Biological life is "merely" a complex set of chemical reactions, a peculiar characteristic of the crust of this planet. I say "merely" not to diminish our value but to put into context that the difference between living and non-living matter is specifically and only a matter of chemistry.

Definitions

So how shall we define life? This is a critical question which also arises from the Copernican Principle. There are many definitions, and the definitions, of course, apply to life as we know it, that is, here on Earth at this time; and they obviously rely on the organisms we have so far observed. Let's consider a couple. A straightforward one is offered by NASA: "Life is a self-sustaining chemical system capable of Darwinian evolution." (360) As we know, NASA is also searching for such life elsewhere in our universe, and it may use that definition if it ever finds something provocative.

A more sophisticated definition was offered in a series of public lectures in 1943 by Erwin Schrödinger, notably one the founders of quantum mechanics. He addressed the question "What is Life?" His answer, "What is the characteristic feature of life? When is a piece of matter said to be alive? When it goes on to 'do something', exchanging material with its environment, and so forth, and that for a much longer period than we would expect an inanimate piece of matter to 'keep going' under similar circumstances." Of course, this is but one characteristic, one that most people would consider insufficient. He goes on to compare the issue to the appropriate physical principle, the well-known Second Law of Thermodynamics we saw in Chapter 3. That is, any matter, left to its own devices, if not influenced by its environment, would continue to "disorganize" until its material reaches a "state of thermodynamic equilibrium or of 'maximum entropy'." Life is characterized by a metabolism which "succeeds in freeing itself from all the entropy it cannot help from producing while it is alive." (451) Life as we know it does maintain itself—actively resisting disorganization—using a flow of energy (until it dies, of course). Of course, this also doesn't affirmatively and fully define "life."

And it may not even be reasonable to define "life" in some unitary way. It's a word we use to conveniently describe some of the things we see around us, and of course, that's useful. We tend to label something "life" when it has the characteristics we're familiar with. But it may not make as much sense as we think in terms of the universe we inhabit. Physicist Carroll observes that there actually cannot be a "correct" definition of life. (78); and the "I know it when I see it" definition just comes up a bit short. Perhaps any attempt to define life becomes a vacuous exercise—as we'll see.

What Is Alive?

You may think it's an easy problem, but deciding what life is and what it isn't is not that simple. We are alive. Rocks are not. Amoebas are alive. Crystals are not, but they grow. Are seeds alive? With bacteria, we generally consider each of them, each cell, alive. Are individual human cells alive? Without extra help (e.g., the laboratory work being done with stem cells), they certainly can't survive outside our body, even briefly, without a medium of nutrients and certain other chemicals, and they don't "go on to do something" on their own.

To clarify, let's consider something that appears to be at the boundary of earthly life—viruses. We really can't ignore this problem. Are they alive? There are more viruses on Earth than all the living creatures combined. Although they are certainly more primitive, they may have actually "evolved" *from* cellular life forms. Whether they are "alive" or not is highly debatable. They are essentially just strands of RNA or DNA along with some other coating substances and sometimes some other minimal structure.

They certainly represent a challenge to our view of "life" and to any definitions we currently have. Are they "merely" biochemical chains? I say "merely" to represent comparison of complexity to us or more "advanced" creatures, not to suggest that they have poorer survival fitness, something which is obviously not the case. They'll probably be around long after we're gone. They don't maintain homeostasis, i.e., a steady state internal environment; they don't show metabolism, in the sense we use that word; they replicate rather than reproduce (actually, in essence, promote their host cells to contribute to the replication). They don't grow. They don't produce or convert energy; they only hijack energy from the cells that they invade. They do adapt to their environments' cells so that they can "survive," and they mutate in order to facilitate transmission. However, without a host, viruses just "sit there" and don't *do* anything.

Viruses do have some qualities of what we consider life, but not others. The consensus of biologists—though not universal—seems to be that viruses are *not* alive. There are also minority views. For example, microbiologists Hugh Harris and Collin Hill state, "The conclusion that viruses are not alive is premature." (205) Stated another way, we just don't know, *which obviously illustrates the problem of defining life in the first place.*

With apologies to Schrödinger, we could think of viruses as both alive and not alive. There is obviously a continuum of living and non-living entities here, especially going back in time. The line between life and non-life, as we conceive of it, is a bit fuzzy, and the border between life and non-life becomes ever fuzzier the more we learn. Perhaps life is not so special after all, and we should be humbler about this. Since life arose from inorganic chemistry, there is no need for a "god of the gaps" to be interposed between nonliving and living matter. Of course, if an omnipotent God created both, as religious apologists espouse, one wonders why it took 13.8 billion years for one type of chemistry to supplement the other.

Cosmic Perspective

We consider "life" to be what we observe here on Earth. Of course, that does seem to be another example of a Copernican geocentric view—that our biological life here has a privileged position in our particular part of space or the universe at large. However, we may eventually find signs of life elsewhere in our solar system. Whether specific evidence of biologic life will be discovered in the near future is an open question, of course. The scientific search continues. Space missions and telescope engineering are constantly advancing. It's certainly possible that, in the next couple of decades, we may find direct evidence, or fossil evidence, or other signs of life, in our own solar system.

Our Universe

And we may find signs of it elsewhere in our own Milky Way galaxy. That's something which, under ordinary circumstances of rational thinking, would clearly be a final blow to the notion of our specialness and Creationism, except, of course, to the outright science deniers.

Earth may not be the only planet in the universe that is supporting life or can support life or has supported life, by any reasonable definition. We may never discover life elsewhere in our vast universe. Or we may. Scientists continue to look for life elsewhere in our galaxy. Obviously, we haven't found it yet, but the search continues. Thousands of exoplanets have been identified; as of this writing, over 5,000. Some theory suggests that nearly all stars in our Milky Way, which is on the order of at least 200 billion, might have one or more exoplanets; and many of these could be in the so-called Goldilocks zone, from the term apparently coined by British philosopher A. C. Grayling as the "Goldilocks enigma." (190) That is, there are likely planets those whose conditions are "just right" to allow life as we know it to evolve—with the possible temperature range, distance from its star, and a gravitational value that could support life as we know it. There's even an incredible number of "rogue" planets roaming the galaxy, which could have supported life at one time. Who's to say that some sort of life didn't once exist on any of those? Technology continues to improve in ways that might enable us to detect signatures of life on other worlds. One of the more exciting developments is the recent launch of the James Webb Telescope (and other pending experiments). This will go even further in this effort. The Webb has actually begun to image some exoplanets and is identifying some of the chemistry consistent with life or habitable conditions for life as we know it. To assume we are alone as living creatures may be quite premature.

Life may have arisen elsewhere and survived for only a short time. It is unclear if life has arisen and evolved elsewhere in a way that can reveal its evidence, and that evidence is discovered by us during a period when our species continues to survive. On the most optimistic projection of our existence, assuming no self-destruction, we will have survived only a limited time to enable discovery of signatures of life elsewhere. And we continue to actually search for "intelligent" life. Then there is the now-classic Drake equation, which estimates the likelihood of intelligent, communicative life existing in our galaxy. It is a probabilistic view, based on a formula which calculates the likelihood of civilized life which can communicate with us—based on a number of variables—but many scientists are encouraged by it. The search for extraterrestrial life continues. The SETI project (Search for Extraterrestrial Intelligence) continues to scan the sky for electromagnetic signs of advanced life. Of course, technologically advanced civilizations may only survive a few thousand years, maybe longer. Do they survive long enough to communicate their existence and before extinction or self-destruction? And does ours exist long enough to "hear" it, i.e., in a time frame in which we receive any signals? Considering such a narrow window, we may never actually uncover evidence of other advanced life. However, that does not mean it doesn't, hasn't, or will not exist. It might be helpful to consider this perspective. Consider this: If an external observer 4.5 billion years ago were asked to estimate whether a technologically advanced human civilization would appear on Earth, the guess would clearly be "no." But it did happen.

Only One Type of Life?

Perhaps our view of life, or even the definition we use, is just another example of the Copernican Principle from which we suffer. The galaxy (and indeed, the universe) is a big place, and the notion that life could have arisen in other forms is something that definitely can't be ruled out.

Our type of life may not be the only type of biologic life in the universe. It just happens that it evolved on a certain planet in certain ways that were obviously compatible with the environmental parameters here. There is no reason to rule out that it could be quite different elsewhere. For example, we animals can only exist—and have evolved for—certain ranges of temperature, pressure, environmental gases, electromagnetic radiation, specific chemistries, etc. However, even on Earth there are plants and other creatures which exist for quite different values of these. What if life, including "advanced" life, evolved somewhere else with the ability to withstand very high temperatures and make use of other parts of the electromagnetic spectrum or perhaps evolve based on other chemistries. The chemistry that led to us might not be the only one. We know that some bacteria, for example, have survived here using arsenic in their nucleic acids and proteins instead of phosphorus, in contrast to the rest of life that we're familiar with. The possibility of life with all sorts of chemical variance can't be ruled out.

There is also the problem of what we on this planet define as life. Our definition of life is based only on what we've observed here, e.g., entities which exhibit growth, metabolism, reproduction, adaptive complexity, and Darwinian evolution. Is it possible that "organic" entities on some other world might exhibit one or two of the qualities of life as we know it and not the others? Would we call that life? For example, perhaps there is something we might call life that does not feature evolution by natural selection, contrary to NASA's definition, for example. Of course, even for biological systems, evolution may not be necessary, and adaptability and survival might occur in ways different from natural selection.

Parallel Universes

Let's take the *big* cosmological perspective. We have life in this universe. But does it exist elsewhere? In other universes? As I discussed in Chapter 3, several multiple universe theories do suggest the principle "anything that can happen does happen," including life, including life such as ours, including human beings just like us. Given the infinity of time and space, the organization of fundamental particles, and the possibility of alternate dimensions or realities, although speculative, this is certainly something to consider.

It is certainly possible that life, as we variously define it, could exist in other universes. Obviously, the fundamental forces of nature had just the right values to allow the development of life as we know it here. But a slight change in the forces, or the values of some of them, might allow for life to develop anyway. There is some reason to believe that slight variations of fundamental forces may still enable life to develop. (250) In fact, there is no reason to rule out that such is possible or that different conditions and/or fundamental properties of nature may exist to subserve some kind of life. Given the "anything that can happen does" principle, matter can organize itself into any conceivable type of life, including various creatures and types of humans. Given infinite time and some other form of natural law or values of the fundamental constants, would matter organize itself into almost any kind of "life?" To extend the argument, as some have, would there be leprechauns or two-dimensional rabbits who can think and talk? Despite infinite time, space, and reorganizing of matter, some things do not appear possible, as they would be self-contradictory to our understanding of their meaning. It's unlikely that any evolution would produce leprechauns that can grant wishes inconsistent with physical laws, or two-dimensional creatures who can create sound waves that can be heard only in a three-dimensional atmosphere, or the monkeys who could write Hamlet, as we ruled out in Chapter 3. But for the universe in which we exist, these things are just not possible—any more than 2+2 could equal 5 or that pigs could fly. Of course, it would have been inconceivable to a human observer—if one existed at the time—that early biological relatives of dinosaurs could fly either, though some eventually did. It all depends on history, the realities which emerge, and how you define the components.

Conclusion

Despite speculation, this just reminds us of the Copernican Principle from which we've always suffered. We are a species that exists only in this time and on this planet. We don't know whether life exists elsewhere. The universe is a very big place. Despite the numbers, the human mind is just not capable of grasping how big it is, in both time and spatial extent. But if you take a long view, we only know that life exists on this planet because we are here (the Anthropic Principle notwithstanding). Technology may help us discover it elsewhere.

Biological life—however you want to define it—may easily arise and be ubiquitous in the universe. And we cannot rule out that other realities may manifest life as well. We can only say at this time that life as we know it is but an interesting feature—albeit a very interesting one—of the surface of this planet. We are left with the conclusion that life as we know it is not necessarily unique or that some supernatural force was necessary for it to arise. Yet, when life as we reasonably know it—or other life which meets reasonable definition—is found on some other world, it will be interesting to observe the mental gymnastics theists use to explain it.

Finally, absurd as it may seem, consider the possibility of "artificial" life. If an artificial being is created, would we consider that "life?" There is even the possibility that self-replicating mechanical, non-organic beings could come into existence, either by our creation or by the creations of AI itself, further muddling our conception of what "life" means. Although this is but science fiction at this time, the possibilities are becoming more real. I'll discuss more about artificial intelligence in Chapter 5.

Evolution

Theists continue to deny the reality of biological evolution and assert that there must be a God who designed the enormous range and complexity of earthly life, including human beings. A great deal has been written about this problem. The empirical truth of evolution—Darwinian evolution by natural selection—is eloquently described in a number of excellent lay works, and I recommend a few here, including Jerry Coyne's *Why Evolution is True* (99) and Dawkins' particularly readable *The Greatest Show on Earth.* (114) Nevertheless, we must consider some of the important facts here.

It is so well-evidenced that Dawkins has suggested that the Theory of Evolution by natural selection is so empirically supported that it should be considered a *Theorem* (113), which is a term used in mathematics to convey something absolutely proven based on accepted mathematical facts.[1] As noted by the American National Academy of Sciences, "The concept [of evolution] has withstood extensive testing by tens of thousands of specialists in biology, medicine, anthropology, geology, chemistry, and other fields. Discoveries in different fields have reinforced one another, and evidence for evolution has continued to accumulate for 150 years." (What's now 167 years.) (362)

Darwin

When Charles Darwin published *On the Origin of Species* in 1859, a revolution of thinking about organisms, including man, was born. Darwin had been educated in the Christian religion, specifically Anglicanism, and also in medicine. His early intent with respect to naturalism was actually an effort to find some evidence of a theistic hand or design in Earth's animal species. What followed, of course, was the painstaking development of the theory of "Evolution by Natural Selection," which remains to this day one of the most well-evidenced theories in all of science. Darwin largely avoided the idea of human evolution, per se, in *The Origin of Species*. But he did address it in the subsequent book, *The Descent of Man and Selection in Relation to Sex.* (106) The works were seminal in the history of natural science. However, as many have rightly observed, and as interesting as the *Descent* volume was, Darwin made numerous errors of fact and prejudice there. He asserted, for example, the inferiority of women and persons of color, being also influenced by the attitudes of the day, without any scientific data of course. Of course, considerable progress has been made in the century and a half since then, and we know a lot more about how modern humans came to be.[2]

Overview

Evolution means unfolding or growth. It has come to signify a broad range of scientific principles, even beyond the development of life forms. There is an incredibly long and complicated history of biological evolution, over hundreds of millions of years. We'll obviously not review that here in any substantial way. Evolution generally occurs gradually, but sometimes more quickly. Briefly, all of life on this Earth began with a very primitive form. No one knows exactly what that did or looked like, but it was likely some self-replicating entity that managed to "survive." As hard as it may be to believe, we are all descendants of that "creature" (see below).

The popular conception of evolution is "survival of the fittest." While technically correct, that might be a bit misleading. The somewhat more accurate description is "survival by the most adaptable." Individuals who are better adapted to their environment have a better chance of surviving and passing their genes on to the next generation. The adaptive evolution of living forms on this planet is largely controlled by genetics. That is, our genes determine structure and function; they are the driving force of evolution, as eloquently laid out in Dawkins' *The Selfish Gene.* (111) Genes vary all the time from generation to generation. And changes to genetic makeup occur pretty much at random.[3] Most of the variability occurring in genes is either trivial or problematic for a species, i.e., it doesn't promote adaptation in any meaningful way. Some variability, however, confers an adaptive advantage. If a bird has a larger beak for eating its prey, which may be certain insects, it may be better suited to survive and pass on its genes for such a bill. The same is true for certain forest trees that may have grown taller and thus closer to sunlight. It's true for certain big cats (e.g., cheetahs) who've evolved a running speed to chase down antelopes, and of course, the antelopes who can run well enough to escape them. And it's also true of primates, who evolved opposable thumbs and larger brains. Nature "selects" for survival the genes which produce a better chance for their host's survival. And the driver is always the environment. Whenever the environment changes—due to predation, climate change, food shortage, invasion of other species, disease, or other external factors—genes that enable better adaptation promote the survival of their hosts. Those that don't tend to "die out." In a nutshell, that's the way evolution works.

Although it remains one of the best-evidenced theories in all of science, it keeps changing of course, as scientific advances provide more data, but 166 years of research only continues to bolster the theory. The modern science of genetics has only enhanced and further supported Darwin's hypothesis. It should be pointed out that evolution, as described here and as universally accepted among biologists, is a strictly Darwinian process. That is, traits of organisms are purely a function of inherited genes. They change over time due to natural selection by the environment of some expressions over others. (I won't delve here into sexual selection, Darwin's other great insight, but that also is a significant factor in the evolution of many species.) There is no evidence for the inheritance of acquired characteristics, as originally proposed by the 18th-century biologist Jean-Baptiste Lamarck. Despite recent genetic discoveries, evolution by natural selection continues to be the clear explanation for how adaptations occur and how species have come into being, including us.[4]

All living things on Earth descended from a single primitive form over billions of years. At some point, plants filled the planet's atmosphere with oxygen, and the first animal life came out of the sea. There have been many "explosions" and mass extinctions of organisms, but evolution has proceeded by creating more and more complex organisms. The Cambrian explosion, over 500 million years ago, led to the explosion of multicellular animal life across the planet. Now, life is everywhere on this planet and in a variety of forms we can only imagine this point. You can't pick up a clod of dirt or a teaspoon of river water or look out the window without finding significant life. It is everywhere, from primitive bacteria to giant redwoods to humans, all produced by evolution. And we have identified probably only a couple of million species living on the planet at this time; there are believed to be countless millions more yet to be detected and classified (There is a wide variety of estimates regarding how many species have actually been identified and named), and this doesn't count all those which have gone extinct. In fact, the consensus is that 99.9% of all the species in Earth's history have gone extinct, and of course, that number will continue to increase.[5] Life is everywhere and of considerable diversity. For example, there are probably hundreds of thousands of species of beetles currently inhabiting this planet (perhaps millions more now extinct), probably more than any other living animal species on Earth. The biologist J.B.S. Haldane reportedly said about the God idea, "If He exists, the creator has an inordinate fondness for beetles." Evolution is a remarkable force of Nature, and it has operated all by itself—without any deistic guidance—for billions of years.

Interesting Examples

We're often amazed at how things in the natural world can come about. Evolution has produced some incredible things, including humans, which I'll get into below. Adaptations that organisms have made are just incredibly fascinating. As Francis Crick, co-discoverer of DNA, once observed, "Evolution is cleverer than you." Over millions of years (but sometimes much, much faster) organisms have found ways to survive and evolve that may surprise even the ardent Creationist (I'll address some of the nonsense of Creationism below). There are countless examples of the interesting things that nature has produced. Here are just a few that I've found most provocative. Consider as you read through these—Why would God do such things?

Evolution has yielded many ways for organisms to reproduce. We're most familiar with sexual reproduction—males and females interacting to produce offspring.[6] But what if there were inadequate numbers of one sex? Well, many organisms, such as the common reed frog, can change its sex "at will" in order to help the species survive. A number of fish and other animals do this as well. The strict male-female dichotomy is but only one variant in the natural world, something clearly anathema to the creationists.

Beetles, as I said, are an incredibly prevalent and diverse group of animals. One in particular has developed a remarkable skill, the Japanese water scavenger beetle (*Regimbartia attenuata*). It is native to much of Asia, but has become an invasive species in many other parts of the world, including the U.S. It is sometimes eaten by frogs. But if it is eaten by certain frogs, it can survive. It sits in the frog's digestive system—without oxygen—for a couple of hours. It manages to turn itself around and proceeds through the frog's digestive system—unharmed—and exits the other end. How ingenious is that![7]

Consider the lowly *Toxoplasma gondii*, a single-celled parasite that invades mammals. It is very common in humans. It's the organism that can cause the infection known as toxoplasmosis. It is generally aborted by a human's immune system, though it can appear vital and dangerous in some weakened individuals. When it invades the bodies of domestic cats, it can reproduce and without any harm to the cat. It generates oocysts (egg-like cells) that are excreted in the cat's feces, waiting to be eaten by other organisms, such as rats, where they develop. But the developed form is neurotoxic to rats. It alters the rat's apparent mental state and behavior such that it loses its fear of cats, who can then more easily secure that prey.[8] The rat is an unwitting and doomed vehicle. The cat makes use of a parasite to facilitate its predation. It is a clever relationship among organisms that evolved only as an accident of nature.

Consider also the "lowly" tardigrade, called a "water bear," one of nature's most interesting creatures. It is less than a millimeter in length, and there are hundreds of species of them. They have evolved to do things which defy our common understanding of complex life on this planet. They've been called the "hardiest animals on earth," living in all sorts of environments and surviving all sorts of extreme conditions. They resist radiation, and they've survived in the vacuum of space (on a space capsule). They've survived extreme temperatures of 150 degrees and near absolute zero. They repair their own DNA. They can survive for decades without food or water, even though they're naturally creatures that dwell in water or very moist conditions; and they show several other remarkable feats we don't see in other living creatures. (62) Unlike more supposedly "clever" advanced creatures, they've been around for at least half a billion years. There are a number of such animals, often called extremophiles, that can tolerate unthinkably harsh environments, e.g., the vacuum of space or extreme heat in undersea vents. For example, there is a microscopic animal called a bdelloid rotifer (and there are hundreds of species of these, too), which characteristically lives for only a few days. However, it has been recovered and "brought back to life" after being frozen in permafrost for about 24,000 years. (473) With apologies to Schrödinger, these kinds of organisms can enter a state in which they're neither dead nor alive. Evolution is truly a remarkable thing.

Then there is the nematode (roundworm) *C. elegans* (more on this interesting creature in Chapter 5). It was recently discovered that these worms can withstand g-forces hundreds of thousands of times greater than humans. (401) And many of these worms, onboard for experiments, actually survived the tragic breakup of the space shuttle Columbia in 2002. (512) Again, evolution is a truly remarkable thing.

One of the most fascinating creatures on Earth is the "lowly" Hydra, a tiny, primitive marine animal that has found a way to forestall aging. Hard as it is to believe, it may even be immortal. It was aptly named after the Hydra of Greek mythology, a multi-headed monster which grew more heads whenever it lost one. The little Hydra has similarly evolved to also be able to regenerate itself. If it's injured or otherwise suffers a problem with some of its cells, it merely grows new ones from its stem cells. These are the kind of cells (and we have them too) that can differentiate into all sorts of tissues, and the Hydra uses them to recover or repair what's been lost. Over a long period of time, the Hydra is not the same creature it used to be, so to speak, since its cells have been gradually replaced over time (We do that too, but very poorly, which is why we die.). It continues to live. Of course, it can succumb to environmental conditions or predation, like most creatures. But other than that, it does seem to be able to restore itself and continue living indefinitely. It may also be that a few other organisms we know about, lacking predation or environmental injury, may overcome their mortality, e.g., certain jellyfish, possibly some corals, and potentially others we don't know about yet. Evolution is incredibly clever. I often wonder why it is these kinds of creatures that evolution has "chosen" to have this power. Yes, perhaps there *is* immortality, though not for us. If God existed, we might ask why the Hydra seems immortal and not the human that He is alleged to have "created" and considers so special.

Evolution doesn't always take millions of years. For example, there is the British peppered moth, a prime example of rapid adaptation to the environment by natural selection. This white, speckled moth sleeps during the day, camouflaged from predatory birds on light-colored surfaces like the trunks of many trees. Pollution from the Industrial Revolution of the 19th century, however, darkened many of the trees, exposing the moths to predation by birds. Genetic variation resulted in turning the moth's coloring black, a new disguise, in just a few years. This occurred for moths living in the polluted cities, while many of these insects remained white in the larger countryside. As pollution abated somewhat and trees began to recover their lighter coloring, the population of the city moths again reverted to white. We can see in this natural "experiment" that the environment drives natural selection, and it can happen pretty quickly.

Sometimes evolution happens fast, in fact, very, very fast. If you don't believe evolution is fundamental to living organisms, consider the following examples. Bacteria evolve into different forms all the time. There's the bacterium now known as MRSA (methicillin-resistant Staphylococcus aureus), which has quickly evolved from another form to evade many of our antibiotics. Every year thousands of people show up in emergency rooms with a life-threatening infection from this. It is very difficult to treat, and many have died from this infection.[9] We hear all the time about *E. coli*, a very common bacterium, some strains of which can produce illnesses in humans. But its survival "instinct" is truly remarkable. If some *E. Coli* "individuals" (i.e., cells) "experience" DNA damage, they will "commit suicide" in order to protect the colony from dangerous rogue mutations. (309) Excusing the personifications, this is an incredibly clever way for any colony of this organism to survive and flourish.[10]

Evolution has done some incredibly clever and potent things to enable survival, as Francis Crick pointed out. I always have at least a private smirk at Creationists trying to explain "God's will" for such things. For those who are a bit frustrated by the "God's will is unknown to us" and "miraculous ways" arguments, one might just as well believe in science.

Experimental Evidence

Most of the evidence for evolution by natural selection comes from methodically examining historical data—from fossils, from genetic evidence, and from systematic observation of what happens over time. But we can conduct specific controlled experiments to show the validity of evolution by natural selection. Here are only a couple of classic examples, of the many that have been done, which illustrate the point.

Consider again the bacterium *E. coli.* Your intestines contain many millions of these little single-celled "creatures." Some of its strains can be harmful, but others are actually beneficial to your health. What is important is that they evolve, and experiments have demonstrated that natural selection is the key. In one classic study, biologist Richard Lensky studied the evolution of a population of *E. coli.* The bacteria were from the same strain (but with a small bit of genetic diversity), and he divided them all into a number of separate flasks. To oversimplify, each day for each flask a very tiny sample was drawn out and squirted into a new flask, which contained a solution of the sugar glucose. Some of the bacteria were better at using the glucose, using it up more quickly (and actually increasing their body size), and others not so much. This was done over and over again, for thousands of generations of the bacteria, and some of the resultant *E. coli* became increasingly efficient in using the sugar. Essentially, they outcompeted their genetically variant brethren. When compared to some previous populations or the original/ancestral populations (which had been frozen, i.e., the fossils), the latest ones proved to be extremely "fit" at using the glucose medium while the ancestral population was not (one of the lines also actually learned to use the citrate that was part of the medium). The experiment has now been conducted by Lensky's teams, confirming the results for a period of over three decades. (304) Bottom line—evolution had created an adaptation and a change in the organism merely because of the effect of the environment. That is experimental evidence of natural selection at work.

Bacteria are very small "creatures," but what about the larger ones, e.g., the large animals we're familiar with? Another classic experiment was conducted on foxes by the Russian biologist Dimitry Belyaev, beginning in 1959, to see if evolution could domesticate an animal. Humans had domesticated animals for their own use for thousands of years, but could it be demonstrated that the process occurred by evolution —evolution by natural selection? He bred generation after generation of silver foxes to see if they could be tamed or domesticated. For each generation, he chose a small number of the foxes for reproduction, those which seemed calmer and less afraid than the rest when humans were present. Voilà! Within six generations of doing this, he had produced foxes that were as tame as any domestic dog. They could be held and petted; they would lick the experimenters' hands; they would wag their tails when people came within eyesight; and they would whine when the people left.

By subjecting a population of foxes for relative tameness, he ended up with a group of fully domesticated animals, essentially with the same type of behavior that you'd find in any household dog. And the experiment has been going on for some six decades, with fully consistent results. (143)

Conclusion

We humans are descended from a single creature, generally labeled the "last universal common ancestor," a "cellular organism" which lived some 4.2 billion years ago. (353)[11] That's right, we have a pretty good idea of what that creature was. Subsequent evolution is responsible for humans, and the same is true of elephants, rose bushes, modern strains of bacteria like *E. coli,* and the other millions of species that inhabit this planet, as well as the 99.9% of all other species that have gone extinct. They have not come about because of some preconceived "design." They have merely evolved because they "work." That is, their current form reflects a long history of natural selection from random diversity, enabling adaptation and survival across time and variously changing conditions.

Again, biological evolution is one of the most well-evidenced theories in science. One might argue that other theories, often more mathematical in their description, are more "scientific," perhaps physics, chemistry, etc. However, they—like all theories—are continually evolving as new data, testing, etc., are done. Evolution is no different. Theories of physics, and indeed cosmology, also continue to evolve and change as our technology and methods of measurement improve. And the advent of modern genetics has only solidified and refined Darwin's theory.

Modern human beings did not spring up magically in a garden somewhere. We also did not arise suddenly from other life forms. There is a long history to us. We did not suddenly arise from a monkey or a chimpanzee, as some have mistakenly cited as criticism of evolution (We actually did not evolve from chimpanzees at all; we merely had a common ancestor millions of years earlier; more on this below). We did not appear because some other primate suddenly "changed its stripes." Evolution proceeds very slowly, sometimes more slowly than we can ever imagine (though there are exceptions, as above). But evolution over millions of years did produce human beings in their present form. To doubt this is no more than willful ignorance. And no god was necessary in this process. Science explains the phenomena in exquisite detail. Of course, we sometimes hear the theological argument that evolution is God's will, but we must take this as but another blind acceptance of irrational dogma and an attempt to deny science.

Creationism and Intelligent Design

Nevertheless, the issues of so-called Creationism and intelligent design still persist, and I'm obliged to comment on the nonsense of these ideas. They are well debunked in a number of excellent lay works. (112, 113) However, we must state that no matter the offering, the notion of creationism has no basis whatsoever. As the National Academy of Sciences notes, "Creationism is *not* science." (362) But we'll look briefly at a couple of the so-called arguments for intelligent design that putatively flow from that idea.

Ever since publication of Darwin's *The Origin of Species* in 1859, we've seen frantic attempts to undermine the solid science it revealed, from Reverend Drummond's "God of the gaps" to modern movements like Creationism and Intelligent Design. The argument from design is the one most often employed by nervous clergy, who cannot accept the well-evidenced idea that life has evolved over hundreds of million years according to natural selection. It just seems too complex for them. God must have designed it all. The modern design argument seems to go back to the English theologian William Paley, who espoused that nature was so complex that it must only have been the work of God. (388) But as Dawkins notes, "The argument from design is the only one still in regular use today, and it still sounds to many like the ultimate knockdown argument. The young Darwin was impressed by it when, as a Cambridge undergraduate, he read William Paley's *Natural Theology.* Unfortunately for Paley, the mature Darwin blew it out of the water." (113)

Creationists stubbornly maintain that the incredible complexity of creatures, including humans, could not have evolved naturally and that some supernatural god must have been the architect. They call this "irreducible complexity." Let's first realize that "irreducible complexity" is merely a comment on human thinking. Some humans are just not capable of, or not interested in, doing or understanding the analyses that reduce apparent complexity into its component and historical roots. Just because you can't or won't conceive of something doesn't make it supernatural. Your knowledge is limited. I have a hard time understanding how a skyscraper is designed and built, or an airplane, or the supercollider at CERN, perhaps the most complex machine ever built. But I know that there are other humans who *can and do* understand such things—in the most detailed and elegant ways. God is not necessary. There are machines and software that are so complicated that they are often fully designed by computers, with minimal direction from humans. I cannot possibly conceive of how that happens, and actually the programmers often cannot either. But to invoke the gap-God because of my limited knowledge would again be ludicrous. The whole notion of invoking a god as a designer to explain things we can't fully understand is wholly unnecessary.

The notion that a Creator designed or promulgated the development of living organisms on this planet does not stand up to any rational or inductive argument. Any god who would have created life as we know it was a pretty incompetent biological engineer. Only a few examples of the blatant design flaws: Why do many birds have wings if they can't fly (emus, ostriches, kiwis)? Why does a bat have solid bones, a relative impediment to flying, and an earthbound emu have bones that are mostly hollow? Why do whales and dolphins have to breathe air? Why does the left recurrent laryngeal nerve, coming out of the tenth cranial nerve, run around the aortic arch and back up to the larynx? It's a poor "design" for us, not to speak of the giraffe or a whale, where the long path is a ridiculous design. Why is the human birth canal often inadequate to the task? Why do so many women die during pregnancy and childbirth, a piece of pretty cruel design if there ever was one? Why do we grow too many teeth to fit comfortably in our mouths? As scientists Randolph Nesse and George C. Williams note, our "design" was both "simultaneously extraordinarily precise and unbelievably slipshod" to the degree that it was "shaped by a prankster." (366) If a God had a sense of humor, there were probably better ways to show it.

A principal Creationist "design" argument is the apparent complexity of life on this planet. There are numerous examples of evolutionary developments which are difficult for many non-biologists to understand. Let's take one of the Creationists' favorite anatomies, the human eye. It just *seems* too complex to have evolved by random variation and natural selection. Nevertheless, this matter has been studied extensively, with clearly compelling explanations. The "eye" began with at least some photosensitive cells, then their later organization into photosensitive organs. It often began to exhibit some "cupping" to better intensify light and so forth, leading eventually to mature vertebrate eyes. In fact, eyes have evolved many times in many different ways throughout biological history in many different types of creatures. Such organs developed to assist an animal in its orientation and later for predation or escape from predators (Some fish have even evolved a "distraction eye" at the other end of the body to confuse predators.). Different kinds of eyes have come about in jellyfish (multiple times), in mollusks like the octopus, and in the long vertebrate lineage. The above tardigrade has a couple of primitive eyes, and even the primitive flatworm planarian has a couple of primitive eye spots. The historical developments have been traced out in specific genetic detail, from primitive species to our own human eyes, including the extensive biochemical and anatomical variations which have enabled different kinds of eyes throughout evolution, which we don't need to get into here. (159) The human eye, and all its vertebrate ancestors, is but one of the many attempts at this by Mother Nature. The subtle anatomical and physiological changes in the eyes of vertebrates, in particular, have been well outlined, and truly compelling explanations, with some studies using genetic and embryonic data to explicate how the human eye evolved—from the primitive hagfish (a creature with "eye" characteristics intermediate between a rudimentary light detector and a camera-like eye which can generate images), up through mammals—like us. (292) That it is admittedly difficult to understand, of course, does not rationalize a supernatural power.

And of course, as remarkable as it is, the human eye is far from perfect. If it was created *de novo*, it was a pretty poor design. We have a blind spot on the retina, where the optic nerve exists. That necessitates hundreds of thousands of additional neurons in the brain working "overtime" to compensate for that problematic design. We have pretty limited peripheral vision. We don't see all that well at night. We can only see a very small portion of the electromagnetic spectrum that affects our lives.[12] Our eyes are wired "backward," with light having to pass through a chamber of liquid to get to the photosensitive cells. Additional mechanisms are then necessary to process the light appropriately. And most of us lose some type of visual acuity as we get older. Our brains make up for most of those shortcomings, but the Creationist view of our eyes is very far from an outstanding creation. It is clearly a product of gradual changes by natural selection over hundreds of millions of years.

Finally, we must consider the Creationists' view of science itself. The methods of paleontology have done an incredible job of establishing various time scales for the evolution of life on Earth. Life on Earth has been evolving for billions of years. It's difficult to see how this would occur on an Earth that's only 6,000 years old. Yes, it is sad that Creationists still think the Earth is 6,000 years old and that God deposited humans and other creatures on it. Creationists even sometimes invoke the notion that the methods of radioactive dating used by paleontologists and other scientists are invalid, as God set up and manipulated radioactive decay in such a way as to fool scientists studying this problem. There's no way to look at this except with a chuckle or as an example of some theologians' delusional thinking.

It is difficult for Creationists to believe that humans are not special creatures. Their allusion to Darwin's findings on this is uninformed. Although he did not comment on this in *Origin of Species*, he did find, in *Descent*, that differences between humans and other animals are "of degree and not of kind." Our specialness began to evaporate. The notion of some "intelligent" supernatural being creating the world and taking particular interest in creating modern human beings is about as absurd as it gets. Science clearly explains how life, including humans, evolved on this planet. The Creationist view of the evolution of humans and other life on this planet would be just laughable if it weren't for the tragedy of so many people believing in it and trying to force it on others.

Our Ancestors

We humans were not placed on this Earth by some divine force. There was no Garden of Eden. We *did* evolve from more primitive forms of life, and most recently from other primate species. We'll see that there is nothing inherently special about us, other than possibly the culture we've created, though not everyone would agree that such has been uniformly positive. Consistent with Darwin's finding, we humans are different from other creatures as a matter of degree, not of kind, as we'll see here. I'll not attempt any comprehensive review here, only some highlights to make the point.

We modern humans are most likely a product of genetic changes that spanned hundreds of thousands, if not millions, of years. We evolved naturally. Our biological name is *Homo sapiens*, which means wise human (Given the very short biological time we've been around, the verdict on such labeling is still out). There were many species of the *Homo* genus that preceded us, and a number of them shared the Earth with us for a very long time. There is a common misconception that we humans evolved from chimpanzees. This is a misunderstanding. Although we share approximately 99% of our genes with chimpanzees, *Homo sapiens* and other species in our group actually evolved from more primitive forms. As E. F. Torrey points out, "It is important to note that hominins did not evolve from chimpanzees as we know them, something evolution deniers misunderstand. Rather, both hominins and chimpanzees evolved from a common ancestor that lived about six million years ago." (519)[13]

We only slowly became human. Here's a brief tour, highlights from the journey, with just some of the creatures that preceded us.[14] In 1974, American paleoanthropologist Dr. Donald Johnson discovered a three-million-year-old fossil in Ethiopia which appeared to have both ape and human characteristics. This is the well-known creature named Lucy, which has long been considered the first human-like ape, *Australopithecus afarensis*.[15] It often walked upright, though it continued to spend time in trees. It was a hardy species, surviving almost a million years, from about 4 to 3 million years ago (We should be so lucky). For a long time, it was thought that this species was the critical branch in the so-called ape-human split.

More recently, the discovery of fossils in 2001 in northern Africa identified a creature who also had features of both apes and humans (*Sahelanthropus tchadensis*); and this was perhaps the species better identified with the split, dating back at least 6 million years. (64) It likely walked upright and had other human-like characteristics, though evidence is still a bit weak in these regards. There continues to be some disagreement as to whether this was actually a bipedal creature, one that always walked upright. (71) Some have argued that it may be a "stem" rather than a direct human ancestor, and others suggest that it was a human ancestor. It was dubbed "Toumaï," meaning "hope of life," in the local Daza language in what is now Chad in Africa. Nevertheless, *Australopithecus afarensis* is still believed by most to have represented the split in the lineage that led to the *Homo* genus from the ancestors of chimpanzees and bonobos. The final verdict is still out, and further discoveries may enlighten the picture. Yet, it is clear that we did evolve from such types of primitive creatures.

We Were Not Alone

From the earliest members of our genus, the *Homo* group, evolution has been a messy endeavor with many branches, some surviving, some not. It took millions of years for *Homo sapiens* to evolve from earlier species, through a variety of intermediate forms, and there are several known *Homo* species that preceded us. Here are just a few for illustrative purposes. The species dubbed *Homo habilis*, dubbed the Handy Man, lived some 2.4 million to 1.6 million years ago. They added meat to their diets and were the first real toolmakers (although there's some indication, still controversial, that *Australopithecus* may have also used tools). *Homo rudolfensis* appeared some 2 to 1.5 million years ago in East Africa. It had a larger brain than *habilis*, and probably did fashion and use stone tools.

We are the only surviving member of our genus, *Homo*. But there are many other species of humans in the family tree included in the genus. Although all our other cousins have gone extinct, several of these shared the Earth with us for a very long time—and until pretty recently. Here's a quick look at just a few of those relatives. Members of *Homo erectus* arose nearly two million years ago, and they didn't go extinct until a little more than 100,000 years ago. *Homo erectus* was the longest-lasting member of our genus. They probably used more sophisticated tools than earlier species, had a large brain, and spent all of their time on the ground. It was probably the first species to have rudimentary speech and the first hominin to use fire. It was a fully committed biped, with a body adapted for modern striding locomotion. It was also the first in the human lineage to disperse outside of Africa, and it had migrated around about half the world by around 700,000 years ago. It also shared the Earth with the species *Homo heidelbergensis*, which lived some 700,000 to 200,000 years ago. They also had large brains and were likely the first to build their own shelters, often using rocks and woods. They hunted large animals with wooden spears. It *may* have been a very close ancestor of modern humans.

And there are others who overlapped with us. Moving a bit closer to modern times, there is the species we call the Callao Man or Ubag (*Homo luzonensis*), named after a mythical caveman. It is a recently discovered tree climber, possibly a pigmy species of human. It roamed an island in Southeast Asia. These cousins may have lived for only a few thousand years, and it appears they went extinct probably around 50,000 years ago. Then there are the *Denisovans*, cousins who ranged across Siberia and East Asia. Not much is known about them, except that they did interbreed with Neanderthals and humans (in fact, their extinction may have been related merely to their extensive interbreeding with us). They coexisted with us and with a creature dubbed the "hobbit" (*Homo floresiensis*). This was a small human which had spread out to southern Asia, particularly what is now Indonesia. And it may have gone extinct, perhaps only 50,000 years ago, associated with the arrival of *Homo sapiens* in its habitats.

The well-known Neanderthals arose as a species probably about 400,000 years ago; and they had a long run, spreading throughout Europe and much of Asia. They were problem solvers, users of sophisticated tools, had families, created art and perhaps music[16] and did other things that we consider quite characteristic of modern humans. They buried their dead, sometimes marking graves with offerings such as flowers. This was the first known time that any human, or other primate, had engaged in such symbolic behavior. (492) Neanderthals had larger brains in relation to body size than humans (but see below). They used clubs and other weapons and fought violent wars with *Homo sapiens* for thousands of years and eventually went extinct only about 35,000-40,000 years ago (though the various reasons for their extinction continue to be debated; see below).

An interesting recent discovery was made in 2013 in the Rising Star cave in South Africa. (136, 556, 557) It revealed another *Homo* species we hadn't known about, the *Homo naledi*. They seemed to have buried their dead, and the species may have lived until about 50,000 years ago. They were small-brained hominins, and they coexisted with us and with Neanderthals. We also shared time and space with *Cro-Magnons*. They are actually considered a subspecies of *Homo sapiens* which arose about 40-50,000 years ago. They coexisted with us and other human species for thousands of years, and they probably existed until well into the late Neolithic period, perhaps into the early Bronze Age; nobody really knows for sure. But this was the time that *Homo sapiens* had already started to develop larger communities and agriculture, and they were still around. In sum, we as humans shared the Earth with our cousins until only few thousand years ago. Perhaps we are unique only as surviving members of one particular genus, and that situation has changed only very recently.

Modern Humans

There are continuing debates and controversy about dates. However, based on the latest discovery of fossils of *Homo sapiens*, it seems we've been around in the neighborhood of at least 300,000 years. (70) Although we've been around for at least that long, what we might consider "fully modern" humans, with our present genetic makeup, likely occurred much later. How this came about is still a mystery, but our genetic makeup is likely a function of long-term interbreeding of different *Homo sapiens* groups. That is, current thinking is that the "single origin story" for modern humans is probably wrong. We did interbreed with Neanderthals, Denisovans, mid-Pleistocene *Homo* groups, and possibly *H. floresiensis.* (25) However, we are not descendants of one unique line of humans. It is likely that we evolved our current form because of early human diaspora, isolation of groups, subsequent evolution and changes in those groups, and ensuing coming together and interbreeding. (449) That is, early *Homo sapiens* spread out across Africa; and various groups, of slightly different genetic makeup, later interbred, leading to our current form as humans.

All this occurred while we were still hunter-gatherers, humans that lived in small bands or tribes for thousands of years. Our brains were a bit smaller than some of our ancestors, but hunter-gatherer *Homo sapiens* were intelligent, and they fashioned more sophisticated spears and other weapons, which may have led to more meat-eating and group cohesion. We learned to cooperate—for hunting but also for defense against other tribes. We began to have more time for socialization and development of culture. We don't have a lot of detail about specific behavior during that period of time, though we do know that there were groups of people who tended to stay together, which was of benefit in hunting and likely child-rearing. We slowly became a domestic species, which occurred through development of social groups, language, agriculture, and culture (180), which I'll examine in a bit more detail below.

Why Us?

We're the only *Homo* species left, the only humans still in existence. We're only special or unique creatures because we've survived to this point in history. The genetic differences among all the human species were amazingly small. *Homo sapiens* did eventually become the dominant species. But if you had lived during the late Stone Age, at that time you would have viewed that such was not at all certain. So how is it that we survived, and other *Homo* species didn't?

Did God make some mistakes in this area? After all, modern humans, as they exist today, have only been the lone species of humans on Earth for perhaps about 35,000 years. That's only about 5% of the time since our possible human-like ancestors started to walk upright. We don't see any sudden "hand of God" or any supernatural influence in how human beings branched out from their earlier ancestors (If you believe that it was God who chose us over the others, no need to read further.). We evolved over millions of years by natural selection, and there was incredible similarity of anatomy and behavior between us and the other human species that roamed the Earth with us. I'll review some of the reasons that we may have survived while all of our cousins did not. There's still no definitive answer to that question, but there are some evidence-based scenarios. Nevertheless, the bottom line is that we're special only by having survived our cousins by a few thousand years. Most importantly, we need to know how we became domesticated since, for our purposes here, our sedentary or domestic lifestyle is what led to the advent of superstitions and organized religion.

How we outcompeted the other *Homo* species of our time continues to be debated. We were not the only species who had language and learned to communicate and socialize. We did make weapons for hunting and fighting, as others did. We probably varied our diet based on availability of prey animals. We likely formed more cohesive groups than other species. Then there's violence. Conflicts between us and other species, especially Neanderthals, are well known. As modern humans expanded their territories, they clashed with Neanderthals, who were very persistent, going into "wars" with us for thousands of years. And that was clearly one factor that helped to reduce their populations.

But there were other factors. We often interbred with other *Homo* species. In fact, we do have genetic evidence that we interbred with both Neanderthals and Denisovans (199), which may have also contributed to their demise. We likely interbred extensively with Neanderthals at various distinct points in our overlapping history. (183) All of us, in fact, still carry a small amount of Neanderthal DNA, up to about 2% of our genome. We should note that other hypotheses for Neanderthal demise, such as migration patterns and other genetic changes, have been recently offered. (280) As for the *Cro-Magnons*, they likely disappeared as such because of interbreeding with us. In fact, as I noted, they are considered a subspecies of *Homo sapiens.* Current terminology is that we who still exist are referred to as *Homo sapiens sapiens,* a designation which implies *more* "wisdom." Whether that means we are even "wiser" than our Cro-Magnon cousins, or other *Homo* species for that matter, remains a question for history.

Then there's climate change, which has been a major driving factor in the evolution of life forms from the earliest days of complex organisms. It was certainly a factor in the demise of other *Homo* species. We know that significant variations in climate occurred just prior to at least three human species disappearing—*H. erectus, H. heidelbergensis*, and *H. neanderthalensis*. (421) With respect to Neanderthals, for example, they were probably less adaptable as they moved south with the last Ice Age, not being able to adjust to other habitats and prey. There have been many significant changes in climate before and after the last Ice Age, when glaciers began receding about 20,000 years ago, so our temporarily privileged position may have been due to a combination of a number of factors, which we'll briefly consider now.

Tools

Did we survive our cousins because we used tools? Doubtful. Across time and species, it does appear that the Copernican Principle applies even to tool use, one of the things we think distinguishes us from other animals. *Homo sapiens*, and the other members of our genus, did not come into being merely because they used tools, and we're not the only animals that do.

We sometimes think of our tool use as reflective of advanced intelligence. Yes, it's advanced considerably in the 300 millennia or so we've been around. But we're not the only organisms that use tools. All of our other hominin ancestors did, of course, including the use of lethal weapons. The use of primitive stone tools by genus *Homo* in Africa dates back at least 2.6 million years to *Homo habilis* (though there is some evidence that *Australopithecus* also probably used stone tools well before that—3-4 million years ago). Some of our ape relatives use tools—gorillas, orangutans, and chimpanzees, and they predate us by a lot. Chimpanzees use some primitive tools, including stones to crack open nuts. They fashion sticks to retrieve insects out of logs. Some chimp tribes even fashion knife-or-spear-like weapons to help kill prey (There appears to be social-cultural learning to all this, with different tribes doing different things, thus not attributable to genetics). Even more primitive primates, e.g., certain monkeys, actually fashion their own gross stone tools to crack open nuts and dig for roots, and even possibly to catch the attention of potential mates. (417)

However, species far removed from us or other primates also construct and use tools. Many species of birds use tools. Corvids (e.g., crows and ravens) are well-known tool users, and some even fashion tools just like chimpanzees do (e.g., the New Caledonian Crow). Dolphins put sponges on their beaks or use large shells, skills they learn from peers, which help them dislodge fish that hide on the sea floor. Elephants use tools in various clever ways—using sticks to scratch themselves, to swat away flies, to acquire out-of-reach food, and many other uses. Recent studies now indicate that even some pigs use tools, notably sticks and bark for digging their nests; and this does appear to be learned behavior. (431) And there are many other species including sea otters, beavers, octopuses, and even ants (yes, they use twigs and other small things they find). We think of non-primates such as crows, dolphins, and pigs as intelligent creatures, even though they are evolutionarily very distant from us. And ants and octopuses are *very* far removed from us. Tool use does seem to be rather ubiquitous in the animal kingdom. It does not make us special, and it's not responsible for our ancestors becoming human.

Language

Did we survive better than our cousins because we had language? Nobody knows. Many organisms can "speak," so to speak. Many other extant species have audible methods of communication, though not true language. We're most familiar with the songs of birds, but many other species communicate with vocalizations—bats, elephants, and others. Bonobos have some actual language which depends on syntax, and it can be argued that the songs of cetaceans (dolphins, whales, and porpoises) may constitute something approaching crude language. Some humpback whales actually learn complex songs from members of other pods.

The development of sophisticated language may have given some species an advantage over others. Language enabled us to communicate with more complexity, binding us together in ways other than the necessity of food acquisition, reproduction, and tribal defense. We were able to share experiences in ways that exceeded our basic needs. Over time, tribes of *H. sapiens* grew in size; and language helped us bond together and create some stability. Larger groups necessitated better ways to communicate, and that may have actually advanced our language skills. As Torrey points out, "The need for large groups among our early ancestors was the driving force behind . . . the evolution of language." (519)[17] Since we had language, we learned to communicate in more sophisticated ways. This helped us in problem solving—two or more heads proved better than one. Many heads might have done even better (of course, the opposite could have been true, as it is often today). Considering more information can be key to solving a problem. Where should we go next? How should the food be kept and distributed? And there were many other problems we had to solve. Communication helped to bond human groups together in ways that some of our extinct cousins may not have availed themselves of, and it did depend on some type of advanced language.

It took a long time for us to be able to communicate in ways similar to how we do now. Language started to develop in humans probably 200 millennia ago, and by 50,000 years ago *Homo sapiens* had developed language such as we know it today, even though it's presently filled with more information than it had back then. But it did not necessarily give us any advantage over the Neanderthals, for example. Neanderthals had speech and language too, having the same gene variant as us. Although group size and culture provided the environment for language, its development in *Homo sapiens* was probably due to genetically driven alterations in brain organization and functioning. Language also seems to have developed independently in many geographies. As Mlodinow points out, "Language evolved as a biological adaptation, aided by a gene that has been present in human chromosomes so long that it has even been identified in ancient Neanderthal DNA. . . . [language] appears to have been invented over and over again, all around the globe . . . each tribe had its own language . . . there is no record that a region has served as a 'cradle' of language from which it spread to previously languageless groups." (350) It would indeed be difficult to ascribe our dominance and survival to language alone, especially vis-à-vis the Neanderthals.[18]

Thinking and Problem Solving

Did we survive better—are we special—because we were smarter? A fundamental, often-cited characteristic of humans is that we can "think." As Einstein once said, "Thinking is to man what flying is to birds. Don't follow the example of a chicken when you could be a lark."[19] It would be difficult to rule out that our intellectual ability to think and solve complex problems was at least a major factor in the development of our species and our survival.

Like all of our characteristics, our intelligence evolved slowly. Natural selection of problem-solving abilities, due to the associated brain capacities, enabled some human groups to achieve a survival advantage, and we developed our complex cognitive abilities while we were still hunter-gatherers. This is the approximate time we became "fully human," what is sometimes referred to as the "cognitive revolution" some 70,000 to 30,000 years ago (203), and the enhancement for actual "complex symbolic thought" was evident perhaps as early as 40,000 years ago. (350)

Humans of that time were anatomically the same as we are now, and they probably had the same level of general intelligence. To illustrate, a baby born into a hunter-gatherer tribe of that time had the same cognitive abilities as today's children. If you were to time-transport that hunter-gatherer baby into today's environment, it would develop the same level of cognition and behavioral skill as any other baby born today would show. That our abilities, behavior, and accomplishments far outstrip our ancestors can't be explained by our native intelligence. They are due to other factors, principally the information culture in which we live. We are not more "special" or unique in comparison to those creatures of 40,000 years ago.

Anatomy

Did we survive because we had bigger brains? We often think of "brain size" as the key to greater intelligence and the greater the possibilities for survival. The bigger the brain, the smarter the creature. But that's not always the case. The evolution of the human brain (and that of certain other advanced *Homo* species) is credited with the development of complex thinking and problem solving. This has clearly been an evolutionary advantage. Brain size, of course, evolves through the mechanism of selection pressures acting on an organism's genome. Natural changes to the genome tend to occur randomly. Those brains which provided better survival value—generally via increased cognitive ability—were selected for, since selection pressures favored those who could better solve problems. But mere brain size is rarely the issue. We mammals are distinguished from other creatures in that our brains are actually larger in relation to body size than other animals, what's technically referred to as the encephalization quotient or EQ, and modern humans have the highest one (orcas and other dolphins are close behind).[20]

A number of *Homo* species, including the so-called dead ends, had large brains. We do know that Neanderthals, for example, had even larger brains than us. Of course, in the end, that didn't make a difference. Some other factors must have been responsible for their demise (possibly interbreeding and conflicts with humans, less stable groups, and other factors). Mammalian brains, especially hominid ones, require a great deal of energy compared to other classes, and saving energy *is* of survival benefit (all organisms must balance energy expenditures in relation to the amount of energy/nutrition consumed). This may have been unfortunate for Neanderthals since brains consume a very disproportionate amount of energy, which may have made them more vulnerable to the deleterious effects of food shortages. There is no way to know if this dynamic played any part in Neanderthal extinction, but it certainly didn't help.

In fact, *Homo sapiens* brains were smaller and have actually further decreased in size as we transitioned from small groups of hunter-gatherers to more localized agricultural communities. (203) That is, as we became domesticated, our brains shrank. This is not an unusual dynamic. For example, domesticated sheep have smaller brains than their wild counterparts. Domestication led to various other characteristics, such as wool types—for our human use, not for any wool characteristic which enabled survival of their wild cohorts. Other domesticated animals, e.g., cats and dogs, but also rabbits and pigs, have smaller brains than their wild counterparts of the same species. Obviously, domestication of humans—and yes, we can call it that—resulted in increased cognitive abilities and output vis-à-vis earlier forms; but we can rightly not ascribe this specifically to brain size, since ours too became smaller. Domestication does appear to reduce brain size.[21]

Problem Solving

It was not brain size, per se, or even the EQ, that was critical to our success. Modern humans do collect more information and consider more behavioral options than other species, but problem-solving ability is not unique to us. Our hunter-gatherer ancestors had the same capacity, even though they did not have access to the same kinds of complex information we do today. And what about more primitive *Homo* species that went extinct millennia or millions of years ago? We know that other extant primate species have well-developed problem-solving skills, as do many species of marine mammals; and they apparently communicate about these with others in their groups. This is even true of many species of birds.[22] Even some insects have the capacity to solve problems, including those in which they have to *learn* the solution (i.e., not innate behavior). For example, bumblebees have even been shown to demonstrate behavioral flexibility and modeling of others in solving problems, even though they do not have much of a brain. (319) It appears that evolution has found ways for organisms to solve problems in their environments in all sorts of ways.[23] But the bottom line is that we are far from alone or special in our use of problem-solving abilities to enable survival.

However, our advanced cognition gave us certain survival advantages. First, we developed the ability to explain things that weren't immediately obvious. This allowed us to solve more complex problems. When the weather changes, shall we wait for the herd of prey to return or try to follow it? We are also a curious species (though there are many others, especially among primates). The generalization of this is the unique human capacity for wonder, the exploratory behavior beginning in early childhood (we'll delve into this more in Chapter 10). It enabled us to seek out more information and solve more complex problems; in fact, it may have led to our own self-domestication.

Agriculture

Modern humans did not come about because we started to farm. Agriculture also did not enable us to outcompete the other species of

Homo that we coexisted with. They had gone extinct thousands of years before. Further, it may not have been intelligence or language that allowed us to become domesticated. After all, we'd existed as hunter-gatherers with language and intelligence for thousands of years before we began to grow things.

The activity of growing plants for food, what's been called the "Agricultural Revolution," actually appeared in the in the Middle East in the late Stone Age, probably somewhere around 9,500 to 8,500 BCE (sources differ a bit on this range). But it also did not occur all at once or in only one place. It also arose independently in other parts of the world, including East Asia, West Africa, and North and South America. The advent of agriculture, however, clearly led to our domestication in different parts of the world. With more consistent food availability, we tended to stay in one place for extended periods, thus enabling the formation of larger and more stable groups and development of societies and cultures. In that context, agriculture was a force enabling group coherence and a communal existence for humans.

But how did it happen? It wasn't designed by any person or group. Humans of 120 centuries ago didn't suddenly become intellectually capable of the rudiments of farming—seed planting and cultivating. It wasn't designed from scratch by a group of humans. Historian and philosopher Yuval Harari, in his seminal book *Sapiens: A Brief History of Humankind*, makes the case when he concludes, "There is no evidence that people became more intelligent with time. . . . The agricultural revolution was history's biggest fraud. Neither kings, nor priests, nor merchants. The culprits were a handful of plant species, including wheat, rice, and potatoes. These plants domesticated *Homo sapiens*, rather than vice versa." (203) It probably happened "by accident." It is reasonable that observations over time shaped humans into making use of the natural history of what happens to plants as they begin to germinate and grow. If a group of people stayed in one place for a modest period of time, they might observe that plant droppings were associated with growth of those edible plants. School children do not routinely conceptualize in advance how a lima bean sprouts. They learn over a few days by observation. That we deliberately designed agriculture, which enabled larger communities and social development, has no specific evidence. Rather, plants in various parts of the world taught us something interesting and beneficial about remaining in groups for a period of time, which benefited our survival.[24] Of course, with less reliance on hunting and meat-eating, such dietary shifts did enable closer and larger communities, which allowed for the development and maintenance of culture. But unfortunately, it also allowed for the advent of malnutrition and the evolution of many infectious diseases. (350)

Group Cohesion/Cooperation

Humans are social animals, and they form themselves into groups, as their ancestors did. Such cohesion was necessary for the appearance of superstitions and later for religion to emerge (see Chapter 7). Living in groups has actually enabled human survival and evolution, but it is not an exclusive human trait. We are not "special" creatures because of it, just one more animal species fighting for survival. Living in cohesive groups is not unique to humans. It exists in other primates and in many other social animals. Wild dogs do it. Lions do it. Ungulates do it, as do marine mammals. It is a survival advantage for those organisms who've developed it. Orcas, for example, bond together in all sorts of ways—from group hunting strategies to mourning rituals and birthing assistance (Yes, some orcas do act as midwives). Even certain species of electric eels cooperate in group hunting activities. Ants and bees do it very well, in highly organized groups. No species of social animals can survive without group cohesion.

Millennia before the agricultural revolution, it became increasingly necessary for small groups of humans to stay together for extended periods of time. As evolution increased brain size and rendered the birth of infants at a less developed stage of maturation, the role of immediate kin and other tribe members became more important—notably for the survival of the infant and subsequent protection and "education" of the child. As Harari points out, "It takes a tribe to raise a human." (203)[25]

Cooperation

Group sizes began to grow, which may have been of benefit in food acquisition and defense. It may also have made it more difficult to move, providing increased opportunity for observing the germination of plants. Increased size also led to more demands for cooperation—for sharing of both food and of responsibilities. Cooperation is one of the behaviors which bind groups together, enhancing chances of survival. Cooperation is a strong trait in most primates, including humans, and the advantages of this are obvious. Such groupings enabled social organization, advantages in locating food, defense of territory, and rearing offspring. These factors certainly enabled cohesion of a group along with evolving social order and protection, especially of the young. It also enabled the sharing of superstitions, the basis for all the later god fantasies and religions.

Leadership

The cohesion and survival value of a group of social mammals is dependent on maintaining social order. Of course, as human groups grew and cooperative behavior emerged, so did hierarchies. Humans, of course, are not the only social animals for whom this is the case; many primates and certain other social mammals show hierarchies—from deer to gorillas, from hyenas to elephants and orcas.[26] What emerges from social hierarchies is that one individual (or in some cases a few) becomes dominant in the group. Such an individual provides guidance, modeling, and punishment of group members for various behaviors not valued by the group (I'll get into the emergence of "morality" in Chapter 9). It has been a pivotal factor in the maintenance of cohesion in human groups, as it has in other social primates and many other social mammals. Dominance and leadership are fully natural evolutionary developments, and they have been critical in how human social behavior and culture evolved.

Human leaders that emerged likely tended to be better at commanding subjugation than other group members. Dominant and submissive behaviors are well known throughout the social animal kingdom based on pretty well-established hierarchies, and this quality tends to be passed down from generation to generation. In fact, in many species, alpha-type males and females produce alphas of the next generation, and they try to enforce bans on reproduction by others. Does "heir apparent to the throne" ring a bell? In most hierarchical species, maintenance of leadership is achieved with physical prowess and domination. As human societies grew, however, this would have been cumbersome and ineffective. Rather, human leaders likely began using symbols and superstitions to enhance their potency—thus, the beginning of some formalization of religious fantasies (I'll detail more about this in Chapters 7 and 9).

Superstition

Imagination

Group cohesion and leadership were necessary for our survival. As such, they also became pivotal to the evolution of superstitious beliefs and religion. However, this would not have been possible without a unique human cognitive capacity, the ability to imagine. *Imagination* enabled humans to behave in ways that were not bound to things in our immediate physical and temporal environment. We can anticipate what "might happen" if we do certain things or certain conditions occur.

When you can imagine, you can also imagine *things that don't actually exist*. As far as we know, humans are the only species which can do that. We have no way of knowing if these are unique skills. We can't ask birds or our pets or other current apes if they do this (and we haven't found ways to test such capacities in other organisms). It's quite possible that some of our extinct human cousins were also able to do this. We survived longer, and it's likely that our ability to imagine and create fantasies led to appearance of superstitions. As we had to generate explanations for actions we couldn't observe, we made them up. We imagined them. Such advanced cognitive skills and abilities to imagine likely did contribute to our survival, but they also led to the *invention* of superstitions and the gods that grew out of them.

Superstitions

For our purposes here, let's consider the nature of superstitions and how they come about. The onset of superstitions seems unlikely unless you consider the evolution of certain fantasies over hundreds of thousands of years. They became well-conditioned in our ancestors and passed down as various religious memes. But what is a superstition? Superstitions are learned behaviors. We are not born with them. They come about because of the accidental association of some events (we'll talk more about superstitions in Chapters 7 and 10). Crossing your fingers *may* be followed by good fortune at some point. Walking under a ladder could lead to some grievous happenings. But there is no necessary connection between such events. They are superstitions.

We humans have always needed to explain things we observe. The ability and tendency to explain events and predict future events, of course, was a pivotal set of behaviors of early humans that provided a survival advantage. It enabled better prediction of events than previous species could accomplish. We were good at recognizing patterns in our environments. If you hear rustling in the bushes, it might be an animal you could run away from or you could kill and eat it, and we learned to tell the difference. If you eat berries of a certain color, you'll feel sick. If clouds start to form, they might give us water. The ability to learn that some events led to others, the power of explanation, helped us survive. As we developed larger brains and the ability to predict more complex events, we grew to believe that everything had an explanation.

The problem arises when this ability or tendency is well-developed and where no reasonable explanation can be found or inferred from the environment and personal experience. Why does it get cold? What makes the river run? Why do people die? There must be some explanation. Humans, by our very nature, have little tolerance for unexplained events. Where no reasonable "explanation" of events was obvious or could be inferred from the available information, based on our capacity to imagine, explanations were "invented."

Also, there must be some force or power behind unexplained events. We knew that our own actions, our own power, caused things to happen. It was personal. Early humans could not conceive that there could exist "forces" which were *not* personal. Therefore, everything that happened must be caused by something or some force with some sort of personal character. They just couldn't always be seen. And so they were *imagined.* There must be some unseen persons or living spirits in the world that caused things to happen. Who caused the sun to come up; who turned the leaves red; who caused death? Fabricated forces or superstitions came into being, and this became what we call *animism*, as I'll address in Chapter 7.

Our ability to think and imagine led to the evolution and dominance of our species, but it also gave us animistic ideas and the tendency to cultivate some elaborate superstitions. And we had sophisticated language that enabled us to share the myths of unseen forces or spirits. Let's take a likely example. The accidental appearance of agriculture provided the context not only for more stable communities, but as a result, it led to more social interaction, communication, and "wonder" about how things worked. That led to the promotion of various supernatural superstitions. Why *did* plants grow? What force made it happen? Since it was unseen, it must have been *super*natural. We did come to know that rain promoted plant growth, and rain came from the skies. And so we looked "up there" in hopes of rain. We anticipated it; we longed for it; and eventually we asked for it. We fervently wished for it—something which much later came to be known as prayer (literally to beg for), and you can't pray to something inanimate. Various spirits thus began to appear, followed by actual gods, especially in the sky. It was a natural process, facilitated by stable communities, advanced language, human imagination, and sharing of superstitious ideas.

Death Rituals

Of course, once established in a group, superstitions were passed down from generation to generation. And they became more elaborate, with ritualized communications and symbols. One of the most important superstitions in human cultural development became death rituals. "Religious rituals" surrounding the dead long predated *Homo sapiens*. Early hominins likely engaged in a number of ritualistic behaviors, but rituals surrounding death of a group member may have been a pivotal one in the appearance of religious behavior and supernatural fantasies. Death is something we all share; as such, death rituals were likely a potent force in the cohesion of human groups, even well before we settled down into agricultural communities.

Probably other *Homo* species had death rituals. Neanderthals likely practiced ritualized burials of their dead, and recent discoveries have provided reasonable evidence of that. (410) The recent discovery of the hominin species *Homo naledi* is also instructive. They appear to have conducted ritualized treatment and burial of their dead. (136, 556) We might consider it to be some sort of religious rite, but at least there may have been a set of primitive behaviors acknowledging death and perhaps commemorating it in some way. This suggests a likely manifestation of some pre-religious superstitious behavior in an early *Homo* species, the value or function of which still remains a mystery. (545) The exact function of the placement of remains in the Naledi's Rising Star cave remains controversial. Nevertheless, it does seem from the evidence that they were specifically placed there and most likely represent some kind of ritual burial.[27] This does suggest some kind of formal acknowledgement and recognition of death, and that it should be treated in some formal way. It may not constitute a "religious rite," per se, by today's definition, but it certainly is a possible cultural precursor of one.

There is no way to know, with current knowledge, whether death rituals in primitive *Homo* species may have been transmitted to other species of *Homo*, or whether burial rituals as superstitious behaviors arose multiple times through evolutionary history. But we do know that such rituals have always been a part of *Homo sapiens'* existence. Various formal *Homo sapiens* burial sites have been discovered dating back at least 150,000 years ago. Apparently, ritualized treatment of the dead became a fundamental superstition in our existence.

But death rituals were not unique social behaviors, and they're not mysterious. They do occur elsewhere in the animal kingdom. Other species exhibit what may be loosely called death rituals, including elephants, giraffes, some corvids (crows and magpies), orcas, donkeys, peccaries or javelinas, and other primates (and perhaps other species we don't know about). Many species actually manifest grieving-like behaviors. (272) Consider the orca, a large dolphin sometimes referred to as a "killer whale." It exhibits a wide variety of sophisticated social behaviors and appears to have advanced cognition and communication, and evidence of culture, per se. In addition to greeting ceremonies, the orcas manifest an elaborate mourning ceremony when one of their group dies. In primates, one of the more interesting death rituals is the one "held" by Langur monkeys. They have been observed to gather around a dead colony member in a funeral-like ritual—silently paying respects and hugging and consoling each other.

One could ask whether death rituals in the animal kingdom are better considered as innate or learned, thus passed down as cultural phenomena. We don't have any reliable data about how such mourning ceremonies may have evolved. It is somewhat difficult to conceive of the possible survival value of such mourning ceremonies for any given individual (which would raise the possibility of genetics). But the notion that this facilitates cohesion of the group and therefore its survival is clearly possible. Although they may have appeared for different reasons in different species, formal mourning ceremonies do tend to occur in species with family groups, low fecundity, long gestation periods, and individual longevity. That is, they have long pregnancies, produce relatively few offspring compared to other mammals, and have long lifespans. One can argue that these factors support the idea of placing increased value on each member of the group—what we might call respect if they were human. The contribution of each group member is significant, and the loss of a single group member is thus a tragedy. It is not surprising then, that the loss becomes associated with a set of behaviors which formally acknowledge it in some animals. Can we reasonably also conclude that they serve some sort of psychological function—shared grieving—and therefore most likely group cohesion? Do they involve some sort of appeal to a deity? If only we could ask them.

Modern humans, of course, do the same thing. The nature of mourning, grieving, and ceremonies are different from culture to culture, from one professed faith to another, etc. But we all mourn and carry out elaborate death rituals when someone dies, including formal funerals. Great loss, which is generally accompanied by the subjective state we call grief, is generally very disconcerting to us and somewhat immobilizing. We find it difficult to function after a great loss. Perhaps our mourning, which includes the intense support of others, helps to lessen the grief and helps to restore more normal functioning. We humans are not special in this regard, and it takes no god to reasonably explain how death rituals arose naturally and what social function they serve. However, when the ritual is artificially connected to some supernatural power, it is a superstition, perhaps the most powerful one in the appearance of religions (as I'll discuss more in Chapter 7). It would indeed be interesting to know if this happened with our extinct *Homo* cousins as well.

Culture

Culture is the set of shared behaviors, communications, and interactions of a group, which are socially learned and passed down in some form generation to generation (We tend to think of culture as including various formal customs, moral codes, and artful and other tangible products of a group, but those particular memes are of "recent" vintage.). The development of human culture is likely a principal factor in how we survived and also how we came to believe in superstitious, supernatural entities.

Other species demonstrate culture as well. Social learning which pervades a group, which is the hallmark of culture, has been observed in many species—from whales to even some birds, rodents, fish, and others. In fact, it's likely that some form of culture develops in most social animals. (425) Orcas and crows are well known for this, and it's particularly extant in primates. For example, you also won't be surprised to learn of chimpanzee culture. Many things that chimps do are learned, not inherited, and vary from tribe to tribe. In one study of chimp tribes, it has been observed there are at least "thirty-eight traditions that vary among those chimp communities," including certain specific types of stone tools for cracking nuts and use of specific medicinal plants. (350)

Culture, in fact, may have driven much of our recent evolution, with advancements in food finding, toolmaking, cooking, transmitting of information, and other shared behaviors. (291) It is unclear exactly how this happened, but there is no doubt that genetically driven increases in cognition accompanied these cultural advances. Of course, other species of humans evidenced cultural development as well, including *H. naledi*, Denisovans, and others. Neanderthals are of special note. Their cultural tendencies and creations were not likely significantly different from ours when we both existed. (222) Recent data even suggest that Neanderthals created products of art even well before *Homo sapiens* migrated into Europe. (13) Our (*Homo sapiens*) success in outlasting the other species of our genus was likely due not to any superior physical capacities or brain size. Rather it was most likely our success in socialization skills, leading to enduring cultures. (204) We most probably achieved more reliable and stable social groups and social cohesion via advancing language communications, advancing cognition and problem-solving abilities, self-domestication by plants (advantages in food security), and cooperative behavior. Such patterns enabled *H. sapiens* groups to remain together for longer periods, thereby developing larger groups and cultural competencies. This enhanced our abilities to endure changes in climate and food availability, to enhance tribal defense, and to more effectively dominate other species. Of course, the byproduct of our cooperative behavior was the advent of superstitious memes that facilitated group cohesion, including the animistic ones.

Since we possibly formed larger and more stable groups than others, we were able to directly enhance the survival of members. As we aged, others in the group provided nutrition and empathy and assistance vis-à-vis injury and disease. Some females began living past childbearing years, which enhanced tribes' abilities to pass down cultural knowledge. This is generally known as the "grandmother effect," something that's still characteristic of many societies around the world today. In fact, such extended families can and do enhance survival of groups in a number of species. For example, other than humans, it has been found that females in only a few species—likely chimpanzees and certain toothed whales—undergo menopause and continue living for some time after that, and grandmothers, particularly orcas, pass down various pieces of wisdom to others in their pods. (63)[28] Such may explain part of their strong cultural traditions and hunting effectiveness. Though there is no known empirical evidence, one cannot rule out that extension of female lifespans was also a factor in humans' ability to survive and flourish as well, including any generational transmission of superstitions.

One of the interesting communal behaviors that enabled *Homo sapiens* social cohesion was communication about things that certainly went well beyond those interactions which were necessary for immediate survival needs. We didn't talk just about food and reproduction. A good deal of communication would have involved reflections on the day, likes and dislikes, reference to others, etc. It's what we've come to know as "gossip." It also would have included consideration of "imagined realities" or myths, since we clearly had the abilities to do this. This would involve storytelling and the evolution of lore, and certainly the consideration of supernatural forces taking on personas. Such supernatural myths may have been the shared lore that was a critical factor in the development of stable "culture" and religious behavior (we'll get into this more in Chapter 7). No god was necessary for this. *It developed naturally.* But it certainly contributed to our unique survival in our genus and took religious superstitions along with it for the ride.

Recent Humans

Most of human culture did not develop in one place. It arose in many places as humans migrated throughout Africa, Asia, and Europe. The diasporas of *H. Sapiens* probably began in earnest about 70,000 years ago, fueled by advances in projectile weapons, changes in weather, migration of prey animals, and growing abilities to cooperate. And there was not just one migration. It probably occurred several times, along with interbreeding with other *Homo* groups along the way. (25)

It was long felt that migration to the Western Hemisphere occurred probably at least 14,000 years ago. However, recent finds suggest humans may have actually crossed the land bridge to North America well over 20,000 years ago, during the last ice age, and had migrated as far south as New Mexico. (54, 529) Further data suggest that at least 16,000 years ago there were additional migrations, probably occurring by both boat and land from Asia. (110); and more recent evidence suggests that humans had migrated as far south as South America over 25,000 years ago. (390) The human diaspora was nearly global by around 10,000 BCE. During those migrations, the parallel cultural development in different parts of the Western hemisphere, and the rest of the world, led to different customs, superstitions, and supernatural forces. The humans who migrated to North America, for example, developed different gods than the ones in the Old World. And so human culture began to develop—again differently in different parts of the world. And this is most likely how various god memes and religions came about.

Still Evolving

No god "deposited" humans as such on the Earth.[29] Humans were not "created." We're the result of natural changes on this planet, and we're still changing. We're only "special" or relatively unique at this brief moment in time. We evolved into our present state of anatomy and culture over millions of years of evolution. But we're not done. Evolution is a continuous process, and it persists. We do continue to evolve.

Darwinian evolution produced our human genetic makeup—through natural selection, sexual selection, and interbreeding with other human species. However, lest you think that evolution of human beings is over, it's not. Even though natural selection has lost some of its clout in the last few millennia, humans continue to sexually select, though this has had no impact on our species' overall survival—thus far. Further, we continue to manifest genetic changes which have enabled adaptations to environmental demands. Let's consider a few examples. Our bones have become less dense over time, perhaps associated with a more sedentary lifestyle. Sitting around doesn't cause your offspring to have poorer bone density, but it is quite possible that selection pressures against denser bones became evident as human locomotor anatomy changed and as food sources for modern humans became more plentiful. (88)

There are more obvious examples. It was only a few thousand years ago that evolution produced the ability of some adult humans to be able to digest the sugar lactose. This was possibly an advantage during periods of reduced availability of foods that were common in certain geographies (though this is still somewhat speculative). It is unlikely that this will prove a survival advantage for our species, since about two-thirds of the world's adult population still manifests lactose intolerance (mostly in the Far East and southern Africa), but it does show how our genome is changing. Skin color is also a well-known example. Ultraviolet light is necessary to enable human bodies to produce enough vitamin D, an essential vitamin. This wasn't a problem for early humans, where there was enough sunlight. However, when humans migrated northward, we couldn't get enough ultraviolet light to produce adequate levels of vitamin D, especially in winter. In that situation, darker-skinned individuals—those with more sun-blocking melanin—were less likely to survive. The result was that evolution tended to select people with lighter skin for survival.[30]

A very interesting modern example is the nomadic Bajau people of Southeast Asia, who've been living primarily on houseboats on the seas between Malaysia and the Philippines for at least the last thousand years. They subsist by spearing fish and gathering undersea foods, diving to a couple of hundred feet, holding their breath for very long periods of time (perhaps more than 5 minutes). How did this happen? Only recently it was discovered that they had evolved larger spleens, which increases a reservoir of oxygenated red blood cells, as well as selection for a better cardiovascular "diving response" (that is a physiologic response characterized by increased peripheral vasoconstriction, which differentially favors providing higher oxygen content to the brain, heart, and lungs), allowing them to hold their breath and work under water for long periods of time. All those abilities have been traced specifically to genetic changes. (234)

There are many other examples, but it is clear that evolution is a continuous process, including in humans. If we are so special and "created," one must ask why we keep changing. The only answer is evolution by natural selection, which requires no god. Obviously, not enough time has elapsed for characteristics like eye color, lactose tolerance, or large spleens to have an effect on our species in general. But other environmental demands may come about in the future—effects which may change us notably—climate change with severe droughts, new and migrating diseases, nuclear war, or others. The bottom line is that evolution, not God, has produced us, and we continue to evolve in response to our environments.

Conclusion

Evolution is a fact, whether some people believe it or not. It remains one of the most well-evidenced theories in all of science. As a theory, evolution is incredibly well supported by observation, experiment, and consistency over time. As theory, it also continues to change over time. That is not to say that it is "wrong" or incorrect in some way. Like all theories, it continues to be refined and updated in the face of new information. It expands to explain more and more observations of the living world. Darwin did not envision the genetic mechanism for evolution, but we now understand much about how this works in the process. He did not envision that evolution could be extremely rapid, for example, as occurs in bacteria when exposed to antibiotics—something which is incredibly real and often to our peril.

And none of this is a God's handiwork. We must accept the fact that Darwinian evolution explains our presence on this Earth. We're also not that special biologically, as the Copernican Principle again dictates. We only appear to have a relatively privileged position because of where we exist on the timeline of species. We are but one species, out of some perhaps 8 to 9 million that presently inhabit the planet (if you don't even count the continually evolving bacteria and certain other primitive life forms). We often fail to appreciate the continuum of life on this planet and our place in it. We are a minuscule blip on a continuum of species, again noting that at least 99.9% of them that have ever existed have gone extinct. We've only been around for a couple of hundred thousand years, a tiny fraction of the Earth's existence (that's only about 0.006%). Rather than being special, we are unique only in the current narrow slice of time. We also represent only one branch or twig on a tree or bush at its outermost reach. But so do all other extant species—from lemurs to eels to planaria to amoebae to the giant redwoods. These are species which have evolved and survived to the present time, and most of them, of course, predate us by eons. Although we currently seem to be at the top of the food chain, are we the dominant species on Earth? Perhaps. But at some point, beetles and bacteria might disagree. Only time will tell. Again, we're humbled by the Copernican Principle.

There also may not be anything "special" about the nature of life itself. It is part of the physical universe we know. "Life" as we know it is a peculiar feature of the earth's crust which arose according to the laws of physics and chemistry. In the scope and scheme of our universe, that's pretty insignificant. And if life—or life similar to ours—is widespread in the universe, that also obviates any specialness for us. All this eschews the need for any "god of the gaps." Considering the infinity of spacetime—even specifically the spatial infinity of our own universe—and that "life" is a matter of a sequence of chemical reactions, we must conclude that the appearance of human beings, and life in general, is not surprising. In fact, it is essentially guaranteed to occur over and over again. This is a purely a statistical matter, for which no god is necessary.

How different are we from our hunter-gatherer or agrarian ancestors? We have the same brains, though we use them in somewhat different ways. And how different are we from our cousins we once shared the

Earth with—*Homo erectus*, *Homo habilis*, and even *Homo neanderthalensis*? Not very much, and the main variable appears to be time, acknowledging that various selection pressures did lead to our survival and the demise of the other *Homo* species. We interbred with some of these, and we still share some of their genes (as we do even with chimpanzees). But of the several human species, we are the only ones still surviving. If we're somehow special as humans, it's only because we're still here. In the end, we must have a respect for biological evolution, how it has operated, and why it has produced us. It is contrary to all religious fantasies of how and why we are here. And we ignore it, and other scientific facts, at our own peril, as I'll discuss in later chapters.

5
CONSCIOUSNESS AND GOD

There is, I mean, no aboriginal stuff or quality of being, contrasted with that of which material objects are made, out of which our thoughts of them are made. —William James[1]

Consciousness is a plain assumption just as unprovable, just as unapproachable, as the old concept of the soul. —John B. Watson[2]

It seems strange for a book on gods or religion to examine issues of human consciousness. That is, why would we be concerned with how our brains work and what we experience as a conscious awareness of ourselves and the world around us? But consider some important questions. Are you a conscious being? Of course. But what *is* that? Is consciousness a factor in faith and belief in God and explaining religious behavior? Does God exist in the human brain? Are we "hardwired" to experience or communicate with God? *To address these questions, we must actually examine the notion of human consciousness itself.* This is going to be a somewhat involved discussion; but it is unavoidable. Contrary to other opinions on this subject, one cannot consider the God question without it.

Let's begin with one of the classic arguments for God—that you are, or can be, *aware* of Him. Most believers know in their hearts that God exists. Here are some typical conceptions:

I am aware of God.
I know God exists.
I can feel His presence.
He talks to me.
I have faith in Him.
He guides me.
God loves me.

All these kinds of statements suggest an awareness or consciousness of a supreme being—presumably. We'll leave for the moment the obvious fact that people simply learn to say such things, which I'll address in Chapter 10. It is generally conceived that "belief" or "faith" in some sort of god is a conscious experience, freely arising in the minds of human beings. If that is the case, we must know what "consciousness" is; where it comes from, i.e., how it arises; and what, if anything, it has to do with human behavior, including faith and so-called religious experiences.

The reports of awareness or experience of God we generally take at face value. People label their experiences in all sorts of ways. In this context, we're talking about the supernatural and omnipotent God that many believe in. We should note that this is different than the experience of wonderment and beauty in the world, or universe, which is sometimes used to suggest that there *must be* a god. It is also different than the often-heard notion that the beauty and wonder of the universe *is* God (the idea most famously associated with Spinoza or Einstein, which I'll talk about more later), or the notion that God is *in* everything. This is not what we're addressing here. What I'm referring to here is the reported *experience* that is expressed as knowledge about, awareness of, or communication with, an omnipotent being with a persona. In order to fully consider this, we must deal directly with a few aspects of the human brain and what we actually mean by *consciousness*.

First, we'll see that, in this vein, no conscious experiences can evince a god's existence or help to understand religious behavior related to that. Thus, it is possible that the subjective reports concerning some supernatural entity can be patently discarded. You may reject this notion out of hand, but in this chapter, let's actually examine what we know and don't know about this. This matter will also be important for our discussion of "free will" and other issues concerning religion, which I'll address in the following chapters. Here we'll consider what consciousness is—and is not, whether there is anything special about human consciousness, and whether it's an idea which helps us understand human behavior, in particular religious behavior.

Not surprisingly, this subject also derives from the Copernican Principle—the belief that we humans are somehow special or unique because of our consciousness. Some humans consider their existence "unique" with respect to any mental quality or capability when compared to other extant creatures, which we'll get into. Are animals conscious? And what about the animals who were our close ancestors and extinct cousins? Remember, our brains are essentially no different from *Homo sapiens* that lived 300 millennia ago and probably not that different from other *Homo* species that roamed the Earth with us for thousands of years.

I will also have to discuss some interesting facts about the human brain. We'll have to look at some reports of "religious experiences" that arise in the conscious brain. We must also entertain the idea of "mind," i.e., that there exists a conscious mind operating "outside" of our material brains. One caution: Although I'll argue against the concept of a non-physical mind, I'm afraid that the terms 'mind' and 'brain' are sometimes used interchangeably, and I'll sometimes do that. Language is again a limiting factor here. Please forgive, though the meanings should be fairly clear from the contexts.

My purpose here is to address the central problem: If there is some conscious aspect to religion, can consciousness itself be demonstrated?

Does consciousness actually exist, or is it an illusion, or is it an ephemeral byproduct of neural activity with no causal influence on our actions? At best, we may shed light on the fact that it is not a useful concept in explaining human behavior, including "religious experiences," faith and belief, and overt religious activity.

The Problem of Consciousness

First, what I'm presenting here is a truly radical idea—that the notion of 'consciousness' is not a helpful idea—in understanding how our brains "work" or in explaining human behavior. I'll refer to various so-called states of consciousness throughout, including faith, beliefs, and other subjective ideas. Of course, there is no way to assess such internal thinking except for the reports and behavior of persons who claim to experience such. I readily admit that I "experience" things, including the world and including my own body (again, we'll assume here, contrary to the extreme solipsist view I discussed, that something besides the human mind exists in our universe), but whether such "experience" is germane to human behavior and religious belief, as we'll see, is an open question.

The whole idea of examining the issue of consciousness might seem somewhat ludicrous on its face. If you ask yourself whether you are conscious, the answer seems obvious. I do not deny that I am conscious, and to deny it would seem absurd. But I can't prove that I am or that anyone else is. Just take the challenge for a moment and try to prove that you are conscious. You can't use circular reasoning (like substituting "awareness" or other subjective words). You can go down the list of synonyms—awareness, feeling, experiencing, knowing, etc.; but you never get to an empirical point where putative consciousness can be affirmed. We talk about consciousness all the time; but strange as it seems, we can't prove that it even exists. We do know some things about how the brain operates when it is "conscious" as opposed to unconscious, for example, when it is asleep or in a coma, or when it is radically altered, e.g., by drugs or a significant disease or injury. Furthermore, we know some things about how some areas of the brain are more important to some behavior than others. Neuroscience has taught us a lot about such things. But this is not the same thing as the notion of self-awareness or consciousness.

What we call consciousness *appears* to be a product, or byproduct, of neural activity in your brain—and not the reverse. It may also be merely a synonym for what's going on in your brain. Despite our personal belief, is "consciousness" actually real? In the end, what we humans may have done to explain our own behavior is "invented" the notion of consciousness. Everybody seems to know what it is, but no one can *tell* you what it is.

The bottom line of the consciousness problem is that it cannot be operationally defined. That is, we cannot "measure" consciousness. We can measure different kinds of "brain waves," which reflect different states of attention, and we can use various techniques to "see" how the brain is operating, but those do not measure consciousness (I'll get into the neuroscience of this below). The problem of definition remains. And if consciousness does not explain human behavior or the behavior of other organisms, what is "it?" It seems to be "merely" a label of a subjective state in search of "itself." Philosopher and physician William James, generally credited as the founder of the modern science of psychology, captured the essence of the problem in 1904.

> I believe that 'consciousness,' when once it has evaporated to this estate of pure diaphaneity [A term not often seen anymore. It is essentially the quality of being transparent, i.e., usage here signifying something that is becoming frail or weak], is on the point of disappearing altogether. It is the name of a nonentity, and has no right to a place among first principles. Those who still cling to it are clinging to a mere echo, the faint rumor left behind by the disappearing 'soul' upon the air of philosophy. (242)

He was obviously prescient (as many others have been), as we've come no further since 1904 in being able to identify or measure what we actually mean by consciousness.

Mind and Brain

For a long time, we have been searching for hypothetical subjective states that we propose to "exist" somewhere in our skulls. The ancients were fond of spirits inhabiting one's "mind." Sigmund Freud proposed the famous triad of the mind: the id, ego, and superego as the various unconscious and conscious entities motivating human behavior. Freud's idea illustrates the problem which we'll see in many other ideas about consciousness and psychology. For various corresponding behaviors, for example, many have proposed the primitive "hindbrain" (our reptilian relic), the brain's limbic system, and a forebrain (highly developed in humans) to account for such "things." But the notion that neuroscience will confirm Freud's entities is totally unfathomable, no matter how sophisticated neural analysis and computing power become. You can't find an "ego" or "superego" in the brain. They might exist in one's "mind," whatever that is, but I wouldn't bet on anyone finding them. (To be clear, you'll see Freud's comments appear in a few of the following chapters. I note here that I'm in no way endorsing any validity of Freud's well-defunct theories on human behavior, pathology, or the treatment of any human aberrations. I merely offer such here and elsewhere as interesting historical illustrations.)

In order to drill down to the issues surrounding consciousness, I need to discuss some interesting facts about the human brain. But again, I will also entertain the related idea of "mind." What is a mind? What is the brain? Are they the same or different? We sometimes use the terms brain and mind interchangeably. Sometimes we seem to use them in overlapping fashion. In fact, the idea of mind has been variously conceived as an organic process *of* the brain. For example, one of the better perspectives comes from psychologist Steven Pinker, who offers the following definition:

> The mind is a system of organs of computation, designed by natural selection to solve the kinds of problems our ancestors faced in their foraging way of life, in particular, understanding and outmaneuvering objects, animals, plants, and other people. . . . the mind is not the brain but what the brain does . . . which makes us see, think, feel, choose, and act. (404)

It is an incisive picture of the issue, focusing on function. Yet, the behavior of the organic brain is generally not the "mind" as generally construed. It is seen as something else, some specific entity, which is also rather intangible or ethereal (including by many psychologists). When we speak of "mind," the connotation is always one of a *conscious mind.* Some philosophers, psychologists, and others do use the term "mind" as a substitute for "brain," even though they may hold that a nonmaterial mind does not exist. It appears to soften the blow to the listener or reader who is offended by the notion that his physical brain is "all there is." But there is clearly nothing wrong with referring to one's brain as mind. It just makes conversations grammatically easier. We never seem to say, "Would your brain allow me to sit here?" or "Have you made up your brain?" And again, you'll see "mind" used casually throughout this book. That's just the way we communicate.

Mind might be thought of as an individual's thoughts and perceptions. These are considered "subjective" entities, meaning they're accessible only the holder. I use the word "subjective" here as a synonym for experience, but only because we don't understand what that really means (I'll get into that in more detail later). Some neuroscientists consider, or define, "mind" to be merely an "emergent" property of brain activity, just as some consider consciousness to be an emergent property. Yet, because it implies various subjective states, as most people use the term, mind is considered to be something that requires consciousness.

Dualism

"Mind" reprises the historical notion of dualism. As Descartes famously put it, *Cogito ergo sum—I think, therefore I am.* But Descartes also went one step further. He split humans into two parts. He proposed a "radical dualism" where there exists two substances in the world, the *res cogitans,* the "thinking substance," and a *res extensa* or "extended substance," representing the physical world composed of matter (There was also a third substance, which was God.). In Descartes' dualism, there is the body and there is a nonmaterial entity containing a soul, operating outside the laws of nature. He also proposed, incongruously, that the seat of the soul was located in the pineal gland, an actual physical structure in the brain. Interestingly, he also denied other animals a soul, even though the pineal gland is present in nearly all extant vertebrate brains. But he did not explain—and no one has since ever explained—how a nonmaterial soul, force or entity can interact with the physical world, including the material brain.

This has now come to be known as the "mind-body problem." It has not been solved by the invention of a "mind" as something real, which is merely a relabeling of Descartes' "thinking substance." The concept of "mind" is variously used to signify an entity or process which exists outside the realm of anatomy and physiology. British philosopher Gilbert Ryle dubbed it the "ghost in the machine." He argued against Descartes' dualism and the idea of a nonmaterial mind, believing that what we call mind and consciousness are merely activities of the brain, foreshadowing Pinker's more descriptive language. But the mind-body problem remains.

Where Did the "Mind" Concept Come From?

The meaning of the word *mind* has only recently come to signify some nonmaterial entity. But how did the idea of "mind" come about? There are various etymologies for the word, but historically it has most directly implied action, e.g., noticing the world, focusing attention, or controlling one's behavior. (548) The practical etymology of "mind" is inescapable—it specifically has to do with human behavior. In fact, we often continue to use the term to indicate some sort of behavior. "Do you mind" suggests attempting to rule out a negative response. "Mind your manners" is an instruction to behave in a certain way. "I don't mind how cold it is today" suggests a tolerant response. To ignore is to "never mind."

Only very slowly over history has the idea of "mind" become something nonmaterial. B. F. Skinner, the founder of the modern behaviorist perspective in human psychology, addressed this issue headon. (You might be taken aback by the name Skinner, as many psychologists and others are, but bear with me. As you might be anticipating, I'll provide a full discussion of the behavioral approach to human psychology and address all the various criticisms of behaviorism, in Chapter 10.) Skinner put the idea of "mind" into an analogous empirical context:

> Mind . . . is the double of the person whose mind it is, a replica, a surrogate, a *Doppelganger.* [à la Gilbert Ryle's ghost] Take any sentence in which the mind does something and see if the meaning is substantially changed if 'person' is substituted. It is said, for example, that 'the mind cannot

> comprehend infinity.' Does that mean anything more than that no person can comprehend infinity? Cognitive processes are behavioral processes; they are things people do. (487)

"Mind" nearly always has a behavioral referent. "To remind" is to be stimulated again, to be prompted by something external. "To mind something" is to behave consistent with expected norms. "To be mindful" is to behave in ways that are careful or considerate of some external standard. To be "out of your mind" is to act very differently from others or your previous self. Such meanings are indeed a clue to how the whole notion of mind came about.

Theory of Mind

Much of the current thinking about "mind," and how it may have influenced human evolution and development, traces back to the notion that we can understand what's going on in the minds of others. This was formalized by psychologist David Premack in his "Theory of Mind." (415) He used it to explain some behavior of chimpanzees, which could also be generalized to other animals, including humans. Simply stated, some of our behavior is based on our presumed ability to guess about the actions or intentions of others. Can we really appreciate the cognitions and emotions of others? It's a very imperfect effort; we don't always know when someone else is angry, for example. Even though this seems commonsensical on its face, we might ask if we are the only organisms that have this supposed capacity. Without belaboring the point, the answer would be "no." Premack made a convincing argument that chimpanzees have some concept or "theory" of mind. Does a dog have a theory of mind? Does a cat or a fish? Where do you draw the line? We certainly know that other primates can sense what we call emotions in their cohorts, and they tend to respond accordingly. For example, let's take some common, intense "states of mind": the subjective notions of pain, fear, or distress, about which cohorts may have a "theory" in dealing with them. In many species (humans, other apes, and other primates), the "theory" recognizes those "states," leading to responses we sometimes group as "empathy." Such responses obviously have evolutionary advantages, as they are often associated with some sort of personal assistance. But do the animals "know" what their cohorts are feeling? Do they "know" when another animal is fearful or hurting? If you have empathy toward others, it might be said that you have a "theory" about *their* minds. When a dolphin helps a human avoid sharks, for example, it is difficult to attribute any such behavior to anything but empathy (It's a bit farfetched to think dolphins know about any of our efforts to protect *them*, but I wouldn't rule it out). Pigs show empathy toward their peers. Meerkats in Africa will risk their own safety to assist a cohort who is ill or dying. How about all the way down to rodents? In one fascinating experiment, for example, it was demonstrated that even mice can overtly demonstrate "empathy" for their cohorts who were exposed to painful stimuli, and they exhibited this empathy only for other mice that they actually "knew." (293) We tend not to ascribe mental events to other animals, but if we take a rigorous approach, we must either endorse a "theory of mind" for them or eschew it as explanation for our own behavior. We should rightly consider a "theory of mind" as the ability to predict someone else's behavior based on our experience with them and our own experience. When we say, "I know what you're thinking," for example, we're really suggesting an increased probability that you'll speak or act in a certain way. It does not signify a nonmaterial mind. It does suggest, and only suggests, that *behavior is predictable*. That is what Premack meant. We do not need to invoke "mind" to understand human behavior.

What is Consciousness?

We do tend to invoke "mind" with the implication of a "conscious mind." We don't usually think of mind as an unconscious state (although some do, at least metaphorically). So, we should rightly ask what it *is*. You could say that it's awareness. And what is awareness? Perhaps it's the quality of experiencing things—the environment and your thoughts and feelings. And what is experience? You see the problem. It is literally impossible to objectively say what we mean by consciousness. Nevertheless, many have tried, and we'll consider some of the efforts here. In the end, we'll see that using the notion of "consciousness" is just a very poor way of understanding human behavior.

Descartes' version proves nothing about this idea of mind or consciousness. We cannot prove that Descartes was conscious, that you are conscious reading this, or that I am conscious writing it. All we know is that we are behaving in ways that we label with the word "conscious," which is a convenient way to communicate about some pretty complex ideas. That's a radical idea, but it's one necessary for us to talk about the problem of human behavior and ideas about "free will" and other subjects, as they relate to religious faith, in the coming chapters.

Definition

According to the OED, consciousness is "the state of being awake and aware of one's surroundings." Of course, we're still left with the problem of "awareness," and the only meaning we can apply here is being responsive, at some level, to external stimuli. That's consistent with the notion that unconscious people are not responsive (usually). Some also consider that the definition must include awareness of ourselves—our bodies and thoughts or experiences—and/or the ability to imagine oneself in the future.

Definitions of consciousness are many and varied, but they have one thing in common. They rely entirely on subjective synonyms and have no objective referents. Again, we can invoke ideas like awareness, self, thinking, feeling, reflection, realization, and many others, but these are also subjective and impossible to define without further subjective synonyms.

What is needed is an *operational definition* of consciousness. That is, to define something, we need to specify the operations we use to measure it. That is, it must be observable or measurable by our senses or some technology that enhances them. But since consciousness, in all our common understanding, is purely subjective, this seems to present an almost insurmountable problem. You may think this is a nonsensical analysis, but accepting the idea of consciousness without an operational definition conflicts with any scientific perspective that we might use to study "it"—though many have tried, as we'll see.

Who Has It?

What does it really mean to be aware and, specifically, to be aware of oneself? Would this apply to dogs or birds? Would it apply to worms? How about artificially intelligent robots? I'll discuss some of these issues below. But there are perspectives advanced by some theorists and philosophers that extend the concept a lot more broadly. Some suggest that there are "levels of consciousness," extending from humans to plants. Some even suggest that consciousness is something that pervades the entire universe, a philosophy often termed "panpsychism." Consider, for example, respected philosopher David Chalmers. He offers that,

> Every system might have some degree of consciousness . . . . every system is conscious, not just humans, dogs, mice, flies, but even [biologist] Rob Knight's microbes.[3] Even a photon has some degree of consciousness. The idea is not that photons are intelligent or thinking. It's not that a photon is wracked with angst because it's thinking, 'Awe, I'm always buzzing around near the speed of light. I never get to slow down and smell the roses.' No, not like that. But the thought is maybe photons might have some element of raw, subjective feeling, some primitive precursor to consciousness. (86)[4]

What? Even the great theoretical physicist David Bohm actually once proclaimed that "Even the electron is informed with a certain level of mind." I don't know of any follow-up or elaboration of this by him, and we don't know what he was thinking. He had been collaborating with an Indian mystic, so perhaps we should take this somewhat metaphorically. Either Bohm's and Chalmer's conceptions are a full abandonment of the meaning of the word "consciousness," as most of us might even consider it, or it's just an attempt to redefine it in some abstract philosophical sense.

Chalmers does admit that it's a "crazy" idea; but he goes on to observe that "It's also worth noting that although the idea seems counter intuitive to us, it's much less counter intuitive to people from different cultures, where the human mind is seen as much more continuous with nature." (86) Of course, this is a cultural observation and explains nothing about the idea itself. The notion of such panpsychism really deserves no further comment here. What I'll go on to consider is only the question of whether there can be any tangible quality to the notion of "consciousness" and whether its existence in humans is pivotal to any evidence or even the experience of God.

Theories of Consciousness

At bottom, being conscious means "to experience something" or "to have an experience." But what does that mean? (We'll take an empirical look at the notion of "experience" in Chapter 10.) A brain cannot experience itself; there are no sensory organs in the brain.[5] And I certainly don't know what that means for you reading this, or anyone else for that matter. My comments on consciousness—and on free will in the next chapter—might be considered radical views. But they constitute a discussion of these issues from an empirical point of view, differentiating them from speculation and philosophy. Nevertheless, we'll examine here some of the actual, possible understandings and proposals about consciousness.

The Easy and Hard Problems

We can think of consciousness as the ability and action of perceiving events in the world and attributing meaning to them, including one's own body: a "sense of self," if you will. Of course, this is not an operational definition, and many scientists and even philosophers would disagree with that kind of definition, believing that consciousness is much more than that. However, it is clear that the human brain does a lot of different things. It perceives the environment—via all five of our senses, and it engages in what's called thinking, feeling, etc. So let's consider some of the additional thinking about what's going on here.

There is a difference between sensing something of the world and processing it. Your eyes can visualize the light reflected by a rose, and your brain enables you to "see" it. The first is a function of your eyes being able to process and translate some of the photons you receive and send information about this toward the brain. The second is how the brain uses that information to create the image of a rose. To drill down on the problem, some neuroscientists (and some AI researchers; see below) divide the question of consciousness into a "hard problem" and an "easy problem." The "easy problem" is how the brain processes information, like the light from the rose, what we generally call sensation. Neuroscience continues to make progress on this. The "hard problem" (as coined by Chalmers), by contrast, is how our subjective experiences—like "a rose"—arise in our heads. It is the brain that creates that picture, what philosophers refer to as "qualia." It is a concept regularly in use today among neuroscientists and philosophers, though impossible to define. Unfortunately, the "boundary" between sensation and more complex "experience" is not something presently amenable to inquiry. We don't know how to find qualia. It certainly would be different for every brain. Even color seems to blur a boundary. For example, what's the boundary between a brain processing a certain visual stimulus from a leaf and the advent of experiencing the color "forest green"? What's the boundary between a noxious sensation and the experience we call pain? (a problem we've been working on seemingly forever) This is a difficult problem for science at this point in history, but we should emphasize the words "at this point in history," as we learn more about the brain. Even further down the tree of processing is how the brain creates what we report to be thoughts, feelings, emotions, etc. It is likely that the "hard problem" will elude us for some time, and the correlates of more complex experience may be even more elusive. I'll address below some of the crude neuroscience that may relate to these issues as we drill down on the idea of consciousness.

Emergence

Many philosophers and neuroscientists consider consciousness to be an "emergent" phenomenon. That is, it "arises" as a consequence of brain activity. Among others, physicist Kaku puts the emergent idea into this context:

> Consciousness is the process of creating a model of the world using multiple feedback loops in various parameters (e.g., temperature, space, time, and in relation to others), in order to accomplish a goal (e.g., find mates, food, shelter). . . . The problem with this picture is that it says absolutely nothing about how this miracle occurs. It brushes all the complexity of the brain under the rug of chaos theory. (261)[6]

Of course, this is not really a definition of consciousness. It merely abstractly describes some "processes," but it's not what would commonly account for what most people call consciousness; and we can't rely on miracles for this discussion, as he notes.

Physicist Carroll disagrees somewhat with any reductionist approach to consciousness, categorizing it as "real." He notes that the reality must be acknowledged "even if we think it is 'just' an emergent way of talking about [it]." (78) He goes on to assert that consciousness is "not an illusion" but something "emergent," just like other phenomena. It is "real" within its domain of applicability, the same way we talk about other realities—tables, or atoms, or weather. This is his view of "poetic naturalism," as I noted in Chapter 2. However, when it comes to subjective phenomena like consciousness, we should remember that this perspective is less a way to model reality than just a useful vocabulary. There is a significant difference in the 'granularity' of various theories, e.g., models of atoms, trees, tables, and love are certainly wildly different. There are many things which we acknowledge exist but are nearly impossible to incorporate into any extant scientific theory, e.g., beauty, awe, love. Such things tend to be subjective concepts, presumably residing in people's heads, and are different for every human being. One can propose models of such things, but they do little in the areas of prediction and falsification. You can't test a theory of consciousness by calling it "real" or valid within its own domain. Carroll does finally admit that, "There is only one, unified, physical world, but many useful ways of talking about it, each of which captures an element of reality. Poetic naturalism is at least consistent with its own standards: it tries to provide the most useful way of talking about the world we have." (78) Unfortunately, this falls short for our discussion of consciousness here.

We'll talk more about the relevant specifics of neuroscience below. But let's assume, for the moment, that emergence is a "real" phenomenon, and that we can go beyond its description as poetic naturalism, or an "effective theory," or an "emergent narrative." In that case, we should know how so-called "emergence" could happen at a more basic level, the operation of the brain—and not a "miracle." There are currently two predominant "theories" dealing with how consciousness might "emerge" in the human brain. These are the global neuronal workspace (GNW) theory and the integrated information (IIT) theory. They purport to understand, at least conceptually, the neuronal basis that gives rise to consciousness. I won't go into these in any detail, but they're worth mentioning, and we'll then see some of the related problematic neuroscience that should be considered.

The GNW theory considers consciousness to flow from information being accessed and amplified from many parts of the brain. (330) There is a so-called global broadcasting and amplification of information by neurons in the brain, which is brought into some type of reality for the individual in the prefrontal areas of the cortex. How? Nobody knows. Integrated Information theory puts forth the notion that consciousness is an irreducible, emergent property of actual neuron behavior. (384) In the brain, there is a complex connection among countless millions of neurons which become integrated to form some higher level of processing, creating a "qualia space" in the brain, thus leading to consciousness. How? Nobody knows.[7] What is interesting here is that both theories fail to explain *how* consciousness may emerge. Moreover, they fail to explicate any specific causal relationships between consciousness and behavior. Further, though the details are unimportant here, IIT explicitly, and GWT implicitly, suggest that behavior can be, in effect, automatic. That is, given the various neuronal activities, the brain can generate even complex behavior without conscious awareness. IIT, in particular, offers that "there can be true 'zombies'." That is, systems in the brain can generate both conscious experience and complex behavior which is, in effect, automatic. There are yet other theories, of course, which don't deserve space here, but none of them explain what we call consciousness. (461) However, no theory has come anywhere near explaining the simplest aspects of what constitutes consciousness.

'Zombie' is a term that's increasingly appeared in current popular entertainment. However, it has been used philosophically to describe people as "automata," that is, creatures who go about their business without any conscious intent. It's been around for a long time. It was advanced by the biologist Thomas Huxley in the late 19th century, noting, "The argumentation which applies to brutes [animals] holds equally good of men . . . It seems to me that in men, as in brutes, there is no proof that any state of consciousness is the cause of change in motion of the matter of the organism." (231) Darwin and Einstein also voiced similar ideas. No, we are not zombies, certainly as they are portrayed in films and TV shows, but as uncomfortable as it may seem, the non-conscious idea does hold some water. Below and in later chapters, I'll discuss more about consciousness as merely an epiphenomenon, as opposed to a cause of behavior, but that is different from a conception of "zombie," though the term continues to be thrown around on this subject.

Neuroscience

I'm now going to journey a bit through some more neuroscience. This may seem off the mark, and there's no problem with skipping all of this without losing the overall theme. My purpose here is to examine whether we can even say what consciousness is and therefore whether it is at all useful in the God problem. To begin, the brain doesn't "consciously" perceive or experience God. The notion that there is a conscious experience of a supernatural being proves vacuous. And when someone says, "I *know* He exists," such a response does not come from any "conscious" awareness. A brain invents the idea, based on what's been learned during its lifetime. Further, we'll see that any theory of consciousness is unlikely to be supportable, and finally—and many will be appalled by the notion—we might rightly consider whether what we call consciousness actually exists. Let's look at some of the science that might bear on this.

If advances in neuroscience can elucidate an empirically based theory of consciousness, perhaps it will discover how *it* "emerges," as the above theories have speculated. The neurological basis of how the brain operates continues to be investigated, but so far there's no reasonable explanation for this. Neuroscience does continue to make progress—albeit pretty slowly (see below)—in identifying neural "events" (generally constituting hundreds of thousands, if not millions, of nerve cells) which are consistent with certain behaviors or mental activities, e.g., moving certain body parts, doing a calculation, and other well-defined behavior. However, these represent very gross correlations between meaningful activity and neuron behavior. While very interesting in their own right, such findings do not represent what we have come to consider as "consciousness," which remains a rather indefinable subjective experience. The exploration of the human brain, whether of states of waking consciousness or self-awareness, cannot explicate anything beyond what is consistent with natural laws, in this case the physical actions of our brains. Consciousness then becomes an epiphenomenon, an interesting "experience" but not a cause of our thinking and behavior. That is, it is a "collateral product," something which seems to exist alongside the actual causes of our actions, as I'll discuss further.

The Human Brain

First, let's look at some basic facts about the human brain. It is the most fascinating organ in all of biology and, according to many, perhaps the most complex thing humans can consider and something we may never fully understand. After all, we do have only a few hundred million years left to work on this.

There are many "structures" and defined areas in the human brain, though the details needn't concern us here. What we're interested in is the neural connections and activity that may lead to consciousness. Our brains are not computers. Functionality aside, the anatomy and operation of the brain are quite different than any computer. Some refer to this difference with the idea that computers have hardware, and we have "wetware." But we both have "software," meaning the methods that the system uses to process information, and our software is continually updating itself, via learning (and certain other influences), just like AI systems do (see below). With respect to our hardware, we have more individual neuron cells in our brains than you can imagine, quite literally. The totality of the human brain, as often quoted, contains on the order of about one hundred billion (10^{11}) neural cells.[8] That number is a large percentage of the number of stars as there are in our Milky Way galaxy.

If consciousness is an ability to store and process information, we certainly have that. The brain stores most of the information it collects, though we, too often frustratingly, can't recall some of it. Have you ever tried to remember a name without success and then later, when seemingly not trying, it "comes" to you? Absent certain problems of aging or significant disease, it's not clear why we "lose" information. How much of it actually disappears, and how much do we lose the ability to retrieve it? We certainly have the potential capacity to store an unimaginable amount of information. The human brain does have almost incalculable capacities.

Mapping and Uniqueness

To explore this capacity, and if a mechanism for consciousness to "emerge" could be theoretically described, we would need a model or "map" of the brain to reveal this. There are, in fact, ongoing attempts to "map" the human brain (analogous to the successful mapping of the human genome, though the analogy is more than a bit strained). That is, we'd need to develop a map of the brain with all the neurons and interconnections among them that we have (the interconnections include spaces between neurons called synapses). Whether this will be successful in any meaningful way can be debated. This potential map of the brain is referred to as the connectome. Mapping the human connectome is an ambitious undertaking, to say the least, but some efforts are underway.[9]

But there are some seemingly insurmountable obstacles. To illustrate the scope and depth of the problem, let's first consider another organism, the worm *C. Elegans,* a nematode, or roundworm, with a "brain" and nervous system. After years of study, its own connectome has been mapped, which includes some 302 neurons and about 7,000 synapses. (160) And now the fruit fly connectome has been revealed, containing about 140,000 neurons and at least 50 million synapses. (139) These are two of the very few "complex" animals so far mapped in this way. But we still have no idea how the worm or the fly engage in *any* complex behavior. Studying the connectome hasn't helped so far, but perhaps some advanced AI will shed light on this problem.

The gross anatomy of the human brain is the same in most of us: a brain stem, limbic system, thalamus, neocortex, hypothalamus, pineal gland (Descartes' seat of the soul), cerebellum, and other areas of the brain. However, the human brain is extremely complex, maybe the most complex entity known to exist. Further, no two brains can be said to be precisely the same. This makes "mapping" incredibly difficult. There are at least 10^{15} neuronal interconnections possible in the human brain, and many of these change over time. No individual brain can be the same at any two points in time. It is dynamic in the extreme. The major functioning of the brain, which develops before and after birth, is a function of genetics, exposures, learning, and other influences. And brains change throughout the life cycle. If connections didn't, no organism with nerve cells could learn and survive. The brain also develops (grows) thousands of new neurons every day, with new interconnections and capacities, and many neurons "die" every day (perhaps unfortunately, in our lives many fewer neurons arise than die). Every brain is unique, and its uniqueness changes from moment to moment.

The human brain can hold information on the order of a zettabyte or 10^{21} bytes, a number just too large to imagine. There are likely also inter-neuronal communications which actually do not rely on synapses, i.e., "extra-synaptic interactions" in which neurons "communicate" with each other, likely with remote chemical signaling. There are also the glial cells, which are involved in neuron health and thus aid in how well information actually flows, but there's also a type of glial cell that may actually transmit information, using the neurotransmitter glutamate. (323) There are also probably at least as many glial cells as neurons in our brains, increasing the complexity manyfold.[10] Even if we could measure all the neuronal and other events in the brain, and track them over time, this would take more computing power than we can likely conceive of.

We also realize that all brains do not work in the same way. There are obvious genetically driven differences in motor, cognitive, and various other brain-related functional capacities in different individuals. They develop uniquely, by definition, and they individually "mature" and undergo demise because of a huge number of learning and physical factors. Analogies to the map of the human genome, even with its variations based primarily on individual base pairs of nucleobases, is grossly overstated, as this is dwarfed by orders of magnitude by the continuously evolving combination of factors in the human brain's operation, principally in the neocortex (there are "only" about 3 billion base pairs in the human genome). Each brain is specifically unique. Even though we know that certain functions tend to use some brain areas more than others, at the level of individual neurons there is no evidence that there is good correlation between one brain and another in various specific functions, e.g., visualizing a flower, saying thank you, swinging a golf club, feeling pain, or falling in love. Micro-anatomies certainly differ somewhat, based on genetic determination, experience, and external physical influences (e.g., diet, drugs, etc.), and there may also be associated differences in structure and functionalities.

There is no reason to believe that each individual brain encodes information in precisely the same way. Patterns of "experience," for example, are certainly unique. To assume that the pattern of such activity (which would be extraordinarily complex) would be the same from moment to moment and person to person would be unjustified. There is no reason to believe, for instance, that my connectome behaves the same way as yours when I learn to add 2+2, or that it "sees" the color purple with the specifically analogous neurons, or that it will be very similar to itself next week in those regards—or the next couple of seconds. The human brain and its "behavior" are unique to each of us and are constantly changing.

The human brain is not a computer. It is not strictly a set of digital circuits. The concepts of hardware and wetware really obfuscate how the brain works. It is not an electronic system; it's an electrochemical system. It is dynamic, *constantly* changing in both arenas. This may account for the less than stellar replication record of certain neuroscience studies (though there are other problems as well), with variability of findings between subjects. Any "map" of the human brain would be obsolete a few seconds after it appears. All this suggests that any "map" would have to be dynamic in the extreme. As I noted, there are a number of "big data" projects underway around the world attempting to map and simulate the human brain, but most experts assert that we are a very long way off from realizing such a goal. (171) The "consciousness" component of this mapping would be even more elusive.

Some Problems

But take a step back. If we could actually, perfectly model the activity of a human brain—with a complete description of neural activity, extra-neural physiology, all their rapidly changing functioning, the appropriate mathematics, and more computing power than we can currently even conceive of—we wouldn't need to invoke the notion of consciousness to explain human behavior, including any god-related matters. Reflect on that for a moment. If we actually knew how the brain works, consciousness itself would be irrelevant, at least scientifically.

Let's look at some of the problems. The search for the biological basis for consciousness is the attempt to find neurological correlates or substrates, as we do for more "ordinary" phenomena. For example, we can identify some workings of the brain when it comes to moving a digit or a limb. We can even grossly identify regions of the brain which are critical in more complex functions, such as balance or speech. Remarkably, we can even identify correlates of some rudimentary "thoughts" (see below). But such approaches fall short on consciousness for a number of important reasons. We do know some things about the neurological basis of sleep, in its various stages, and are clear about what happens to an EEG when someone wakes up—going from unconscious to conscious. That's easy. But it may be an insurmountable challenge to identify a state in a human where there is an "awake condition" without self-reported consciousness.[11] The notion that data derived from any subject can identify "consciousness" as different from a state of unconsciousness without controlling for sleep, coma, etc., is probably no more than a fool's errand.[12] In addition, the notion that we can report meaningful activity from an unconscious state, e.g., sleep or coma, is overblown. Also, the notion that a basis of consciousness can be informed from dreaming is probably not possible. The images of dreams which occur during REM sleep do not arise in a conscious state. They are only reported when a person is conscious, and the relationship of that to what actually occurs during REM sleep is ambiguous, being mixed with other information as the person awakens. We still don't even know the purpose of dreams. The only thing we can be confident of is that dreams are not a medium between us and the gods, as ancient humans believed.

Finally, it probably makes no sense to suggest that there are "easy" and "hard" problems of consciousness. The brain is not made up of discrete functions with locations of neurons that subserve perceptions and "higher" level activity in a formally detached way. We only know that certain functions rely more on some particular areas of the brain, e.g., speech processing in the temporal lobes, perceptual-motor functions in parietal lobes, etc. The brain is dynamic, with areas of perception, cognition, emotion, and action all interplaying in ways that are beyond any technology we've currently conceived of, and they certainly work in ways that are at least a bit different from individual to individual. You could theoretically do a fine-grained and temporal analysis of every neuron and every synaptic interaction in the brain and the rest of the central nervous system (CNS)—if there were ever adequate computing power to do this—a very unlikely scenario—and you could do this for given individuals over some small or large amount of time. But you would still not know any more about consciousness than you do now.

Consciousness and Emotion

There is incredible resistance to the idea of rejecting consciousness as a tangible or quantifiable phenomenon, whether it "emerges" from neurological activity or not. The "search" for it continues unabated and well-funded (though we may never "find" "it," the search will undoubtedly reveal many interesting things about how the human brain works). Have we taken a constructed idea and then tried to go about finding it in reality? Descartes was pretty confident that the seat of the soul was in the pineal gland. Most of today's neuroscientists are not seeking that. Will we find consciousness? How about joy or purpose or beauty? Even though progress is being made in identifying gross areas of the brain that may be critical in certain functions—recognizing objects, speech capacities, etc., mapping complex subjective experiences seems out of reach. Let's take the example of emotions, which have been a particular focus for brain scientists for a very long time. Conventional wisdom is that a particular structure in the brain, the amygdala, serves as the seat of strong emotions, such as fear or pleasure, but, as neuroscientists Ralph Adolphs and David Anderson also correctly argue, it's very far from being that "simple." (4) It turns out that interactions among many cortical and other areas of the brain are necessary for subjective states such as emotions, and we even lack a reasonable definition of what we mean by the term "emotion." There is no indication that so-called emotions are "unitary," that is, contained in any specific brain structure or area. For example, what is generally called fear, one of the most basic "emotions," is not "housed" in the amygdala, where it has often been assigned. Subjective emotional states are partly due to activity there, but they are more a function of innumerable interconnected brain areas and activities. (98) Language and individual experience, of course, are significant limitations to identifying any part of the connectome which may subserve particular "emotions," as all persons experience and label things differently. And there is no hard evidence that strong "emotions" such as fear specifically cause or drive behavior, at best being covariates. In fact, what we may experience as "emotions" may be merely the subjective covariates of the brain's attempt at energy conservation in certain situations. As Kaku points out, "The brain saves energy by making snap judgments in the form of emotions" (261), making the activity unbelievably complex. It would be inefficient to process lots of threatening information linearly, e.g., for fear. The brain can do this essentially all at once in situations that demand it. And what about joy or anger or anxiety? If we can't understand or "map" even basic emotions, consciousness itself would seem out of reach.

Mind Reading

If we could "read" someone's mind, perhaps we would "know" an individual's consciousness. Other than the feats of magicians, could it be done? No, we cannot actually "read" a nonmaterial "mind." But some progress *is* being made in neuroscience to measure or reflect what someone may report to an experimenter as "having in mind."

To begin with, there is good progress identifying neuronal complexes associated with certain motor responses, e.g., movement of some fingers. In open brain surgery, for example—such as when done for surgical treatment of intractable epilepsy—a surgeon can provide a specific stimulation, causing the patient to move her index finger. Computer interfaces which read brain activity, quite wonderfully, have also enabled stimulation that allows people to move various body parts which have been disabled. These things are a far cry, however, from the brain activity which might underlie complex motor activity like kissing or hitting a golf ball.

Experiments with sophisticated EEG techniques have been able to expose reports of crude images or "thoughts." There have been a number of experiments in which a subject is trained to think of an image from a small number of discrete options, e.g., house, cold, tree, yes. (141) Then the neuroscientist repeatedly records activity of the brain while the subject is thinking of the particular object or action. That is, patterns of neural activity are measured in response to given stimuli, e.g., "think of a box" or "imagine a circle." The private experience of the individual can be "decoded" from the recordings of brain activity. Then a subject can engage in some self-selected choices of a small number of the options, and the live EEG data can be "read" by the experimenter (actually some type of artificial intelligence) to reveal what the subject is presumably thinking about. "You were thinking about a house." Of course, this would not be the same thing as the experience that a person is "having"; more reasonably just reporting. And yet, we can see that the neurological covariates of some "thoughts" or words are clearly measurable, though only once the subject's brain is well-trained on the system. It is a modified form of biofeedback, if you will, not what we might informally call "mind reading."

The technique of "reading" thoughts or emotions has also been attempted with functional magnetic resonance imaging or fMRI. MRI is a technology used to study medical problems, but fMRI can track circulation in the brain; it can "see" what is happening in the brain by tracking blood flow to various areas (It's actually much more complex than that, which doesn't concern us here). "Mind reading" with this technique is something which has been increasingly reported in the popular press. (493) However, the current status of fMRI findings—how certain groups of neurons or regions of the brain "act" in correspondence to stimulation, behavior, thinking, or "emotion," based only on regional blood flow—is in its infancy. In a recent study, fMRI was used in a small sample of subjects to generate signals which *did* conform to language and meaning. fMRI was able to reasonably identify patterns, for example, of the implied content of spoken narrative stories or silent videos. That is, using machine learning, the experimenters were able to identify, for example, which of a few stories were being experienced, at least at a better than chance level, though only after the brain was trained on such stimuli. (513)

These are truly exciting developments, and replications and technical advancements may well allow the electronic "reading" of complex cognitive brain activity. However, they are a far cry from "reading" one's "thoughts" which haven't been actually trained on a system. Patterns reflected in EEG or fMRI used in similar studies must be specifically trained on each subject. Specific "thoughts" or "emotions" cannot be diagnosed or revealed when originated by the subject, as opposed to being proactively trained in experiments, and there are numerous problems with methodology and replication of these.[13] Some progress has been made in generalizing the locale of prompted thoughts from one subject to another. However, results of experiments attempting this have been poor. There is no evidence of an experimental ability to read "thoughts" or "emotions" which haven't been previously trained on at least some subjects. The level of accuracy is still crude, and, most importantly, patterns of neural activity are different for each subject for the same output or recognition task. All brains are different in how they perceive and code information. Again, a flower for one brain will not code the same way for another person, not to speak of complex thoughts or emotions; and flowers and dogs will not be coded the same way today as they will be next week.[14] But considerable progress *is* being made on reconstructing complex thoughts. For example, artificial intelligence, using brain training and sophisticated EEG, has even been able to reproduce music from the brains of subjects, e.g., a particular song, though far from perfectly. (43)

Nevertheless, technology and methods will improve over time to increase our understanding of brain function. But such knowledge, however detailed, cannot reveal what we call consciousness—only the neural correlates of any rudimentary "experience" or thought. To presuppose that we encode information and perform tasks (neural activity) in precisely the same way even on two different occasions is unjustified. Thus, efforts to create a static picture (or even a dynamic picture) of the human brain or of any particular brain must fail, almost by definition (by the time you could complete the measurement, it would be different). From examining a scan, I cannot tell you if your right shoulder is hurting, or that you're thinking about Hawaii, or that doing calculus is difficult for you. That is not to say that we can't further understand how information is generally processed, and, more specifically, identify how damage generally affects brains. Modern electroencephalography, or functional MRI, examination of individual or "small" groups of neurons, computer simulations, and several other promising techniques have been used to make great strides in neuroscience. We can even assess the behavior of some individual neurons in live lab animals. But specific "mind reading" from EEG and fMRI has certainly been oversold thus far. And any hope of "reading" consciousness seems forever elusive.

Conclusion

Neuroscientists are searching for the mechanism by which consciousness "emerges" from the activity of the brain. I wish them luck. This may be the most difficult problem ever undertaken or a fool's errand. One can go far down the rabbit hole in trying to understand consciousness but end up nowhere. The problem again comes down to definition. Since we don't know how to define consciousness, we don't know when we've empirically "got it." Philosophers and neuroscientists vigorously disagree on such things. Neuroscientist Christopher Koch (277) takes great umbrage with philosopher Daniel Dennett (120) and others who deny or avoid the idea of qualia or similar subjective experiences. Koch argues for the continued search for the "neuronal correlates of consciousness (NCC)" as a subjective state. Theoretically, you can fully map brain activity and never come up with an answer since we still can't provide a definition for consciousness, but the neuroscience pushes on.

There is also an ongoing debate about whether machines are or could be conscious, which I'll discuss more below. All this may render the search for the neurological basis of consciousness a dead end. The matter is well illustrated by a wager placed between Koch and philosopher Chalmers in 1998. Koch bet that the neural basis of consciousness would be fully explicated by 2023. Obviously, it didn't happen, and Chalmers won the bet. (224) Another bet for 25 years hence was taken, but I'd put my money on any neural basis still being elusive.

Quantum Consciousness

Some scientists think they have a different type of theory for this problem, based on quantum mechanics. Much like the above neural ones, there is no empirical support, but we should at least briefly review this since it has received a good deal of attention in recent years, both in professional journals and the popular press. If there is a determined universe and consciousness is not a useful way of explaining human behavior, are we, in fact, zombies? No, we're not. However, there has been a never-ending effort to rescue the idea of a freely operating conscious mind from a determined universe. Quantum mechanics, as we discussed in Chapter 3, according to some physicists, leaves room for some sort of randomness in the universe and now, by extension, some sort of consciousness.

I point this out since it is relevant to issues of faith and belief, which I'll address later. I also entertain this subject because some have actually linked quantum mechanics to the idea of some kind of human soul. (201) This would seem unbelievable on its face, but the search for some sort of science to reify Descartes' idea and some sort of immaterial spirit never seems to end.

First, there is merely the recycled Cartesian argument that consciousness must exist, but now we have theories that it is presumed to arise from brain activity consistent with quantum mechanics. That is, because some sort of "observation" is necessary for reality to emerge (e.g., in the form of particles) from the haze (superposition) of quantum phenomena, we observers must be conscious. It is reminiscent of the defunct notion of some of the Enlightenment thinkers and solipsists, as we reviewed previously—that reality consists of only what one can experience. As mathematician Edward Frenkel notes about such theories, "It's as though it has resolved to prove the famous Bishop Berkeley maxim 'to be is to be perceived.'" (172) Nevertheless, I need to give this some serious attention, since related quantum concepts have also been proposed for the related concept of "free will," a critical matter for the god problem, which I'll take up in Chapter 6.

One of the most talked-about theories in this effort has been proposed by the Nobel Prize-winning physicist Roger Penrose. In *Shadows of the Mind* (394) he proposed a theory of "quantum consciousness," now extended to a so-called "Penrose–Hameroff theory" of consciousness. (200) Since this appears to be a current, often-quoted theory, we'll give it some space here.

It goes something like this. Some living organisms are eukaryotic (animals, plants, fungi, protists), while others are prokaryotic, like certain bacteria.[15] In eukaryotic cells, there are structures we call microtubules, which provide assistance in cellular structure and platforms for intracellular transport and several other functions. To oversimplify, the theory proposes that quantum effects do emerge in brain tissue, in the microtubules, and it is argued that the quantum superpositions there are reduced by "quantum gravity," leading to the emergence of consciousness. (200, 394)

But There Are Some Problems

The link is not explained. First, there's the definition problem. Penrose considers the simultaneous reduction of superpositions of quantum states in many neurons of the brain to be, in effect, consciousness. However, no actual definition of consciousness is offered. In fact, Penrose, quite admittedly, doesn't define what he means by consciousness, noting, "I shall not attempt a full definition of 'consciousness' . . . even though my arguments are directed towards the goal of ultimately comprehending this phenomenon in scientific but non-computational terms . . . When all is said and done, we must ultimately rely upon our intuitive comprehension of its meaning." (394) This is, indeed, a strange assertion from a physicist.

But let's say, for sake of argument, that we could get past this. There are a few other problems. The notion of collapse of the wave function—or the simultaneous reduction of superpositions in all the relevant neurons in the brain—leading to "consciousness" seems again to appear no more than magically. Then there is the notion of "quantum gravity." That such hasn't been verified, or even a reasonable theory about it yet offered, seems to be a problem, to say the least. Penrose does go on to admit, "It should be stressed that there is no accepted theory at present—there is not even an acceptable candidate" for quantum gravity. (394) Also, the Penrose–Hameroff "theory" also essentially denies determinism, a problem in its own right. We can assume that the universe is an indeterministic or random system, a serious problem. Or we can subscribe to "classical" quantum theory that allows for collapse of the wave function by observation, though who would the observers be here? Or perhaps other theories of quantum mechanics will prove out. But none of these solves the problem of consciousness.

An additional problem is that the theory is based on the concept that, fundamentally, there is some random action in fundamental particles in neural cells and therefore the phenomenon is "noncomputable." Absolute predictions (theoretically) cannot be made. Others disagree with that premise, of course. (117) Electrons in the brain, for example, repeatedly interact with other particles. Otherwise, there could be no physiological activity of neuron cells (such as conducting impulses or communicating with other neurons), which is how the brain actually works. This is the process of decoherence, as I noted in Chapter 3. Physicist Carroll notes, "Important processes in the brain are well described by the approximation of classical physics . . . . Every particle in your head is constantly being jostled by other particles, leading to an ongoing process of 'collapse' (or branching of the wave function, for fearless Everettians like me." (78) To wit, observation and superposition are irrelevant in the brain's internal operation. "In Everett's Many-Worlds Interpretation . . . everything continues to smoothly evolve according to a deterministic set of equations." When a macroscopic observer interacts with the environment, this results in the system evolving from "one universe in quantum superposition" to "two separate universes." Consciousness plays no role. "The fact that observers happen to be conscious [so to speak, his assumption] plays zero role; measurements can be easily [his adverb] carried out by nematodes, video cameras, or rocks." (78)

What Else Is Conscious?

Human brains are not the only organs with cellular microtubules. The behavior of electrons in eukaryotic cells is critical throughout the body—in the lungs, kidneys, colon, skin, and other organs. We would certainly not propose that any quantum states of electrons existing in those organs form a basis for consciousness there. It is unclear, therefore, why they would do so in the brain. Certainly, Penrose would not advocate that such structures, in and of themselves, are sufficient to give rise to putative "quantum consciousness" in such organs. Some would say the heart could be an exception, but of course, other than the apt metaphor, there's no evidence for that.

Other than electrons moving around in chemical reactions in microtubules, what are the other conditions would be necessary for consciousness to emerge—an actual nervous system, sufficient brain size, complex learning, awareness of self, etc.? What about other organisms? Are those quantum effects sufficient for dogs, birds, fish, jellyfish, planaria, etc.? (I'll talk more about animals below). Let's leave for the moment Chalmer's idea that everything in the universe has some type of consciousness. Some have suggested that there are elements of the Penrose–Hameroff theory which would suggest a "type" of consciousness even in plants, which I just cannot bring myself to detail here. (32) Do rose bushes or trees have consciousness? What about fungi? Many plants and fungi communicate with each other through a vast array of signaling systems. Are they "aware" of what they're doing? There is even recent speculation that plants may manifest a "diffused consciousness" throughout their anatomies (72), analogous to the apparent diffusion of brain processes in the octopus. The implied continuum—over species and over evolutionary time scales—creates an absurd problem. If quantum effects in microtubules is the key, where do we draw the line on consciousness in living things? This presents two interesting alternatives. This either goes against any reasonable consensus of what most people mean by the term, or else the idea of "consciousness," so to speak, in living organisms is nothing special, *including for humans*.

Conclusion

In sum, most physicists tend to disagree with the notion of quantum consciousness. Physicist Carroll notes that "almost no modern physicist thinks that 'consciousness' has anything whatsoever to do with quantum mechanics. There are an iconoclastic few who do, but it's a tiny minority, unrepresentative of the mainstream." (78) That is certainly true. However, we should point out that sometimes minority views prevail in the long run. Nevertheless, no empirical evidence exists at this time for the quantum idea. In reviewing the quantum consciousness hypothesis, two eminent biologists, Bernard Baars and David Edelman, point out that "In science the burden of proof is on the proposer, and this burden has not yet been met by quantum-level proposals. While in the future we may discover quantum effects that bear distinctively on conscious cognition 'as such,' we do not have such evidence today." (24)

Again, the definition problem. The notion that any quantum effects are responsible for a subjective state which has no meaningful definition provides no reasonable understanding or prediction of human behavior. The attempt to explain "consciousness" this way falls short, and as heretical as it may seem, it provides no empirical evidence that such a phenomenon actually exists. And we certainly can't use it to support the notion of a soul (something which is necessary for the God idea), as some have attempted. (201)[16]

Behavior and Experience

When you have faith in God, when you believe in His existence, when you pray, does consciousness have any role in those activities? Most people would say "definitely." But what does consciousness have to do with human behavior anyway? I'll address the science of human behavior in Chapter 10, but let's consider here a few problems with this particular reasoning.

As I've shown, we can't define consciousness (Perhaps Penrose is right and there *is* no way to define it.). We don't know where it comes from, and "its" role in what we do can't be described. As I noted, we should quite rightly consider consciousness as a "collateral product," an epiphenomenon that has nothing to do with our actions. If we knew the instantaneous action of all events in the brain at any one time and therefore could predict its output, we would not need the notion of consciousness to explain and predict behavior, including faith. This makes "consciousness" an almost irrelevant idea, at least scientifically. It may someday appear associated with brain activity, but it is not the cause—of your faith or any of your religious endeavors. That you have an experience you call "with God" or any subjective "knowledge" about God becomes irrelevant. And when you pray or go to church or read the Bible, those activities and behaviors are determined, much as their basic chemistries are; and consciousness, if it exists, is simply along for the ride.

The only way to get around that is to employ some sort of mysticism. That sort of philosophy, of course, merely reprises Descartes' dualism, with a corporeal body and an immaterial soul. Of course, to explain how nonmaterial entities interact with the physical world only sends mystics and theistic philosophers into intellectual knots. Or it requires Chalmers' wild notion that everything in the universe has some sort of consciousness. Therefore, such concepts are of no use to us here.

Given that, why would we go down the rabbit hole of trying to find "it"? We've essentially given up trying to operationally define other subjective states, realizing that words we use do not necessarily imply some coherent physical representation—beauty, sorrow, hope, etc. Even if one could "map" the activity of a connectome and its momentary dynamics associated with those ideas, they would be notably different from person to person, moment to moment, etc. And the reason for this is crystal clear—we are all different, with different genetic, environmental, and behavioral histories. And the brain, most uniquely the neocortex, develops as a function of experience, and therefore its operation is also unique. To look in the brain for "hope," for example, would be a hopeless exercise. There are better ways to examine such human behavior. I will not deny that I have internal "experience." But what that means is totally idiosyncratic. My "consciousness" is not a predictor of anything. It is not a useful concept in understanding human behavior—or the behavior, for that matter, of monkeys, fish, or plants . . . or photons.

Most—in fact, nearly all—of what we do, we do "unconsciously." That is not to say we are unconscious. It merely means that we are not specifically aware of doing it. We might more correctly call it "nonconscious." When you breathe, you do it automatically, occasionally with effort. When you smile, you don't focus on the muscular mechanics; you just smile. Putting aside the zombie notion, we should realize that consciousness, per se, is not relevant for the generation of most behavior. Let's take the example of a pretty complex behavior—typing. When you type quickly you usually don't "think" about it. You just type. Even though you are aware of the fact of your typing, it seems automatic, and essentially it is. Such was not the case when you were learning to type. Every stroke was deliberate, and you made many mistakes. But acquisition of the skill moved the behavior from its components into a complex set of responses that required no "awareness." The acquisition and use of language is a prime example. We generally don't "think about" the grammar or syntax we're using when we speak. We just speak (though perhaps we should pay a bit more attention to it at times). Even "simple" behaviors like eating are done without particular focus. You don't think about the flexion of your elbow when you eat. What behavior is done consciously, and what is done without any awareness? And where do you draw the line? We even multitask, focusing on one behavior while doing another without much attention. Most of what we do is done "automatically"; we don't really "think" about it. Consciousness, again, is a superfluous idea and not the "driver" of behavior.

The typical rationale for consciousness, and not an empirical one, is "I know I'm conscious because I have experience." Of course, that's just a tautology and a reprise of Descartes, and it doesn't make it a tangible phenomenon. Humans are capable of imagination; we can fantasize and daydream. We can "experience" things that don't exist, from fairies to past lives. Brains can create all sorts of realities in our perception that can't be validated with some external reference. Hallucinations are another example (which I'll discuss further below). We can imagine or experience nearly anything. "Martians are talking to me." This is no different from saying, "I know there's a god because I experience him" or "God talks to me." One is not necessarily more ludicrous than the other. Experience is *not* evidence or even a reasonable indication that consciousness is a useful concept—or certainly that a god exists, even though the "reality" of the person's experience cannot be denied.

Artificial Intelligence

Are we God's unique creatures because we're supposedly conscious? Unlikely. Does it take intelligence to be conscious? Unlikely. We'll consider some other creatures (below), but let's first consider the rather fantastic question of whether consciousness can "arise" as artificially created. If that is so, we humans are certainly not all that unique. I'll not review in any detail the idea of artificial intelligence, a broad and intricate subject way beyond our mission here. But another way to examine the concept of consciousness *is* the idea of artificial intelligence. Historically viewed as a science fiction notion, AI is already in our everyday lives. Whether it is AI on your phone, a self-driving car, "chatting" with your bank's computer, or Google anticipating your preferences or giving you "answers" to some of your most complex questions, AI is all around us. The AI question will be critical in how we look at the idea of consciousness. Further, this subject will be critical to the notion of God.

Do machines think? Are they conscious or sentient? This seems absurd, but don't dismiss it out of hand. Remember, you can't prove that *you* are conscious. If consciousness is something that "emerges," and it is purely a function of the interaction of specific entities (e.g., neurons and certain brain chemistries), then we must consider whether complex interactions among artificial entities could do the same. Computer scientists call these neural networks, parallel in function to our own brains (These are actually mathematical systems, not specifically analogous to biological brains.).

Advanced AI systems have been the fantasies of science fiction for quite a while. The notion of AI and robots has been around for a very long time (The term robot appears to have come from a 1920 play by Czech playwright Karel Čapek, 'Rossum's Universal Robots,' with the term *robots* signifying 'forced labor.'). In the *Iliad*, Homer gives us the character of Hephaestus, the blacksmith to the Greek gods. He fashioned golden robots for the gods, handmaidens "who" could learn and reason. They even had souls. And the historical and current science fiction literature is replete with them. What is relevant here is that we are coming closer and closer to the science fiction of "machines" which think and act in ways which are eerily similar to us.

Learning and Intelligence

We can extend what we've addressed about neuroscience to the rapidly evolving world of computers and their capability to behave in ways that seem to mimic human thinking, at least at a primitive level at this time. This is called *artificial intelligence* and is a relatively recent concept. It refers to complex assessments or behaviors emitted by machines—whether sophisticated data processing and analysis, or being able to solve complex problems in ways that seem to mimic human behavior and cognitive capacities. At this point in human history, sophisticated computers have been able to do all sorts of complex analyses. Consider IBM's Watson, a very sophisticated computer system which, just a few years ago, easily won games on the television game show *Jeopardy*, playing a couple of the long-time champions of that show. Answering the questions (or questioning the answers, so to speak, as is the format of the show) required not only vast knowledge stores but the ability to quickly appreciate human idioms, humor, and connections among information and types of reasoning that we engage in. Watson was able to solve complex problems (what we call intelligence) and do it even faster than very bright humans. And AI has made incredible progress since then, which I'll discuss further.

Turing Test

But great intelligence, in and of itself, is not consciousness. Humans are more than stores of information and problem solving. One way to tell if an artificial system is conscious is to interact with it and assess whether it "seems" human. This is known as the Turing test, proposed in 1950 by British computer scientist and mathematician Alan Turing. He proposed a theoretical test of whether a machine can exhibit intelligence and other characteristics adequate to convince an observer that it is human. (520)

It is theoretically possible to construct, or at least envision, a physical (non-biological) device which could mimic human thinking and verbal behavior (though actual, observable behavior would require another level of unimaginably complex output systems), including continuous changes based on physical exposure and learning. The only difference in "scaling up" such a system—to mimic a human brain—would be the formation of new "tissue" and the dying of other ones. But one can imagine such functions being built into the system. But that is also not the point. Any system which can be analyzed and theoretically replicated would be indistinguishable from the model it was based on. But would we say it has consciousness? Again, whether it would be "conscious" in the sense that we use that term is immaterial, as we cannot assess this in other presumably sentient humans either, nor in other animals. Such a machine could assume consciousness for itself and report its own self-awareness.

So far, no machine has actually passed the Turing Test. Many philosophers assume it would be impossible, even if the output was disguised. American philosopher John Searle disagreed with Turing's premise. He noted that even if such a machine could pass the test, it can't be shown to have "mind" or intention. (454) A computer, even though it can process information to simulate human thought, does not actually "understand" what it is doing, so the argument goes. But what is "understanding?" And how do we really know that the system doesn't understand? If we mean by understanding that there is something over and above the mechanical system, such would be incompatible with the laws of physics. If we mean that answers and other functioning can be placed in a larger context—perhaps an entire "life" of experience—then what is lacking is just that experience by any tested machine.

The current consensus is that the Turing Test is presently inadequate for the task of establishing consciousness. Neuroscientist Gary Marcus describes how some primitive machines can fool examiners with specific responses. There need to be significantly enhanced criteria for such a machine, including better natural language, ability to process and understand ambiguous information, and have a wide basic knowledge base. (327) However, in recent years those criteria may no longer be limiting, as current AI possesses or can instantly access and abstract more current and historical information than any human possibly could; and their "large language models," as they're called, do enable machines to understand and communicate in ways that seem eerily human. Perhaps the Turing Test may be too easily tricked. The Loebner Prize for passing the Turing Test was established in 1991, but so far, no entry has fooled a panel of judges into believing it was human. At this point, leading experts in the field of neuroscience and consciousness have not been able to agree even on what constitutes a reasonable test for AI consciousness (165), though some in this debate have proposed that such test should be applicable to newborns and animals as well. You see the problem.

Deep and Broad Learning

AI systems are gaining more and more of the world's current and historical information, and "they" have access to it instantaneously, even if it's not "on board" at the time (You'll note the temptation to refer to such systems with personal pronouns.). The capacities of AI are often projected based on what's known as Moore's Law (in which the number of transistors on a chip doubles every two years, though it does not appear to be holding as well as predicted, and most feel it will end or slow down to a crawl over the next decade). Physicist Kaku notes, "Given . . . Moore's Law, some say it is only a matter of time before machines eventually acquire self-awareness that rivals human intelligence." (261) He equivocates intelligence and self-awareness, which is generally unjustified, of course, but the point is well taken. Nobody knows if Moore's Law will hold, and it doesn't seem that it will under current technologies. Nevertheless, if quantum computing becomes practical, he may have a good point. Kaku goes on, "Moore's Law cannot last forever. In fact, we can already see it slowing down now. It might flatten out by the end of this or the next decade; but the gap may be filled by . . . quantum computers, molecular computers, nanocomputers, DNA computers, etc." AI systems may not be limited in ways we have previously envisioned, coming ever closer to human "self-awareness."

We know that systems we call artificial intelligence "evolve" capacities on their own, having not been programmed for such, and often in ways that even their designers don't well understand. As AI acquires more information and capacities, it becomes able to essentially learn to do other, nonprogrammed things on its own. That is, it *learns* by gathering information and by trial and error or "reinforcement learning." (The analogy to human behavioral learning is striking, as we'll discuss in Chapter 10.) It has new capabilities, some things that weren't there before. Neural networks "evolve," which is what is known as "deep learning." The AI system is said to develop, on its own, "emergent abilities." Sound familiar?

As it learns how to learn and form concepts, AI becomes intelligent in ways that we don't well understand, and it is able to solve novel problems. This is what researchers call artificial general intelligence or AGI. It is the ability to form concepts, reason, and solve a wide range of problems for which no specific antecedent learning can be identified. This is specifically analogous to what we call general intelligence in humans, i.e., the ability to think and reason beyond the level of specific problems, the capacity to solve various types of problems, and to self-correct in a way that leads to even more efficient problem solving. When the English psychologist Charles Spearman came up with the concept of "general intelligence" in 1904, this is the type of reasoning that was behind it.[17] Such AGI is now being modeled by machines, in which algorithms build upon themselves by acting toward an abstract goal, e.g., "make my business more efficient" or "why did Shakespeare write in the style he did?"[18]

Analogy to Brain

Our brains can process incredible amounts of information, though we can use only a small amount of it at one time. And our brains are incredibly slow compared to artificial intelligence. Despite the apparent potential of the brain, we can hold only one thought at a time, and the brain actually processes information and makes decisions at only about 10 bits per second, even though our senses can process information about the world at around a billion bits per second. (564) If we need to get more information, we pause to consider or we actually look it up. AGI can store and use more information than we can; it can do it incredibly faster; and it doesn't need to store it all. It can be stored "elsewhere," and, of course, it has incredibly rapid access to most of the information in the world that's somehow available.

Most neuroscientists conclude that the functioning of the human brain could not be currently simulated, even if it could be aped well enough to pass a Turing test at this time. The human brain is not merely a collection of circuits. However, at its most basic level, it is clearly an information processor. So let us imagine that, at some time in the future, an artificial system could be created to *function* the way the human brain does. Imagine a continuously changing connectome, with the ability to learn from experience, store and retrieve as much information as a human, integrate with past experience, continually update its algorithms (microsecond by microsecond), dispose of (and refabricate) "tissue" in ways analogous to various kinds of CNS cells, and consume, store, and use energy.

While the brain might not be perfectly modeled and constructed in terms of neuronal activations, electrical gating, chemical processes, etc., the dynamic information and functioning could be—theoretically. Such a system—or "being," if you will—would "behave" in ways eerily similar to us. Whether we enable adequate immediate environmental inputs (sensation) and outputs (motoric activity, movements, etc.) in such a being is another matter—capacities that many fear in such a "being." It is what happens centrally—in the "brain"—that is relevant here. This is obviously a wild simplification, and we have pretty limited understanding of the human brain itself at this time. But as an analogous functioning entity, would the output of a human-mimicking brain device have consciousness? If it passes the Turing test (at least an adequate one) to convince an examiner, would it be *conscious*?

Current AI

Is your smartphone already conscious? Often it seems so. Does your computer anticipate your behavior and needs, like completing your thoughts or sentences? I make no comments here regarding the wisdom of more and more advanced AI in our lives and its future in our culture. That is a subject unto itself. But we can certainly observe that technological advances—tools—have always been the hallmark of the developing human culture. Since our *Australopithecus* ancestors might have begun using stone tools in the Olduvai Gorge in Africa millions of years ago, and since we evolved as a species into the late Stone Age, tools have always been a driver of human cultural development. It is unclear how much AI will penetrate our lives, and what attempts will be made to "slow it down" or "manage it," as some have suggested. Yet, history is clear that there has never been a successful human effort to reliably curtail tools from becoming more and more sophisticated, often being replaced by other, even more advanced tools, or others, for the same tasks, and being actively used. Tools, if you will, seem to have an inevitable march of their own.

We have no choice but to deal with the question. Current AI is able to "behave" in some truly remarkable ways. It can independently drive a car (though there have been problems with this lately). There are also numerous applications that are significantly advancing science, technology, medicine, etc., which the public is only now becoming aware of. For example, it analyzes medical imaging tests and diagnoses illnesses. It can diagnose certain kinds of cancer (sometimes better than physicians, e.g., dermatologic or breast cancers). It has achieved some amazing results, such as early diagnosis of Alzheimer's Disease or beating radiologists in diagnosing lung cancer, designing new drugs, and other remarkable achievements.

It can write some extraordinary things, all on its own. It can write original essays, screenplays, and other texts, and in a prescribed style, and produce remarkable technical and creative writing (a growing problem for educators). It can write some pretty good poetry. AI can write an original essay on the societal value and risks of AI. It can write an intriguing play in the style of Shakespeare. It has created original art and pictures from only verbal prompts, and it can create some beautiful art, even mimicking certain styles of the Masters. It is also creating fully novel art pieces in its own styles, which people can't discriminate from human-created art, including some highly regarded paintings. (153) You also can specify a style, and the system goes on to "create" something. The resultant artistic piece is a remarkable original composition. (407) That some AI is now producing art and music and even substantial videos (that are hard to discriminate as artificial constructions) should be a provocative hallmark in what we think of as consciousness.

We're all familiar with AI's speech recognition. But there's also facial recognition, lip reading, and affect recognition, actually detecting human emotions. It can understand what a person is saying—beyond the actual words—and converse with you about almost any subject, using natural language. It can learn much more sophisticated behaviors. You can "chat" with AI about a wide variety of subjects. It has also demonstrated the ability to simulate real people using audiovisual dialogue and behavior indistinguishable from the real people it's aping, so-called "deep fakes." And now, some systems are even providing online psychotherapy independent of human backup. (Whether such a trend is reasonable, effective, or ethical is questionable at this time, and some tragic outcomes have occurred. The future of this is unclear.) All this is not to say that we're on the verge of a system which can pass the Turing test any time soon, but recent progress should convince you to merely imagine that it could be done, possibly in the near future. Would that make it conscious?

Conscious AI?

So, could AGI systems actually "have" whatever we call consciousness? Would a computer system even closer to our own capacity have consciousness? AGI forms general concepts (it learns how to learn and abstracts), which seem to be distributed throughout the system. One might say that it shows an "emergent property." Again, sound familiar? Now multiply AI "experience" billions of times for billions of environments and tasks (I merely use "billions" metaphorically since it's unknown how far quantum computing and any successors will go). Is this the beginning of a conscious "organism?" With deep learning, we really don't know all of what goes on with an AGI system and how it generates "behavior," the so-called "black box" problem.[19] Computer scientists are reluctant to address the issue of consciousness, yet there is actually no way to rule this out for some AI systems and by the same logic we use for human consciousness. Many terms are used to disguise the problem—

artificial consciousness (AC) or machine consciousness (MC) and even synthetic consciousness (SC)—to get around any coherence with the idea of human consciousness. Nevertheless, the question does remain. Some researchers have proposed that artificial systems could clearly have "experience," that which we consider essential for human consciousness. (117) Even the notion that machines with AI can dream has been recently offered (387), and we do know that they hallucinate—sometimes they generate ideas or responses that are impossible in reality. At bottom then, what is actually required for a machine to be conscious? Some researchers argue that consciousness can "arise," so to speak (though neuroscientists use the equivalent term "emerge" about brains), where there is both information processing and self-monitoring. Cognitive scientist Stanislas Dehaene and colleagues summarize the argument:

> A machine endowed with [both information processing and self-monitoring capabilities] would behave as though it were conscious; for instance, it would know that it is seeing something, would express confidence in it, would report it to others, could suffer hallucinations when its monitoring mechanisms break down, and may even experience the same perceptual illusions as humans. . . . Although centuries of philosophical dualism have led us to consider consciousness as unreducible to physical interactions, the empirical evidence is compatible with the possibility that consciousness arises from nothing more than specific computations. (117) [20]

What about self-awareness? Is the machine aware of what it is doing? It's sometimes said that self-awareness for a machine is the wrong standard since it essentially has "knowledge" of everything it is and what it is doing, awareness being undefined. However, some computer scientists have publicly offered that some systems might be "self-aware," being able to recognize itself, and have a "notion of self." (539) Self-recognition is a reasonable criterion, provided it is not fed pictures. This is often cited as a criterion of human consciousness as well (Note that elephants, dolphins, and many primates satisfy this. I'll deal with animal consciousness below). Though true machine recognition and self-awareness has not yet been unequivocally demonstrated, there are suggestions that this may be on the near horizon. (33, 84)

Let's do a somewhat fantastical thought experiment. Let us suppose that a computer could be constructed which perfectly mimics a human brain and its functioning, with a similar or analogous arrangement of billions of neurons and an almost unimaginable number of changeable, neuronal interconnections, or neural network. Whether this will ever happen is an interesting matter of pure speculation at this time. However, if we envision such a device, would we possibly envision a capacity for self-awareness and consider it "conscious?" If we knew all the antecedents which led to our own capacities, we could make no better argument for our own consciousness than we could for that computer. Dehaene opines that machines are "not there" yet (117), but we really don't know and can't ask them—any more than we can "ask" chimpanzees or elephants.

Remember that no matter how intelligent AGI becomes, that's not consciousness. Intelligence and consciousness are not the same thing. Intelligence can be operationally defined, e.g., the ability to solve problems. Humans can solve more complex problems than apes, which are better problem solvers than planaria.[21] But consciousness is a different animal. We have no way to operationally define it, and thus, we cannot say whether an AI is conscious or not. However, if you teach an AI machine what consciousness *is*—that it is tantamount to conceptual learning, experience, and self-awareness—would it answer you in the affirmative when asked if it is conscious? Would you believe "it"? Would you consider it the same "type of consciousness" that you claim to have? Why or why not? Are we convinced, or doubtful, or skeptical, or in denial about whether some AI is already conscious? So far AI hasn't developed the relevant reporting mechanisms. But this may be merely a technical problem.

Does a sufficiently advanced AGI system "think" or "know" that it exists? If thinking is whatever an entity says it is, perhaps we must believe it.[22] Would it affirm the standard *Cogito ergo sum*? Remember, other than this Descartes axiom, you can't prove that *you* are conscious. If a gibbon or an octopus or, in Chalmer's view, a photon, can be conscious, why not AI? Perhaps it wouldn't manifest jealousy or gratitude or anger or achieve Nirvana, but those are human traits. There is no reason to believe that AI could not—or hasn't already attained—some sort of consciousness.

What Would It Be?

If a machine becomes conscious, what does that mean? We certainly would no longer be the unique conscious characters we consider ourselves to be—special to ourselves and some God who supposedly created us. We're probably not there yet. The problem, of course, is that even if you enable a machine to have the sum total of a human experience, the machine will not have experienced it in a way that a real human does—throughout time, relationships, physical changes, successes and failures, etc. Beyond programming, unless the ontogenesis (development) of a "typical" human can be replicated, there would be no convincing evidence that human thinking and behavior could be truly mimicked. For our purposes here, however, we'll just naively assume that this could be approximated to the degree that an AI system is self-aware, endorses that it "feels" it is conscious, and is cognitively reflective to the same degree as a human without having been specifically programmed in that regard.

It might "think" and behave much as we do, and of course, many have envisioned such computer systems encased in shells, what have been called robots or cyborgs. The modern science fiction literature is replete with such beings. Let's take a couple of examples. Take Arthur C. Clark's *2001: A Space Odyssey*. The computer system aboard the spaceship resisted attempts to turn it off and attempted to take over the mission, as "he" (named Hal) saw it. Did Hal have consciousness or free will? Was Hal more concerned with its own survival? Is a survival instinct purely and only a function of "biological" evolution?" Then there's the provocative example from the television series, *Star Trek: The Next Generation*. A principal character is Commander Data, an android being. In the episode "The Measure of a Man", a trial is held to adjudicate whether Data has the right to determine his own future—in this case, not to be disassembled for study. If he *tells* us, as he does in this episode, that he is self-aware, has a right to make his own decisions, and embraces his survival instinct, on what basis could we disagree? Data had "evolved" beyond the scope of his programming (as has current AI, though perhaps to a lesser degree at this point in history). Who are we to say that any real-world instances of a Data behaving that way would not be conscious? Would we deny it to him and ascribe it to a cat, merely because the latter has wet neural networks? Science fiction is just that, but we should consider such seriously when we try to understand the idea of "consciousness." (Remember, science fiction has often become reality.)

There's another, related notion that's pretty fantastical—humans putatively being able to "upload" their consciousness into a non-organic machine. If the entirety of a human brain's history and function could be uploaded, would that machine be conscious? The often-discussed science fiction notion of this actually being possible and creating some sort of immortality is not really relevant here, but just trying to imagine such might be instructive. The issue is considering whether a virtual, inorganic "clone" of oneself, with its entire history of behavior and experience, would be conscious. Whether a being (brain) evolves from a "few" lines of code or from a perfect template of you, to envision such things casts doubt on the whole notion of human self-awareness and consciousness. A cybernetic creature composed of silicon and other elements can claim to be fully conscious, and it would be difficult to doubt it. Yet, he or she can no more define or reify the idea than you can. This is leading us somewhere of great importance, so stay tuned.

Again, the Definition?

If a machine could conceivably attain consciousness, then we would again ask the question, "What is it?" As we've reviewed, there really can be no operational definition of consciousness. And if AI attains "it," we may still not know, since AI may not share the neural nets, algorithms, and other operations it refers to as consciousness. And it may not know itself.

We certainly can't assume that it is the same type of consciousness that humans might claim. Being afraid of the dark is an unlikely emotion for AI systems as we might conceive of them. Falling in love seems out of the realm of possibilities. And the desire for sex and reproduction seems unlikely (though some would disagree). But the reason for this is simple. It's not really conceivable that a machine could be programmed and learn in the way humans develop throughout their lives. Nevertheless, if we consider consciousness as somewhat variable (though perhaps not as variable as Chalmers suggests), there is no reason to doubt a self-reporting conscious AI.

We could ask the related question, "What does it 'mean' to be conscious?" We might ask a machine, as science fiction muses, a number of other, probing or philosophical questions: "What is the meaning of life?" "What's the best form of government?" "Should we genetically engineer humans?" "How do we balance economic resources for earthly human needs with space exploration?" "What should the human species do to survive?" "What is the one true religion?" And personal questions: "How do you feel?" "What's your favorite color?" "What do you think your dreams mean?" "Do you love me?" We can project a future in which the responses to those questions are no more believable or unbelievable than those we get from each other today. The closer we come (theoretically!) to such a being, the more the notion of consciousness tends to lose all meaning.

Morality and AI

Now let's assume that some AGI *is* conscious. This is a bit off the point, but we can't ignore it entirely. A number of interesting questions arise. One of the things that we identify as making us human and special —not just intelligence or even consciousness—is some sort of illusion that we have free will and have consciously generated our own morality. We'll get into these subjects, in subsequent chapters. However, it may be helpful to consider that if some advanced AGI is "conscious," will it exhibit and claim the notion of free will and profess its own "thoughts" and espouse its own morality? Would it exemplify the same moral compass as its creators, or of the culture in which it emerged, or something that it evolved on its own? Would that change with experience? Would it be guilty of "sin" if it violated its own code?

Interesting problems also arise when we might "find" consciousness in some artificial systems. How would we respond? How would we treat them? With what rights do we imbue them? Shall they be the same as for humans, and do we even have the right to make such declarations? We certainly have a poor track record of all humans having equal rights. Would we differentiate and reward or punish a robot judged as "good" (e.g., Star Trek's Commander Data) vs. "evil" (2001's Hal)? Would we reserve the right to "turn it off"? Could we assume that it would not suffer if "injured" or punished? How would we feel about it reproducing? (Note that some advanced systems already design other ones.) Would it be "life?" Would we send it on a one-way mission to outer space?

These are perplexing questions, in a sense. However, we humans *are* contemplating the idea of some AGI becoming conscious. That's obvious, since the other question frequently pondered is, "Would some conscious AGI system be a threat to humans or take vengeance on us if we treat it 'inhumanely'?" Would it actually be an existential threat? Perhaps our fear is a bit too conditioned by science fiction movies, but the fear is real, including for some scientists who've seriously contemplated such matters. As Stephen Hawking once said, "The development of full artificial intelligence could spell the end of the human race." (82)

This may seem a bit off the mark, but how do we hedge against this? In 1942, the great science fiction writer Isaac Asimov proposed his famous three Laws of Robotics, as follow:[23]

> A robot may not injure a human being or, through inaction, allow a human being to come to harm; it must obey orders given it by human beings except where such orders would conflict with the First Law; and it must protect its own existence as long as such protection does not conflict with the First or Second Laws.

The argument about whether sufficiently intelligent AIs would *actually* have moral values or be a threat to humankind is not really relevant for our purposes here. What is relevant is that if we considered them conscious, we would also *consider* them a threat. Thus, Asimov's laws ring prescient. We reserve the notion of consciousness to ourselves; any other beings who might "possess" it make us very uncomfortable. We must have rules—for them. Such also threatens our putative uniqueness as human beings—the Copernican Principle again. This is all fantastical, of course, but contemplating such questions allows you to consider what you really mean by consciousness.

AI God

All this is leading up to an even more interesting question. We could ask a conscious AGI, "Is there a god?" The answer would be interesting, to say the least. So far, if you ask that of your smartphone, the answer is unsatisfying. If AI can be conscious AI, does it believe in a god? Does it "know" in its heart that such a being exists? Is this absurd? Maybe. But bear with me. If we assume that AI can be conscious, and if we assume that it has at least all the information that human beings have developed (and probably a great, great deal more), and, therefore, it is intelligent enough to ponder the nature of the universe, we cannot rule out that it has considered the problem of where everything came from. We humans have done that, and many assume the answer is an omnipotent supreme being. There is no reason to believe that a supremely intelligent and conscious AI wouldn't also consider the problem of creation as well. And what would its answer be?

A truly advanced AI system, of course, might be the "individual" that achieves an understanding of all the concepts of cosmology, relativity, quantum mechanics, dark matter and energy, etc., and perhaps yet-undiscovered laws of physics. Perhaps it will solve Einstein's quest, and the rest of physics, for the so-called Grand Unified Theory. Maybe the evolution of conscious AI *can* provide an answer to this, assuming it will actually still be a relevant question by that time. Given the obvious incompleteness of our understanding of the universe, it might come up with a good answer. But I'll bet it wouldn't be the God of the Judeo-Christian Bible.

Let's do another thought experiment. Let us assume that a truly advanced cybernetic creature exists and claims to be conscious, gives all indication of self-awareness, passes an advanced Turing test, including provocative discussions with humans about, let's say, philosophical issues, and understands at least as much as we do (probably more) about cosmology. If it does claim to believe in an omnipotent personal god, what does that actually mean?

Perhaps it conceives of a god in its own image, a machine operating on its own algorithms. Absurd? In Chapter 3, I mentioned that some cosmologists have seriously considered the idea that our universe is indeed "merely" a program running on some "supercomputer," and that it is "merely" a simulation. (11, 386) If that's the case, it's not too difficult to conceive of the notion that the "real god" is a "machine" and that conscious AI is closer to understanding the god problem than we are. The AI would then be within its "philosophical" rights to consider that biological life is "merely" an intermediate form in the evolution of artificial life. This is just wild speculation, of course; but the logic remains. We created God in our image, of course, and AI could also do so on its own. Is this an outlandish conclusion? It was the eighteenth-century French political philosopher Baron de Montesquieu who offered a prescient view of this, "If triangles had a god, he would have three sides." The logic remains.

Here's the bottom line of all this. *Merely being "conscious" of a god, or the idea of a god, or consciously conceiving of a god, is meaningless*, and the notion of consciousness offers no explanation of human behavior, including ideas about any supernatural being. And, when you get down to it, it makes no more sense to consider a human's belief in a certain "personal god" than it does for a robot to believe in a cybernetic god. We "create" all sorts of things in our own image.

Animal Consciousness

Can we assume that we are the only living organisms that have consciousness? We've considered whether an artificial entity can, or likely will, achieve consciousness, something that can't be dismissed out of hand. Closer to us are other animals. Are any of them conscious? If so, perhaps there's nothing special about the idea and nothing which is fully unique to us humans. This will obviously have implications with respect to any notions about gods.

Who is to say which animals are "conscious" and which are not? Are there different levels or types of consciousness in animals? Regardless of any instinctual drive to do it, is a cat aware of the fact that it is its own paw that it is cleaning? Is a dolphin midwife conscious of what she is doing? (Dolphin females do assist in the birth of others in their pod.) Does a gray squirrel "understand" why she is nursing her young? Does a chimpanzee remember, from time to time, what it was like as a child? There is no way to know, of course, since there is no empirical test for consciousness, and we cannot ask these creatures.

Our Genus

Is it unreasonable to think that consciousness per se—as most people conceive of "it"—arose suddenly and distinctly in modern humans? Did human consciousness, whatever that is, evolve only as *Homo sapiens* came into being? Or can we speculate that it "existed" in earlier *Homo* species or human ancestors? Is it reasonable to assume consciousness for other more "modern" species, e.g., Neanderthals? They certainly solved problems in ways not dissimilar to us. And what about our earlier ancestors? Was *Homo habilis* conscious? What about *Homo erectus*? Some scientists think so. Torrey observes, "*Homo erectus* was also self-aware. . . . [and it] was the most successful hominin species that ever inhabited this planet, surviving for approximately 15 times longer than our own species has so far survived." (519) Then there are some older ancestors like *Australopithecus* or the Toumaï (*Sahelanthropus tchadensis*). Then there are the other extant apes—chimps, orangutans, etc. You get the picture. Might we attribute consciousness to them? They certainly behave in ways that suggest it. After all, when you see a chimp mother gently caressing her baby, you have to work hard not to infer some sort of conscious experience for her.

Intelligence

Is some sort of intelligence a prerequisite for consciousness? Most would not equate the terms. As I've pointed out, intelligence is not consciousness. Nevertheless, we do assume, rightly or wrongly, that the more intelligent a creature is, the more likely it is to be conscious. So let's examine whether there is any logic to the notion that consciousness and intelligence are somehow related.

How you define intelligence determines how you measure it. We tend to "assess" intelligence by human standards. And the standards are pretty variable. Intelligence being a function of performance on an intelligence test, often cited, is a pretty narrow and circular idea. In the most general sense, intelligence is the ability to solve problems. And consider that the problems other animals need to solve are very different from most of ours. The famed primatologist Franz de Waal argues that we should view animal intelligence based on the ability of organisms to solve problems *they* need to solve in order to adapt and survive, what he terms "evolutionary cognition." (131) Mice are not very good at counting. Horses are not very dexterous. Gorillas wouldn't do very well on a vocabulary test (though we do know that some have learned to sign a great many words). Most of us humans wouldn't have the requisite intelligence to survive in the wild for very long. We usually consider "higher organisms" more intelligent than "lower" ones (in point of fact, there is really no good biological/ evolutionary reason for the term "higher organism." We tend to use the term as being roughly synonymous with "intelligence," though the meaning of that term is generally couched in human standards.) Strictly used, intelligence implies the ability of an organism to behaviorally adapt to its environment and solve the problems that it needs to. In that sense, it may have valid usage only within, not between, species.

Yet, several nonhuman species are capable of solving complex problems—orcas, elephants, crows, ravens and other birds, octopi, and more—even though their problem-solving mechanisms may be somewhat different from ours. A couple of examples: In order to get at some food, an octopus can hold a jar and open it (and it can even open a jar from the inside), and it has a number of other remarkable talents. It can use tools. It can solve certain puzzles. It's been known to escape its tank, steal fish from another one, and return to its own and re-close the lid. (Remember, none of these things exist in an octopus' natural environment.) They also "play" and have "personalities." (332) They can even learn to solve problems by observation. We're rightly tempted to infer that the octopus "reasons things out," suggesting both consciousness and advanced intelligence, as we typically define those. But some are uncomfortable using these adjectives with an octopus (though not all of us), which has only about half a billion neurons and a "brain" mostly distributed around multiple sites in its body. And then there are birds. Very complex problem solving has been observed in some species of birds, like parrots and corvids (ravens, magpies, and crows, for example). Actually, ravens seem to show comparable intelligence to chimpanzees and orangutans. (403) If intelligence somehow "grows out of" consciousness, then birds and octopuses are clearly conscious too.

We tend to look at intelligence in other animals in a rather unified way. That is, each member of a given species seems to manifest a similar level of intelligence. Beavers all go about their work in a similar fashion. When bees adapt their signaling for location of nectar, they do it in similar ways. Thus, we're inclined to attribute their problem-solving capacities to some innate character or trait, not to some conscious intellect (though we may be quite wrong about that). Nevertheless, with some creatures, we see some variability in how problems get solved. For example, some tribes of chimpanzees solve food acquisition differently from others; thus, we consider the species to have some advanced intelligence. Intelligence, so to speak, is variable among humans, at least for the capacity to solve problems that we as a species seem to value. Such variability may have assisted some *Homo* groups to outcompete or out-survive others; but such variability does not translate to some sort of consciousness which we might possess and which chimpanzees, beavers, and bees do not. All animals learn to solve certain problems relative to their needs, i.e., they are intelligent in such contexts. To assume there is some particular "level" of intelligence along the evolutionary spectrum which requires consciousness would be absurd reasoning.

Self-awareness

Most scientists consider self-awareness as perhaps the critical component of consciousness. Many animals do seem to be aware of themselves. It is not necessarily unique to humans, and by extension, pre-humans, etc. Consider various nonhuman animals. Several of them can recognize themselves in a mirror as a distinct and unique individual different from their kin, a classic method used to assess self-recognition in humans and animals. Many species can pass this "mirror test"—elephants, chimpanzees, orangutans, bottlenose dolphins, orcas, even some corvids, e.g., certain magpies. There is even demonstration that a particular fish species, e.g., the cleaner fish, can pass the mirror test. (275) If we are "self-aware," how does one rule that out for other species? We look at this as somewhat remarkable, since each member of those species looks very much the same to us. Yet, many of them do recognize themselves as unique individuals. Many would claim self-awareness for their dogs, even though we can't ask them about it, but dogs don't pass the mirror test. There are even some suggestions that some ants, believe it or not, may have some ability to be aware of themselves as opposed to others of their cohort (73), even though few would ascribe conscious self-awareness to them. Humans don't truly recognize themselves in a mirror until at least 18-24 months of age (10), though there has been some variability in the literature on this. We might ask if such infants and toddlers are "self-conscious" before that time. Most parents would certainly say "yes."

Do the self-recognizing animals "have consciousness" and other creatures not have it? Perhaps the notion of self-awareness is best seen as being on a continuum, across species and across time. Are some plants conscious, as Chalmers would have us believe? It's unlikely we'd attribute that capacity to a rose. Some would argue that the idea of self-awareness is not exactly the same thing as consciousness, though it does seem to be a critical piece. All this, again, casts some doubt on the notion of consciousness itself.

Behavior

We cannot discount that other animals exhibit behaviors which we consider to reflect consciousness, whatever that is. Actually, we tend to ascribe consciousness to other organisms to the degree that their behavior resembles our own—in social interaction, complex problem solving, and the ability to delay behavior in pursuit of an outcome (presumably reflective of advanced thinking). For example, we think of human conscious intent when it comes to behavior such as the ability to "delay gratification." Some animals can pass what's become known as the "marshmallow test," bypassing a reward for success on a task for a delayed, more favored reward later. The marshmallow test experiments, done on young children, were designed to assess ability to wait for a better treat rather than consuming one quickly (It was supposed to predict later developmental intelligence, but it has been widely criticized based on socioeconomic and other design flaws. (365) Nevertheless, it is the often-quoted model for investigating the idea of delayed gratification, a capacity we once thought was purely human.). We know that crows can pass the marshmallow test. (502) In fact, it's now been found that even the cuttlefish, a mollusk, can pass the marshmallow test, sometimes delaying gratification even up to two minutes, but of course for other "treats," not marshmallows. (450) Is this consciousness?

We also attribute consciousness to other creatures to the degree we recognize "human" behavior in them, i.e., we anthropomorphize. We like to see "apparent" emotion, as chimpanzees clearly manifest. We like to see friend behavior, as our dogs are "best" at. We like to see complex "caring" behavior, as dolphins manifest with their calves and other pod members. We would tend to attribute consciousness to a gorilla; we might not to a fish; and we would certainly not to a fern. We "see" consciousness in primates and other "higher" mammals, for example. And there is evidence that the constellation of behaviors in great apes suggests that they are "consciousness." (282) Of course, this is not a "validation" of consciousness —any more than for me and you—merely assumed correlates of behavior and presumed cognition that we tend to equate with the idea of consciousness. Perhaps there are degrees of consciousness in various organisms. Even Darwin argued for a continuity of such mental experiences between animals and man in *The Descent of Man*.

When you consider the totality of an individual's behaviors, you get an "impression" of consciousness. We might even start to say "who" instead of "what" or "it" when describing an organism, much as we do with people, as we do sometimes with the other great apes. That is, we start to infer what's called "personality." Humans act in certain ways, e.g., being extroverted or introverted, acting aggressively or passively, being contemplative or impulsive, etc. However, there is some evidence that animals *within* a given species do the same. We're familiar with the pattern that some members of a group are dominant and others submissive. But it seems to go further than that, with many animals seemingly manifesting individual "personalities." That is, behaviors within a group are sufficiently variable that particular characteristics can be assigned to the individuals. In studies of dolphins and, most recently, in both chimpanzees (6) and orcas (522), patterns of behavior have been found to the degree that certain personality traits have been classified in them—such as extraversion, agreeableness, conscientiousness, neuroticism, and openness (there were also inferences of dominance in these species, as expected). Provocatively, this is the same basic cluster of personality traits that usually emerge in testing on humans (I'll have more to say about this subject in Chapter 10).[24] It is difficult to suppress the idea that so-called personality, and individual character, are absent in other, nonhuman animals —another blow to the notion of a fully unique human consciousness.

Is there self-knowledge in other animals, i.e., awareness of some personal agency? Consider, for example, the rather ingenious beavers. They sustain themselves and raise their young with a very complex set of behaviors. They fell trees, construct dams, and build incredible domiciles. You might assume that all these behaviors are fully innate, but that's not the point. Does the beaver "know what he is doing"; i.e., is he "aware?" Is there any change in planning and execution when things go wrong? That's what neuropsychologists call "executive functioning," the ability to solve problems and make decisions which affect more remote consequences. We might consider that the beaver is "aware" of what he's doing. But what about other creatures? Bees perform incredible dances and other communications with each other. We look at such activity as innate, though now it appears that some of it may actually be learned (377); but we generally don't consider them conscious. But maybe we only deny that out of some species arrogance. Perhaps beavers and bees do "know" what they're doing.

What about "consciously" taking care of yourself? Many animals actually medicate themselves for illnesses, in a process known by the fancy name "zoopharmacognosy." There is considerable evidence of this in many species, especially for removing parasites. (124) Much of this is believed to be innate. Cats, dogs, and deer eat certain plants to "treat" parasites or other diseases by inducing vomiting. Deer have also been observed to treat itching of their new antlers by rubbing them against the bark of willow trees, and we know that this material contains a precursor of the substance from which we get aspirin. But other species have come up with medical solutions which seem to be more "conscious," learned, or cultural. What about apes? A recent study found that some orangutans in Borneo chew up certain plants to make a topical analgesic, which actually acts as an anti-inflammatory agent, a rather intelligent and clever set of behaviors. And such behavior is not universal in orangutans, as it has been found only in a small geography, suggesting it is a learned behavior. (354) Would we deny the ape's intelligence and consciousness? Recently, chimpanzees in Gabon have been observed orally mashing up insects and then applying them to wounds—their own and others'. (329) So far, it's not apparent that chimpanzees do this as a species at large, so it does appear to be cultural. Could we deny that this suggests consciousness, intelligence, and perhaps empathy? When animals show planning, self-care, and care for others—again, empathy—it is difficult not to ascribe consciousness to them.

Many ravens and crows socially cooperate and solve complex problems. They take some time to consider alternatives and solutions, often cooperate with each other in the endeavor, and even communicate concepts to their young. (216, 532) Watching such behavior, one must conclude that they possess both intelligence and seeming self-awareness or consciousness. One could ask if this is any different than the criteria we apply to the notions of consciousness and intelligence in young children, or even ourselves. Perhaps this is a different "kind" of consciousness. But this takes us again down the rabbit hole of what such terms really mean.

As I noted in Chapter 4, many species engage in elaborate rituals, especially for the dead. Other *Homo* species also did this. Death, for the survivors, can include a powerful set of rituals for a number of other species; we cannot discount the fact that there may clearly be awareness and consciousness in these organisms, including the emotions of "sorrow" we know only too well. It is quite likely that many animals grieve, and related behavior has been observed in many species. (272) But we do not often grant these other intelligent creatures the emotions of sorrow or grief. Nevertheless, the observations that animals might be manifesting these emotions is difficult to separate from the notion of consciousness.

Imagination

One of the putative features of consciousness is imagination. Humans can certainly imagine—construct ideas and images and project them across time. If we assume a reality for consciousness, we must assume the ability of animals to recall themselves in the past and "imagine" themselves in the future, something we call *autonoesis*. If we can imagine ourselves next month or next decade, or next minute, we might also rightly assume that Neanderthals did too. We might reasonably assume that elephants and chimpanzees can imagine themselves a few hours or days ahead, maybe more. And squirrels might imagine themselves a few minutes or hours in the future—when and where should I create a false burial site to place my nuts to deceive others today? For elephants, to imagine a water source might project a great deal longer. Elephants do *remember*.

Another type of imagination comes out of dreams. If dreams imply some sort of consciousness, then we must also consider the fact of this happening in other creatures (and AI too, as I noted above). Most people make the likely reasonable inference that their cats and dogs have dreams. We can see them behaving, while they sleep, in ways that suggest certain actions to us. But what about any "lower" animals? For example, REM brain waves, indicative of dreaming, are observed in mollusks such as cuttlefish and octopuses. (233, 331) Not surprisingly, they also show dependence on complex learning and memory, all indications of what we consider consciousness in humans. Dreaming may even occur in birds, reptiles, and in some insects and arachnids. The capacity to generate images, whether waking or sleeping, can't be denied as a marker of consciousness. Thus, if consciousness exists, there is little to justify holding it only for ourselves.

Which Organisms?

Nevertheless, we tend to ascribe consciousness, like the theory of mind, to some *Homo* species and to a few other primates. You might think that comparison to more "primitive" organisms is unfounded, even absurd. But is it? Since there is no acceptable definition of consciousness, who are we to say what *it* is and who has it? Remember the worm *C. elegans*. Might we assume that its behavior is at least partly a function of consciousness? For either answer here—yes or no—you see the problem.

The parallel evolution of many traits across species belies the notion of correlating complex behavior with consciousness. Consider bees again, who have sophisticated social behavior, rudimentary learning, abstraction, mathematical abilities, and problem-solving capacities despite a rather minuscule neuronal anatomy. (276, 377) They are remarkable creatures. They sleep and perhaps dream, they can actually play (137), and they may be the only extant creatures other than humans who use very sophisticated symbolic communications. Consciousness?

Unless you consider modern man (whatever that is) to be the only conscious animal, itself a questionable proposition, it raises the question of where along the creature continuum you want to label something as "conscious." Is it primates, mammals, corvids, octopus, squirrels, worms, etc.? Some have suggested different "types" or "levels" of consciousness throughout the animal and other biologic kingdoms. And we recall Chalmers' photons. But this only illustrates the non-usefulness of the concept. It does not explain anything.

The search for the neurologic underpinnings of consciousness continues. If, in fact, this actually becomes a fruitful endeavor in the coming decades (or centuries), we should look at whether this suggests—or evinces—consciousness in other organisms. One wonders whether other animals with well-developed nervous systems, i.e., neocortices, have what we call consciousness. In this context, Torrey compares human brains to other creatures. We "have 25,000 to 30,000 neurons per cubic millimeter of brain cortex. By comparison, whales and elephants have only 6,000 to 7,000 neurons per cubic millimeter of brain cortex. . . . Glial cells also underwent evolutionary changes . . . [with] enhanced ability to integrate information." (519) If we assume such integration is necessary for consciousness, as some current neuroscience theories offer, then whales and elephants might be conscious. Most of us think they are. However, the density of brain tissue, level of integration, or even a neocortex may not be relevant to the question. In fact, in reviewing the extant literature, an international group of scientists from various neurological disciplines issued the following summary statement that concludes,

> The absence of a neocortex does not appear to preclude an organism from experiencing affective states. Convergent evidence indicates that non-human animals have the neuroanatomical, neurochemical, and neurophysiological substrates of conscious states along with the capacity to exhibit intentional behaviors. Consequently, the weight of evidence indicates that humans are not unique in possessing the neurological substrates that generate consciousness. Nonhuman animals, including all mammals and birds, and many other creatures, including octopuses, also possess these neurological substrates . . . . Furthermore, neural circuits supporting behavioral/electrophysiological states of attentiveness, sleep and decision making appear to have arisen in evolution as early as the invertebrate radiation, being evident in insects and cephalopod mollusks (e.g., octopus). (320)[25]

We should probably realize that the notion of humans as uniquely conscious is probably unfounded on the basis of neurology. You can't draw a line between creatures who are not considered conscious and those who are. Some have even envisioned a gradient along phylogenetic lines, with invertebrates, for example having a much lower probability of being conscious (but consider again the cuttlefish, which is a mollusk). The hedge some neuroscientists make is for "levels" of consciousness; but it's unclear what that means. We might assume monkeys, for example, have consciousness. Is that a different type or level of consciousness? They certainly act like people in many circumstances.[26] Let's leave for the moment Chalmers' idea that photons might have some consciousness. But do plants have consciousness? Some plants do communicate with others via chemical signaling, but few would imbue even remote ideas of "consciousness" to them. There are even some preliminary data suggesting that some plants "scream"—emit ultrasonic pulses—when stressed, e.g., tomato and tobacco plants, though the functionality of this remains uncertain. (270) Incredibly, even some plants can engage in associative learning—like Pavlov's dogs—to "decide" in what direction to grow. (117) (I'll have more to say about this in Chapter 10.) Does all this suggest a kind of consciousness? Most of us would say no, but again, can that be proven? No.

Conclusion

Perhaps there is nothing "special" about consciousness at all. Perhaps all living creatures with brains experience some type or level of consciousness. But does it require a brain? In that sense, we might consider consciousness on a continuum. Of course, the logic is difficult to extrapolate to fish, amoebas, trees, and photons. We're left with the questions of when, and in what species, some level or type of consciousness evolved. Does a plant "know" that it is growing toward the sunlight? Some would pose that this is a *reductio ad absurdum* argument since some organisms cannot possibly be conscious, but who knows?

We seem to reserve concepts of consciousness, "mind," free will, etc., to ourselves. It is a shame that we do not have any common language with orcas or elephants or certain other species, or we might have a very different view. Our concept of uniqueness in the animal world is one of unjustified hubris, if nothing else. Both across current species, and certainly with our ancestors, the properties of putative consciousness and awareness cannot be reasonably discounted. These characteristics exist across time and species.

Whenever we see an animal doing something we value, we do consider "what must they be thinking?" We "see" consciousness, intelligence, and reasoning. When a crow solves a very complex problem, when an elephant comforts a cohort, when a dolphin midwife assists her sister, when a chimp treats another's wounds, we do think we must have something substantial in common with those other creatures. We're not unique or special, and we attribute a good deal of self-awareness to them in those behaviors. When you look at your dog, do you generally infer some sort of consciousness? When you look at chimps hugging each other, the feeling that there's consciousness is hard to ignore.

It would be folly to assume that human "consciousness" makes us unique in the animal world. A recent interdisciplinary panel (including renowned experts in biology, ethology, psychology, neurosciences, philosophy, and others) concluded, "The empirical evidence indicates at least a realistic possibility of conscious experience in all vertebrates (including reptiles, amphibians, and fishes) and many invertebrates (including, at minimum, cephalopod mollusks, decapod crustaceans, and insects)." (372) The notion that a god created humans and gave only them the singular ability to think and experience—including things like love, hate, fear, and even reverence, which is analogous to what some call worship—makes no sense. The tired argument that our human consciousness (other than the information we process) is qualitatively different from what any other animal experiences has no basis. Nothing about our consciousness or putative sentience makes us special in the animal world across species and time. *The Copernican Principle of human specialness takes another hit.*

Religious Experience

Now let's examine the intangible idea of consciousness as it relates to "religious experiences," something that believers often rely on. If you can't prove there's a God, and if you haven't actually witnessed any miracles (no one has), at least you *know* there's a god—you can feel Him, you can be moved by Him, you just "know" in your heart that He exists. You may at least "feel" very differently when you imagine or consider God or any of His supposed powers. There are many types of what people call "religious experience." And they often use such subjective experience as affirmation of God. This is not evidence, of course, but it is regularly cited as such. Many people claim they have an awareness of God, or "experience" Him in some way. Stated another way, people report that they are conscious of God. The only "truth" which arises here is your feeling, of course. We wouldn't deny that such a feeling exists for you; but it would be ludicrous to assume that translates into any validity of some supernatural reality.

Perception

Experience is unique to each of us and unknowable to others. All we know about others' experience is what is said or otherwise communicated. What we experience is sometimes referred to as perception, and each of us perceives external events in our own way, based on our unique capacities and histories. We can see a sunset or hear a symphony. We can be moved by the beauty and grandeur of the universe, or a dance, or a flower. But it is different for every person.

If we hear God or "feel" His presence, those are perceptions. No one can doubt another's experience or perceptions, though they may not translate into any valid external reality. The faithful report that they perceive God. Believers often report that God "reveals" Himself to them in various ways. But perceptions, of course, can be deceiving. When someone reports a so-called revelation, for example, we can infer one or more realities. The revelation is either (1) actually made up or fabricated, or (2) there is some misperception of certain realities or objects, or (3) there are actual hallucinations, or (4) the report is an outright delusion.

Hallucinations

Our brains can play tricks on us. For example, we can see or hear things that just don't exist in the real world. These are hallucinations. They can be experienced by persons with certain severe mental illnesses, but such are far from exclusive instances. People hallucinate for a variety of reasons. It's not just a problem of schizophrenia or other psychosis. Hallucinations are more common than are generally acknowledged. In fact, likely 5 to 10% of "normal" populations have experienced auditory hallucinations (without any drug induction). (295) They can occur sometimes in "seeing" a departed loved one. They sometimes appear with dementias. Serious pain or agony can enable hallucinations, and they can be disturbing or frightening. There are hypnogogic or hypnopompic experiences which seem real—images seen or heard in the transition to sleep or waking. Extreme fatigue will do it. So will sensory or sleep deprivation. Elderly people hospitalized for a long time can experience hallucinations. And of course, hallucinations can be induced by various drugs.

We understand that such "perceptions" may be subjectively real to the perceiver; but they do not reflect some external reality. We dismiss them out of hand or ascribe them to a "misfiring" brain. As hard as we may try to convince the subject that he does not "see" or hear what he sees or hears, reality for him is different. He may later admit that the hallucinations didn't reflect reality, but they were certainly a reality at the time he had them. What we call hallucinations *are* real to the experiencer.

Some people's religious experiences are actually hallucinations. Many people have reported visual perceptions of religious figures or experiences of God actually speaking to them. Where the subject of religiously associated images or communications is concerned, we are asked to make an important discrimination. Only some of us can see Jesus in a slice of toast, which is often just an interesting visual perception. But if Jesus then speaks to the perceiver, we're forced to make an assessment. Is this just a fleeting idea, is it a brief hallucination, or is it a substantive and serious communication as perceived by the individual? The latter is often consistent with a severe mental illness, and we sometimes psychiatrically hospitalize people who repeatedly insist on the latter. For context, divine "communications" usually don't result in dangerous behavior—unless, of course, someone uses God's alleged word to substantiate a religion and take hostile or violent action against its nonbelievers. Has this ever happened in human history? *Who* God talks to is also relevant. Given that no "miracles" have ever been verified, it is curious how some perceivers of divine words are considered prophets inspired by some supernatural being, and some are just "crazy." This just becomes an issue of status, timing, leadership, and other conditions of the culture in which it may occur. But when an individual hears the voice of God and exhibits some other unusual behavior, e.g., returning the communication, we might initiate some type of intervention. Why the discrepancy? Context is everything. Taking God's word to "speak in tongues" is okay if it's in the right setting, but doing this in other contexts will result in other humans actively avoiding you.

Delusions

If someone appeared, perhaps on TV, and announced that he was Jesus Christ returned, what would people think? This is what Christians have been eagerly waiting for! But he would be rejected anyway—thus the ludicrous nature of such beliefs. In any case, 'Jesus' would likely be hospitalized or put on some antipsychotic medication (unfortunately, his delusion would persist; he would just be less anxious about people not believing in him).

People can believe all sorts of things. Some people believe they have been abducted by interstellar aliens. Others believe that the government or a technology company has planted chips in their brains. Others believe that

"God told me to run for office." We might cringe a bit at that, but I would put observers of that behavior into four different groups—those who accept the comment at face value, those who see it metaphorically as some sort of inspiration, those who see it as mere political leverage, and those who snicker. However, with such statements as "God told me to jump off the bridge" or "I am Jesus returned", we tend to hospitalize such people for their own protection. They suffer from a *delusion*, i.e., endorsing a different reality from that perceived by the rest of us. Many people are under a delusion of the existence of God.[27] But we do question people who claim to be in direct communication with God or that God has directed them to do something. If God tells someone to always wear white, we pretty much ignore it. If they assert that God commanded them to kill their children, we're less tolerant, as this is likely a delusion, something which has, in fact, happened (we'll get into the actual issues of responsibility and some "abnormal" mental states later). Context is everything.

We also tend to ignore endorsement of God while rejecting other fantasies. If God is "real" to the otherwise sane believer, we are inclined to accept the experience as a valid one. There are other fantasies, of course. We expect children to outgrow their perfectly valid experience of believing in Santa Claus. Who are we to say that such is not a valid experience for a child? Yet, it is no more fanciful that the one about God —period. We're less tolerant when adults endorse fairies, or leprechauns, or other human-created fantasies. Delusions are such because they're disconnected from reality, and this applies to God, angels, fairies, Santa Claus, and Mother Goose.

God in the Brain

The "experience" of a god is clearly created in the human brain. The reasons for this are many, but it is specifically the action of our neurons that lead to such religious experiences. Kaku even cites the example of a split-brain individual, a patient who has had the two cerebral hemispheres of the brain surgically disconnected from each other (such surgeries have been occasionally performed to treat otherwise intractable epilepsy). The result is that the two hemispheres of the brain *do* show very different capabilities and behaviors, including generating responses when stimuli or questions are visually presented to one or the other hemisphere by some experimental procedure. The subject claimed he was an atheist, but when a response was provoked from only his right brain, he claimed that he was a believer. (261) It was a compelling example of how the brain actually creates or denies a god.

Some argue that we are biologically primed to experience God; that our brains have been specifically prepared by God to experience Him. A related phenomenon to this is that many people with temporal lobe epilepsy sometimes have "hyper-religious" experiences or even report the intense feeling of a personal god during seizures. Building on such reports, experiments have been conducted attempting to artificially simulate those experiences. Psychologist Michael Persinger claimed some success in promoting the so-called "god experience" by stimulating the brain with a weak magnetic field. (400) Others have conducted similar experiments, though success is questionable. Nevertheless, the procedure became known as the "God helmet." Several years ago, neuroscientist Andrew Newberg and colleagues at the University of Pennsylvania summarized the literature on this neuroscience and the idea of "god in the brain," i.e., that God was represented in our brains, or that we were specially programmed or evolved to "experience" God. (368) Although the authors devoted a great deal of text to mystical, religious, and spiritual matters, the inescapable conclusion was that there is no way to discriminate such a "god experience" from other "mystical" experiences, or even more common ones. Whether such experiences can be labeled delusions, hallucinations, or merely thoughts is not of much importance here. Neuroscientist David Sharp concludes, in his review of the work, that for a unitary supernatural being, "The case is not yet proved at the levels of neuropsychology and imaging," an understatement if there ever was one. (464) The "search" seems to have died out in recent years, though occasional studies appear. Neuroscientists Mario Beauregard and Vincent Paquette used fMRI in an attempt to reveal neurological patterns in the brains of nuns while they reported being in a state of "union with God." (39) Not surprisingly, no particularly unique pattern emerged. Although encouraging further studies in the field, medical researchers Newberg and Lee essentially admit to the obvious when it comes to a god. They assert, "That a neurobiological correlate exists does not specifically refer to the causal mechanism of such phenomena. That is, if the brain activity changes during a mystical communion with God, it is not clear whether the brain activity caused that experience or responded to that experience." (369) In addition, other studies in neuroscience have not been able to demonstrate that *any* "belief" and "disbelief" are differentially located in the human brain, whether such are relevant to religious thinking or more mundane matters. (209) Assuming the existence of a god, of course, negates the whole investigative enterprise; and the implication that it is possible to have an "experience" without brain activity is rather ludicrous. It does not matter that you may have a "spiritual" experience that may suggest a god to you. Your consciousness is an epiphenomenon correlated with activity in the brain. It is irrelevant to the question of the existence of any supernatural entity. Being "conscious" of something doesn't confer any external reality to it. The notion of finding God in the brain or that humans are somehow hard-wired to communicate with It should obviously be cast into the dustbin of experimental history.

Conclusion

As we've seen, the idea of consciousness is not definable. We don't know what "it" is and even suffer from the idea that we just can't define "it." I cannot prove that I am conscious, or that you are in reading this book. All of this comes down to a philosophical question. Does consciousness exist as a discoverable reality, or do we use that word merely to communicate something or "try" to explain something which is otherwise opaque to us? If consciousness does exist as such, "it" will always be inaccessible until it is defined.

Theories of consciousness, principally how "it" putatively "emerges," have offered no insight. That we may employ a word to summarize a pattern of neural circuitry and operations doesn't make it real or useful—though no such meaningful pattern has obviously ever been "discovered." The only possible conclusion is that consciousness is a word we use to describe our biological information processing that we cannot actually describe adequately with present measurement tools. Perhaps our subjective experience *is* merely the activity of our neurons—and nothing more, excepting those who think a "mind" or some other intangible spirit exists outside the laws of nature. Despite provocative theories, the search for how consciousness emerges from the activity of our neurons hasn't yielded any actual results, as philosophers have predicted. (303)

Asking the question "what is consciousness?" is almost the same as asking "does consciousness exist?" The better question is, "Why do we use the fabricated idea of consciousness to explain things that have not yet been otherwise explained?" Of course, there is a long history of this type of reasoning in human behavior, including in science. There is nothing wrong with using provocative ideas to stimulate investigation and research, of course. Einstein was renowned for such. But "theoretical" concepts, used over and over again, often take on palpable qualities that are unjustified, without *any* evidence. Thus, trying to develop a theory about this idea may always reach a dead end. You may criticize all this as a *reducto ad absurdum* argument. However, that is where we are. In the end, invoking the idea of consciousness provides no real advantage in explaining human behavior or ideas like mind or worship or free will, which I'll address in Chapter 6.

On Behavior

Our veneration of consciousness may be a bit premature. The notion of consciousness is irrelevant to the generation of human behavior. It is, at best, "merely" a byproduct of the brain's activity. We are left with the idea that—at least so far—the notion of consciousness offers us no further explanatory power for human behavior than labeling sleep as being "unconscious."

It is not necessary to invoke the notion of consciousness, as William

James famously asserted, to explain our actions and "choices," which operate according to the controlling laws of behavior (Chapter 10). You may never be able to accept this any more than you will endorse that free will is an illusion. But our personal, subjective impressions are not science and do not demonstrate the validity of either consciousness or free will, as I'll discuss in the next chapter.

When you get down to it, the idea of consciousness is a "psychology of the gaps." It is another concept presumably used to explain human behavior. Unfortunately, it does not, and, by its subjective nature, cannot. We must accept the fact that our behavior does not depend on the notion of consciousness and that it is difficult to imbue "conscious" motivation as the source of our action, whether it is decisions, motor behavior, problem solving, or *belief in some sort of deity*.

Copernican Principle

The pursuit of human consciousness is, in another way, a search for how humans are special, in defiance of the Copernican Principle, which has extinguished our specialness in every other way. There is no evidence that consciousness, if it exists, is a feature unique to human beings. We cannot consider ourselves special creatures in this regard.

We do tend to resist ascribing consciousness to other creatures, reserving it for ourselves; and we do this, in part, because it makes us feel superior or special, which we are not. We do ascribe some sort of awareness to other living things, though only to the degree that they seem to act *like us*. Some of them "seem" aware and sentient. If that's the case, should we be slaughtering millions of creatures who may have such consciousness, and do show some indications that they do—cows, pigs, whales, etc.? Does consciousness confer some sort of personal rights? We seldom slaughter apes anymore (though this does occur in some parts of the world). We could even ask the question, "If apes are conscious, in fact, might they be considered 'persons'?" Should they be granted human rights?[28]

We may also ask the question of whether consciousness, or at least any pattern of neural activity "correlated" with it, confers an evolutionary advantage to humans (or to certain other creatures). We just don't know. Thus far, sophisticated intelligence clearly has. But its value over time may be quite limited. Most readers of this will bristle acutely at such an idea, but human history is still pretty short.

If we accept Chalmers' view that everything in the entire universe has some sort of consciousness, it again loses all meaning. In that context, if there are gradations or types of consciousness, as opposed to that of a dog or an oak tree or a photon, we're certainly not that special. This again leads us to the notion of artificially intelligent "beings." Since consciousness cannot be ruled out in such machines, now or in the foreseeable future, we really cannot consider ourselves fully unique because of the fact that we're biological creatures with some sort of self-awareness or sentience.

On God

Which brings us to the point. To use the notion of consciousness to affirm the existence of a god is wholly unjustified. Being "conscious" of God or "knowing" there is a God or having "faith" that God exists is not evidence or an explanation of anything. We cannot rely on the notion of consciousness with respect to a supernatural being any more than for any other human behavior. Humans do not believe in a god because we've consciously decided this. Consciousness cannot explain our behavior; and therefore, belief in gods, faith behavior, and other reports of so-called god experiences have nothing to do with whether any god exists. When people endorse the idea of a supernatural being, they are merely reporting a fantasy created by a dynamic human brain. No "conscious" idea is evidence of anything—but merely a figment of imagination. We'll discuss more about how such fantasies come about in later chapters. For now, the bottom line is that consciousness provides no insight into theistic religion. In sum, the "conscious experience" of God is not an indication or evidence of any supernatural being. It is literally nonsense.

6

NOT SO FREE WILL

All theory is against the freedom of the will. —Samuel Johnson

Human beings in their thinking, feeling and acting are not free but are as causally bound as the stars in their motions. —Albert Einstein[1]

Two roads diverged in a wood, and I—I took the one less traveled by, And that has made all the difference. —Robert Frost (The Road Not Taken)

Robert Frost summarizes what we all feel. We have choices. Shall I turn right or left? Do I say, "I love you?" Can I choose one future or another? Nothing seems more important than our choices, and we see our decisions as matters fully of our own volition, our own "free will." We all "feel" we have free will. And our decisions, however unimportant or life-changing, seem to just arise quickly and spontaneously in our minds. As the poet T. S. Eliot observed,

In a minute there is time
For decisions and revisions which a minute will reverse.
. . . Do I dare to eat a peach?
(The Love Song of J. Alfred Prufrock)

I have that feeling; so did Frost and Eliot; and so do you in reading this. There is no escaping it. However, we'll see here that, in the most basic sense, free will is really just an illusion. If you're taken aback by that statement, and most people are, read on. *This issue is a critical one in considering the hypothesis of whether some personal god exists.* In order to examine this idea, we'll have to consider the various arguments about free will and how it relates to the God problem.

If we live in a determined universe, there is no free will. That is, our behavior and everything else in the universe is explainable (or potentially so) with the mathematics of natural law. And if you believe in free will, the only conclusion is that you believe it comes from a supernatural being in a universe which is *not* fully determined. That is, the laws of nature might apply to most events in the world but not your ability to freely make some decisions that are totally independent of them, including belief in God. *Without "free will," the concept of a supernatural being that reigns over the world becomes meaningless.* If we are able to make "free" decisions—ones not fully controlled by our physical and biological histories—then we have the "choice" as to whether or not we embrace a god meme, as well as the capacity to make a myriad of other important and unimportant decisions in our lives. If free will does not exist, then there is no actual choice, and "belief" is merely another learned superstition. In fact, one of theologians' prime rationales cited for the existence of God is that humans *do* have free will—and that this character must have come *from* God. We must apparently choose between good and evil, decide whom to worship, and decide to follow the dogma of one or another particular religion. Since free will cannot possibly be explained by science (by definition) and, according to the theological argument, since it exists, it must come from somewhere, then it must be granted to us by God. This is the classic argument in a nutshell.

When we decide to do something, to make a decision, or have an idea, we believe that we are acting fully of our own volition. This is what is commonly called "free will," and it is usually ascribed to one's "mind," a nonphysical structure that is the seat of our being, our identity, our "soul," as Descartes and many others have conceived of it. Of course, these are abstract and untestable ideas that do not bear on the actual question of whether we can "freely" initiate things separate from the chemical history of our brains, which have been shaped by genetics for millions of years, as well as our personal histories since conception. In this chapter, I will show: how this view is an illusion, why our behavior is not technically "free," why the idea of "free will" is in conflict with a scientific understanding of human behavior, how the notion of free will may have naturally evolved, some reasons why the illusion has persisted, and yet why there is still no reason for dismay about this.

Most serious scientists appreciate that human beings, like all creatures great and small, behave according to the laws of nature and that no external or non-material capability—such as free will—exists. We continue to conduct our lives, however, with the illusion that it does. The crux of the argument, for most people, is that since they "experience" choice, there *must* be free will. Of course, just because you "feel" something doesn't make it valid, although the feeling itself may be, as we discussed for consciousness. Distinguishing reality from perception is the key to appreciating that the notion of free will is just an illusion. You don't have free will just because you think you do.

We humans have a natural antipathy to the idea that "free will" doesn't exist. The human motivation, if you will, for endorsing the idea of free will is that there is a natural aversion to the notion of determinism. It is just psychologically unacceptable for us to embrace the idea that all the events in the universe from the beginning of time—including our thoughts and actions—are determined by the laws of physics. It just "seems" impossible. Further, if there is no free will, how can we be expected to make "decisions" and have any control of our own actions? It just seems subjectively impossible that this is the case. We fervently "believe" in our ability to act fully outside the constraints of our biology and circumstances. To deny this is admittedly a psychologically uncomfortable idea, but I hope this is clarified and softened a bit by what follows.

Respect for science demands that concepts like "free will" are not valid. By definition, this is an idea whose character is not, even potentially, able to be examined by observation and scientific method. But a great deal depends on what you actually mean by "free will." If you mean that you think and behave in ways unrelated to your biology and experience, I hope you'll be dissuaded from that notion here. If by free will, you mean that it is "you" who makes decisions and you are aware of making them, there are some good things to say about that, and I'll address this further.

You might assume that in writing this book, I am actually illustrating the validity of the "free will" idea. It may not exist as such; but it is still part of our lexicon. After all, with each word, phrase, sentence, idea, etc., that you encounter, you may believe that a choice has been put on the table and into the computer. If I say, "now consider that . . . ," you may or may not do so. If I say, "I reject that notion," you may infer a voluntary decision on my part. If I state things a certain way, you may not agree—*your* decision. But all this is a limitation of our brains and our language. We tend to communicate in ways that assume a purely voluntary choice on the part of a writer or speaker and on the part of a reader or listener, i.e., some notion of free will. We say that "I want to do something or other . . . " or "I chose a different path . . . " or "I refuse to . . .", etc. However, this is just the way we talk. It does not mean that free will actually exists. Without belaboring the history of language and grammar, this is just the way we think and communicate. It is also a reflection of how dearly we hold onto the concept of free will. But set that awkwardness aside as you go through these arguments.

Many other authors have eloquently addressed this problem, and I would refer you to them for some additional, provocative perspectives. (121, 208, 445) This is one of the oldest debates in the history of human thought; and I'll admit a tinge of hubris even to tackle the question. But here goes. One cannot prove the existence of "free will." And if free will as such does not exist, if it is an illusion, what's responsible for this firmly held belief? We may not have the free will we think we do, as I'll demonstrate in this chapter. But it does go to the heart of the matter: *Free will is an illusion, as is the God that it implies.* If you "choose," read on.

History

The debate on free will extends back in history to the ancient Greeks (and perhaps even before that). There have been many attempts by philosophers and theologians throughout history to rescue free will from a determined universe. The ancient Greeks subscribed to the notion that natural law was the guiding force of human actions. Most thinkers of the time were atomists, believing that all matter and persons were composed of atoms, as Democritus had suggested, which obeyed natural law. But these laws were actually the powers of the various Greek gods, who, in their often-arbitrary wisdom, directly influenced human actions. However, they also allowed for *some* independent thinking. There was room for humans to have some choices. Aristotle was a proponent of this position. People were allowed to make some decisions and not others. We could freely act in some ways, some of which might lead to punishment (which also led to the notions of moral responsibility, which I'll discuss later). However, our destinies would be up to the gods: for example, the three Fates of Greek mythology, controlling birth, the overall course of life, and death. That human destiny was controlled by the gods and that, yet, we could make some decisions freely was certainly a major conundrum. It was never resolved by the Greeks, and as we'll see, this confusing dilemma has been the bane of theologians and philosophers ever since.

Belief in Free Will

We humans believe that we have free will, i.e., the ability and capacity to make choices and engage in thinking and behavior fully independent of the physical influences which have acted upon us—from genetic history to physical exposures (from congenital to present) to developmental history to the lifelong histories which shape the contingencies acting on our thinking and behavior. Whether in rudimentary behavior or "deliberate" actions, we believe that we can think and act freely, avoiding or ignoring anything that has influenced us in the past. In a determined universe, we still believe that we can act in ways that are not controlled by the laws of physics, which has led most scientists to consider free will to be an illusion.

Simply stated, people believe in free will because it suits them. The "belief" in free will—illusory as it may be—is actually conditioned in us. We grow from early stages of life to believe that we have free choice in our thinking and behavior. We resent any idea that we don't have free will. That we are conditioned to think and behave in certain ways is not a palatable idea, as I'll discuss in Chapter 10. Rejecting the notion of such conditioning leads to the vehemently endorsed idea that we freely make decisions. It is a "belief" which is difficult—if not impossible—to shed. Nevertheless, the belief is likely to have a vital function, as I'll discuss further, and it remains incredibly strong despite the fact that it is really an illusion.

This may be a bit difficult to appreciate, but it's pretty much the same as other beliefs we develop in life. We believe in all sorts of things. We believe in many things we are not able to verify. We believe in gods; we believe in a heaven, a hell, or some kind of afterlife (depending on what culture you grow up in). Some of us believe in a variety of "miracles;" some believe that we carry the memory of past lives; and many of us believe in angels, the Tooth Fairy, and Santa Claus. We cannot verify such things, but the beliefs hold strong, both for individuals and as cultural memes which may have existed for millennia.

We are conditioned to believe in free will, as we'll see. But the belief is somewhat variable. Many experiments have been done on the various factors which contribute our belief in free will. I'll not get into the weeds just yet; but there are a number of factors. People tend to believe in free will the more options they have in their lives, the more they reflect on their choices, the higher they rate their ability to make choices, how much deliberation or forethought goes into a decision or choice, whether making choices they consider moral, and their perceived ability to make choices independent of external pressures. (158, 352, 507) I'll expound on these concepts below.

Let's look at the idea of free will in a bit more detail. The common understanding of this is that we can think, choose, and act as we wish or "want." That is, we can do whatever we want, with "want" being something that spontaneously arises within us. Our actions are completely self-initiated. What we "want" is not determined by our histories or any laws of nature. When you exert your "will," you're actually referring to moving something forward, to causing something to happen, what psychologists sometimes refer to as motivation. We talk about "willpower" or "strength of will" as though it were a commodity. Our internal motivations are sacred. We revere "will" as an ultimate human motivation, e.g., as in our "last will and testament." But when you get down to it, motivation is just shorthand for a complex set of behaviors or functions in the brain, without regard to how they got there. To exert your will boils down only to suggest that there are actions that are or can be forthcoming. And what about "free?" Free of what? That's where the rub comes in. Do you mean free of the immediate circumstance, your behavioral history, your decision-making abilities, your genetic propensities, or everything that has ever happened to you? Most people consider free will to be the latter. Our thoughts, choices, and behavior are "free" of any physical influence and history whatsoever. I can "choose" one road or the other; to turn right or left, to eat a peach or not. It is a completely "free" decision.

From a scientific perspective, an operational definition of "free will" is actually *not* possible. That is, we cannot come up with any ways that we can measure "it." That there is incomplete understanding (massively so) of the neural complexity of outputs in the brain and the physics which subserve them actually does not translate into a possibility of free will. Consider this: If one proposes an "operational definition" of free will, it therefore does not exist. That is, if we could measure "it," it would cease to be independent of the laws of physics. The common understanding of free will—that it is independent of the laws of behavior and physics—cannot be addressed empirically—by definition.

If by "free will" you mean initiating an action or thought which is "free" of what you currently understand or can identify about your own life, then, yes, you have free will. But that is not what is normally considered free will by most philosophers, theologians, and lay people. The effort to squeeze free will out of biology does remain logically impossible. There have always been attempts to rescue free will from a determined universe; and we'll discuss some of them; but these have been fruitless. Can you act freely—outside the laws of nature? Of course, this is logically absurd unless you conceive of "degrees" of free will. Not really a valid argument. It's reminiscent of the ancient Greeks, who were allowed only a bit of free will by the gods, who determined everything else. If you endorse a universe obeying natural laws, what you're implying with the idea of "free will" is actually a miracle.

God and Religion

Which leads me to the God notion. Since free will is a pivotal topic for theistic religion, we have no "choice" but to address it here. (Do bear with me. As you can already see, language itself is a significant constraint in this discussion, as it is throughout the book). If there is no free will, then the issue of whether we are "free" to believe in a god or not is irrelevant. And if this question is irrelevant, there is no god, at least in the sense of a personal God such as the one espoused by the Abrahamic religions. And if free will actually exists, some particularly awkward problems arise, as we'll see.

Where Does It Come From?

Either free will exists or it does not. If it does not, you are not "free" to believe in God (or not), and then the notion of a personal deity makes no sense. If you do accept the idea of free will, you must account for where it comes from. And if we can make decisions and emit behavior totally free of our complete genetic, environmental, and behavioral history, i.e., in a determined universe, from whence does this ability arise? It *must* come from some *super*natural source—by definition, and of course, the only source cited by theologians is God.[2]

When pressed, the theist's answer is always that God *lent* free will to humans. The answer of "loaning" or "borrowing," of course, is not actually an answer. Are we supposed to give it back at some point? Would God have lent free will to humans just to see what happened—since our actions would be temporarily "free" of His influence? That is a rather cynical view of a presumably omniscient (see below) supernatural being and makes no sense. Then there is the Christian doctrine of "predestination," i.e., that God has foreordained, if you will, what will actually happen to all humans—our actions, and particularly whether any are "saved." Again, this is incompatible with free will, which He is supposed to have lent us. If one's ultimate fate is pre-destined, choices have little meaning. Of course, Catholic doctrine, for example, allows that you can still have free will so that you can choose between good and evil (though there is still some controversy about Hell, which I'll discuss in Chapter 7), which is also incompatible with the notion of predestination. Predestination and being "free" to make choices on "moral" issues are inherently incompatible ideas, even within religious doctrines. All this makes the whole theological notion of free will meaningless.

Choice and Belief

We might consider—Can we choose to believe or not, to choose one or another god, to follow one religion or another? We could also ask—If there is a God responsible for humans, (for example, the one most Christian apologists believe in), then why did He create the conditions for a whole variety of god-beliefs, often in conflict with each other, and provide the alleged "free will" to choose one over any other? To be sarcastic (or not), was it just to allow for some sort of "voluntary" worship in palliating His own insecurity or to entertain Himself with violence and wars?

Do we *all* have a choice about belief? There's the old notion that some of us might not be informed about our choices. That's the plaintive cry of the evangelist. Uninformed humans must be taught the "truth," which of course varies a good deal depending on which sect one follows. God Himself admitted that He had done a poor job of inspiring belief, for example, as noted in the fable of having to send His only begotten son to rectify the problem. This seems to be a crude method of dealing with the issue for an omnipotent being.

Then there's the "ability" to believe or not. Some humans are just not that capable of some sort of philosophical belief—lacking certain language, information, or intellectual ability, for example. It is unclear why any so-called God would endow some with that ability and not others. That would be a truly bizarre picture. I won't get into the fabricated, weird notion of whether salvation is granted to such people *de facto*. I won't also address the bizarre notions of whether or not other animals are "saved." Many theologians hold that they are not so entitled. The notion of God endowing souls only to humans does show some blatant disregard for the natural world that He supposedly created. However, religious acolytes should wonder what happened to the pre-Christian humans. They did have their fantasy views. And what about our biological ancestors? Did Cro-Magnons, or Neanderthals, or even *Homo erectus,* for that matter, have the free will to believe in God? We're straying a bit here, but it just illustrates how bizarre some of the religious notions are on this topic.

Omniscience and Omnipotence

It is often argued by theologians that, despite God being omnipotent and omniscient, despite what happens in the natural world, God endows humans with a soul or spirit, and this gives rise to the ability to act freely. Of course, this is just one more "gap"-filler without any evidence or even coherent logic. It is variously known as the "paradox of free will" or "theological fatalism." The problem has been addressed many times through history. The Catholic Church, other faiths, and religious philosophers have struggled with this for a couple of millennia, but without any convincing solutions whatsoever. Here it is simply stated: If there is an omniscient god, then It knows what you are going to think and do. Thus, there is no free will. If you have free will, then any god is *not* omniscient; that is, it doesn't know what you're going to do. This is not the God Christians espouse. You can't have it both ways—the notion that God has given you free will, and that He knows how you will exercise it is just literally nonsense. St. Augustine, the fourth-century theologian and philosopher, argued that God's omniscience *was* compatible with humans' ability to freely make decisions, but he never resolved the paradox. Nor has anyone else. It was just an assumption. He just made it up. In fact, the idea wasn't part of any classic theological tradition or writings. As psychologist and theologian Rodger K. Bufford points out, "Freedom in the sense of human will as an uncaused cause of human actions is not a biblical concept." (66)

More specifically, if God is omnipotent and omniscient, he knows the future—what is going to happen—and He is the *power* behind it all. We cannot "choose" what is already chosen. This obviates any meaning of the idea of free will. If God knows what's going to happen—and He will make it happen—you have no free will. This enables two possibilities: One, no god exists. Two, the Christian God is not omniscient. Of course, if He does not know how you are going to exercise your free will, he is also not omnipotent. That is, if He controls everything, there is no room for people to influence any events. But if He is finite in His abilities, humans have free choices. This is again eerily similar to the ancient Greek gods or even more modern polytheisms, where supernatural beings have power over some things and not others. And of course, if He is finite, the troublesome question of who or what created God comes into play (which I'll address in Chapter 8).

And let's consider this: Why would a supposedly omnipotent being promulgate the laws of nature, use them for nearly all human functions (as well as for the rest of the universe) and yet exempt a certain small set of human choices from them? So they could choose whether or not to believe in "Him," and to reward or punish them for their choice, as theologians always claim? It seems like a pretty sadistic game for an all-knowing, all-powerful, loving personage to play. Of course, all this is merely a fantasy in the minds of religious apologists.

St. Augustine's mystical idea—yes, we can call it that—remains nonsense. You can't have it both ways. There is great resistance to this reasoning, but there is no escaping it, and the pretzel logic that theists engage in to overcome this problem is just embarrassing. God grants free will but exempts Himself from knowing. God's power is supposedly greater than the reasoning we mortals use. God *again* works in mysterious ways. And so forth. If this isn't the most tortured type of logic imaginable, I don't know what is.

Conclusion

Theistic religion relies heavily on the notion of free will. This is particularly characteristic of the Abrahamic religions. The absence of free will, as this is construed, closes the door on any personal god. Albert Einstein once offered, "I am a determinist. I do not believe in free will. Jews believe in free will. They believe that man shapes his own life. I reject that doctrine."[3] The whole notion of free will is supposed to lend credence to the idea of God; but it makes no sense from logical or philosophical standpoints. Religious philosophers and apologists have been engaged in various mental and philosophical gymnastics to resolve the absurdities of this for centuries, but no reasonable propositions have ever emerged. The theists' principal argument about all this, of course, is that God's behavior doesn't have to make sense to be true. Think about that one.

In conclusion, we must assert that if there were a god who created the universe and its natural laws but also endowed humans to exercise free will outside of those laws—so that they could voluntarily choose to worship him—*we can only infer that He was a pretty insecure, not-so-omniscient and not-so-omnipotent being.* Or perhaps the whole God idea is just nonsense.

Freedom

At this point, we should also be clear about the confounding of free will and the notion of "freedom." When you say you have free will, you are assuming that you can act fully independently of any historical or contemporary influences, that is, the ability to act without any known, or theoretically known, causes. Freedom, on the other hand, is your ability to do something, relatively speaking, for which there are few physical constraints, little coercion or likelihood that you'll be punished for doing it. I'll drill down on this a bit more, but one of the best definitions of freedom I've heard comes from the Janis Joplin song, "Me and Bobby McGee." *"Freedom's just another word for nothin' left to lose."*[4] It is purely the lack of external contingencies or constraints on behavior that we call freedom.

As French philosopher Jean-Jacques Rousseau observed in the opening of his *The Social Contract*, "Man is born free, and everywhere he is in chains." (433) Although he referred mainly to larger social and political matters, his meaning was clear: We are naturally "free" creatures, in his estimation, but we are coerced into behaving in ways that conform to the expectations of others, especially the powerful. The "chains" which act upon us limit our "freedom." We have innate freedom which is forever compromised by the control of our social obligations. Of course, it is our social obligations—the need to get along with others in order to survive—that provide inescapable influence over our behavior. Freedom here is a social and political problem. It acknowledges that "choice" is often a victim of social obligations. Einstein clearly saw the dynamics of freedom similarly: "By freedom I understand social conditions of such a kind that the expression of opinions and assertions about general and particular matters of knowledge will not involve dangers or serious disadvantages for him who expresses them." (150)

Freedom and Constraint

People have "freedom" within whatever constraints pervade their lives. The notion of "freedom of speech" is one of the more notable examples. In America, our Constitution prohibits the government from impeding the exercise of free speech. We can speak, write, sing, dress, and communicate in any way we like without being constrained by governmental authority. However, it is not unlimited. We cannot yell "fire" in a crowded theater, incite violence, or in certain circumstances, publicly defame others—without consequences.

But your ability to "freely" act is also constrained by all sorts of things—the capacity to do certain things, the opportunity, and the consequences that may follow. Your freedom of speech is limited, for example, if you are a school child, or you are incarcerated, or you're writing under editorial approval. This is not a matter of free will. Let's take some mundane examples. I am free to select either an apple or an orange from the store. I can wear whatever shoes I want. The choice of peanuts or walnuts is a nearly "freedom"-based decision, usually limited only by your own history (cost or allergy may constrain you). Shoe choice is also rarely limited by external conditions (though there are certainly exceptions). Walking or running is a free choice. That is freedom.

The more options a person has in thinking and acting, the more she feels free. Where the options are numerous, we have a great deal of subjective impression that we have freedom. Given other similar circumstances, for example, children have fewer behavioral options than adults. Highly educated people have more options than others. Wealth creates more options than poverty. And so forth.

You are also not "free" to make decisions outside of your capabilities, so that is a limiting factor to begin with. You are not "free" to do things beyond your ken. A young child cannot freely choose to do calculus or compose a symphony, although we do know of exceptions. Humans of average physical stature and fitness can't freely choose to play professional basketball. How "free" we are depends on our circumstances. Such are limitations to freedom, not free will. I'm free to solve any equation I want. I can say any Mandarin phrase I want. But you can only speak languages to the degree that you know them. You may "want" to solve Einstein's relativity equations, but few of us can. You can certainly go to the library and get any book you want, but it might not be there, thus limiting your opportunity (an increasing problem for young people in some parts of America). We are constrained—to lesser or greater extents—by our talents, capacities, and opportunities.

Behavior

Freedom is also dependent on the consequences of behavior. Janis Joplin's hit song well describes the issue. We *feel* free when "there's nothin' left to lose." This can be interpreted in many different ways. But for our purposes, this is a poignant metaphor for the contingencies of human behavior. Where there are no further potential punishing or negatively reinforcing consequences of behavior, one is seemingly "free" to engage in whatever one desires. That's not technically true, of course, but it does suggest the implicit understanding that our behavior is generally a function of the contingencies put on it (as I'll illustrate further in Chapter 10). We can do whatever we want when we perceive that our actions will not result in any troublesome outcome or loss, and we then feel free to act without deliberately considering the consequences. Where we perceive that we're unlikely to be hurt further than our present or future circumstances, we perceive freedom to act in all sorts of ways. It is not free will; it is the lack of constraints and anticipation of troublesome consequences that enable "freedom." There are obviously also "degrees of freedom" in human actions, so to speak.[5] We are "free" to engage in any particular action to the degree that it is possible—that any action is physically and behaviorally possible and that we are unlikely to be punished for it. But again, this is "freedom," not free will.

Belief

Which brings us to God. What is the "freedom" to believe in God? History is replete with constraints on the belief in all sorts of gods. It is one of the principal reasons for discrimination, hatred, violence, and war (which I'll discuss further in Chapter 7). But there is another dynamic. You may have the freedom to believe in a god, even though you do not possess the free will to do it. You may declare that Jesus loves you; you may pray in a mosque, if one is available; or you may sit home and do nothing on Saturdays. There are constraints and consequences for such behavior and beliefs. They may be trivial or severe. They may result in reinforcement, comfort, praise, fellowship, etc. Or they may result in discrimination, hatred, or death. The freedom to believe in a god is highly variable, depending on your circumstances. However, we should be clear that such are the opportunities and constraints of belief, not free will.

Cosmology

Let's start again with the larger, very much larger, perspective. We accept that the universe behaves in a determined fashion. That is, it operates consistent with the laws of nature, as we discussed in previous chapters. Events are determined according to those laws, and this again is what is called determinism. This implies a specific unfolding of events, though not by a specific agent, e.g., God—just that the evolution of our universe is theoretically predictable.

"Choices" are a function of the evolution of the universe—beginning with the Big Bang. The whole notion of believing in a supernatural entity evaporates in the context of essential physics, as we discussed in Chapter 3. If the universe or multiverse is "determined," then the notion of faith or belief is meaningless. The thoughts and behaviors of humans are as equally determined by the laws of nature as any other events; and you are not free, in a cosmological sense, to "choose" to believe or not to believe.

All "theories" of free will (and they are not actually theories in any scientific or practical sense) make the assumption called "indeterminism," either explicitly or implicitly. That is, all events are not strictly determined. Natural law asserts that the past is the predictor of the future. To assert otherwise is the idea of indeterminism. The notion of free will specifically demands the notion of indeterminism, that there are some events that do not follow the laws of nature. Some have argued that indeterminism can exist along with determinism. The history of philosophy is replete with commentary about this, as we saw with the Greek gods. Are some events actually subject to determinism and some not? Should we believe that some indeterminism, putatively characteristic of human minds, is also characteristic of flatworms? When does determinism come into play, and when does it not?

Some serious thinkers disagree, of course. Physicist Kaku offers that, "Strict determinism is destroyed by a subtle combination of quantum effects and chaos theory. In the end, we are still masters of our destiny." (261) However, as we'll see (below), quantum effects I discussed in Chapter 5 are clearly unlikely to influence the actions of individuals or neurons in their brains, rendering activity at the cellular and higher levels of human behavior theoretically predictable, consistent with determinism. To fall back onto chaos theory, as he suggests, also doesn't clarify the problem. Chaos theory is merely a statement of very large, dynamic outcomes from some initial conditions, which makes prediction difficult, at least in any practical sense, for us humans. But it is rendered moot when all influencing events are known, i.e., the calculational impasse problem again. Quantum effects and chaos theory don't rescue free will from a determined universe. We do not live in utter chaos and randomness, nor in partial chaos and randomness. Thus, the conclusion is clear—free will and a universe which unfolds according to Natural Law are specifically incompatible.

Time and Causation

We tend to believe that we "cause" things to happen, and this reinforces the idea of free will. That is, we assume a temporal "cause and effect" relationship that really doesn't exist. That is because time does not "flow," as we've seen. As I discussed in Chapter 3, "causes and effects," don't really exist, as such. We merely infer them as human beings. The universe has been evolving in a particular way, with all the "events" we observe (and of course those we don't) coming about according to the laws of physics which emerged at the Big Bang. What we observe is that one event seems to precede another, and we infer from this that the first one is the "cause." In fact, to oversimplify, both types of events—as they are different in our perception—arise out of the entirety of how the universe has evolved to any point in time. It is the so-called "arrow of time" which describes our perception. If we believe that we have caused something to happen, it is only because the thought of it has "occurred" to us before the observed action, what we label as an "effect" (though not always; see below). But in actuality, both the "effect" and our thought of our initiation are merely events that have evolved with the universe and are perceived by us as different in "time." The notion that you *freely cause something to happen* would be inconsistent with the fundamental laws of nature and, quite logically, with the notion of any free will.

Cosmic Inflation

If you consider the idea of multiple universes, as we reviewed in Chapter 3, then such notions have profound implications for the concept of free will. Let's consider a couple of the possibilities. Though this again seems a bit esoteric, it is important to the notion of free will as illusion.

Let's consider the model that our universe is an infinite, ever-expanding spacetime without bounds or temporal limits, as is the current consensus. Again, in such a scenario, in any sufficiently large region of space, any arrangement of particles necessarily repeats, and this occurs infinitely. In this situation, what "you" are is not unique. The unlimited number of particle arrangements in various regions of an infinite universe allows for this. You exist an infinite number of times with the same history, the same features, and the same thinking and behavior. You can also exist with different histories, thinking, and behavior. This seems ludicrous on its face, but the cosmology does suggest this possibility.

The "choices" you make are made in parallel an infinite number of times, as are the alternate choices for any particular scenario. There is nothing unique or free about any given decision or behavior. In some patches of the universe, the particles are arranged a bit differently, which would lead to different choices for the same or similar conditions. In fact, your choices would be extremely variable. Any and all decisions are made by "you" and all your "clones," and where you are somewhat different, you make other decisions. In some cases, "you" make the same choices; in others you make different ones. In some you turn right; in others you turn left. It depends only on how particles are arranged at the time and place. Any and all choices are made. This all flies in the face of the idea that your choices represent something unique or fully voluntary. This is yet another example of how the idea of free will is meaningless in the grand scheme of the cosmos.

The Big Bang gave rise to our universe, but it also gave rise to an infinite number of other universes or bubbles. In each of these, there is also an infinity of spatial regions in which "choice" behavior can be emitted. Thus, the choices you make are not only replicated in our own universe but also in other universes—at least those with the same fundamental constants and other characteristics as ours. In either case, it is easy to see that anything that can happen does. In the largest sense, the notion of "choice" becomes an irrelevant idea. Any thought or behavior occurs an infinite number of times. "You" just happen to experience the one that "you" emit here.

Many Worlds

Returning to quantum mechanics, either the wave function collapses, or it doesn't. If it does, then in reality we can have the phenomenon of decoherence (see below), which allows behavior—of electrons, neurons, and brain activity—to become fully predictable (I'll address quantum mechanics further below). And since any neurogenic activity leading to a behavior or "choice" can theoretically be predicted, the notion of free will disappears.

If the wave function doesn't collapse, we have alternative theories which purport to explain the quantum measurement problem. The most popular notion of interpreting this appears to be Everett's Many Worlds theory I described in Chapter 3. In the Many Worlds interpretation of quantum mechanics, wave functions don't actually collapse; they merely split into different universes. That is, with any observation or interaction with fundamental particles, the universe splits into some other reality. Anything that can happen, does happen, just in different realities. For example, when something happens or a human makes a "choice," the result that we are aware of is only a function of what we can know in *this* universe. Alternate events or the other "choices" do happen "elsewhere." We just can't perceive them here.

If you choose to go left instead of right, that choice is the one observable in this universe. But the "choice" to go right may also occur, just in some other reality. If any selection, choice, or binary decision is possible in your self-identified universe, the other choice simultaneously exists in other, parallel universes. The wave function branches into different realities. That we might experience one instance of a thought or choice, for example, does not mean that it is somehow unique. All the other possibilities of those choices *do* occur simultaneously; we are just not aware of them. We might refer to the "choice" you make in this universe as the "reference universe," i.e., *your* reality. But an infinity of other choices or behaviors is possible and does occur in other "realities," that is, in other "branches" of the wave function at the time. In any scenario, all "choices" *are* made. Since you have made each and every possible choice, you have not made *any* choice! There being no "choices," the free will concept doesn't hold up. This all seems absurd, but it is what the theory tells us.

You may also believe that you have made a choice "at random." If you chose to go right, you could have chosen to go left. But you didn't—giving you the *illusion* of free will. However, the "randomness" that you assume underlies your choice doesn't exist. To reiterate, physicist Max Tegmark notes, "There is no true randomness in the cosmos, but things can appear random in the eye of beholder. The randomness reflects your inability to self-locate." (358) The Many Worlds theory suggests that you simultaneously make all possible choices, but you are only aware of the one you observe or experience "here"—in this branch of a particular wave function. He goes on, "Whenever observers are asked a question, make a snap decision and give an answer, quantum effects in their brains lead to a superposition of outcomes . . . each of these alter egos is unaware of the others and notices the branching merely as a slight randomness: a certain probability of [any given behavior]." (517) Therefore, the notion that you have made any "random" choice of alternatives fades with the reality that all choices are actually made.

Although he may not have meant it literally, William James, philosopher, physician, and founder of American Psychology, noted the following, "The world of our present consciousness is only one out of many worlds of consciousness that exist." (241)[6] Is it possible that "realities" exist in many "branches" and that we experience only one of them at any point in time and space? The Many Worlds interpretation of quantum mechanics would certainly be consistent with that notion; William James may have been incredibly prescient about this issue. If that is the case, it is again inconsistent with any notion of free will.

The Multiverse and You

In the various multiverse scenarios, the meaning of "you" disappears into an infinity of replications or branches in different realities. This seems a bit esoteric and "off the mark" with respect to what most people consider free will. However, it is perfectly consistent with what modern cosmology is suggesting. We are left then with only two possibilities when it comes to the issue of free will. One is that in each and every iteration of "you," you are making an identical choice, in which case every one of you would have to be exerting the same "free will," an absurd notion. The other is that an infinity of choices is both available and actually made, in which case the whole notion of choice becomes irrelevant—all choices exist and are made. Again, no free will. If all this now seems to fly in the face of your own experience and understanding, so be it. But remember that your "commonsense" experience is not a solution to such a problem. We have extended the problem of free will to what science can tell us about our place in the universe, or universes. In each model of an infinite universe, of infinite replications, or infinite branches into different realities, anything that can happen *does* happen—somewhere at some time. As science writer George Musser notes, "Parallel universes are not some exotic idea out there in the cosmos. Our body and brain are little multiverses, and it is the multiplicity of possibility that endows us with freedom." (358) In this context, the notion of free will again just evaporates.

The Quantum Argument

The notion of quantum mechanics is the latest effort enlisted to try rescuing free will from a determined universe, much as it has for consciousness. Since this has gained some popularity in recent years, and since it appears to be currently the last bastion of any serious thinking supporting the notion of human free will, we'll dive into this a bit. The argument is that quantum mechanics, with its probabilistic nature, avoids a strictly deterministic universe and thus allows for free will. As is generally stated by proponents of this view, quantum mechanics "leaves room" for free will. We've shown how the Many Worlds interpretation is at odds with this. However, if we assume that the Copenhagen interpretation is the correct one, that the wave function does actually collapse (for example, with measurement), perhaps there are other phenomena we need to consider.

Decoherence

A realistic approach to this problem is the phenomenon of *decoherence* I discussed in Chapter 3. If the wave function does collapse, we know (theoretically, at least) what behavior will occur—whether an electron or other fundamental particle can be found here or there, not just anywhere. A decoherence may occur with observation, i.e., interaction with photons (which is how our eyes or cameras work); or it may occur via interaction with other fundamental particles or other aspects of nature. If this is the case, then decoherence in effect is not merely or only a function of observation, and specific predictions can be made. That is, any quantum phenomenon in superposition (or the "blurred" or probabilistic nature of it) may encounter interaction with other entities in its environment, e.g., other particles. Where organisms are concerned, this may also include atoms, molecules, etc., not just observation. As such happens, the coherence of the wave function is disturbed, and the probability of any event can be (theoretically) calculated. Quantum effects become irrelevant in practice to the behavior of such things as molecules, neurons or more complex systems, e.g., humans. As physicist Carroll asserts, "When we aggregate many particles the overall behavior becomes fantastically predictable (at least in principle)." (78)

Brain Basics

Here's how it happens. The behavior of neurons in the human brain depends largely on something called an action potential in part of a neuron, called an axon. It is an electrochemical phenomenon that we describe as an impulse. Through some additional chemistry in the synapses, each neuron triggers other neurons to be activated. Further, a neuron or nerve cell is essentially a binary system; at any point in time the axon of a neuron conducts an electrical impulse—it either "fires" or it does not. It either conducts an impulse or it does not; it operates as either "on" or "off," nothing in between. And when it does activate, that cell stimulates others in a complex network of neurons. That is how our brains actually operate. (There are other things that influence brain activity—circulating chemicals, action of glial cells, etc., but these are not to the point here.)

Each neuron in the brain has an axon, which conducts the electrical impulse. That impulse is generated by the movement of certain ions, charged particles, which obviously involves the action of electrons. Quantum mechanics describes the behavior of electrons in general. However, the electrons in those ions can't do their neurological job without interacting with their environments. In this case, there is a series of interactions of those electrons with certain other entities in their chemical environments, resulting in decoherence. Thus, the nature of the impulse becomes obvious, as does the behavior of the neuron itself. Any quantum uncertainties are reduced to certainties and are (theoretically) predictable. They are canceled out in the chemistry of the impulse. Thus, we know when a neuron will fire; we can track, to a certain extent, the certainty of the impulse and predict how that affects its neighbors (all at least theoretically). As I noted in Chapter 5, and as Carroll asserts, "Important processes in the brain are well described by the approximation of classical physics. . . . Every particle in your head is constantly being jostled by other particles, leading to an ongoing process of 'collapse'." (78) Thus, the workings of each neuron, and the brain in toto, and its output at any one time are predictable—though remembering the calculational impasse, only theoretically. Nevertheless, such possible predictability fully rules out any randomness or "room" for free will to emerge.

It may also be helpful to look at the quantum uncertainty argument for animal brains. You say that your dog "wants" to go outside. Is that free will, and is it subserved by some quantum uncertainty? And we can say—whether you see it as nonsensical or not—that a roundworm "desperately wants" to get out of the sun (remembering Chalmers' notion that everything in the universe is supposed to have some sort of consciousness). The same quantum states exist in such creatures as they do in humans, whether we call them randomness, probabilities, uncertainties, or superpositions. Nevertheless, the behavior of such creatures *is* predictable too, per the dynamic of decoherence in all of the animal's neurons. Neuroscientist Sam Harris puts it well, "The indeterminacy specific to quantum mechanics offers no foothold: If my brain is a quantum computer, the brain of a fly is likely to be a quantum computer too. Do flies enjoy free will? Quantum effects are unlikely to be biologically salient in any case." (208) One could argue, of course, that the fly can learn something (as all animals do) and therefore initiate a "decision" favorable to its own survival. But such "decisions" do not arise out of some quantum uncertainties. They are a function of knowable brain activity. Same for fish, octopi, and all other animals. (I won't go into other phyla of organisms here, e.g., plants, fungi, and bacteria, as the quantum free will argument hasn't been applied to them, as far as I know. Their only recourse is Chalmers' universal consciousness approach, though it's unclear if he would actually assign free will to them.) The quantum neurological argument just doesn't hold up for humans any more that it does for other organisms. Human behavior, including our "choices," is not controlled at the quantum level. Our behavior, as mediated by our brains and other parts of the CNS, obeys the laws of physics for larger objects. The quantum neurological argument just doesn't hold water here. Again, no free will.

Room and Randomness

Is there still "room" for free will in a determined universe, as some have argued? We already noted the consensus that there is no true randomness to the universe. We must conclude that quantum uncertainties or probabilities do not leave "room" for free will. But the bottom line is this. If we *perceive* any randomness, what fills in that "room?" There could be—and probably is—a more fundamental theory of nature which explains the apparent randomness of quantum events that we infer, as Einstein doggedly argued for. Or the apparent randomness is not really random. We may not be aware of "realities" which complete the better prediction of events because we can't actually access them, as is described by the Many Worlds theory. (517)

It should also be pointed out that "room" for free will, of course, is not the same thing as evidence for it. That there may be apparent "uncertainties" or probabilities underlying some of our perceived reality does not lead to the conclusion that free will must occupy any "uncertain space." This would be a leap too far for any logic. Contrary to Penrose and others (below), this is not a valid argument. It is no more than filling in a "gap," as science doubters have done for evolution and many other things that seemed to defy empirical explanation. "Room" is merely a synonym for "I don't know yet, so there must be free will in the gap." This is not the same thing as the quantum mechanical assertion that the fundamental nature of reality is probabilistic, being in some "haze" or in one potential state or another. This is not a "gap" in knowledge but rather a still-evolving area of physics. "Room" for free will is not the same thing as evidence for its existence.

Thus, any conception of uncertainties in quantum mechanics cannot be reasonably used as an argument for free will. The behavior of our bodies—and our brains—like everything else we actually observe in the world, per modern physics, is explainable with definite probabilities in the action of particles, molecules, neurons, and other structures, again, of course, theoretically. Thus, quantum uncertainty cannot be used to create some philosophical vacuum for free will. Further, quantum principles apply only to the realm of fundamental particles, not to the explanation of other, larger phenomena—where the other laws of nature apply—from general relativity to those of the biological, anthropological, and behavioral sciences.

Penrose–Hameroff Theory

Let's return to the related, so-called Penrose–Hameroff theory, which I began describing in Chapter 5, and which has been offered as an argument for free will. (200, 394) In this theory, it is argued that there are operations in the brain that are inherently "noncomputable." It is proposed that the theory "identifies discrete conscious moments with quantum computations in microtubules inside brain neurons . . . [This] can cause temporal non-locality, sending quantum information backward in classical time, enabling conscious control of behavior." (200) What? We need not elaborate on what that might mean because there is no evidence for such. Penrose also argues that an active aspect of consciousness is "the feeling of free will." (394) However, he flatly eschews any "definition of the word 'consciousness' here either (and certainly not of 'free will')." He goes on, "While I do admit to there being passive and active aspects to consciousness that are sometimes distinguishable, the passive having to do with sensations (or 'qualia') and the active with the issues of 'free will,' I take these to be two sides of a single coin." Penrose does admit, "It appears that neither classical nor quantum physics, as presently understood, allows room for the non-computable behaviour of the type required, so we must look elsewhere for our needed non-computable action." (394) However, he fails to make any convincing argument for finding it anywhere or how to go about even looking for it. Despite these authors providing elaborate technical analyses and models, there remains no empirical evidence for this theory. In sum, there is no evidence for the supposed conscious production of free will under this particular quantum proposal. (24)

Conclusion

Nevertheless, philosopher Daniel Dennett poses the pivotal question, "How can the indeterminism of quantum mechanics be harnessed to give us a clear, coherent picture of a human agent exercising this wonderful free will?" (121) Of course, if there were a "coherent picture," it wouldn't be free will. Again, if there were a coherent picture, we could operationally describe it, and it wouldn't be *free* will. Despite the pleas of Dennett and others, most physicists merely do not even attempt to address this question. It would be a fool's errand. In a popular commentary, physicist David Kaiser astutely notes that, for scientists working on quantum mechanics, "Obviously, we aren't going to try to prove that humans have free will." (260) To wit, since free will, as it is commonly understood, would lie outside the laws of nature, it *cannot* be proved.

That quantum uncertainty is a currently accepted principle and that observation or measurement may be necessary to "bring forth reality" in some sense (as we commonly know it), i.e., fundamental particles out of some quantum haze or superposition, does not enable a logical leap to considering free will a reasonable idea. Physicist Greene "seriously doubts" quantum uncertainty leaves any room for free will. "If Schrödinger's equation is all there is to quantum mechanics, then quantum physics, in its entirety, is as deterministic as classical physics." (192) I don't want to beat a dead philosophical horse here, but a couple of further summaries should illustrate the points. First, physicist Carroll:

> Quantum mechanics has nothing to do with the question of free will. . . . The fact that we can't predict the future isn't the same as the idea that we are free to bring it about. Even in textbook quantum mechanics, human beings are still collections of particles and fields obeying the laws of physics. For that matter, quantum mechanics is not necessarily indeterministic. Many-Worlds is a counterexample. You evolve, perfectly deterministically, from a single person now into multiple persons at a future time. No choices come into the matter anywhere. (79)

And neuroscientist Harris:

> It has been fashionable, for several decades now, to speculate about the manner in which the indeterminacy of quantum processes, at the level of the neuron or its constituents, could yield a form of mental life that might stand free of the causal order; but such speculation is entirely oblique to the matter at hand—for an indeterminate world, governed by chance or quantum probabilities, would grant no more autonomy to

> human agents than would the incessant drawing of lots . . . . Although we can find no room for it in the causal order, the notion of free will is still accorded a remarkable deference in philosophical and scientific literature, even by scientists who believe that the mind is entirely dependent upon the workings of the brain. (206)

Why free will might be sometimes accorded some "deference," however, might be a bit off the point. As I've said, sometimes we just know how to talk about ordinary events and behavior with common lingo. Again, "room for free will" is not evidence of any free will existence. Quantum mechanics doesn't rescue free will from a determined universe. Unpredictability, uncertainty, or implied randomness is not the same thing as free will in any common definition of this. As physicist Paul Davies points out, "Quantum uncertainty cannot create freedom." (109) The whole notion of trying to rescue free will with quantum probabilities is no different than other "gaps" in which explanations are merely invented. By extension—and this is the point—this closes the door on an idea of humans being "free" to believe in anything, including anything supernatural.

Neuroscience and Awareness

If quantum mechanics does not explain free will or even "leave room" for it, and if we continue to fervently believe it, can there be some way of looking at the brain that gives us some useful information about this matter? Can there be any test of the brain's operation which helps us even understand the problem? Bear with me. Some of this is truly provocative.

The human brain can be thought of as the proximate cause of most of our behavior. That is, it includes the last series of events that may precede our overt actions (I use "cause" here in the somewhat common meaning. Recall that, at the most basic cosmological level, there are no specific causes of one event in relation to another.) The meaning here is merely that there is a rather direct connection or correlation between neurological activity and the emission of one's private or overt behavior. In this regard, we rightly should consider neuroscience in its infancy at this point. However, some behavior can be predicted on the basis of recorded activity of our neurons, as I've suggested. This again raises the question of awareness in relation to what we actually do. Are we aware of a fact that any particular behavior is coming? The question becomes pivotal for the whole notion of behavior or choices as "voluntary," a key to the whole notion of free will.

The Classic Awareness Problem

One of the basic tenets of the idea of free will is that our intentions precede our actions, leading to the inference that we decide what we will do before we do it. Thinking precedes behavior. If that is not the case, it is less likely that free will is in any way relevant to our actions. In fact, there have been a variety of experiments on the human brain which address this issue, that is, the possibility of "awareness" preceding some purely "voluntary" actions.

Awareness is not a particularly useful concept in explaining human behavior, as I've suggested in the last chapter. However, there is also a body of experimental evidence from neuroscience that bears on this point. In some now-classic experiments, physiologist Benjamin Libet and colleagues recorded subjects' brain waves while they were being asked to voluntarily move their fingers or wrist whenever they "chose" to, while the subjects also viewed an oscilloscope with a dot rapidly moving around the image of a clock face. (311, 312) They were asked, after the voluntary movement, to note the exact clock position or "time" at which they made their choice. The result. It turns out that some neurons in the motor cortex of the brain showed activity for specific finger movements about 350 milliseconds (that's about a third of a second) before the subjects reported the awareness of a decision to move the finger. That is, the brain started the "voluntary" movement before the subject was aware of it. Subjects reported that they had become "conscious" of the decision to move *only after the brain had begun to initiate the act*. The data cast some serious doubt on the notion that the movement was fully initiated by some conscious decision process. Such pre-awareness activity in the brain is generally referred to as a "readiness potential," the activity of neurons in the area of the brain involved in the movement. That is, the brain's readiness potential for the movement began to appear in a specific way preceding any urge or awareness of the future action by the subject. In this light, we can't consider the actions as fully "voluntary." A number of similar experiments have been done, using EEG, and they have produced results consistent with this.

More recently, this type of experiment has been repeated by recording the activity of even smaller specific groups of neurons in the brain. This has been done when neurosurgery is performed to intervene in patients with uncontrollable epilepsy, for example. In one such experiment, the Libet-type study was done with researchers implanting electrodes directly into the brain, measuring the activity down to the level of only about 1,000 neurons, an incredibly tiny area of brain tissue. (175) When subjects were asked to press a button, it turns out that the "pressing action" of some of those neurons occurred between 0.7 seconds and 1.5 seconds before subjects estimated that they made the decision to press. In another study, very sensitive fMRI procedures were used for a slightly more complicated task. (494) Subjects were asked to fixate on a screen while a stream of letters flowed across it. When they felt the urge, they were to press either a right or left button with the right or left index finger, along with the instruction that they were to remember the letter that had appeared when they made their decision. It was found that "decisions" were recorded in certain parts of the brain (the prefrontal and parietal cortex), remarkably up to 10 seconds before subjects were actually "aware" of their decisions. This readiness potential is sometimes construed as the brain "preparing" for a decision (or "accumulating" neural activity for a choice). Nevertheless, it does indicate that the subjects are unaware of the brain initiating the decision process before it "came to mind," so to speak.

Some studies have addressed the neural activity which precedes supposed conscious awareness of a decision, where the decision itself is even more complex. Neuroscientist Uri Maoz and colleagues, for example, found that subjects' preconscious awareness was lacking for certain arbitrary tasks (such as the above), while this was not the case for more complex decisions. In one experiment, readiness potentials were recorded while subjects were asked to make a financial decision. They were to make a choice of what nonprofit organization a donation would be made to. (325) If the subject had to make a decision about which of two organizations to donate to, readiness potentials seemed to be absent. But if the "decision" was merely to donate, regardless of either target, the readiness potentials were again apparent.[7] This does support what's apparent about choices. The "simpler" the action, the less pre-behavior awareness we have. The more complex the decision, the less likely an isolated, specific neural action can be identified, and we're more likely to merely label such actions as "voluntary."

There has been a lot of controversy surrounding these types of experiments. It is not clear what pre-action neural activity represents and what types of "decisions" may conform to these types of findings. Such decisions are generally interpreted in the context of "conscious" or non-conscious experiences, though these concepts are impossible to define. We only know that for some actions, there is often preceding neural activity that subjects do not report an awareness of. Experiments with highly sophisticated recording and data analysis have "muddied the waters" of any readiness potential in such experiments but have not clarified anything about ideas of supposed "free will" that may underlie subjects' responses. (453) At this time, we are not able to isolate any sort of "readiness potential" for very complex decisions. Can you imagine trying to assess what brain circuitry is operating before throwing a curve ball, or deciding on a job, or saying "I love you?" Obviously, neuroscience is very far from such capabilities.

There has also been much consternation in the philosophical community about whether Libet's experiments and others that followed actually relate to the concept of so-called "voluntary" action. (429) Pre-awareness findings are certainly a threat to the idea of free will. However, the notion that "awareness" does not always precede action in these experiments has not been debunked. It remains that choice of "voluntary" action is a complex process, and "awareness" is often unnecessary in this regard. In sum, Libet and successors have demonstrated that "conscious choice" is not the free and specific cause of our actions. Further, the human brain is a complex organ (that we do not well understand), but the notion that there is any reliable binary system of "conscious" and "nonconscious" operations in generating behavior we call "free will" would be impossible to justify.

Of course, these experiments tend to show that our conscious awareness is not necessary for the emission of behavior, even when we think that it is. *This is not a spectacular revelation*. We do a lot without really thinking about it—in fact, nearly all of what we do. But such studies do show that specific brain functions for acting can often well precede conscious "decisions." It has also been argued that "decisions" are not made all at once. They come about gradually, including what experiments have measured preceding a recorded decision time. (274) Nevertheless, it is clear that the experiments demonstrate that conscious awareness does not necessarily precede other neural events subserving the behavior. To call those preceding neural events "unconscious" offers no further clarity. The point is that subjects did not make choices that they were "aware" of before the brain started to produce the response. One can say that some things are conscious and others unconscious or that "awareness" exists on some kind of continuum. But this again really begs the question. Actual awareness does not need to precede behavior, which comes about for reasons other than voluntary or "free" choice. Some decisions are made in rather sudden fashion, while others require more time. The Libet-type experiments are very simple examinations of decision processes, and more complex ones haven't been done, that is, for readiness potentials that are not obvious (to our current state of measuring technology). It is not surprising that specific readiness potentials can be identified in the former and not the latter. Neuroscientist Patrick Haggard summarizes the findings on this:

> These data give the impression that conscious intention is just a subjective corollary of an action being about to occur. Such models agree with previous accounts that voluntary actions begin unconsciously and enter into our conscious experience only when medial frontal activity has reached a given threshold level of activity. In this sense, the current work is in broad agreement with a general trend in neuroscience of volition: although we may experience that our conscious decisions and thoughts cause our actions, these experiences are in fact based on readouts of brain activity in a network of brain areas that control voluntary action. (198)

In fact, the neuroscience is clear. What is interesting here is that experiments have demonstrated that our behavior can be activated, and often is, without our conscious awareness, even when we *think* we are completely, "voluntarily" making a relevant decision. The meaning of *voluntary* here generally implies "free" action, but technically it suggests only that such actions are a function of some areas of the brain as opposed to others. Such brain activity constitutes collateral events, occurring after or coincident with response initiation, and not the actual cause of our actions.

The Open Brain

It is possible to artificially stimulate the brain, i.e., without going through our senses. Again, this can be done during open-brain surgery, which is sometimes done for people with certain neurological disorders or tumors. The patient remains awake, and the surgeon stimulates certain areas of the brain, producing a variety of effects.[8] Depending on where the stimulation is given, a variety of sensations, experiences, or movements can be produced. For example, in one classic experiment, people undergoing brain surgery for tumors were given local anesthetic, and electrodes were used to stimulate various areas of the brain. (126) (There are many such experiments, which will not be summarized here.) If some small stimulation was given to a certain area of the parietal cortex, the subjects actually reported feeling a strong desire to move certain body parts, e.g., feet, arms, hands, chest, or even lips. That is, there was an "intention" or urge to move, but there was actually no movement of those body parts. Interestingly, if the electrical currents were made stronger, the subjects reported that they were, in fact, moving those body parts, *even though they actually hadn't* (recordings of electrical activity in the relevant muscles confirmed that no movement had occurred). *It was an illusion.* When a different region of the brain was stimulated (the premotor cortex), the opposite effect was found. With that stimulation, the patients moved all their analogous body parts (hands, arms, or mouths); but they reported no awareness of doing those actions. They had no idea their hands or mouths were moving. In such studies, awareness or consciousness of action was not necessary. As we see, movement can occur without awareness, and awareness is not necessary for movement. The brain is a complicated organ, but such experiments do demonstrate that "voluntary" initiation or conscious free will is not a necessary part of the equation.

Conclusion

Neurophysiological studies, however imprecise, tend to expose the complex workings of the brain, including the possibility of reported awareness in relation to actions. Some show awareness preceding actions, others do not. The difference may be the complexity of any decision; any neurological "readiness" may be much harder to identify in highly complex decisions. Our behavior exists along a continuum, from actions that are reflexive to some being associated with reported sudden awareness to some related to years or decades of cognitive planning—and everywhere in between. Neuroscience has not yet exposed the role of particular brain functioning, timing, or location for various levels of decisions and actions, but the notion of voluntarily or freely emitting behavior cannot be demonstrated.

Actual intentions do not originate in consciousness, but rather *appear* in consciousness. This makes any awareness essentially a covariate, not a cause. If decisions are not "conscious," then free will is an absurd idea, since we consider our decisions to be fully conscious activity. The bottom line is that actions we take, including verbal behavior (both public and private), are determined by neurological events which precede our reported awareness—by milliseconds or much longer—and that our "thoughts" (if such can ever be defined neurologically) are collateral products, not causes, of our actions. The proximate causes of behavior are biological; the "mental states" which may accompany them are not the causes of the action; they're just along for the ride, generally in other areas of the brain. This seems like a pretty heretical statement, but empirical data do tend to support this, and there's no evidence to the contrary. *Again, no free will.*

Mind and Brain Again

If you still endorse the idea of free will, and if what we know about neuroscience seems to rule this out, there is no alternative than to questioning where "it" comes from. We've covered the theists' idea that it comes from God. This necessarily also brings up the idea of a nonmaterial "mind," as I've discussed. Is "mind" somehow separate and distinct from the brain, partially related, otherwise controlled, etc.? Is there some unique existence to it? We feel that there *must* be a mind superior to a brain. *After all, free will requires it.*

Of course, the idea of separating humans into body and soul, made most famous by Descartes, gave Enlightenment impetus to the belief that there was some sort of human mind that could operate outside the laws that governed the physical body. And that mind makes all the most important choices in our lives. It has fundamental power over how we feel and act. The "mind" can make any decision it "wants." And it seems critical to every complex choice or decision we make in life. As Milton wrote in *Paradise Lost*, Lucifer says that "The mind is its own place, and in itself can make a Heaven of Hell, a Hell of Heaven." It is independent of circumstances. If there is not a nonmaterial mind, we have no choices, and, as the popular notion goes, we are "merely" victims of the world around us. That is part of what makes the free will idea very difficult for us to reject.

Language

We use the idea of mind almost cavalierly. Without considering whether such a thing actually exists, we use it to describe our actions and the world around us. I don't mind if you sit here. I have a plan in mind. I'm doing mindfulness meditation. Never mind that. We use the concept of the mind to describe the behavior of others. He doesn't mind me. We're of two minds on that. He's minding the store. We also use the term for nonhuman organisms or inanimate objects. The phrase "mind of its own" is a well-established idiom in our lexicon. A runaway machine, an errant golf shot, or a plant that grows the wrong way, for example, is often described as having a mind of its own. (You'll notice that such language regarding "mind" actually describes overt behaviors, as we reviewed in Chapter 5) This is an informal variant of what we call a "psychologism," the philosophy of "explaining" nonpsychological phenomena using psychological principles (though "mind" is not really one of them). It seems like we anthropomorphize inanimate objects, but it's often just a convenient type of communication. We don't really imply that inanimate objects and primitive organisms have "minds." Nevertheless, it also reinforces for us the idea of a nonmaterial mind, as it's pervasive in our language.

To clarify, using the concept of "mind" also suggests that decisions or actions can be freely initiated in a nonmaterial mind. On the other hand, if I say that your brain makes a choice, you may look quizzical or deny it outright. To say, "I've made up my brain," sounds ludicrous. Nevertheless, as I previously apologized for, the term "mind" appears frequently throughout this book. It is part of our lexicon, something we use to economically communicate, but please remember that it doesn't imply a nonmaterial soul or similar idea.

Brain

The operation of the human brain and its capacities—as we discussed regarding the connectome in Chapter 5—is "mind"-boggling. What remains is to be discovered is how these connections develop and operate. The inputs are likely genetics, congenital history, nutrition, physical/environmental influences, and influences from interaction with the social and other external environments, which we call learning. That we cannot well appreciate this complexity is, again, the problem of a calculational impasse, limited by our present theories and measurement and computing technology. But the complexity does not support the notion that the brain possesses any special character which allows it to initiate its own "voluntary" action outside those potentially understandable influences and trillions of neural interconnections. We are not "free" to act outside of this realm.

Some have suggested that we might consider the idea of "mind," as opposed to brain, as the genesis of free will. This is unjustified. Even the notion that mind is an "emergent" phenomenon is highly questionable. Some have suggested that we should merely be looking at this problem with different levels or layers of analysis, i.e., a different type of theory, even if we secretly consider that the mind is not really a "ghost in the machine." However, as neuroscientist Michael Gazzaniga suggests, the more we understand about the brain, the less "free will" will be seen as a causative agent. (94) As Greene puts it, "One's physical and mental characteristics are nothing but a manifestation of how the particles in one's body are arranged." (193)

In the end, many neurophysiologic studies counter the notion of "free" decisions. But they actually do not address free will, per se. They only purport to expose reports of awareness associated with decision-making. They don't formally address the issue of free will. If this is free of the laws of nature, then brain studies will never find it. If "free" means something neurological, as yet unidentified or poorly defined, then further study is certainly appropriate in understanding the brain. And if neuroscience eventually maps or explains the "illusion" of free will (how marvelous would that be!), that does not mean that free will actually exists —only the illusion. Neuroscientist Harris summarizes only what we know and goes a step further, again countering the notion of free will:

> There is no question that (most if not all) mental events are the product of physical events. The brain is a physical system, entirely beholden to the laws of nature—and there is every reason to believe that changes in its functional state and material structure entirely dictate our thoughts and actions. But even if the human mind were made of soul-stuff, nothing about my argument would change. The unconscious operation of a soul would grant you no more freedom than the unconscious physiology of your brain does. (208)

Consciousness Yet Again

The issue of "mind" is closely related to the idea of consciousness. It is the "conscious mind," so to speak, that is considered the vector for free will. Without mind and consciousness, it doesn't exist. In the last chapter, we dispensed with this idea of consciousness as any possible explanation for human behavior. Whether there is ever any pattern of neural activity that is associated with "consciousness," or if such pattern is ever discovered to "emerge," is beside the point. Such a finding is unlikely, of course, but people continue to claim that they make "conscious choices," affirming the idea of their free will.

Nevertheless, some have tried to squeeze the idea of free will out of a physically operating brain. For example, philosopher Eddy Nahmias takes exception to the Libet model of choice behavior. (359) He acknowledges that there are certain neurologic antecedents to our decisions and actions that we are not aware of, but consciousness or awareness is not seen as a unitary phenomenon. He notes that there are "degrees" of consciousness and willfulness in human thinking. Therefore, "free will" can still emerge, even though we're not fully cognizant of the contributing factors. Of course, if we were able to document all the contributing factors, there would, in fact, be no "free will" left. Nahmias' argument just doesn't hold water.

Artificial Intelligence

Let's take a rather extreme equivalence, or so it might seem. We entertained in Chapter 5 the possibility that some AI system could be self-aware. Without specifically answering this, consider the question—can an advanced AI system have free will? This seems like a nonsensical question, though interacting with a current chatbot or similar system with AGI sometimes can seem like there's free will "in there." With advanced AGI, there is often little or no understanding by us humans of how such a system often makes decisions. It learns and makes decisions based on its "experiences" (i.e., the training it may have been exposed to, as well as the algorithms it generates for itself, as described in Chapter 5). Consider —an AI system might be programmed, or generate its own programming, to "believe" it can make free choices. That is, an AI system can be "trained" to reflect on its own designed decisions (its self-generated algorithms) to infer that certain decisions it makes which are independent of all its previous direct learning should be considered "free." Or, more intriguingly, given its knowledge of human behavior, it might infer that the decisions it makes which don't rely specifically on its experience *are what actually constitute its "free will."*

Isn't that exactly the point? This is an intriguing thought experiment, and I'm not aware that anything like this has been carried out. But it does go to the heart of the matter. This is precisely what we humans do—label our processes as awareness and our choices as free will. When we can't explain our thinking and behavior based on accumulated experience, we label it free will. In most of the decisions you make, you are not attentive to the "algorithms" in your brain that are brought to bear leading up to the decision. As you read this, for example, you are not "aware" of the natural algorithms (so to speak, the activity of neural circuitry in your brain) that lead you to recognize letters, read words, and even construct the meaning of this sentence. You are not "aware" of the algorithms that lead you to make various decisions, no matter how complex. You might "consider" various previous experiences in deciding what job to accept; but *all* the "deep learning" which leads to such choices is much too complex to consider at any one point in time. That would be paralyzing, and evolution has prevented that sort of thing from happening (see below). You consider the decision "free." The AI analogy seems absurd on its face, of course. But since *both you and an AGI system* make decisions from information processing, "deep learning," and development of algorithms, there is no reason that an AGI system wouldn't be as justified in labeling its various decisions as "free will" as humans are. Take another moment to reflect on that. It may be psychologically uncomfortable to ponder this (and as Steven Hawking feared, pretty threatening). When you consider the matter at its most basic level, however, decisions and other behaviors we label as "free will" are just immediately unexplained complex actions growing out of deep learning, some sort of processing, and a "lifetime" of perceptions and experience.[9] This thought experiment, of course, becomes rather philosophical since we are "seemingly" far from the ability to conduct such a trial. For now, we are left with the obvious conclusion that complex information processing does lead to the subjective impression—the illusion—of free will.

Behavioral Perspectives

We've seen that the notion of free will is not supportable—on the basis of quantum mechanics, neuroscience, or ideas about mind and consciousness. Nevertheless, we humans do endorse the concept of free will, continue to "believe" in it, and vigorously object to any notion of it being an illusion. In fact, the notion of "free will" is perhaps even more resolutely held by us humans than even the idea of God, remembering also that the former is necessary for any theist's conception of the latter. Why there is such incredible devotion to the idea does seem to require further explanation. In short, why do we endorse "free will" as the cause of most of our actions? In formulating questions, as I noted in Chapter 2, the even better type of question is—what are the conditions which lead us to claim the free will ability?

To endorse free will, or make choices, or cite motivations for our actions—all of this is human behavior, no more or less. And the factors responsible for such thinking and behavior can theoretically be explicated. This is the science of human behavior. I'll discuss behavioral science, per se, in Chapter 10, including the various environmental contingencies that shape our behavior. For now, I'll lay out some of the behavioral reasons that we continue to so strongly espouse the idea of free will.

Free will is often couched in the language of "choices." But what does that mean? In the largest perspective, choice is a behavior. Yes, you can choose—you make choices. But "choice" is more of a metaphor than something you do free of the events which have preceded any choice behavior. Such behavior or choice is then theoretically predictable, and if it is predictable, it is not free.

Voluntary Behavior

Related to the notion of free will is the concept of "voluntary," as I've alluded to. We prefer to invoke purely voluntary motivations for our behavior, i.e., that our choices and actions are specifically intentional and deliberate. To acknowledge that our complex behavior is not fully voluntary—that it is controlled by known or knowable environmental and physiological factors—is a great threat. The notion of *voluntary behavior*, again, however, glosses over the fact that the brain operates according to the laws of chemistry and physics. All this information processing and the inference that we voluntarily initiate behavior is "potentially" explainable with such science. Again, that we are not presently able to completely do this is a "calculational impasse," not an argument for free will. When behavior is predictable, it is the laws of nature that are responsible, not a free will outside of those laws.

We tend to deny, or at least take for granted, all the experiences that lead us up to a so-called voluntary decision or action. Our subjective experience is that we are "free" of those antecedents, leading to the belief that we have free will in emission of any behavior. When we say "I want" to do this or that, we seldom reflect on the history and conditions that brought such impulses about (save impulses that are purely or "mostly" unconditioned, like hunger or pain avoidance). But this free will impression, of course, is merely an illusion. Harris summarizes this well, "The *feeling* [emphasis added] of freedom arises from our moment-to-moment ignorance of the prior causes of thoughts and actions." (208) Of course—and this is the crux of the matter—if we were fully aware of prior causes, we actually couldn't function. Life is "long" and much too complex, and we rather automatically ignore nearly all the antecedents to our behavior. Thus, we are left with the impression that we've acted voluntarily.

Summarizing the larger view, as I've reviewed, mental operations themselves are not the causes of overt behavior. They are only viewed as such temporally—they *tend* to come before the behavior they're attributed to, as we perceive them. But they are not the cause. What we may experience as "awareness" or "intention" is a collateral product of what happens in the brain, which leads physiologically to the emission of behavior, but it is not the cause of any behavior in question, as the neuroscience above suggests. This is not a novel insight. The matter was cogently summarized by the great English biologist T.H. Huxley even before the twentieth century:

> Our mental conditions are simply the symbols in consciousness of the changes which take place automatically in the organism. . . . to take an extreme illustration, the feeling we call volition is not the cause of a voluntary act, but the symbol of that state of the brain which is the immediate cause of that act. We are conscious automata. (231)

Automata (or zombies, as we discussed) is an unfortunate term for this reality, carrying pretty negative connotations, but it is clearly on point. We have the subjective awareness of making choices and decisions. However, behavioral science and neuroscience dictate that such thoughts/feelings are merely coincident to the process of initiating a behavior or even other thoughts/feelings. We are so often influenced by the seeming potency of our own thoughts that we are convinced that they *must* be the cause of our actions.

To reprise some basic biology, the concept of "voluntary" is sometimes used to describe actions which arise in the central nervous system as opposed to the autonomic nervous systems of our bodies. However, this is mostly a discrimination for convenience and anatomical study. It is actually not fully synonymous with the ideas of "voluntary" and "involuntary." This calls into question the whole notion of what behaviors are guided by supposed free or voluntary choice and which are not—under an infinite variety of circumstances. We certainly acknowledge that reflexes and certain autonomic behavior are not guided by choice—knee jerks, temperature control, breathing, heartbeats, etc.[10] Most people will acknowledge that the vast majority of our actions are "automatic" in the sense that they occur without our thinking about them, i.e., prior deliberate thought—from ordinary motor behavior (walking, sitting, etc.) to routine activities. For example, how many times do you stop yourself and try to remember whether you moved forward to get out of the chair, or flushed the toilet, or even closed the garage door? You did, of course, but you also obviously did it without entertaining a specific decision process. If you did, you were almost certainly only minimally aware of it. Did you take all the pills you were supposed to this morning? Do you stop to consider whether to swallow the morsel you just ate? Of course, you *could* stop and spit it out, but if it's good, you don't have to pause and make a decision about swallowing it. Most of the actions we take in our daily lives are done without specific awareness and "conscious" decisions preceding them. Are we "free" in all those behaviors? Even if you claim they are "voluntary," such actions are determined by personal history, and by physical and social environmental conditions at the time and place they are emitted. The concepts of awareness and "voluntary" are of no help in understanding the assumption of free will. To draw a red line between actions which seem automatic and those which are preceded by actually thinking about them is a specious argument for free will.

Contingencies of Behavior

We are conditioned to believe in the "reality" of free will; and we continue to maintain the validity of this concept despite the fact that it is an illusion. What maintains that behavior? Much of human behavior is due to its external influences, both historical and contemporary. We refer to these as the *contingencies of behavior*—mostly those effects from a person's social or interpersonal environment. That is, our behavior is contingent upon the environments in which it occurs and/or has occurred. Some of our behaviors are reinforced or rewarded; others are not, and over time our behavioral repertoires are "shaped" by our histories in this regard.[11] It's not that simple, of course, as I'll address more in Chapter 10, but such is an important dynamic in shaping how we behave.

There are a number of dynamics to highlight here. First, sometimes the contingencies responsible for a choice or decision are clear and compelling, and therefore we don't give much potency to free will; at least we don't use it to explain our actions. We don't talk about free will as the genesis of sudden anger, or moving our legs to walk, or using a fork to eat—even though we might label it as free or voluntary in retrospect.

Second, behavioral contingencies exert an influence on our perception of free choice to the degree that they are balanced or unbalanced. Where there are drastic differences in reward or punishment for taking a certain action, choice seems to be very limited. To swallow a mouthful of food usually presents a striking imbalance. Swallowing is rewarding; not swallowing is uncomfortable. We don't typically think of such decisions as purely "voluntary," even though one might label them that way in retrospect. Where risk and reward are relatively balanced, choice becomes a more "considered" set of actions in our brains. Should I take a job for more money and leave my more secure position? Should I eat that fish even though I'm a bit put off by the odor? Do I dare to eat a peach? These kinds of decisions are generally made with the perception that they are again self-initiated or voluntary. And this is because the various potential consequences can be reasonably envisioned and risks balanced, leading to an impression of free will.

Third, where the contingencies of behavior are very remote, we tend to think that we've freely chosen. When we say, "I love you," that's certainly ascribed to our own volition. Even though we might retrospectively understand the long history of enabling conditions, it's not usually productive to do so. Recanting all the history and reasons you say "I love you" is usually neither possible nor productive. It is more than convenient to consider this as purely voluntary. When you make a "choice" to marry, to wear a certain dress, to shun an acquaintance, you believe that you made those decisions freely—even though there were external contingencies that clearly influenced them. Even if we understand the contingencies, it is not very efficient to lay them out. When I say, "I want ice cream," it is not economical to explicate the personal history of what goes into that behavior. We just accept that it is a function of all the experiences we've previously had. The various consequences of your behavioral history are very remote, and you merely consider your choices to be free ones.

Fourth, where behavior has a long and very complex history, we don't invoke notions of free will. You do not invoke free will to explain why you've just read these words, (if you got to this point) though you might for picking up the book in the first place. You wouldn't invoke free will to explain the specific behaviors involved with painting a sunset or throwing a baseball or playing an oboe. They are just too complex, with long histories, well beyond our ability to analyze them. These are well-learned behaviors, and we acknowledge that it is the learning or conditioning which enables the responses; but to analyze them would be nearly impossible; and to see each component as "free" wouldn't make sense. We see them as just "voluntary" behaviors which we have freely chosen to engage in.

Early Development

We are not born with an idea about free will, and essentially none of our behavior at birth might generally be considered "voluntary."[12] So how does the illusion of free will develop? Young children are naturally curious. Curiosity is a large set of mostly innate behaviors which enable rapid learning about the world. In fact, curiosity has been an evolutionary advantage for primates and many other animals; it enabled learning that is crucial to survival. Young infants emit many seemingly random behaviors. In addition to those, there is a tendency to be curious and to emit novel behaviors. Parents like to see lots of these sorts of behavior from their infants. A good deal of this behavior may be innate; and as specific behaviors rapidly become evident in young children, they are rapidly and enthusiastically reinforced by parents. "*You* kicked your legs . . . wonderful." "*You* picked up the spoon." And novel behaviors and curiosity are later prompted. "What do *you* want? . . . Good choice." Thus, when novel behaviors are emitted and reinforced, or "choices" are made, the illusion that "It's *my* decision" becomes strengthened in the child. Parents who reinforce decisions made by young children generally do so without any reference to the prior causes of the child's decision. Thus, the concept of free will—albeit an illusion—is shaped. The child grows to believe that she is the unprompted agent of the decision and the behavior. The more this is reinforced, the greater the illusion of free will. "*I* made the decision." Young children learn to respond to increasingly complex stimuli, commands, and expectations. In doing so, a developing child must consider more and more pieces of information in order to respond. "What do you want to eat?" This takes time, allowing for a child's consideration of options. Even when time is very short, e.g., "Where's the green block?" there's opportunity to focus on alternatives. Such temporal delay and focus reinforces the idea of voluntary choice. In addition, a parent imbues the child with the expectation that there *is* choice in considering the information and emitting a response in a seeming "infinity" of behaviors. This creates the illusion of voluntary choice, since that is how it is labeled and reinforced. "*You* made a great choice." The self-concept of acting freely thus develops, i.e., the child believes he is making voluntary choices about his own behavior. Of course, punishment can also play a role. "*You* can't do that." And some punitive actions from a parent may follow. In short, behavioral contingencies rapidly shape the illusion of voluntary behavior or free will in the child, and it occurs very early in the child's development.

Thinking and Emoting

Although the development of human "reasoning" was based in part on experience with cause and effect, so-called logical thinking associated with most ordinary human behavior is generally ascribed to free will. "I decided there were better options." There is other cognition which seems more rapid and is often described as impulsive or irrational. As I hinted in Chapter 2, this dichotomy has been brilliantly explicated by the work of psychologists Amos Tversky and Nobel laureate Daniel Kahneman.[13]

There is rapid thinking, for which we seldom use the term free will, and there is slower thinking, for which we consider choices that are deliberate, voluntary, or generally consistent with free will. We see in Tversky and Kahneman's experiments that people often make "irrational" decisions (based on their unique histories and biases) where better consideration of circumstances and data would lead to more effective outcomes for a person. It is a matter of how any relevant information is processed and considered. When information is not well-considered—thinking fast—we tend to rapidly discriminate between options; and where the outcome of such a choice is positive, we may call a decision "intuitive." Where it isn't, we often label it as a "snap" or impulsive decision. But what we mean is that in the first instance, relatively speaking, the person has merely used stored information to make a decision, whereas, in the second, the person has not. However, where choices are more thoroughly contemplated, we tend to invoke volition regardless of the outcome. We attribute "free" choice to a behavior only to the degree that it is "considered," i.e., the amount of information and reasoning effort that precedes the behavior—thinking slow. This reinforces the illusion that we can make free choices, even though our histories dictate the actual choice behavior.

However, there is clearly a need for rapid decision-making, including actions which are sometimes referred as emotions; "emotional behavior," so to speak. This has had an evolutionary advantage. For example, it would be pretty risky to have to "consider" all relevant observations and data in making rapid decisions where survival is at stake. One such decision, for example, is what we call "fight or flight" behavior, the tendency to deal with a threat by fleeing or confrontation. In its purest forms, it is an acute survival response and is not generally labeled as voluntary or a choice born of "free will." Running from an attacking lion isn't exactly a choice. However, "fight or flight" behavior has become generalized in modern humans as responses to stimuli which don't present threats on a survival level but merely on an interpersonal or other mundane level, and they can signal "emotional" components. We do still contain the biological basis for such behavior, but it has become part of the way we make decisions, as Kahneman has laid out. Nevertheless, certain "fight or flight" behavior, despite its biological roots, is often labeled as voluntary. This is because the "fight or flight" impulse has become conditioned to many other stimuli, mostly interpersonal ones, in our lives. "I'm afraid of what will happen tomorrow." "You better not come over here." "He didn't have to yell at me." "No way can we invite that guy." Emotional behavior is often seen as "freely" emitted, even though it has both individual and biological roots. When we experience anger and heart palpitations with a choice, for example, we still see the choice as voluntary behavior. We label it that way in order to perceptibly move it away from any evolutionary history, where we would have trouble assigning the label of "free will."

Timing and Complexity

Even though consideration of multiple types or pieces of information can take but a few seconds, that delay is often significant enough to reinforce the perception that the "decision" to engage in some type of behavior is "self-initiated," even though considerable experience preceded that "final common pathway." In fact, our experience of free will appears proportional to the time it takes to initiate any particular choice or set of behaviors. The "decision" to cough to relieve some acute distress, for example, is relatively easy, has been well learned, and does not involve a great deal of cognitive processing. In fact, at times it may require a modicum of cognitive processing, but we generally do not invoke a concept of free will to explain it. The "decision" to eat one thing or another may involve more cognitive processing, and it may take a bit longer (at least for some people), though only in seconds. If you make a decision about buying a car, choosing a career, or choosing a partner, for example, that takes a good deal more time, consideration of alternatives, and cognitive processing. The longer it takes to come to a "choice", and the more complex the matter, the more opportunity there is to perceive that one is making it voluntarily—independent of the operation of the history which actually subserves it. We invoke the idea of free will to the degree any choice takes time. Above other factors, this may be the one which most often imbues us with the impression of free will. *The longer it takes to make a decision, and the more complex the decision, the more likely we are to label any choices as free will.*

Postdictive Thinking

The notion of thinking, as I've said, is often invoked as supposed evidence that we have free will. "I thought about doing something, and then I did it." The thought is presumed to always come before the decision or action. We reviewed some extant neuroscience on this, but for now we just know that this is not always true. This is a bankrupt argument for free will, of course, but we should consider another factor about "decisions" in this context.

When you ascribe an action to a previous thought, you may actually have the sequence wrong. That is, you may cite the reasons that you spontaneously made a choice, but those reasons may come to your awareness only after the behavior is emitted. We saw in neuroscience that choices can be made before a person is "aware" of them, even though they thought they had made a "voluntary" decision. However, in a practical sense, people often tend to label their choices as free because of experiences that come *after* the choice has been made. This is known as a "postdictive" influence and has been demonstrated in a variety of studies. (38, 537) In various experiments, subjects are asked to quickly make choices and remember what choices were made. For example, in some of these experiments, subjects are asked to choose among a series of identical objects flashed on a screen before any object changes color. (38) Afterward, the subjects are asked to identify which object had been previously "chosen." As you might suspect, the object identified is much more likely to be the one that did change color, even if it had *not* been chosen. The subjects' recollections were influenced by other things that occurred to them after they made the choice. As one experimenter put it:

> 'Postdictive' illusions of this sort are typically explained by noting that there's a delay in the time it takes information out in the world to reach conscious awareness: Because it lags slightly behind reality, consciousness can 'anticipate' future events that haven't yet entered awareness, but have been encoded subconsciously, allowing for an *illusion* in which the experienced future alters the experienced past. (37)

This, of course, is what happens in everyday life. We often engage in seemingly voluntary behavior without "thinking" about it and then reflect on it as if we had reasonably chosen to act in certain ways. We might "choose" a flavor of ice cream and only thereafter—however short the time—consider that we loved it before. Or put on a certain shirt. Or select a TV program. We often ascribe our choices to some free action, but only afterward come to realize why we did what we did. Nevertheless, the illusion remains.

The Miracle

What do we actually mean by "free will?" Again, we are generally referring to decisions or "choice behavior" for which no physical or behavioral explanation is known or can possibly be known. We use the labels free choice, voluntary, and similar concepts when we cannot appreciate a driving force or contingencies of the behavior for the choice. It is as though a minor miracle has occurred. Other "miracles" are described with the phrase "god of the gaps" to provide "explanations" for things we don't understand. Similarly, *free will can be seen as the "gap" for human behavior*. When we can't explain the behavior, we invoke the miraculous idea of free will. A choice is made completely freely without any historical or other present external contingencies.

To the degree we understand the generation of behavior, the less we assign pure volition to the actor. We often label thoughts or actions as "free" or the exertion of free will where we do not quickly appreciate the contingencies of behavior which have enabled them. Free will is invoked as the "uncaused cause" for behavior. It is no less a "miracle" than the ones we ascribe to a god. This may seem too extreme a viewpoint. But again, there is no scientific way to explain free will—by definition. It defies the laws of nature; and we do use it when we don't understand. It is no more than an ethereal "gap" that we employ to categorize our own actions. It is a miracle.

Continuum

Human behavior exists on a continuum—from the autonomic to the reflexive to what we might describe as habitual to the deliberate to the very delayed and complex. Although these types of behaviors are often under the control of different mechanisms or combinations of mechanisms, it is helpful here to consider any behavioral repertoire along a continuum. The often-used binary discrimination of voluntary or involuntary is not always a helpful perspective. Again, where do we invoke the notion of voluntary choice along that continuum?

We don't invoke "voluntary" or free will to explain pulling your hand off a hot stove. Do we "choose" to alter our brain waves when we sleep; or to cry as we are born? Do we choose to enable our kidney functions? We might "explain" a choice of shoes with free will. And we nearly always offer that the choice of mate is a voluntary choice. Where do we draw the line on "voluntary" behavior? It is arbitrary, of course, differing from one situation to another, one behavior to another, one culture to another, one person to another. Volition or free will is just not a useful way of looking at our actions. It is merely a convenient label we use when we reach certain points along a huge continuum of human behaviors.

Conclusion

Human behavior is a function of its history, pure and simple. The contingencies which have shaped our behavior lead to nearly all our choices and other actions that we emit at any particular time. Biologist Robert Sapolsky summarizes the issue well:

> Someone behaves in a particular way. Maybe it's wonderful and inspiring, maybe it's appalling, maybe it's in the eye of the beholder, or maybe just trivial. And we frequently ask the same basic question: Why did that behavior occur? If you believe that turtles can float in the air, the answer is that it just happened, that there was no cause besides that person having simply decided to create that behavior. Science has recently provided a much more accurate answer, and when I say 'recently,' I mean in the last few centuries. The answer is that the behavior happened because something that preceded it caused it to happen. And why did that prior circumstance occur? Because something that preceded it caused *it* to happen. It's antecedent causes all the way down, not a floating turtle or causeless cause to be found. (445)

This deserves a bit of emphasis and context. We can acknowledge that the brain is the proximate cause of our complex behavior. But only that. Everything that happens to you and everything that you do (whether you are aware of it or not) combine to produce your behavior. No matter what changes occur in the brain, no matter what algorithms it generates for itself, all our actions are a direct result of the contingencies of behavior in your history—from early development to the happenings of recent minutes and seconds to your lifespan. Free will plays no part.

Philosophy

I would be remiss if I didn't give a little space to philosophy here, since the notion of free will has been a hotly debated subject throughout history. Plato and Aristotle were supporters. Plato believed that the soul was the source of free will. Aristotle thought that humans could be rational and therefore engage in some voluntary choices. The Christian tradition in this regard was epitomized by St. Augustine, who espoused a self-determining power for humans independent of the world around us. Much of the thinking involved the necessity of free will in order to act morally and, in many instances, to discriminate good from evil—as a route to salvation. No more need be said on that latter topic here. And most of the Enlightenment thinkers conceived of some version of free will. Of course, the "paradox" of the free will problem was never resolved by any of them. Nevertheless, philosophy continues its search for some perspective to bring apparent credence to the matter.

Compatibilism

There have been many attempts to resolve the issue of free will existing in a determined universe. They share some commonalities, but the most often currently cited is what's generally known as *compatibilism*. It is the idea that, despite the universe being determined, we have freedom to make decisions. That is, for any given situation or where a decision is made, the classic aphorism is offered, "*You could have done otherwise*." This is an idea which has run through history in one form or another. John Locke was an early proponent of this view, and it is the one most espoused by philosophers today. Philosopher Christian List, for example, summarizes the position—given the world as it is, in fact, "different alternative actions can be possible for the agent." (316)

This is the now classic philosophical argument about free will. Despite countless volumes written about this, the "could have done otherwise" approach holds no water. Like all philosophy, it is based merely on an assumption. It *assumes* you could have done otherwise. But in fact, *you did not*. You could have turned left instead of right, but you didn't. You could have eaten a peach, but you didn't, or you ate a banana. You could have worshiped God, but you didn't. Yes, you could have done otherwise, but you didn't. The notion that "you could have done otherwise" is rendered ludicrous by the fact that "you didn't."

And you couldn't. We could also pose the question, "Could you have been a different person, a person with different characteristics?" Yes, you could have; but you aren't. If your parents didn't meet, *you* would not have been born. You can wind the proverbial clock back as far as you want. In the end, this boils down to "could the universe have been different?" Of course, it could. But it wasn't. You can ask the question, "Could the Big Bang have been any different?" "Could the strength of gravity have been different?" Yes, it could have been, but it wasn't. If it was, you wouldn't even be here to ask the question. The multiverse concept tells us that in fact it could have. But in this universe, it is what it is. At bottom, posing the "could have" question is not an argument for free will.

Taken another way, if one can "freely choose" among alternatives at any one point, that is synonymous with the idea of choosing which of various futures can be obtained. If I turn right, that leads to some particular outcome, e.g., going to the store to buy orange juice. If I turn left, I get lost or delayed and, e.g., and there's no more orange juice at the store. Then my future, all futures, are changed by my choice. *In effect, if "choosing" one future over another is possible, we must throw out the laws of nature.* The future is predicted by preceding events, not you as a "free" agent. The future *is* determined by the past. This goes to the heart of the problem. You might again consider this to be a *reducto ad absurdum* argument. However, the factors determining any thought or behavior, including your existence itself, are determined by how the universe has evolved to that point, as I've already discussed. To suggest that you are "free" of that and can willfully alter the future reality of the universe requires some other imaginary force. It is but a science fiction fantasy. One cannot acknowledge that the laws of nature have brought your brain to a certain temporal point, the point at which there are a small number of possible alternatives, and then, according to philosophers, the "magic" happens and you make a decision—and that creates an alternative cosmological future. That is what you'd have to believe to enable some sort of validity to the concept of "free will"—that any of us can change the future "at will."

Some other perspectives on this: A leading philosopher on the issue, Dennett's view of compatibilism argues that despite a deterministic world, we can still make complex and moral decisions since we have evolved the abilities to do so. (121) He argues that, "Compatibilism . . . free will and determinism, are compatible after all." He makes an exception for circumstances where people are coerced, though that is a pretty ambiguous term. This is like saying that we have free will about some things and not about others, a rather ludicrous position. That we are granted only certain free will abilities, while everything else is determined for us, of course, only harkens back to the Greek gods. At this point, the idea of having "some" free will is only a matter for science fiction (For a good example, see Philip K. Dick's 1954 short story, "Adjustment Team."

(134)) Also, if we've *evolved* to make "moral" decisions, as Dennett suggests, then we've also evolved to make "immoral" ones, as we see all around us. This makes no sense, logically or biologically. Products of evolution are amenable to empirical inquiry; free will is not. We've addressed the issue of complexity, and "morality" (a set of pretty complex behaviors) provides no solace for free will. He ignores the fact that we are conditioned to behave in various ways, including those that lead to complex and "moral" decisions, which themselves are highly variable, based on culture and other circumstances. But that also is not free will. Dennett's idea applies more to the notion of personal "freedom," not the idea of "free will" outside the laws of nature. Of course, this still begs the question: What accounts for those abilities and how are they exercised, matters which are not addressed by compatibilism? Dennett celebrates our "freedom" as human beings, but that is not free will. This is still the inconsistency and muddiness of the compatibilist view.

Similar to other philosophers, Christian List offers that, "The best way to defend the compatibility of free will and determinism is to recognize that free will is not a physical phenomenon, but a higher-level phenomenon on a par with other familiar higher-level phenomena such as beliefs, desires, and intentions. If we are searching for free will at the level of fundamental physics, we are simply searching in the wrong place." (316) Of course, we should ask, "What really is a higher-level phenomenon?" Do beliefs, desires, and intentions arise from some non-physical force? Are we reinventing Plato's or Descartes' soul? It would be ludicrous to assume that beliefs and intentions, for example, do not arise from the actions of our brains. That we do not fully appreciate how that happens is not an argument for free will. Compatibilism, again, can't rescue free will from the laws of nature.

We should note that there are a few other "theories." For example, philosopher Liam Clegg has offered what he calls "protean free will." (91) To simplify, since human behavior is so variable and unpredictable, free will becomes the only explanation. The concept seems derived from its biological use. Protean behavior is often used as a description of the variable and unanticipated actions of prey animals, for instance, in trying to elude predators. It is essentially the adaptive unpredictability of behavior. Perceived unpredictability in this regard has likely been of survival value in certain organisms, but it is doubtful that such behavior involves some considered, deliberate decision processes on the part of an animal. It is not evidence for human free will. It is a vacuous philosophical argument.

Philosophers have been struggling with this problem for a long time, but various philosophical gymnastics have not rescued free will from a deterministic universe. No convincing solution to compatibilism has ever appeared. As philosopher Shaun Nichols points out, "Philosophical conflicts over such concepts as free will and consciousness often have their roots in ordinary intuitions, and the historical debates often end in stalemates." (376) And of course, they always will—which is why such ideas as compatibilism are destined for the dustbin of intellectual history. The debate is a sterile one. Again, the notion of free will is merely a "gap" filler in search of some fabricated philosophical rationale.

The Problem

Where does this leave us? If compatibilism is a dead end, perhaps you might subscribe to some theory of quantum mechanics that may "allow room" for free will. As unlikely as that is (and allowing "room" isn't the same thing as providing evidence for it), science would still insist that you must explain (1) what free will is, (2) where it comes from, and (3) what physical phenomena are responsible for such behavior (whether it is thinking or actions)? The only answer to number one is that there is no affirmative definition of free will. Proponents can only suggest the self-contradictory notion that it is something inherent in human beings which cannot be explained by science. If it is some undefined, fundamental characteristic of humans, such is no explanation. The only proposed answer to number two is that it comes from God, of course, an unverifiable and fanciful idea. And for the answer to number three, we must conclude that any empirical explanation contradicts the notion of free will itself. If behavior can be explained by physical phenomena, then actions observed outside of this realm are not possible. As science continues to drill down into the fundamentals of quantum mechanics—or perhaps someday discovers a unified theory for all the forces of nature—perhaps we will understand quantum effects and probabilities at a more fundamental level, as Einstein fervently held, and as current physicists and cosmologists continue to pursue. But this does not mean that the notion of free will is valid because human behavior is not fully explained. This is a "gap" hypothesis with no foundation, as we reviewed. The idea of free will's existence based on incomplete scientific understanding of workings of the brain, provides an opportunity for philosophical musings but adds no understanding. The bottom line is that we use this idea when we can't immediately find other reasonable explanations.

Theory?

Can there still be a "theory" of free will, theory in the most general use of the word? Science does have something to say about human behaviors, including supposed voluntary actions or choices. But we generally do not have enough information to precisely predict them. This again goes back to the problem of the calculational impasse. It has been stated in a variety of ways. Stephen Hawking's perspective, for example: "In the case of people, since we cannot solve the equations that determine our behavior, we use the *effective theory* [emphasis added] that people have free will. The study of our will, and of the behavior that arises from it, is the science of psychology." (214) "Effective" here means practical, not well-developed or well-evidenced. It is obviously a conversational convenience, not an actually useful theory. It is not even a reasonably good theory, since prediction based on free will is impossible, both practically and theoretically, as was admitted. How do you predict something if the options are "free?" And if "free will" operates outside the laws of nature, it cannot be included in some "theory," by definition. We're a long way from solving the equations of human behavior. As Greene's calculational impasse suggests, we do not even have the requisite data, but "free will" can't fill the gap.

Many have tried to redefine free will to make it amenable to scientific inquiry. Of course, we can redefine anything to make it measurable, though that's unlikely to solve the problem. For example, neurobiologist Björn Brembs has argued that even invertebrates—like flies, cockroaches, and worms—emit complex behavior and have the ability to make different responses to "identical" situations or conditions. He concludes that "Abandoning the metaphysical concept of free will does not automatically entail that we are slaves of our genes and our environment, forced to always choose the same option when faced with the same situation." (60) Of course, he neglects the obvious point that there is no such thing as an "identical" situation or condition—history *is* a contributor to all behavior and no two situations are precisely identical, including the fact that an organism itself becomes different from moment to moment. Eschewing the emotionally loaded notion of "slaving" to our genes doesn't bolster the argument. He suggests "re-defining the familiar free will in scientific terms rather than giving it up, only because of the historical baggage all its connotations carry with them." But it is that baggage which nearly everyone dearly holds onto, most notably philosophers. And the idea that there can be a scientific definition of free will is rather absurd. Philosophical arguments about free will shed no light on the issue. Based fully on assumptions, they fail to demonstrate in any way that it is a reasonable idea.

Animals

If humans have free will, are we the only creatures on Earth that have "it"? Are humans distinct in this putative ability? Do other animals have free will? We sometimes ascribe this power to certain other creatures—apes, dolphins, corvids—where higher-order cognition is evident. We have no way of knowing that other animals hold the same conceptions of their own desires and actions as we do, i.e., that they consider their choices are "free." Is this a ludicrous question? Perhaps. If only we could ask them.

A Few Examples

We tend to ascribe free will to other species to the degree that they act like us. When you look at the behavior of a chimpanzee, a dolphin, or an elephant, do you get any impression that there is free choice behavior going on, that freely considered decisions are being made? We are not able to communicate with them, so we don't know their own impressions. Nevertheless, these types of animals are some of the ones we've assessed to have higher levels of intelligence, something we arbitrarily and incorrectly equate with the capacity for free will. We could extend the analogy back through time and across other species. We've talked about how clever other animals are, especially some birds and mammals, solving pretty complex problems, demonstrating empathy, and engaging in all sorts of behavior we would call voluntary when done by humans. Do these creatures have free will? If we do, why can't they?

We might also consider the honeybee, which learns a great deal and even communicates complex data to its peers, a very critical social dynamic. But is there any free will there, or do we ascribe all of this to genetically determined behavior? Most people would view it as the latter. It was once felt that such communication was purely innate, but recent research confirms that it actually must be socially learned. (432) Social learning and impressions of free will do seem to go hand in hand. Who's to say that the honeybees don't show any voluntary choices here? We generally ascribe behavior based on such social learning, in humans for example, as tantamount to free will. And now we know that some honeybee cousins, the bumblebees, can engage in what can now be considered play. They can manipulate things with no obvious or envisioned consequences. And the play behavior is quite variable (not stereotyped); it seems to have no survival value; it isn't at all related to functional bee behavior; and it's not to reduce stress, as young bees seem to do it more. (137) This sure seems like free will. Most people would have trouble assigning free will to the bees, but all this social learning and play certainly "seem" like it.

Jumping way down the phylogenetic tree, consider again the case of *C. elegans*, the primitive creature with about a thousand cells, whose entire connectome has been mapped. It can develop food "choice" through learned behavior. Few would say that this learning results in "free" choice, but we have no idea at all where exists evidence in the connectome for this behavior. It isn't explained by any biology we know of. Thus, should we consider it free will?

Our Genus

Let's also consider our so-called more primitive cousins. Did free will only appear with our particular species of *Homo*? *Homo erectus* appeared millions of years ago and somehow tamed fire and could orally communicate. These behaviors certainly reflect complex thinking, planning, and decision-making. Were there conscious choices here? Was that free will? What about *Homo habilis* or Neanderthals? If you could ask them about free will (they did have language), what would they say? How about our subspecies, the Cro-Magnons, who lived among us until perhaps the early Bronze Age? Would we deny them free will? Do you draw the line only at modern humans? That would make no sense whatsoever.

A Continuum?

Does free will require a sufficiently complex brain or nervous system? Do we support the idea of free will in dogs? Fish? Chimpanzees? Planaria? Amoebas? We might ask, "Is there some sort of continuum for free will?" That is, perhaps animals can make some choices which are independent of historical and social learning, and some which are not. Does free will require some sort of consciousness? If that view holds, the entire concept of free will seems to evaporate, much as we saw for consciousness.

It is more economical, and clearly more scientific, to see our "choices" as a function of environmental and social learning. Organisms differ across the phylogenetic tree in their ability to make decisions. They have become more and more capable of complex "choices" and actions as phyla have progressed over hundreds of millions of years. The relevant continuum is also well analogized by the classic principle of "ontogeny recapitulates phylogeny." That is, humans develop more and more capacity to do things as they grow and mature, beginning *in utero* and extending for a few years after birth, quite analogous to how species have evolved and progressed in this way over the eons. We do not attribute early infant suckling to free will. We often label as "free will" the somewhat older child's food preferences. And showing empathy with others is yet more advanced. You get the picture, and it doesn't necessitate the idea of free will. As we look at other branches of the evolutionary tree and individual human development, it becomes clear that the *seeming* ability to "freely" make decisions, as we construe it, is only a function of the complexity of an organism's behavior.

Either "free will" exists outside the laws of nature for all sorts of living things, or it is a patently absurd idea. We could ask, "Where on the continuum of living creatures does free will emerge?" Hawking and Mlodinow sum up the matter well, using the well-worn *C. elegans* analogy:

> If we have free will, where in the evolutionary tree did it develop? Do blue-green algae or bacteria have free will, or is their behavior automatic and within the realm of scientific law? . . . but what about the roundworm called *Caenorhabditis elegans*—a simple creature with only 959 cells? . . . it too has a definite preference in food and will either settle for an unattractive meal or go foraging for something better, depending on recent experience. Is that the exercise of free will? (214)

Do we ascribe free will only to ourselves? That we attribute "free will" to ourselves uniquely is scientifically illogical, of course, and perhaps just our "species arrogance." The Copernican Principle again. If

God gave free will uniquely to humans, you'd have a good deal of trouble explaining why chimps care for each other emotionally or medically, or why our *Homo naledi* cousins had ritualized death burials. And if you extend the problem back in time and across living things, you might question the phenomenon altogether. Do Chalmers' "conscious photons" have free will? The only logical alternative is that free will is a useless concept in explaining our behavior. In this context, we can close the door on free will, much as we did on consciousness.

Illusion of Free Will

Nevertheless, the effort to preserve the notion of free will runs deep in our lives, even though it is an illusion. It does not actually exist, yet we hold that idea as strongly as any other. We *must* have free will; otherwise, how could we go on? So let's devote some attention to the *illusion* itself. We seem to know that free will exists because we "experience" it. Free will is *real* to the beholder. It is a perception that can't be shaken. The only empirical explanation, therefore, is that the illusion must somehow exist in the operation of our brains. Yet, the human brain can create all sorts of illusions, perceptions that don't conform with external reality. Our brains can trick us in a variety of ways, and we come to experience a number of things that just aren't valid. Over time, or sometimes suddenly, we thus come to believe them to be true.

Hallucinations

Let's again return to this issue. That we *perceive* we are "free" to take actions or make decisions is an illusion. Illusions are real to the holder, but so are hallucinations. And we must take them seriously. There are many reasons that hallucinations can occur, which I've discussed. These are reported perceptual experiences that some people have which don't conform to reality as it's generally conceived at the time by most others.

Hearing people who aren't there is a hallucination. So far, there's no evidence that ghosts are real. Getting messages from Martians is pretty unlikely. Though such experiences may be real to the holder, we don't validate them, and we tend to medicate or physically confine people who make these claims too assertively. They are perceptions without any external referents. They are illusions, and therefore, we don't place any stock in some "reality" they may represent. Yet, seeing or hearing things that "aren't there" does reflect activities going on in the brain, even though we don't well understand them at the present time.

Neuroscience

In fact, our brains can do some weird and incredible things, including some other well-known illusions. Let's consider a powerful example—phantom pain. That is pain that people experience in missing body parts. Part of an arm or a leg might be missing, and the perceptible truth of the lost appendage is real. However, some individuals who have had an extremity amputated feel pain in the extremity which is no longer there (particularly if the amputation began with trauma). The pain is real, often agonizingly so. To clarify, it's not just the pain that's sometimes felt in the stump of the lost extremity. It's a clear experience of pain in the anatomy that's missing. It's pain in fingers or toes, for example, that no longer exist. Our brains create this experience, which we can only view as illusory—even though the neural activity that creates it is real. We don't know exactly how this happens. Your brain merely creates that reality, just as it creates various hallucinations (though the specific mechanisms may be different).[14] The human brain is a powerful organ, and it can develop an experience or belief in almost anything. That it can create the illusion of free will is not at all surprising.

We might rightly ask, "What is it about the brain that creates the subjective impression of free will?" There is no way we can presently test for the biology of the illusion itself. Nevertheless, there have been some provocative experiments in neuroscience which seem to bear on this issue. One of the most impressive studies was conducted by neurologist Ryan Darby and colleagues, using brain imaging, in an attempt to localize possible components of the phenomenon. (105) They examined 28 patients with what's called "akinetic mutism." That's a disorder caused by damage to the brain in which people lose the *desire* to move or speak, even though they remain motorically capable of such. (This is not a type of aphasia, a disorder in which speech or language is actually disrupted by brain damage.) Using MRI and CT imaging, the researchers were able to identify a network of neurons which were associated with this malfunction. (In this case, areas connecting to the anterior cingulate of the cerebral cortex.) That is, damage to the network was interfering with the *desire* to act—what we otherwise might consider a component of "volition." They also examined 50 patients with brain damage resulting in what's called "alien limb syndrome," which causes people to use a hand or arm to touch or grab things or make other "seemingly voluntary" actions *without* the subject actually intending to do so. Often the individual is not able to control a seemingly autonomous arm or hand, for example, without grabbing it or sitting on it. The limb seems to act of its *own* "volition." For instance, a person may try to unbutton a shirt without any subjective desire to do so, and he would not be able to stop the action without grabbing the "alien" extremity with the other hand. These are truly strange, complex behaviors—not reflexes—but they do sometimes occur in people who have suffered certain types of brain damage, generally certain types of strokes or tumors. The researchers concluded that a network of neurons connecting to the precuneus, a portion of the parietal lobe of the brain, was responsible for the lack of "agency" or "responsibility" for these people's strange behaviors. Interestingly, this is an area of the cortex which has also been linked to descriptions of one's "self," what some have labeled as our sense of "self" or "self-consciousness." They also repeated the imaging tests with people who had certain psychiatric disorders associated with a disturbed perception of free will. These are certain so-called "conversion disorders" (where people report not being able to move a body part, even though all their relevant anatomy is intact), catatonia (where a person just doesn't move much), and nonepileptic seizures (seizures without any relevant brain abnormality). Not surprisingly, neuroimaging abnormalities in these cases were found for areas similar to the results for patients with the above types of brain damage. Although these findings are not perfect, they do suggest that our ability to perceive the initiation and control of our own actions, the perception of acting freely, do correlate with rather specific networks in our brains.[15] Our ability to act with some sort of "intention" and to *believe* that we are doing so voluntarily do seem to have neural substrates.

Volition, per se, does appear to have some organic vectors, not dependent on some nonmaterial mind, and neuroscience may eventually explain the private holding of the "free will *illusion*." For now, we do know that the *perception* of free will *is* purely an epiphenomenon, somehow arising from what is encoded in our brains. The perception is real, though the inference of "acting freely" continues to be an illusion. This is not a surprise, of course; but the more we understand about how our brains work, the less we may be convinced that some other freely operating non-physical mind is at work.

Conclusion

The matter of the illusion is cogently summed up by physicists Hawking and Mlodinow:

> Though we feel that we can choose what we do, our understanding of the molecular basis of biology shows that biological processes are governed by the laws of physics and chemistry and therefore are as determined as the orbits of the planets. Recent experiments in neuroscience support the view that it is our physical brain, following the known laws of science, that determines our actions, and not some agency that exists outside those laws. . . . It also seems reasonable to conclude that the outcome [of innumerable brain actions] is determined in such a complicated way and with so many variables as to make it impossible in practice to predict. . . . It is hard to imagine how free will can operate if our behavior is determined by physical law, so it seems that we are no more than biological machines and that free will is just an *illusion*. (214)

Evolution of the Free Will Illusion

Of course, since the illusion itself is real, and since it obviously becomes somehow encoded in our brains, we must ask, "How could this have come about?" "How is it that humans came to believe such a thing?" We generally don't believe plants or mushrooms or planaria have free will, and they're unlikely to have an illusion of it. We might be convinced that dolphins or chimpanzees might have some idea about free will of their own. What about our more recent ancestral cousins? Do we believe that Neanderthals had an impression of free will? They certainly created weapons, produced art, and had some form of advanced linguistic communication. What about *Homo habilis* or *Homo erectus*? Did these hominins *perceive* that they were the autonomous agents of their own actions?

Returning to Hawking and Mlodinow's challenge, "If we have free will, where in the evolutionary tree did it develop?" (214) This was obviously rather sarcastic, but we could actually entertain the notion of where the *illusion* of free will developed. There would not be a fossil record of such things, and we cannot ask other species about their perceptions. Nevertheless, given the existence of the illusion, we can reasonably speculate on its history. Where did it come from, and did it help us function? Or is it no more than a meaningless epiphenomenon?[16] That is, might the illusion itself actually have conferred some benefit for early *Homo sapiens*? It is unlikely that the illusion provided an actual competitive survival advantage for humans, but it may have functioned in other important ways. I'm assuming here that Neanderthals and our other cousins would have had at least some rudimentary illusory perception of their own behavioral autonomy. However, any such cognition where *H. sapiens* was superior would not likely have been a factor in any other species' competitive disadvantage. Their extinctions were likely due to other factors, as I've outlined.

It's safe to say the *illusion* of free will did not spring forth in humans all of a sudden. We can admit that amoebae and tardigrades probably didn't have such an illusion. (Actually, I'm not too sure about those remarkable tardigrades.) It must have developed over the course of evolutionary history —particularly of our biological order, genus, or species; and there must have been reasons for this. Thus, we are left with the challenge of how this may have happened. What benefit did it enable? This is a very difficult but potentially not insurmountable problem. There is obviously no fossil record which would reflect such human perception; it would have evolved millennia before we learned to write, paint, or produce any other instructive artifacts. All we can do is consider it relative to how it may have related to the challenges early humans faced, the increasing capacities of the human brain, and the analogy of how children develop it.

We reviewed above how children are shaped into the conception that they make autonomous choices. If we look at a more basic level, consider this. Brain development and capacity are critical in this process. The human fetus develops brain capacity both before and after birth. First, it takes many weeks for the brain to mature, to the extent that a child can grasp complex ideas, e.g., recognizing oneself (as noted above), using symbols like gestures and words, and especially using concepts like "me," "I," and similar ideas. Second, evolution has also granted us (perhaps our genus, at least), the capacity to *imagine* and, with that, the capacity to imagine our own autonomy. These dynamics are most likely the underlying factors in development of the free will illusion. As human cultures "matured" and language became more efficient, it is quite reasonable to assume that the free will idea was more and more easily shaped. Complex language was certainly a key. Such is the most likely overall explanation for the evolution of the free will illusion—imagination and complex language.

Brain Again

Yet, there are still some other specific factors to ponder. The consideration of options, often complex ones, is enabled by an advanced brain, which developed in humans over hundreds of thousands of years (and millions in our ancestors before that). I alluded to this briefly above, but now seriously consider the complexity of brain activity that precedes any given thought or behavior. To be overly general, this includes processing an array of complex stimuli as well as using a history of information, sometimes a lifetime of it. In making a decision or emitting any complex behavior, any active perception or consideration of even a small part of such a huge amount of information would be paralyzing. The human brain (and that of most other mammals) has evolved to attend to what is important at the time and to ignore or suppress irrelevant history and activity. In fact, the dominant activity of the human neocortex is *inhibitory*. That is, it keeps things from happening that are unnecessary at the time, whether it's movement-related or some appearance of some "stored" information. Otherwise, there would be a processing of a seeming "infinity" of data. The neurologic chaos that would otherwise occur would render a brain immobile. Evolution has taken care of that problem. We only attend to information that is necessary at the time.

If you are not able to inhibit a lot of the relevant information for any choice or decision, it is more difficult to emit considered responses, and you are more likely to have difficulty with a decision. The analogy of being paralyzed is apt. The other alternative is to emit a choice or other behavior which seems to ignore much of the relevant information. This can result in rather sudden responses, often labeled as impulsive or sometimes bizarre. It's well known that some drugs can induce such behavior, and they do this, in part, because the brain's ability to suppress irrelevant information becomes impaired. That is why people in such conditions report bizarre experiences. We also often see this in persons with certain types of brain injury or impairment, who often have difficulty suppressing behavior because they have trouble squelching irrelevant information.

The fully functioning human brain "hides" the history of prior causes and uses only the information necessary to make decisions. When the decision is complex—even choosing a beverage, let's say, enormous amounts of information and history need to be suppressed. That gives the individual the impression of a "free" choice. That would not be possible unless our brains had evolved to differentially suppress "noise" and selectively access the relevant pieces of information. It is likely that evolution produced brains in humans (and perhaps other creatures) that were better and better at suppressing irrelevant information and focusing only on what was necessary. Of course, then, for any choice, the *postdictive* mechanism I discussed comes into play. The individual perceives that he has used all sorts of historical information, reinforcing the idea that he had actually considered such things and thus made an autonomous choice, something our advanced brains began to allow. And the illusion of "free will" became reinforced.

Problem Solving

Over evolutionary time, humans have needed to make decisions about more and more complex problems. Given humans' physical capacities when complex thinking had evolved, survival would have been dependent on such problem-solving abilities. Planning and social coordination, for example, would have been necessary to seek and kill prey, grow food, develop social relationships, etc.

As we formed larger groups, we became increasingly able to solve more complex problems—when to move the troop, whether and how to discipline young and noncooperators, or who should accompany you on a hunt. As we became more sedentary, even more complex problems arose—where and how to distribute grain in a community, what roles individuals should have, and thousands of other problems. For any decision, there were usually a variety of options. That is, there are "degrees of freedom" in making choices. When these were many, as they arose for Neolithic humans, much more information was required to consider alternatives and come to a decision. Complex decisions require abstract thinking and more processing time. This again led to the illusion of making voluntary choices, i.e., independent of the individual's history. The options grew exponentially with larger and more stable groups, and, at some point, humans began to use their capacity to imagine—to consider—options that weren't immediately available. I noted that the more options you have, the "freer" you feel. Being able to "consider" more options and take actions would certainly have been useful in early human problem solving, reinforcing the perception of acting "freely." Again, no fossil record is available on this hypothesis, but it has been widely discussed in the psychological and philosophical literature. (35, 500)

To wit, the use of intelligence, and a processing delay—which we all continue to experience when confronted with a difficult problem (Kahneman's "thinking slow")—allows for the perception that the solution is *self-generated*, i.e., we believe we have *created* the solution. Though we have no specific empirical evidence, the perception of "self-generated" solutions would certainly have reinforced the idea that "*I* did it." As more and more problems became solved in this fashion, we developed the perception that our thinking and behavior could be independent of the complex historical and current environmental stimuli in our lives. Thus, over time we developed the ability to solve complex problems and to believe that self-generated ideas were the solutions. Thus, the *illusion* of free will may well have evolved in humans because it enabled more of this seemingly autonomous problem solving.

Human Culture

Evolutionarily, early humans needed to emit more and more complex decisions in order to solve survival-related problems. We began to see choices as our own "uncaused" volition. That assumption likely also proved an efficient way to communicate. It makes no sense to outline a relevant history for a choice or decision, for yourself or others. "You planted the wheat over there because you saw your parents do it, you understand that it will grow, that other plots or soil aren't as productive, that it can be eaten, that planting other seeds doesn't work out as well, that such behavior has been reinforced by others, that it's part of your responsibility in the community, etc." You merely chose to do as you did. And the causal agent became your own, uncomplicated volition. You just freely *decided* how to plant. It's just easier to talk that way.

Humans began forming larger and more stable groups; intelligence and seemingly self-generated problem solving were part of this development. Behavior in such groups is what we call culture, and it is the phenomenon which was pivotal in getting us here. The seemingly autonomous solution to problems in human cultures—the illusion of free will—was pivotal. How this concept was communicated in prehistoric groups remains unknown. However, early humans eventually developed some sophisticated language, and the maturity of our language enabled communication of ideas that no other species had. We were able to *label* our behavior in increasingly abstract ways. And this laid the foundation for the advent of "free will," *as a concept*. It was likely central to our cultural evolution, as it led to both perceptions of autonomy and responsibility for group members, and we referred to it in such ways.

We could say that "free will," or whatever language referents were used, became cultural memes. Such a notion of free will was then passed around members of the group and down to subsequent generations. Children were conditioned to believe in their own autonomous agency, as I discussed. That is, each generation did not have to discover that it had created solutions independent of antecedent stimuli. The *idea* itself was learned, and problems in childhood that required more complex solutions became associated—through parental labeling in a variety of ways—with the idea of self-generation. Of course, it was also reinforced by the application of rewards and punishments for various behaviors—since such are doled out with the assumption that individuals are the voluntary initiators of their actions, i.e., "responsibility" (more about this below and in later chapters).

The free will meme became more and more refined in early human cultures. Ascribing free will—your own volition—became an integral part of human cultural development—from early agricultural communities or before, throughout the late Stone Age, into the Bronze Age. As far as we know, the notion of free will may have been first *actually labeled* in some way in the ancient Sumerian Code of Ur-Nammu in Mesopotamia (c. 2100–2050 BCE); and it is noted in the subsequent Babylonian Code of Hammurabi (c. 1755–1750 BCE). (294) The illusion of free will was becoming codified. It became part of advanced human culture, and it thus has continued to be used, revered, and variously codified in law ever since.

Responsibility

A closely related concept is the idea of "responsibility." We can't talk about free will without addressing this. If there is no free will, we certainly can't ignore the issue of responsibility—whether humans are in fact responsible for their own actions. This is an age-old question, but it does deserve some space here. If there is no free will, can we and should we hold people responsible for their actions and punish them accordingly? This is obviously a philosophical question, and in a practical sense there is no satisfying answer.

What do we actually mean by *responsibility*? This often goes to notions of *morality*, which bears on the issue of God and which I'll discuss more in chapter 9. Definitions of "responsible" vary but here are two often cited: "Being the primary cause of something and so able to be blamed or credited for it." (OED). "Liable to be called to account as the primary cause, motive, or agent" or "being the cause or explanation" (Merriam-Webster). In general, it is assumed that a person is responsible for her actions if—and generally only if—the action arises freely from the individual herself—autonomously—without any untoward historical or contemporary influence. Where this idea might come from, no one knows. But despite its long history of being so construed, it places responsibility, per se, beyond the reach of natural law. At this point, I'll just observe that the notion of responsibility and "free will" do feed on each other; they are two sides of a philosophical coin. If you are "free," you are responsible for your actions. If not, you are not. Gradations of this, of course, (if that makes any sense) continue to be associated with philosophical, professional, and legal debates, which I'll touch on a bit.

History

The notion of free will has probably been around a long time; I did speculate on whether our hominin ancestors might have had such a belief, which would support sanctions against the young and other tribe members.

In the 4th century BCE, Aristotle's classic *The Nicomachean Ethics* elaborated on the concept of voluntary and involuntary actions. (18) Aristotle believed that we are responsible for so-called voluntary actions, i.e., our intentions, based on those behaviors that involve our use of our own human discriminative powers. We were not responsible for involuntary ones, which were those that involved particular circumstances and obligations. Which are which has never been clear. Obviously, the definitions problem does go back to ancient Greek culture, and so-called legal codes did change considerably over the centuries, based on what was considered "voluntary."

Even well before Aristotle, the *Code of Hammurabi* clearly promulgated the assumption that man was free and autonomous in his actions and could and should be held responsible for them. It's most famous, of course, for its prescription of very Draconian punishments. (Note that Draco himself, a Greek nobleman, would not promulgate his laws for about another millennium subsequently.) Some have speculated, quite reasonably, that the notion of the Ten Commandments and other ancient specifications, laying out what we're *responsible for*—were largely appropriated from King Hammurabi's *Code*.

Behavioral

Of course, none of the ancient texts addresses the pivotal question: Where do we draw the line between voluntary and involuntary? This is a difficult discrimination, as Aristotle certainly experienced. As we've seen, the differentiating of voluntary and involuntary behavior is often one of convenience, not born of any critical analysis. At bottom, responsibility essentially means "respond-ability," i.e., the behavioral capacity and action which is independent of some level of historical or external current conditions. It is, in fact, tantamount to the idea of free will in your actions. We use the term "responsibility" to the degree we cannot, or tend not to, identify the contingencies which shape or enable any particular behavior. That is, we attach responsibility to our actions in relation to our understanding, or more rightly *ignorance*, of the conditions which led to any behavior, much as we do for free will in general. If we can thoroughly account for why a person does something aberrant, for example, we tend not to hold them responsible or accountable for it. If it is seen that you "have no choice" in an action, we don't hold you responsible. That is, your behavior has been fully determined by external contingencies. We don't hold you responsible for tripping and falling down. However, if you wore sandals to traverse rocky ground, we might hold you responsible for that decision. We generally don't hold you responsible for sneezing and coughing, though we might do so if you eschewed a dictum to stay home from work when you were ill.

We continue to struggle in trying to define "personal responsibility." But it is a hopeless task. The only approximation to this is the tendency to act in ways that are consistent with others' impressions of your character and history. If a one-year-old child throws his toy, breaking an expensive vase, we deny personal intention and responsibility because "she doesn't know any better." If a disinhibiting brain tumor, for example, causes you to leave the stove on and burn down the house, we "might" minimize your "responsibility" for that action. But a fully competent adult doing such things would be considered responsible for those actions.

It is also interesting that we acknowledge some mitigating circumstances when people emit unwanted behavior, but we assume free will when behavior is highly valued. For example, we might acknowledge that some aberrant behavior in a child is sometimes due to a parent's history of abuse and not punish it. On the other hand, when the same child does well in math class, we assume it was freely done—ignoring the years of education and parental support—and tend to reward it. The common view of responsibility for any action is based *only* on our own views of the person and circumstances involved.

The God Rationale

Which brings us to God. Theistic religion *must* have a notion of free will in order to rationalize humans' responsibility to make the *right* decisions in life. As we noted, the theists' argument for free will is that it was given by God—or at least loaned to us. And it supposedly enabled humans to make voluntary choices, and we thus have a *responsibility* to choose between "good" and "evil." It is the so-called moral argument, which, more than anything else, is what allows us to punish others. The 19th-century German philosopher Friedrich Nietzsche summarized the problems well, "Today we no longer have any pity for the concept of 'free will': we know only too well what it really is—the foulest of all theologians' artifices, aimed at making mankind 'responsible' in their sense. . . . Wherever responsibilities are sought, it is usually the instinct of wanting to judge and punish which is at work." (373)

We even ascribe some of our behavior to supernatural guidance, and we invoke the notion of God to avoid responsibility. "The Lord told me to run for office." "It's God's will that I can't allow my child those antibiotics." "We need to spread the word, and it is God's will that I collect all this money." God apparently directs human behavior and simultaneously holds people responsible for their actions. The hypocrisy is more than stunning. Einstein summarized the problem well, "If this being is omnipotent, then every occurrence, including every human action, every human thought, and every human feeling and aspiration is also His work, how is it possible to think of holding men responsible for their deeds and thoughts before such an almighty Being?" (150) We might even say that He—not human actors—was fully responsible. We can only marvel at theologians' bizarre attempts to come to grips with this issue.

Neuroscience

In common practice, we generally endorse the concept of free will, even though the operation of the human brain is inconsistent with the notion. Nevertheless, we sometimes use neurological findings which might bear on the concept of responsibility. When your brain is significantly impaired, we tend to eschew the notion of responsibility for certain unwanted actions. Harris lays out a variety of scenarios which illustrate the problem. (208) You might have frontal lobe impairments and impulse control or judgment problems. Such things can happen with a tumor or traumatic brain injury. Or you might have an acknowledged congenital intellectual impairment. Or there might be some other kind of brain dysfunction. There are many ways your brain can become dysfunctional. Athletes who've suffered CTE (chronic traumatic encephalopathy) often have great difficulty with life because of the damage, and they have occasionally committed violent acts or suicide. There are many medical conditions that have a good deal of influence on your behavior. For example, there are various dementias that, when severe, lead to major behavioral aberrations. Brain tumors can seriously affect your actions. In these types of scenarios, we might eschew concepts of responsibility for problematic behavior. This is because we might possibly understand, through medical or psychological assessment, the conditions in the brain that make someone act in aberrant ways. Without impairments, however, our behavior is still seen as a function (at least the proximate cause) of what's happening in our brains. Whether there is damage or not, the action of our brains is still the "responsible party." We only tend to consider personal responsibility for our actions *when we don't understand* the neurology of what's happening, and we tend to assign less responsibility when we do. It seems rather arbitrary, and it is, given our primitive knowledge of how the brain actually works, but that's the current view of our culture on notions of responsibility.

Punishment

If there is no free will, in that absolute sense, why would anyone be punished? Good question. Lack of free will obviates the whole notion of punishment based on the presumed responsibility for an individual's behavior. However, we should first point out that punishment for unwanted behavior is natural. We have an evolutionary tendency in this regard and are also conditioned to do so, of course, but there is rarely realization of this fact. Many other social animals punish other group members for unwanted behavior, particularly behavior which might be hazardous to an individual or the group. Animals punish for risky child behavior, noncooperation, maintaining dominance, and various infractions of group norms. Chimps do this, as do dolphins, meerkats, elephants, and many others. Evolution has provided us with punishment as something which confers a survival advantage. It serves a purpose for safety and survival, and it is an integral part of human and other mammalian cultures. It is a natural phenomenon, and it doesn't depend on the notions of free will or responsibility.

Punishment is a fact of life, and it works, although generally only in the short term and not as well as other behavioral actions (as I'll discuss in Chapter 10). So why do we punish? Sometimes it does happen for safety and survival reasons. A young child trying to cross a busy street. Speeding in an automobile. But we also punish for unwanted behavior which has little to do with physical survival. It depends on where you live, your social "status," what culture you come from, and other variables. Using foul language. Poor school grades. Wearing the wrong clothes. Buying the wrong lettuce. Anything unwanted can be and is punished—from minor aberrations of behavior and infractions of legal and moral codes, to saying the wrong thing, all the way to murder. You might also be exposed to the threat of eternal punishment for doing things that nearly all people do, which we'll get into in Chapter 9. Don't even *think* about evil ideas or sins. And recall George Orwell's dystopian novel *1984*, which brought to life the idea of a "thoughtcrime." Merely thinking forbidden ideas made you a thought criminal subject to enforcement by the "Thought Police." In Philip K. Dick's science fiction story, *The Minority Report*, merely conceiving of a murder can get you arrested for the crime. This harkens back to the issue of criminal intent, something our culture often punishes if it goes too far, as evidenced, e.g., by sharing a nefarious idea with someone else. As we continue to endorse the idea of human free will, the idea of responsibility for these types of transgressions will always be with us.

We also punish for reasons other than the particular suppression of some unwanted behavior. We have the reliable impulse of needing to blame others for aberrant behavior. The more we punish, and punish for various reasons, the more it becomes a habit. We can't blame ourselves as a culture; thus, the natural tendency is to attribute responsibility to the putative offending targets. The belief in free will allows us to do that. In various studies, it generally occurs that persons who consider themselves libertarians, for example, those who believe strongly in the autonomy of the individual, tend to be more punitive in their attitudes toward others' transgressions than people who see themselves as determinists. (215) Although the correlations are far from perfect, it does seem that the more you believe in free will, the more you hold others responsible for their actions.

Criminal Justice

How about serious offenses to a society? Do we need notions of free will, personal responsibility, and punishment to enable social and cultural stability? It's an age-old philosophical question without any cogent answers. Nevertheless, societal laws are always based on such notions. The notion of responsibility holds that people must account for their actions, as if they could voluntarily choose behavioral options completely free of any current or antecedent influences. Thus, the ideas of God, free will, and personal responsibility are intricately woven into our criminal justice system. Enforcement of the rules of behavior—most notably in criminal justice—depends on the notion of responsibility, based fully on the notion of free will. In fact, this is specifically laid out in American court decisions. One of the most notable and cited cases in this regard is a 1978 U.S. Supreme Court decision, *United States v. Grayson*. (524) It established that the notion of "free will" is, in fact, the basis for our criminal justice system. It emphasized that a foundational principle of the judicial system is "the freedom of the human will and a consequent ability and duty of the normal individual to choose between good and evil." Of course, good and evil are pretty vague notions (though American definitions of these rely heavily on Judeo-Christian tenets). And the concept of "normal" individuals is a rather useless concept. We are all different based on our genetic and behavioral histories. Nevertheless, courts almost always use such ideas to judge aberrant behavior, based on the notion that we do have free will, and that it takes major variations from the typical condition of most humans to weaken any notion of personal responsibility.

Since free will is assumed, we generally hold people legally responsible for their actions based on their "state of mind," which is where responsibility is felt to reside. Did he intend to do it? Modern Western legal codes are based on the notion of some form of free intentionality. The idea is that an individual's state of mind preceding an action is the basis on which we judge responsibility; it is the foundation of our legal theories and how we dole out punishment.[17] The notion that we can reach into a person's mind and retroactively expose intent is an age-old problem, and one for which there is no good answer. If we are creatures of our genetic and environmental histories, the notion of mind-reading of intent becomes meaningless. To diagnose that an individual was acting "freely" comes down merely to a matter of opinion; and, not surprisingly, legal opinions in such matters vary greatly—often depending on the role of the person whom you ask.

The criminal justice system encodes mind reading with what's referred to as *mens rea*, i.e., having a certain state of mind, particularly of an intent to do wrong. Legal systems consider this in assessing guilt and doling out punishment. It is generally based on whether a person has a specific intent or purpose, whether there is awareness of right from wrong, and whether or not an act was deliberate. Again, it assumes free will—with qualifications—and necessitates mind reading. A particular type of this "responsibility" problem involves whether or not an individual is "insane." To oversimplify, you may have free will, but we generally don't hold you responsible for an act you may have committed if you are judged insane as being part of the reason you committed it. Of course, this issue continues to be hotly debated. The generally cited legal standards are derived from what is known as the M'Naughten Rule, based on an 1843 English court decision. It basically says that one is not "responsible" or "not guilty by reason of insanity" under certain conditions—if, at the time of the act, the individual was so deranged that he did not know the nature or quality of his act, or, if he knew the nature and quality of the actions, that he was so deranged that he did not know that what he was doing was wrong. The implied derangement is generally based on any so-called "disease of the mind." Courts and mental health professionals continue to disagree on precisely what this means, on what constitutes a mental "disease," on how that might affect a person's actions, and on what sort of professional assessment can inform on such issues.[18] Different psychologists or psychiatrists easily disagree on whether a given accused person is, in fact, legally insane and whether her insanity was a contributing cause of a criminal action. In point of fact, criminal proceedings often include "dueling experts" testifying on precisely those issues. Free will and responsibility are always assumed unless there is agreement, decision, or adjudication that free will should be thrown out in favor of the mind-reading exercise. It is just "mind-boggling" that one expert can infer that an individual is acting freely, and should be held responsible, and another disagrees. There have been a number of attempts at refining this problem which I'll not into here.[19] Nevertheless, the subtleties don't seem to work, and free will becomes a meaningless idea, leading to merely opinionated views of a person's private mental activity and responsibility.

There is a whole forensic area devoted to this, of course, within the fields of psychology, psychiatry, and criminal justice, and there is an exhaustive literature on the extant scientific data and views on neuroscience and criminal responsibility. (324) I'll not delve much into that here. Yet, our knowledge about neuroscience and the underpinnings of responsibility is very poor. It's why you can find any number of qualified mental health professionals who adamantly disagree on the notion of "responsibility" in any particular case. Unfortunately, we can't even ignore that forensic mental health assessments are clearly influenced by prejudices and sources of funding.[20] Your free will and responsibility are assumed unless you're lucky enough, and perhaps wealthy enough, to get a good label and report.

If you just walk through a prison and talk to the inmates (most of us can't do that), you'll find some pretty horrifying life stories. The criminal justice system almost never considers such things—extreme poverty and malnutrition, serious abuse as a child, drug addiction, lack of education, repeated head injuries, etc. We hold you criminally responsible for your offenses, unless, of course, some mental health professional can attach a psychiatric label which presumably forces you to act in a certain antisocial/criminal way. For example, it's been found that criminal psychopaths show several EEG abnormalities in various cortical and limbic structures compared to non-psychopathic criminals (271), though this parameter is never considered in actual forensic cases. Also, in one major recent study, MRI and CT scans of the brains were conducted in a large number of male inpatients in a high-security mental health facility, essentially a prison for psychiatric patients. (554) Significantly more "structural brain damage" was found in the forensic inpatients than in the control subjects. This was the case even though none had been previously determined to manifest signs of any organic brain syndrome, as such are commonly considered by mental health professionals. Unfortunately, mitigation based on such findings is rarely considered. We prefer to use personal responsibility for meting out justice.

That is not to say that there are not good reasons for incarceration for certain crimes. But that is often a matter of public safety, something which has evolved as beneficial to the culture. However, such consequences are often more a function of retribution and vengeance than punishment that we think will be effective. And we do know that incarceration as punishment is fairly ineffective if not accompanied or followed by some sort of rehabilitation. In sum, the criminal justice system holds people responsible for their behavior unless it can find adequate external and historical explanations for your behavior. Of course, when it can't, it merely resorts to the unsupportable notions of free will and responsibility. The notion of practicality aside, this reflects our limitations in understanding human behavior. We have filled in the "gap" in our understanding with free will and its cousin, personal responsibility. Of course, the answers to more efficient methods of control of unwanted behavior are well known and do not depend on people eschewing concepts of "free will" and responsibility —just a more educated and informed society capable of appreciating the values of prevention and effectiveness over vengeance. The overreliance on free will and responsibility are certainly impediments to this. As biologist Sapolsky once said, "The world is really screwed up and made much, much more unfair by the fact that we reward people and punish people for things they have no control over . . . We've got no free will. Stop attributing stuff to us that isn't there." (419)

Conclusion

We generally hold onto the notion that we must have free will to know right from wrong or to engage in any other moral behavior. This is an antiquated notion, of course, based on religious tradition, but most people, even the secular, endorse it. In fact, the whole notion of free will is maintained, in part, by our tendency to hold people accountable for their moral or amoral behavior, as Nietzsche suggested. And this seems universal. In one large study, a variety of laboratory experiments and large-scale surveys were used across several nations. It was found that the belief in free will is almost universally closely tied to individuals' tendency to hold others morally responsible for wrongful behavior. (89)

What any culture "chooses" to do about socially aberrant or problematic behavior is another story. The application of responsibility and punishment for the maintenance of social order (being able to act for the benefit of the community by providing suppressive consequences) will always be with us. They're part of how we got here. Although Western societies base the issue of punishment largely on the concept of free will, it's again helpful to look at these separately. Reward and punishment are effective methods of behavior control (though often not applied well or fairly), which don't depend on any notions of free will, responsibility, or voluntary action. Societies evolve to protect themselves from deviant or dangerous behavior, with various behavioral norms depending on the relevant culture, and to promote growth. This is the basis for ethics or morality, which generally holds that people are freely responsible for their actions. However, the laws of behavior operate whether one subscribes to notions of responsibility or not, much as the laws of physics do the same (Chapter 10). The notion that one must eschew appropriate behavior controls and shaping because of the philosophical concept of "free will" is an untenable and scientifically absurd position. The only reasonable way of coming to grips with this problem would be more humane methods of dealing with aberrant behavior and criminal justice. We already well know that shaping positive behavior—also known as prevention—is more effective and cost-effective than punishment by incarceration. This doesn't require any debate or consideration of "free will" or notions of personal responsibility, merely a more educated and informed public. I take no position on whether this will happen. But the problems, quite obviously, stem from one powerful cultural tradition—belief in God.

What If People Thought They Didn't Have Free Will

The question has been posed many times, mostly by philosophers. We just *have to have* free will. Otherwise, how could we function? How could we go on? Many have tried to envision what would happen if scientists somehow convinced us that free will does not exist. (461) It's an interesting question, albeit without any possible answer. This is an old argument, and actually quite pointless, which really doesn't go to the heart of the problem, but I'll address it briefly here. What would happen if science somehow debunked the idea of free will? Harris offers that "If the scientific community were to declare free will an illusion, it would precipitate a culture war far more belligerent than the one that has been waged on the subject of evolution." (208)

I think that science being able to do this is pretty unlikely in our lifetimes—or that of human civilization. Science couldn't actually do this, of course, since "free will" is something outside the realm of what can be empirically studied. And then there's the notion—increasing these days—that a significant portion of the population doesn't believe in science anyway. The attitude toward science, particularly in America, would lead to a large percentage of the population just rejecting the idea, of course. Moreover, the belief in free will (albeit illusory) is a very powerful superstition, like the God meme and many others, and would be highly resistant to extinction. People will continue to "believe" in free will, for all the reasons which have shaped the belief in the first place, which we reviewed—science be damned.

There have been "experiments" with people putatively showing that beliefs or perceptions about free will significantly affect behavior. Widely cited studies in this area suggest that if people didn't believe in free will, they would be less likely to support punishment for crimes, but they might also behave less "ethically." (527) They might also be less helpful to others and more aggressive. (36) Unfortunately, studies in this area do not hold up very well to replication, and generalization to real-world behavior seems doubtful. Behavioral scientists cogently make the argument that, if the public comes to increasingly doubt the existence of free will, this may reduce retribution as a motive to imprison people. (461) In psychological experiments inducing "suggested disbelief" in free will, per se, this does tend to seem to make people less likely to be vindictive and favor retributive punishment, though the trend seems to hold mainly for women. (80, 462)

The related question often asked is, "What would you do if the world was known to be purely deterministic?" That is, what if "choices" were already determined? It's essentially the same question, of course. But how would "moral" behavior be affected? People do tend to reject determinism as a cause of our behavior. However, in various experiments, when challenged with specific issues of moral responsibility, people strangely often see determinism as not inconsistent with moral responsibility. (375) It is quite unclear why this happens. Punishment may be a function of how significant a transgression is considered, but there is no clear relationship in the attitudes people hold about the relations of free will, morality, and determinism. The idea that legal and moral standards may change as the result of scientific work on free will is rather fantastical. Punishment does have some minimal effects itself, but it'll be a long time before moral and legal principles are significantly plucked from the clutches of religion and retributive justice.

Despite any scientific evidence or opinion to the contrary, people will continue to believe they have free will. This is because the illusion is compelling. The illusion of free will didn't happen by accident. Our behavior about this has been conditioned for many millennia and ingrained in our culture. We do not want to believe that we can't make free choices, and cannot believe otherwise. Physicist Paul Davies admits that the illusion of free will "may still be a fiction worth maintaining," since many current social systems depend on it, while acknowledging that "we must believe in free will—we have no choice." (109) The awkward grammatical self-contradiction notwithstanding, we will go on unfettered by it. Any increased (though incomplete) understanding we have of the ultimate nature of reality, including the operation of our own brains, will not constrain us from thinking and behaving as we always have, including our belief in free will.

You and Choice

There is a related question, and it's the one that always comes up. If there is no free will, do I *really* have no choices? Do I then have no potency over my own life, no real reason to act, no meaning in my life? I would say *no*. Despite the deterministic nature of a universe, following the laws of physics for each of us, I'd offer that there's no reason for dismay.

The crux of the matter may be what you consider to be the meaning of "I." Do you mean something besides the sum total of the biological processes which operate within you? You generally consider that "you" are your mind, perhaps more honestly your brain, and that other parts of "you" are referred to at some arm's length. You refer to my arm, my heart, my indigestion, etc., as though they were possessions, not things specifically comprising your "self." What you think of as "I" *is* your brain, but we clearly "feel" that, when a decision or choice is made, it is "I" who makes it. We don't say, "My brain made it." When you say "I", even though it is your brain and only your brain that is doing it, you don't focus on that. It is "I."

What "you" are is a function of your history and relatively current conditions. You cannot be different than you are at any given moment. And that "moment" is where any action or decision takes place. But when we take a closer look, there is actually not even a "moment" or present, only history (There is actually no such thing as a "moment," only a past; the future, while determined in the cosmological sense, for you is only an idea). Whatever occurs to you has already happened. You cannot separate yourself from this history. The decisions you make are only such in retrospect, whether this is a long-term issue or a few milliseconds, as the above experiments demonstrate. But your perception is always of "I," and in your present moment as you perceive it. Nothing that can shake that.

In any situation where you might consider your choices purely voluntary (remembering that this is a matter of practical degree), your brain does formulate an intention, a subjective event which is instantiated by the chemical actions of your brain, and this is followed, *as you perceive it,* by the actions you then take. In that sense—which is eminently genuine to you—you do have the perception of choice. "I" typed on a keyboard to write this. "I'm" going to sleep now. Free will *is a reality to you* even if it is an illusion when considered in the framework of a determined universe. "You" determine your choices. "You" make choices based on your accumulated history and current functioning—which is no less satisfying even if you ponder some cosmological framework. There need not be any feeling of powerlessness in that context.

You continue to make decisions and choices, and to emit "voluntary" behavior. You process information from your senses and your memories. *You* put all those data together (often without focusing on them), consider options, and make "choices." *You* appreciate the beauty of a flower without regard to any illusions or biological underpinnings. *You* decide what to wear. *You* enjoy a Mozart concerto. *You* decide whether or not to eat a peach. All this is without any regard for the history of the universe. For now, the personal belief in free will is a given, and it is unlikely to be disturbed. And when you go through the process of making a decision, whatever the neural reality, it is still a choice *to you*. And that's certainly enough.

Conclusion

In sum, "free will" cannot be rescued from a determined universe. Free will is clearly an illusion. Nevertheless, we do "feel" it. That's okay, as above, but that's not evidence of the matter. The Copernican Principle becomes relevant here. We believe that there is a natural world out there, but somehow we are able to make choices and behave in ways that do not follow the natural law. We are so special that we feel exempt from it. Also, the common belief is that we alone employ "free will" as the genesis of our thoughts and actions, though we've sometimes ascribed a bit of free will to certain other creatures. Thus, we come down to the conclusion that either every living thing may have some degree of free will or the concept is meaningless. Obviously, it's the latter. You may again regard this as a *reducto ad absurdum* argument, but it is unavoidable.

Human Behavior

The classic free will argument, of course, is that there is no adequate explanation for why people say and do what they do. That we cannot adequately explain all human behavior is not a convincing argument. Nevertheless, free will philosophers continue to rely on the complexity of human behavior to support the "free will" idea. Remember, invoking free will as an "effective theory" provides no explanation. It is merely conversational convenience or a "philosophy of the gaps."

Human behavior is extremely complex, given our level of understanding. That we do not fully understand how decisions and actions come about is, of course, not an argument for uncaused behavior. Invoking an abstract and untestable idea to fill a knowledge gap in no way demonstrates its validity or reality. It attempts to bridge the gap between what science can't adequately predict and whether there is something beyond the physical laws of nature that helps us do what we do. It is a philosophically and logically untenable idea. The only way to overcome that is to reprise notion of the soul, as Descartes did and others continue to do.

Perspective

Neuroscience has already taught us a great deal about the human brain, and this will certainly continue, most likely at an ever-increasing pace. We are learning more and more about how the brain actually learns, how it initiates behavior, and why it "goes wrong" in various ways. As these scientific pursuits enable better predictions of human behavior, they do not diminish the worth of the human "condition." They merely give us a better understanding of how we, as humans, operate. The more we understand about human behavior, and the neurologic mediators of it, the less we may rely on free will as a pivotal meme for our actions.

Does the logical and scientific contradiction of free will somehow demean our human dignity? No. In practice, the realization that free will is but an illusion will not affect how we behave. "You" will continue to make choices on a day-to-day, moment-to-moment basis without worry about determinism, quantum mechanics, or your entire genetic and behavioral history. Our understanding of human behavior clearly doesn't change the rich history (and future) of humankind in a way that minimizes such things as love, compassion and other emotions, and our great artistic, scientific, and technological achievements. These will obviously continue unabated (unless some events suppress them, of course). That there may be behavioral explanations for our actions doesn't impair the dignity or the value of our lives. Those qualities are unique to each individual, precisely because each of us has a unique history. Our enjoyment of our creations, reflections, and each other do not diminish merely because we have an increased logical and scientific appreciation of how our brains and our physical universe operate. And the enjoyment and dignity of our lives need not be compromised in any way because we get over the God superstition.

Accepting, at least intellectually, that free will does not exist might presumably lead to a nihilistic worldview. It need not. We still make decisions. We still function in the ways we desire. We still appreciate and value our temporary lives. There is nothing about the possible scientific denial of free will that diminishes anything about your behavior or experience—except possibly the acknowledgment that we might be more aware of the actual contributions to our own behavior—and possibly to renew our understanding of social justice issues like responsibility, deviance, and punishment. At this juncture in history, we seem merely to "need" the free will meme. Similar to physicist Davies' perspective, when philosopher and journalist Christopher Hitchens was once asked whether he had free will, he ironically admitted, "I don't think I have any choice but to believe I have free will." Despite the apparent self-contradictory language (again, always a limitation in looking at this problem), I think that sums it up well, at once admitting to the illusion and providing oneself with psychological comfort. Humans believe they have free will because not to believe so would be unacceptable.

Will our perception of free will change as science marches forward? Neuroscientist Michael Gazzaniga offered that, "I think we will get over the idea of free will and accept we are a special kind of machine, one with a moral agency which comes from living in social groups. This perspective will make us ask new kinds of questions." (94) I don't see that happening in the run of our species, but I could be wrong.

On God

To conclude, regardless of the personal dynamics of free will, here's the bottom line about God. Since free will does not actually exist, the idea of a personal god becomes ludicrous. You may hold that either the universe is determined and evolves according to the known (and yet to be discovered) laws of nature, or you may believe in something supernatural which is in control of all reality—and you, by definition. There is no other alternative. There is no sense in which both of these things, or parts of these things, can coexist.

The notion that God must exist because you believe He gave you free will and expects belief in return is the most patently specious argument on this subject. And the notion that you'll be held responsible for such belief or disbelief is just absurd. *You do not have free choice to believe or not to believe. That behavior is already determined.* That there *is* a deterministic universe without free will means that any belief in a fantastical, supernatural god is already set. Some people believe, and others don't. Faith becomes irrelevant to any legitimate notion of God.

You continue to act and make decisions "freely," knowing that your history and the laws of physics are the driving factors, along with the knowledge that no god is there to interfere with or supervise your fate. Nevertheless, for a variety of reasons, our culture continues to promulgate various god memes, which has led to the formation of organized religions. I'll get to that presently.

7

RELIGION: ITS EVOLUTION AND PATHOLOGY

My own view on religion is that of Lucretius. I regard it as a disease born of fear and as a source of untold misery to the human race. —Bertrand Russell[1]

I cannot conceive of a personal God who would directly influence the actions of individuals or would sit in judgment on creatures of his own creation. —Albert Einstein[2]

I do not believe in God because I do not believe in Mother Goose. —Clarence Darrow

For this book, I did not set out to talk about religion, per se, meaning the varieties, pathologies, and "sins" of organized religion, at least as it has come to be over the past 6,000 years, particularly the last 2,000 years or so. Countless books have been written on this. Nevertheless, I've found it difficult to totally avoid the subject, and I need to provide some context for what we've been talking about—whether any supernatural being exists. I shall not attempt to delve into much detail of the hypocritical, cruel, barbaric, and genocidal character of organized religion, principally the Abrahamic ones (Christianity, Judaism, and Islam) and their sacred writings in any significant detail, though again, it is impossible to avoid these things completely. So, we'll have to take a brief tour into the origins and nature of religion in order to show that religion is a purely human-constructed activity, not inspired by some divine being. And we'll explore how it operates in its own interests, reinforcing superstitions for its own survival, and creating all the well-known dynamics of discrimination, hatred, violence, and war.

Religion, at its core, is a method of social control. It enforces moral standards of behavior, though it often does not do a very good job of this. Religion, especially organized religion, is a pivotal social problem for our species. Religions are powerful social movements, and we can't ignore how people coalesce around these superstitions in order to maintain social control and discrimination against those who profess different beliefs. It is often how we identify ourselves, and it is set of traditions that keeps the God meme and other myths alive. Such is the cultural "pathology" of organized religion.

The notion of God—the God meme—is a very modern invention. If we take the theists' fantasy about this, we could ask, "When did God appear, i.e., when did He make Himself known to Earthlings?" Remember, the universe we know had been around for some 13.8 billion years, the Earth about 4.5 billion years, human beings probably only around 300,000 years, and communities of people for just a few millennia. To reiterate, why didn't God reveal Himself to *Homo habilis*, or *Homo erectus*, or chimpanzees, or fish? Why did it take perhaps a quarter million years of our *Homo sapiens* existence to begin to "realize," that is, fabricate, the notion of gods? The human diaspora was nearly global by around 10,000 BCE, and various gods and spirits had been emerging. But let's take a step back and consider the evolution of religion in the first place.

Animism

Religion is a set of behaviors and traditions formed around a group of superstitious beliefs. To see how this developed, we need to take a brief look at how superstitions arose in early humans. Building on what we began in Chapter 4, let's consider how ancient humans developed such ideas. If a rock rolls down the hill and hits you in the knee, you'd rightly blame the rock. The rock is the cause of your injury. But what made the rock roll down the hill? There must have been some force. Early humans just couldn't see it. And what made it rain? And what made the sun come up? And what made the trees grow? How in the world can a bird fly? Can you imagine what primitive humans—without much knowledge—must have thought about major happenings in their environments? Just look up at the sun or the moon, or watch a flowing river. Try to suppress your knowledge about such things, and try to imagine what they are and why they do what they do. You realize the seemingly awesome powers there. And think about your emotional reaction to great thunder and lightning. It is easy to see how early humans would have been moved by the power of these things, the overwhelming changes in the environment and the fears associated with them. If you don't know anything, even though you have a big brain and maybe because of it, you searched for causes. But you couldn't find any.

Humans (and perhaps other or more primitive apes) do learn about relationships between events in the world by observation. It was clear to early humans that everything had a cause. If I do something, something else specific happens. You know that *you* can create change in your environment; *you* are the force or cause. Early humans didn't know that other creatures and other things—rivers, trees, clouds—didn't have something inside them that caused their actions or movements. Early humans were clearly able to discern a difference between living and non-living things. This may have been a bit difficult, but they certainly "saw" some sort of power in everything that moved. They probably saw something in themselves that was different from most of the rest of the world. But there was a certain similarity with many animals. They also saw living and dead plants and some seemingly inanimate things that also moved—streams, falling rocks, rain, wind, etc. They all must have had some forces that moved them or created some sort of change. You just can't see what made these things happen.

The unknown force which causes things to happen is often referred to as "agency." For ancient humans, philosopher Daniel Dennett proposed that there existed a "human hyperactive agent detection device" (HADD), in which a human reacts with concerted attention to various events in the environment. Since forces were unknown, this led to the general attribution of agency in various living and nonliving things in the environment by early humans. (123) He describes the evolution of religious behavior as derived from such agents' *intentional system*, that is, the tendency to ascribe intention to other objects in the world. "Whenever an animal treats something as an *agent*, with beliefs and desires (with knowledge and goals), I say that it is adopting the intentional stance or treating that thing as an *intentional system*. . . . At the root of human belief in gods lies an instinct on a hair trigger: the disposition to attribute *agency*—beliefs and desires and other mental states—to anything complicated that moves." (123) The wildebeest or the tree must have some kind of intention to do what it does. There was some "life force" in all those moving things, something that gave them the intention to do various things.

Since no agents could be seen (or known by any other means), they had to be invented. And those inventions became various *spirits*, things that were gradually imbued with their own energy. Such *animism* was the root of the idea of forces, powers which were initially perceived as purely natural. Everything had some living force inside it. Thus, the advent of unseen forces evolved into actual spirits. But what *were* they? Primitive *Homo sapiens* did need some way to communicate about such things. Since early humans had well-developed language, it's obvious that assigning some sort of identities to those spirits made communications more efficient. Spirits became personas with various unseen powers. Foraging and hunting humans—and later domesticated humans—undoubtedly developed particular named spirit entities for all sorts of things—for various plants and animals, the wind, sun, etc. The spirits made things happen. Some inhabited the birds. Some resided in the buffalo. Some made plants grow. Some caused death. Early humans looked to these spirits to explain what was not otherwise understood. Per our need to explain things we don't understand, the "gap" had to be filled in. And thus, elaborate anima and superstitions began to appear. It wasn't long—a few thousand years—until those spirits became reified with variously named personas, depending on what primitive culture they came out of. You can't talk about a great force for long without giving it some identity and particular character.

Unimaginable powers would certainly have been the perceptions, and they had to be spirits with powers beyond anything humans or animals could muster. They would have to be something else, spirits beyond the natural world humans were observing—something *super*natural. The notion of supernatural forces responsible for the sun, the moon, the rain, death, etc., had to be "identified," and labeling them with various personas was a perfectly natural outgrowth of our explanatory need.

Evolution of Religion

The evolution of religion, per se, directly followed from the beliefs of such animism. It was a natural process. Humans employed animistic notions to explain the world around them. Obviously, we are not able to go back far enough into paleolithic history to find evidence of actual spirit memes. Nevertheless, we do see some anthropological evidence, as superstitious anima persist even to this day. For example, as Torrey suggests, "A belief in animal spirits is widespread in the world. In some cultures, it is believed that the animal is the ancestor of humans; such an animal is called a totem. . . . [in some cave paintings we even find] half-human, half-animal figures." (519)

In Chapter 4, I touched on the evolution of superstitious anima. But consider some possible, very primitive ones. They were very powerful, e.g., causing the death of a tribe member or the buffalo that was speared, creating wind that felled trees, making the sun come up or making it rain. The relevant dynamic here is that spirits of such power were obviously to be feared. Can you imagine early humans' fear of the spirits who could cause death or the sun to rise? The shaping of such powerful spirits was a central factor in the evolution of various "gods" and various primitive religious behaviors, and it was likely predicated on the great fear of spirits who existed beyond the natural world that humans perceived. It is natural for humans, like other animals, to fear those with much greater power and to become submissive to them. The spirits with great power would have been intensely feared and respected. What followed was obviously the evolution of supernatural memes and humans' intentions to revere and "please" them—worshiping and making efforts to appease them, even leading to animal and human sacrifice (I'll get back to that later). That's a condensation of several thousand years of human history. Nevertheless, such traditions persist to this very day. We continue to venerate and worship imaginary supernatural spirits, though our propitiation of them has become a bit more symbolic. Yet, supernatural worship and subsequent organization around superstitious beliefs was clearly the natural evolution of early religious behavior.

Our Ancestors

Since the evolution of religion was perfectly natural, we might ask whether our ancestors engaged in what we might call formal religious activity? If religion is the communal sharing of superstitious beliefs, it certainly did occur. Let's take a prime example, ceremonial burial of the dead, as I introduced in Chapter 4. This is obviously an important dynamic in how organized religious behavior came into being. Formal *Homo sapiens* burial sites have been discovered dating back hundreds of millennia. It seems we have always buried our dead, and ceremoniously so.

There is a natural human desire to forestall and avoid death, as there is for all other creatures. The question of mortality—whether death is "the end"—led to various rituals and helped to shape religious behavior, including the notion of an afterlife (which I'll detail more below). It is likely that Neanderthals buried their dead (562), and there is some new evidence that there may have been ceremonial aspects involved. (410) Even the small-brained hominins, the recently discovered *Homo naledi*, likely deposited their dead in a communal grave, discovered a few years ago in the Rising Star cave in South Africa. (136, 556) Though there is some controversy about this, the site does suggest some rather "modern" ritualistic behavior associated with the burials, including some artificially fashioned objects being included in the site. We refer to such things as "grave goods." We do know that burying the dead with "grave goods" was important to early humans and even other *Homo* species. As Torrey points out, "The most common reason for burying personal items in a grave . . . is so that those items will be available for use by the deceased in the afterlife." (519) This is clearly religious behavior. Cro-Magnon humans certainly seemed to manifest some complex rituals. They buried their dead with various grave goods—ornaments, shells, even pendants or necklaces, and tools, in preparation for the afterlife. But remember, these are a subspecies of *Homo sapiens*, and thus, a pretty recent variety. Whether we learned religion from them or vice versa is unimportant. It's just further evidence of organized religious activity in early humans.

The notion of an "afterlife" was critical to the development of religion and belief in some sort of deity. Superstitious animism enabled invisible forces or spirits to exist, which powered everything in early hominin existence. The spirits certainly had the power to take a life; therefore, they had the potential power to preserve life in some ethereal fashion, which must have been the greatest power of all to early humans. It would have led to the ultimate in reverence, leading to more formal "worship." It is likely that death rituals were the beginning of worship of some sort of god, and the sharing of this among tribal members was the beginning of what we call religion.

Group Cohesion

In order for early humans to survive, they must have banded together in stable groups (and as I previously noted, it was likely one factor in how we outlasted our ancestral cousins). For religions to evolve, groups must stay together and share ideas over a long period of time. Such social cohesion was a factor in enabling religious memes to develop and thrive.

Humans "knew" that there were unseen forces operating on their lives. We just didn't know what they were. But as our imaginative capacities and communication abilities developed, and as we settled in one place for longer periods, we began to talk with others about things that didn't actually exist. Fictions of spirits and other fantasies took on more palpable qualities. They were undoubtedly shared using sophisticated language, leading to collective reinforcement of such superstitions. Various myths became the sum and substance of shared communications and rituals. It is likely that such behavior functioned to facilitate communication, reinforcement of behavioral norms, and social harmony within small groups. The predominant anthropological perspective does point to collective rituals as the genesis of early religious behavior. Such behavior would have reduced stress and reinforced shared values, norms, and beliefs, and the rituals promoted what was the primary effect of religion, group solidarity. (495) These became what we know as traditions, and cultural traditions became the basic architecture on which religious traditions evolved and were maintained. Without "communities" and shared traditions, including religious memes and rituals, organized religions themselves could not have developed and been maintained.

Further, in order for groups to remain stable for long periods, leaders must emerge, as we've reviewed. This is not unique to humans among the primates, or in many other mammals for that matter. Leadership enables group cohesion and thus the increased survival potency of the tribe. The pattern of hierarchical organization of the group is a strong tendency in advanced primates, and strong leadership is a model humans inherited from our primate ancestors. In other primate species, leadership is made rather explicit, and physical dominance is the preferred method. A primate leader is usually one member of a troupe, usually male, and his power is essentially inherited (We continue to have this today in some human cultures as the inherited maintenance of royalty—symbolic in some places and the exertion of specific power in others).[3] Maintenance of dominance, and therefore leadership, is preserved by power, including often by fighting off ambitious rivals.

In early humans, leadership may have derived not strictly from physical power dominance but from other skills that enabled group survival, e.g., hunting skill. Physical dominance or skills became less important over time as cognitive capacities became more important, e.g., communication abilities, knowing where the prey was migrating, primitive planting skills, and assistance to others. Principal among these would be communications about various spirits or anima that were the actual perceived power over humans. Undoubtedly, some individuals in a tribe were better than others at embracing and sharing these superstitions or religious memes. Individuals employing such dynamics would enjoy enhanced prestige within the group. Such leaders were looked up to as if they had more knowledge with respect to the supernatural forces. Hierarchies began to form around religious superstitions, with particular leaders or shamans at the helm. Those who best promulgated the spirits were probably seen as authorities or leaders in any tribe, and the greater the employment of religious memes, the more power could be ascribed to the individual. They were more revered and undoubtedly employed various religious "signals" (behaviors) to facilitate the message and, therefore, religiously based group cohesion. Such leaders have likely been important factors in the evolution of religion. We know that even today the signaling of religious memes to other members of a group tends to increase cooperation in the group as well as in-group favoritism. (380) For a current example, an interesting study looked at the religious/ritualistic behavior of groups of Christians and Hindus in South India. It was found that people who engage more in religious rituals tended to be perceived as more hard-working and physically strong, having good command of moral dicta and a commitment to helping others, all components of emerging leadership. (412) Thus, religious traditions and the leaders who modeled them became a way to further organize the tribe, facilitating its cohesiveness and endurance.

The Appearance of Gods

Eventually, the various anima or spirits evolved into more specific personas or gods. It's unknown, of course, when this began to happen. Torrey offers, "It seems doubtful whether gods existed among early *Homo sapiens* 100,000 years ago. . . . the concept of a god would not have occurred to hominins prior to about 40,000 years ago, and the gods themselves would probably not have become fully visible prior to about 10,000 years ago." (519) (We do take the meaning of "visible" here as metaphorical.) This would likely have had to occur sometime around the appearance of agriculture, with the domestication of plants—at various times in different parts of the world. Agriculture enabled larger groups to form and to settle in one place for longer periods. Such obviously allowed for more imagination, speculation, and sharing among community members about the forces which were in control of the things they saw around them in various cultures in different geographies—which is one reason history has given us so many different gods.

Further, we should remember that we humans do tend to anthropomorphize things that move or act. Torrey suggests this as a component of religious evolution. We "attribute human agency to inanimate things or events. Thus, we assume that thunder and lightning, floods and droughts, the rising of the sun and the phases of the moon must all be caused by some superhuman or divine power. Such pattern-seeking theories have given rise to multiple theories of the origins of gods and religions." (519) We depended on such "beings" to support us. For example, one of the things we had trouble explaining was the behavior of other living things in the environment. They had spirits moving them, probably named at some point, which is what organized and animated the material universe of early humans. However, we certainly couldn't communicate with them as we did with each other, and they didn't have the same awesome power as the sun, moon, etc. We didn't have any good channels to connect those living things, e.g., help with finding prey or growing food. The solution had to be invented, and the invention was to ask the gods for assistance. As Harari notes, "The gods' main role was to mediate between humans and the mute plants and animals. Much of ancient mythology is in fact a legal contract in which humans promise everlasting devotion to the gods in exchange for mastery of plants and animals." (203)

Further, there were many spirit personas to explain all the other significant events in early humans' lives, but, just as humans themselves, they needed to be organized in some way. Thus, as Torrey notes, it was also "necessary to establish a hierarchy among the competing spirits, just as it was necessary to do so among the human leaders of the various hunter-gatherer groups. Out of the hierarchy of spirits emerged the first gods, which were simply high-ranking spirits with advanced degrees conferred by a consensus of the people." (519) And some of them became spirits of great power, eventually a "person" with supernatural abilities, including over the fates of humans. Philosopher and psychologist John Teehan has described such a being as having enormous powers (being invisible and immortal), as leading to the idea of a god as a "full access strategic agent," one who could enforce moral judgment upon *all.* (515) And a specific god or gods began to emerge. Further, each human settlement had its own panoply of gods. But the expansion of trade and communication eventually necessitated that local spirits be replaced by shared deities, giving rise to the agreed-upon polytheistic gods—gods whose missions could be appreciated across communities. (203) And organized religions were born.

Group Selection

Did some groups of human hunter-gatherers survive while others did not because they manifested certain social or cultural characteristics? Did some survive because they invented better weapons, because they organized themselves better, because they had more stable groups, or because they shared some cultural memes and behaviors which made them more adaptable? One of the theories proposed for this is called "group selection." That is, groups of humans, as opposed to just individuals, may have survived specifically because they had some set of behaviors which made the group more adaptable and thus enabled better survival—as a group.

It has often been maintained that one of these group behaviors or cultural traditions was religion. Did religion survive because it enabled human tribes who practiced it to survive better than others? We've outlined some of the functions that it may have played in early human cultures. But why did it persist? The group selection hypothesis has been advanced by many. Even physicist Kaku opines on this evolutionary theory of religion as "increasing the chances of survival of early humans.

Religion helped bond bickering individuals into a cohesive tribe with a common mythology, which increased the chance that the tribe would stick together and survive." (261)

The anthropological theory of "group selection" has been variously advanced over the years. The idea has been around for a long time. It has come in and out of favor; and there have been recent attempts to revive various forms of it. (426, 549) It continues to be a controversial theory, though it has been advanced over the years by some to suggest that some human tribes survived and others did not because they had one particular cultural characteristic—religion. (549) In this theory, religions are so-called "adaptive units," and such units can be "selected for," in Darwinian evolutionary terms. As Torrey notes, "The most common argument put forth by the adaptationists is that gods improve the survival of the group. According to this theory, 'ancestral societies with culturally widespread God concepts would have outcompeted societies without such concepts, given the cooperative advantage of believing groups.'" (519)

In taking a Darwinian perspective, it would be necessary to postulate that some genetically driven traits were at work for this. That is, is it possible that there is some "genetic" underpinning for this kind of behavior? In this context, psychologist and neurobiologist Jay D. Glass offers an explanation for evolution of the god meme and religious behavior based on analogy to animal behavior, principally chimpanzees. Whether there exists a god or not, he suggests that we are genetically programmed to follow the alpha male of the troop, the supreme being, as this enables improved chances of survival via protection and resources and that "we are compelled to put our faith and trust in a Supreme Being in just the same way as we are compelled to breathe." (184) Therefore, the emergence of a god meme would have been of survival value itself. The possible mechanisms for this have not even been reasonably speculated about, and this theory has been substantially debunked. (31) The notion of humans possessing a genetic tendency to believe in a supreme being, including those based on analogies to human or nonhuman dominance hierarchies, has not shown any real promise. That tribes of humans were favorably selected for survival, or outcompeted rival tribes, based on a hypothesized genetic tendency to follow the alpha male—and certainly a fabricated one—does not seem a reasonable hypothesis at this time.

Unfortunately, the archeological record is very meager for examining such things, and specific genetic markers or other hard data are still lacking. But formal Darwinian group selection would not be necessary for certain cultural behaviors, such as memes or superstitions—e.g., those for gods or their preceding anima—to be reinforced and persist. Memes which are strongly associated with potent leadership in a group are the ones more likely to survive—as memes are prone to do. That doesn't mean that they operate as competitive "adaptive units" of group behavior. Memes can appear, often with similar characteristics, separately in different groups. The "group selection" hypothesis remains a skeptical idea, and no good empirical evidence has emerged. In addition, some geneticists have questioned the validity of group selection, per se, noting that biological evolutionary theory, of course, applies only to individuals.

It's more likely that superstitious beliefs developed alongside other social and cultural traditions which helped humans to survive. As Dawkins notes in *The God Delusion*, religion "doesn't have a direct survival value of its own, but is a by-product of something else that does." (113) We've outlined some of those byproducts—group cohesion, leadership, shared rituals, and cooperative behavior. That is, as groups bonded together for longer periods of time, various memes were more likely to be shared, reinforced, and become part of the tribe's culture. Thus, religious symbols and practices could develop socially and independently in various groups of humans. And they did so, in their own unique ways, in different parts of the world. Such may have enabled more stability and thus the survival of groups; but it's unlikely that religion enabled groups to outcompete each other for territory or resources.

The human diaspora had reached around the world thousands of years before there is evidence of agricultural revolutions, when humans "settled down" and began to develop more cohesive cultural norms. Thus, the idea of tribal competitiveness via religious behavior doesn't ring true. That some primitive cultures may have survived because of shared behaviors, cooperation, leadership, or supernatural memes is not evidence of competitive group selection. Such are passed down culturally, not genetically. And it is likely that all *Homo sapiens* bands or tribes manifested such types of behavior, which may have differed only with respect to details. Cultural evolution based on myths developed independently in many parts of the world. There were different gods, rituals, and traditions. The god superstition, writ large, may be the most instantiated one of all time, but it likely arose independently in various human groups. Humans in the Middle East had different gods than those in northeastern Africa, or in the Far East, or in the Western Hemisphere, for example, and those populations had obviously diverged many thousands of years earlier.[4]

Finally, let's give some credit to Darwin about this problem. He seemed to support the *social evolution* of religious concepts, not something to be considered a driver of group fitness. This was not an endorsement of any god; he merely asserted that belief in the supernatural is a social construct. In the *Descent of Man*, he offered that,

> The belief in God has often been advanced as not only the greatest, but the most complete of all the distinctions between man and the lower animals. It is, however, impossible, as we have seen, to maintain that this belief is innate or instinctive. . . . The idea of a universal and beneficent Creator of the universe does not seem to arise in the mind of man, until he has been elevated by long-continued culture. (106)

This was not a new idea, but Darwin was one of the first scientists to debunk the idea of God as anything more than a meme of primitive human groups, not a genetic trait. Religions developed as "cultural evolution," not a Darwinian system of group survival via competition or fitness.

Conclusion

We must conclude that religion is a natural phenomenon, and that it came into being through natural and cultural processes. Religious memes and practices are learned behaviors, shared socially and passed down culturally. They don't have any genetic drivers, and they didn't have any effect on some putative competitive survival of human groups. The memes of religion—particularly the god ones—are human inventions which arose as people began residing longer in particular places and eventually settling down into communities. In this context, I cannot summarize the evolution of religion better than Albert Einstein did.

> Long before men were ripe [what a clever metaphor], to be faced with such a universal moral attitude, fear of the dangers of life had led them to attribute to various imaginary personal beings, not physically tangible, power to release those natural forces which men feared or perhaps welcomed. And they believed that those beings, which everywhere dominated their imagination, were psychically made in their own image, but were endowed with superhuman powers. These were the primitive precursors of the idea of God." (150)

Do Animals Have Gods?

If human religion developed as a natural phenomenon, and if more primitive *Homo* species showed signs of this, then we might rightly ask what other animals have religion. Before we go further, this seems like a nonsensical question, but consider the following: It is not as absurd as might first appear. To wit, if we ask, "What is religion," then the question seems potentially answerable. Religion is a set of ritualistic behaviors which have no immediate or near-term survival value—for food acquisition, mating, or territorial defense, for example. They are merely superstitious behaviors shared by individuals in a group. They are rituals which tend to enable group cohesion, but they do not, in and of themselves, individually foster survival.

Humans are not the only species to engage in such rituals. A number of other animal species, especially some primates, exhibit ritualized behaviors which do not appear to have anything to do with a functional advantage. Let's take the example of chimpanzees, another primate species we know a lot about. The well-known observations of Jane Goodall are particularly instructive. Goodall observed that some chimpanzees engage in various ritualistic behaviors. Some tribes of chimpanzees engaged in what were called "rain dances." At the start of the rainy season, when there is heavy rain, some of the adult males will engage in charging displays, drag branches around, slap the ground, vocalize pant-hoots, and emit other displays. They've also been observed to engage in "waterfall dances," in which the chimps will stand up, sway around, throw rocks into the waterfall, and even get into the water (something they otherwise are quite averse to). (185, 543) (Does any other species "sway around" during a religious ritual?) These actions do not appear to have any reinforcing value; they don't generate food or mark territory, and the participants don't differentially compete for mates. They are mere rituals. In another study, chimps have been observed to construct specific stone accumulations or "shrines" at *'sacred' trees* [emphasis added], which bear similarities to those constructed by some indigenous West African peoples. (289) All these behaviors are only observed in some chimp tribes, not others, suggesting that they are cultural or learned customs. They do not seem to involve any dominance or food-seeking behavior or any other utilitarian purpose. We can't ask them if they have any supernatural significance, but they are clearly rituals.

Other, more primitive primates also engage in ritualistic behavior. A troop of capuchin monkeys living in Costa Rica is instructive. Some of its members sometimes sit around in a group and shove their fingers into their peers' eyes or noses while repeatedly swaying around (swaying again). They also engage in rituals with "sacred objects," e.g., sitting around focusing attention on some non-edible parts of plants. The behaviors appear to be acquired "via social learning", and only some of the troop members do these things. There does not seem to be any reinforcing or utilitarian purpose to those rituals. (398) Are they merely social bonding exercises? Are these sacred rituals? Do they suggest religion? As we've defined it, they certainly do. Remember, ritualistic behavior associated with artificial memes *is* religion. That's what we humans do.

One of the most poignant of religious rituals of animals is the ceremonial behavior on the occasion of the death of a group member. As I noted in Chapter 4, many animals exhibit death rituals, including elephants, giraffes, some corvids (crows and magpies), orcas, donkeys, peccaries or javelinas, and primates other than us. We don't know the meaning of their ritualistic behaviors, but it would be shortsighted to deny that there is some sort of grieving taking place with these ceremonies. (272) These species do have numerous other ways to manifest social bonding. Thus, upon the death of a member, it is reasonable to infer that the animals' rituals are an acknowledgement of the death, along with whatever psychological loss might accompany that. You might balk at the notion of "psychological loss" in this context. I do not suggest a "theory of mind" between us and such other animals. I merely suggest this as an inference based on behavior. In many animal species, a mother will continue for a long time to look for a missing offspring or try in vain to revive a dead one before "moving on." Some orcas have been observed to "carry" a dead calf along for weeks. To suggest that there is no *sense of loss* when such an animal dies would indeed be shortsighted. Thus, it would also be shortsighted to rule out grieving as an underpinning of those animals' death rituals.

Are human funerals any different? Yes and no. We have more sophisticated cognitions, language, imaginations, symbols, and traditions than other animals. But the coming together upon the death of a community member and the performance of ritualized behaviors is consistent with what these other animals do. It's unclear, of course, whether such animals consider that these rituals have any higher significance. If you could "talk chimp or monkey," perhaps they would suggest some type of belief for the ritual. We have no way of knowing. We only know of one commonality. Since the death of a group member in higher primates is often a significant loss, religious rituals are likely to relieve the stress shared by the group—in them and in us.

Do the animals envision some higher power involved? We humans, as well as some of our ancestor cousins, developed superstitions, and we ascribed to those superstitious powers greater than our mortal selves. Actually, anything with superior unexplained power is a god to some (which I'll return to below). That's how animism likely started in primitive *Homo* species. The notion that some animals would not perceive powers greater than themselves would be shortsighted, including those powers responsible for death. The notion that it is any different for humans, just because we have cognition enabling an understanding of the universe and a fantasy about a being that may have created it, might be seriously questioned. If an animal can conceive of powers greater than itself as a mortal being, perhaps we're not that special. We don't know of such animal superstitious "thinking." We only know that humans often think that way, and we behave in eerily similar ways to some other animals upon death of a group member. That is religion—for us and for them.

Ancient Civilizations and Cultures

As religion developed as a natural phenomenon, let's take a look at religion through the lens of developing human culture and civilization. Humans became the dominant species on Earth and took gods and religions along with them. Evidence of organized religion goes back to the late Stone Age, as bands of humans formed themselves into somewhat more stable communities. It's likely that even before stable agricultural communities arose, some formal religious traditions began to develop.

Archeological evidence suggests that construction of what is considered the world's oldest temple, Göbekli Tepe ("belly hill" in Turkish) in what is now southeastern Turkey, began about 11,500 years ago. It was likely the first human "holy place." (519) It was built by hunter-gatherers and appeared about 7,000 years before the Great Pyramid of Giza. It apparently was a religious sanctuary where people buried their dead. There were carvings of menacing animals (scorpions, snakes, wild boars, lions), which the people may have worshipped. There were religious rituals performed in groups, knitting them closer together, which some have suggested was "the real origin of complex Neolithic societies." (350) In fact, it is now understood that in building this structure, "only a sophisticated religious or ideological system could sustain such efforts." (203)

Not far away, just a few hundred miles west of Göbekli Tepe, on the plains of what is now central Turkey, there arose the settlement of Çatalhöyük (Chah-tahl-hoo-yook), a proto-city of about 30-40 acres. It was built around 7,500 BCE. People were apparently drawn to this site by various superstitious beliefs. Many religious symbols have been recovered there, as well as evidence that people buried their dead underneath the floors (possibly to keep them near). (350, 371) But in ancient Mesopotamia, the first identified community we might call a "city" was probably Uruk, the ancient Sumerian city-state near what is now Basra, Iraq, settled at the latest around 4,000 BCE. There, various religious traditions began to emerge. Religion became more mature, as not only a system of beliefs but also for the imbuing of the gods with very special discrete powers. "Fear of the gods" was a mechanism for enforcing obedience in Urukian society, which was led by a priest-king, who had that special role in the community. He was the mediator between the people and the gods. As various communities or cities arose in Mesopotamia, each one had its own god or goddess, with the executive power of the priest-king using his divine powers to enforce the rules and hold the society together. Citizens, of course, had many roles in the cultures, and various gods needed to be invented to help with the relevant responsibilities. Thus, Mesopotamians had many gods, for all sorts of roles. "There were gods for everything—literally thousands of them—including a god for brewing, gods for farmers, scribes, merchants, and craftsmen . . . . each city state had its own head god, but also a whole court of subordinate gods who played roles such as gatekeeper, gardener, ambassador, and hairdresser." (350)

In ancient Mesopotamia, there was no distinction between church and state, and the kings of various communities would assume "divine prerogatives." As Torrey observed, as far back as "2,300 years ago, the Greek philosopher Euhemeros of Macedonia said that 'gods were originally human rulers who were gradually deified by their subjects. . . . As one moves up the continuum from spirits to deities, one acquires more supernatural powers." (519) For example, Nram-Sin, the ruler of the Akkadian Empire, which followed the Sumerian culture, who reigned c. 2254–2218 BCE, did declare himself to be a god; and Shulgi, who ruled two centuries later, "was worshiped as a god during his lifetime." (519)[5] As the first Babylonian Empire emerged, c. the eighteenth century BCE, the strength of the gods in everyday life was becoming manifest. The great Babylonian king, Hammurabi, was directed by the god Marduk to "bring about the rule of righteousness in the land, to destroy the wicked and evildoers." (350) Hammurabi's *Code* was supplied to the people in 1750 BCE, about a year before his death (I'll discuss the *Code* a bit more in Chapter 9). The Pentateuch, or Torah, as written down, wouldn't begin until hundreds of years later. Early human civilization was well on its way to various invented gods, the establishment of religious traditions based on various superstitions, and the imbuing of authority in various kings and other rulers to enforce religious law based on divine inspiration. Our species, in effect, through its own cultural evolution, began constructing the idea of particular gods and organized religion.

Early Gods

From the dawn of civilizations, humans have invented their gods. Every early culture is replete with mythological gods that controlled the events and fate of humans. Ancient religions were primarily fertility cults. That is, they were characterized by various rituals in which gods were worshiped in an effort to ensure reproduction—for agricultural needs, animal reproduction, or for the ability of man himself to reproduce. Early communities began to formalize religions for such; to do this, gods had to be invented and given reign over various activities of nature and of human life, as the ancient Mesopotamians had started off. This is a voluminous subject, and I'll touch here only on some notable examples.

Hindu gods began appearing in the Bronze Age, probably no later than around 2,000 years BCE, in the Indus River Valley of southern Asia. There was the creator god Brahma, but there evolved gods for most everything else—literally millions of other gods. The countless, though largely unnamed, spirits obviously grew out of the dynamic of animism. Even today, in that culture, certain animals continue to be sacred, e.g., rats, elephants, tigers, the well-known lotus, and a huge number of other plants.

Other religions, per se, arose independently in other parts of the world. These evolved in ancient Egypt and Babylonia in the interval from approximately 800 to 200 BCE. In ancient Babylonian culture, many gods began appearing, probably around the eighth century BCE. There was the god Anu, the creator god as well as a number of gods for all sorts of functions. There was Marduk, the Babylonian national god, if you will, who created humans and often assumed for himself the power of the other gods.

In ancient Egypt, there arose in the fourth to third century BCE, various gods, some of which I'll refer to later. There was the creator god Atum. He/she (was "bisexual," having both male and female qualities) gave rise to Ra, the sun god. Ptah was the god of craftsmen and architects, the spouse of Sekhmet and the father of Nefertum. Nefertum was also known as the "son of god" and the god of beauty and healing, among other things He was considered the father of the sage Imhotep, best known as architect of a pyramid, who was a mortal but became deified over the centuries. There was the Ennead, which added Atum's children to the mix —Shu and Tefnut, and their descendants (including Geb and Nut, who fostered their children Osiris, Isis, Set, and Nephthys). They lorded over most everything in the world.

Gods began to appear in ancient Greek culture in around the fourth century BCE. The Greeks eventually consolidated their gods into 12. With Zeus at the top, they controlled the fates of all reality, including humans (though recall Aristotle's position that the gods allowed for some human free choices). They all lived on Mount Olympus, even though they were immortal. And of course, as Roman civilization became dominant, a very similar panoply of gods, albeit with different names, became the objects of worship. Human culture was beginning to organize its superstitious beliefs in the supernatural.

Polytheism

A huge array of gods permeated ancient cultures. The phenomena evolved from the earlier superstitions of spirits inhabiting everything humans were concerned about. During the Bronze Age, there were hundreds, if not thousands, of gods and goddesses worshiped throughout cultures of the world. There were gods for everything, from many gods for the sun to gods for agriculture, fire, wind, and the underworld. And there were gods for more ordinary cultural phenomena like wine, blacksmithing, fertility (and chastity), medicine, shepherds, hunting, houses, clothing, and most other aspects of everyday life. Unfortunately, this was not a very usable system. Independent worship for everything doesn't seem to be an efficient way of getting what one wants. You might spend all day praying.

Polytheistic religions are less common today than they were 4,000 years ago, at least in the West. Polytheism does persist in some religions today. We still have religions with many gods. Not to single one out, but Hindus still do acknowledge that there are several million gods, not just the Brahma, Shiva, and Vishnu we hear about. After humans migrated to the new world, perhaps as early as 25,000 years ago, native bands began to develop their own groups of gods. It's unclear how much of this was brought over the land bridge or other migratory routes, but humans in the Americas began to worship all sorts of gods. From the Incas to the Aztecs and various other North American tribes, large numbers of gods were worshiped, with many persisting to this day.

Monotheism

Polytheism is not a very coherent, workable, or practical system. Eventually, supernatural powers had to be collapsed into one, or perhaps just a few gods, and the number of gods diminished over the centuries. For most cultures around the world, polytheism gave way to monotheistic gods—though notably different ones in different cultures. There is no evidence of belief in a deity arising just once and then "the word" spreading throughout the world. Each culture developed its own monotheistic gods. Monotheistic gods (note the self-contradiction) served as unitary powers, often becoming omnipotent, and they were more useful to enforcement powers of the "divinely inspired" monarchs.

This transition occurred uniquely in different cultures. It was certainly a more efficient system, i.e., worshiping one god for everything. "The first monotheistic religion known to us appeared in Egypt, c.1350 BCE, when Pharaoh Akhenaten declared that one of the minor deities of the Egyptian pantheon, the god Aten, was, in fact, the supreme power ruling the universe." (203) Worshiping "It" became the state religion. Nevertheless, the edict did not outlive the Pharaoh, and the old panoply of gods experienced a resurgence. As civilization developed and spread, the idea of a singular, omnipotent god increasingly took hold. It was certainly more economical to have one unified supernatural power, though the idea of gods for various functions did linger on for some time. In fact, some versions of polytheism persist, again most notably in Hinduism, where there are countless supernatural beings. Although Hindus worship some "primary" deities, e.g., Brahma the Creator, Vishnu or Krishna the Preserver, and Shiva the Destroyer (interesting that this is also a "holy trinity"), they acknowledge that there are literally millions of gods.

The so-called Abrahamic religions, with a monotheistic god, seemed to appear with the advent of Judaism around the fourth century BCE, give or take, in an area currently known as Israel. The lore that it originated millennia earlier with the Moses character, of course, is only an unsupportable legend. The God Yahweh proved to be a potent force. "He" was a "useful" monotheistic deity, and probably more saleable to the population than His predecessors. (Although Judaism embraces a monotheistic God, ancient Jews weren't necessarily comfortable with this.) As a vengeful, punitive God, He likely was effective in enforcing "morality" and cultural traditions. The superstition became so powerful that—voilà—it persists to this day.

Much later, Christianity began as an esoteric sect of Judaism. It was the dominant force in human culture throughout the Middle Ages, and its sects (yes, we can call them such) constitute the most widely practiced religion today. Though it is generally viewed as a monotheistic religion, it has a clearly polytheistic quality. It endorses a Holy Trinity—the Father, Son, and Holy Ghost. In addition, "Christianity . . . developed its own pantheon of saints, whose cults differed little form those of the polytheistic gods. . . .The Christian saints did not merely resemble the old polytheistic gods. Often, they were these very same gods in disguise." (203) For example, St. Brigit of Ireland (the Irish goddess of spring, fertility, and life), was adopted from Brigid of the Celts. Quite obviously, history repeats itself as much in religion as any other aspect of human life. Only the names change.

Christianity was also a model "for another monotheistic religion that appeared in the Arabian Peninsula in the seventh century - *Islam*." (203) In fact, much as Judaism was founded on the basis of a mythical historical figure, and Christianity was founded on the mythical Jesus character, Islam was founded on a messianic figure, Muhammad.

There used to "be" many gods. Over history, in various cultures, there may have been thousands of different gods with different powers in different cultures. Of course, no one of them is more valid than any other. They all rely on superstitions passed down through their respective lore and are maintained by factors which are purely cultural. We have no more reason to accept the idea of a monotheistic god than we do for polytheistic ones or for an animistic philosophy. We're just more comfortable with a monotheistic god; it makes prayer more efficient. The various labels used for It allow for better sanction of discrimination, hatred, and war—all very mortal, though powerful, rationales and motivations for collapsing supernatural forces into one being. Of course, there continue to be various monotheistic gods around the world (again, an interesting concept). Monotheists have promulgated a much more efficient system, but as Christopher Hitchens astutely put it, "They are getting ever nearer to the true, round figure." (211)

Myths

Of course, the gods which have appeared over the millennia are merely myths. But there are countless other myths which underlie religion. Having other myths merely enables religions to "appear" substantial. There's much more detail in these, which I'll only briefly review. It is not difficult to see how religions use myths to reinforce belief. They are shared memes with some "shock value," often involving miracles, which tend to create social coherence in a group. What is a myth? Per James K. Feibleman, American philosopher, "A myth is a religion in which no one any longer believes." With all due respect, billions still do hold onto religious myths as if they were facts. Nevertheless, their perpetuation is an essential force in the maintenance of organized religions. Without them, religious systems would collapse. That's right. Organized religion can only exist on the back of myths.

A huge panoply of myths began to develop in many cultures. There were myths about certain animals, the power of the god-kings and later the prophets, the origin of the world, a son of a god, virgin births, a catastrophic flood, divine punishment, and literally thousands more. I'll review just a few of these. Most organized religions share the same kinds of myths, with different labels, personal figures, etc., especially the monotheistic religions. For example, they often employ a supernatural figure who created the world, concepts of an afterlife or death and rebirth, the idea of a divine, yet mortal, prophet or messiah, and many others. What is behind these apparent commonalities? Some would argue that these are all the phenomena handed down by a supreme god, and that they diverge only by slightly different names, stories, and interpretations. Of course, the argument of coherence of myths from a unitary supernatural being is merely another "gap" fantasy. It is more likely that such myths grew out of commonality in superstitions among early humans, and the further concordance of religious myths across cultures was likely due to what might be called "parallel cultural evolution," with the development of similar memes in different geographies of extant humans. This has been systematically studied, and we can reasonably infer that certain major myths, sometimes termed "mythemes," did spread and change somewhat with prehistoric human migrations, along with other aspects of culture. Primary myths of various cultures do seem to have grown out of a small number of superstitions. These spread around the world and appear to have taken on various forms in different cultures beginning at least 15,000 years ago. (119)

Despite this, of course, different religions still maintain that *their* "stories" are more valid than others', which has obviously led to numerous problems throughout history. Either cultural variations account for the evolution of different religious myths, or there are many, many gods, or one god has played a cruel game with the various cultures of Earth. Obviously, the first explanation is the most likely one.

Creation Myths

Of course, the most common of myths are the ones "explaining" how our world came into existence, the origin or creation stories. But why *are* there creation myths? As I've discussed, we humans do have a need to explain things, and the origin of the world, of course, is at the top of the list. We need to have a cause for all of existence, and, lacking an immediate explanation, we invent a few. In fact, all cultures, both modern and ancient, have mythical world origin stories as well as myths about the origin of humans.[6]

Ancient cultures created "theories" of the origin of the world, generally relying on the extant supernatural beings of each culture and often relying on some of its mundane aspects. For example, the Norse felt that the god Odin (and some of his relatives) created the world, and it was created out of ice, a characteristic feature the Norse people dealt with. Each human culture throughout history had its own "true" story. To illustrate the point, here are just a few other examples. (Extensive illustrations are cataloged in other works. (300)) Early Egyptian mythology relied on the god Atum, the creator. He was responsible for the chaos of the universe, creation of the actual world, and, according to legend, had the power to destroy it all and bring it back from its chaos or "nothingness." (The modern theory of a cyclic universe comes to mind.) Atum gave rise to Ra, the sun god. The sun was likely the first major phenomenon in human perception that needed to be explained, and Sun worship, of course, persists to the present. Ra became the king of all the other gods in Egyptian mythology as well as the father of mortal Egyptian kings (yes, a god siring some mortals). And there were many, many gods—male, female, and gods with both characteristics, responsible for nearly everything in human life, from war to hunting to craftsmen. But it was Atum that got it all started.

There are a number of origin stories told by the ancient Chinese. An early one does not include any creator god. That culture believed that the world (and humans) has always existed and that the universe is "self-generating." (356) (Recall again the various modern cosmologists' agreement on this.) Another ancient Chinese legend suggests that there was a vast primeval vapor, an undifferentiated chaos which contained "all possible realities." (Any foreshadowing of the quantum Many Worlds theory or M-theory?) Out of this came the *yin* and *yang*, the "forces" which led to creation of our world and humans. (300) Yet there are other ancient Chinese philosophies that do include the idea of particular gods. There was the goddess Nüwa, for example. She molded humans out of earth, and she shaped differed kinds of humans. The aristocrats, for example, were made from fine yellow soil, but commoners were just made out of "brown mud." (203)

In ancient Hindu lore, there was the creation myth of the primordial giant named Purusha. He is a primeval and possibly eternal being, having no beginning and no end. He split into the sun, moon, and various castes of humans. The gods—whoever they were—dismembered him; his head became the sky, his feet became the earth, and his breath became the wind. This is how the known universe came into being. The sun came from Purusha's eye and the moon from Purusha's brain. Of course, many other entities derived from Purusha. For example, warriors, or Kshatriyas, came from his arms, the Vaishyas, or peasants and merchants, from his thighs, the Brahmins, or priests, from his mouth, and the Shudras, or servants, from his legs.

The ancient Greeks believed that before there was a recognizable universe, there was only "chaos," a kind of infinite void, the primordial randomness. According to the poem "Theogony" by Hesiod (c. 700 BCE), the gods and the universe spontaneously arose out of the preexisting chaos, including most notably (and first) the female god Khaos (the goddess of air), then gods of the Earth and the underworld and the goddess of love (Eros). What followed was the well-known polytheism, with Zeus at the helm and all of his subordinate gods with different powers, many of whom had both mortal and supernatural qualities. We're familiar with Apollo, Ares, Poseidon, and other major gods, as well as "minor" gods like those for wine (Dionysus), the underworld (Hades), and agriculture (Demeter). But there were many others. Before the panoply of gods was collapsed into a usable number, the Greek mathematician Thales of Miletus believed that the gods were everywhere in the universe; they were in "everything," not surprisingly, somewhat consistent with the Hindu perspective.

Skipping ahead to the New World, the Mayans (of what is now Central America and Mexico) thought reality was initially only the sky, the sea, and what they call the "Makers," actually two gods, who created almost everything else, including all the animals. They created the first humans, possibly from the earth or from wood (interpretations vary), but that effort failed—they soaked up water and dissolved. That was unsatisfying, of course, and the Makers eventually created humans out of maize, the principal crop of that culture. Another interesting New World culture is the Lakota of North America. They believed that there existed a world before this one. The creator was unhappy with it and rained upon it until all life was destroyed, save a crow. Sound familiar?

The so-called origin story of the Abrahamic religions is perhaps the most well-known and popular one today, the Old Testament story of Genesis. We're told that God created everything, including humans, from scratch or out of nothing, and it took 6 days, followed by rest. Rest? That's a rather strange dynamic for an omnipotent being. I would point out that the theists who offer that all this should be taken metaphorically are just insecure about their positions. As we've seen, the Biblical account was far from the first attempt by human cultures to conceive of how our world came into being. It's not really an original or novel story. We see voids or chaos in many origin stories that well predate the Old Testament. It's not unlike the ancient Egyptian or Aristotelian notion of an eternally existing, formless chaos that needed to be massaged in some way to "create" the world. History again repeating itself. Shermer also points out a related glitch in the Biblical account of the "creation" story.

> The Christian theologian argument that God created the universe *ex nihilo*, or 'out of nothing,' based on the English translation of Genesis 1:1 that 'in the beginning God created the heavens and the earth.' . . . is misleading. Recent scholarship has suggested that the Hebrew word for 'creation' in Genesis 1:1 is bara (ברא)—a verb that more accurately translated means to 'separate' or 'divide.' Genesis 1:1 should read, 'In the beginning God separated the heavens and the earth.' Separated from what is not indicated. (470)

This does appear to be the consensus on the language, but that's not the full story. Whoever put down some putative original language, and whatever scribes provided edits over the millennia, were guided purely by their own intuition and other extant writings. The Biblical account merely invokes a supreme being to organize the eternal chaos, as many other origin stories had addressed with their gods throughout ancient history. Nothing new here. The Biblical account is merely poor plagiarism of other, extant stories. It thus also leaves unresolved where everything came from. There are many other creation myths, of course. We merely see creation myths being reinvented in various human cultures. Of course, none is more valid than any other, including the Judeo-Christian myth of creation, a bane for modern true believers. The notion that any particular supernatural creation story is true seems laughable, but they persist to the present day. They are all "true" to their believers. We believe only what we're taught. These, again, are cultural memes, nothing more. That there is any need for a "creation," of course, is again strictly a fantasy, based on human needs for practical explanations. It remains farcical to consider any of the origin stories as anything more than memes which hold religious groups together.

Other Myths

There are many other religious myths, of course, passed down over thousands of years. The list would be extremely long, but let's consider just a few of the more popular ones. Many religions have myths of *resurrection* or *death and rebirth.* This is not new and well predates the fable of Jesus' resurrection of Lazarus and his own death and rebirth. Many religions have stories of mythical figures who died and were reborn. Dionysus, the Greek god of wine and fertility, falls into this category. He was killed by the Titans but brought back to life by his grandmother. The Egyptian god Osiris was killed by a rival god and his dismembered body scattered around Egypt, but the goddess Isis collected and buried them, which brought him back to life. Cultures have embraced these myths with the sincerest of belief. The "death and rebirth" myth, of course, is consonant with the cycles of nature; death coming in winter months and rebirth in the spring, from which such myths arise. Rebirth, of course, is also famously epitomized by the Hindu myth of reincarnation. Being reborn into something better, based on your current life, is a fundamental goal. It is a sacred principle of Hinduism. "Death and rebirth" has been a religious myth for literally thousands of years.

Many religions embrace the myth of a *catastrophic flood*, an enormous deluge, often sent by a god dissatisfied with humanity. It was the job of the deity to replenish or revitalize the Earth. The Biblical myth of Noah and his flood is only a recent one (For the omnipotent Judeo-Christian God, as I've noted, this is a pretty bizarre failure.). God is alleged to have directed Noah to load his ark with two of each kind of animal—male and female, separated so they couldn't reproduce. Of course, given the millions of animal species, this must have been a pretty big boat. Then there are all the animals that reproduce asexually. And if God wanted to replenish the Earth, what about all the plants and other kingdoms of living things? I'm being a bit snide here, but this is just to illustrate the nonsense of the legend and the bizarre ideation of the theologians who interpret this even metaphorically. Catastrophic or apocalyptic floods are also legends in the mythology of the Norse, Chinese, and many others. It is noted in Hinduism, along with warnings to build a giant boat. The Greeks had a similar mythology, as did the Egyptians and the Babylonians, as noted in the *Epic of Gilgamesh*, which I'll have more to say about below. The flood also appears in the Quran, and similar stories have been told in Buddhism and by ancient Hawaiians, Nigerians, Aborigines, and many others. In the new world, the Cheyenne, and perhaps other tribes, have similar lore.

Then there's the myth of the *virgin birth.* Many a prophet or god-like character has been born that way. Virgin births characterize various Greek gods, like Dionysus, for example, the god of wine and various other mortal pleasures. There was Mithra, a Persian god, whose birthday, by the way, is December 25. The Aztecs' Quetzalcoatl was born of a virgin, interestingly symbolized on a cross, and arose from his death (or hers, or both). We can't forget Romulus and Mercury of Roman mythology; Ra and Horus of Egyptian mythology; and many, many others. Even figures largely considered to be fully mortal, e.g., Genghis Khan, Plato, Alexander the Great, Pythagoras, and Caesar Augustus, claimed to have been born of virgins. And there are various miraculous births in Native American, Japanese, and other cultures. If someone claimed such a birth today, most cultures would shun or more likely confine such a person, but it has been a popular theme throughout history. The Jesus fable is only a very slightly more modern one. If it wasn't purely a "conspiracy theory" (a concept we'll revisit in Chapter 11), the whole notion was obviously just plagiarized from some other culture.

Prophets and Messiahs

A predominant myth in many cultures has been the prophet or messiah. Messiahs, supposedly inspired by gods, are the putative saviors of humans. Prophets merely get messages from some supernatural god and spread those words. Of course, throughout history particular myths have imbued some prophets and messiahs with both deistic and mortal qualities in the same mythical figure.

Prophets or messiahs are generally constructed from the developing lore and are more rightly considered memes than actual people—or gods for that matter. Almost all religions have divinely inspired prophets or messiahs. These all well preceded the Christian character of Jesus, which was obviously plagiarized from more ancient myths. As merely a sampling, we see such figures in the ancient religions of Hinduism (Kalki), Taoism (Li Hong), Zoroastrianism (Saoshyant), and Buddhism (Maitreya). Siddhartha Gautama, the Buddha, probably did exist, though there's some controversy about this. It's unclear exactly when he lived; but he's variously considered a prophet, a mortal human being, or, per some Hindus, an avatar of the god Vishnu. Legend describes a prophet called Moses, a supposedly mortal character, but there's no evidence that such a person ever existed. Orthodox Jews are still waiting on their messiah.

Islam does recognize the Jesus character or Isa Masih, among other references, as the messiah in the Quran. He is of a virgin born and performed miracles, but is mortal—not a god or son of God (unclear how he got miraculous powers). Muslim lore proposes that the text of the Quran was given to the Prophet Muhammad by God or the angel Gabriel (The actual human authors of the Quran remain unknown). However, contrary to popular belief, there is little evidence that the prophet Muhammad actually existed. In a comprehensive examination, archeologists Yehuda Nevo and

Judith Koren review the argument and conclude that in all likelihood the Muhammad figure was merely fabricated to give credence to the text of the Quran. (367) An all-too-typical scenario.

The Jesus Myth

Let's take a prime example, the Jesus character. I'll devote a little space to this since this is the meme that most Westerners are familiar with. Christianity is a set of organized religions that pervade most populations around the Earth and its central protagonist—the Jesus figure—deserves some comment here. The myth obviously grew out of the messiah lore of the various religions which preceded it, as I noted. As an example, let's just take one parameter of the legend, the related fact of his "birthday." It is, of course, celebrated with the winter solstice, i.e., in late December. But this again is borrowed from a much longer history. Earlier pagan traditions of the winter solstice were "celebrated" around this time of year. It is marked by many religions as the symbolic time of death, e.g., of the sun at its nadir, and the beginning of rebirth. And long before the Jesus character, birthdays on that part of the calendar were attributed to many gods, including the Egyptians Horus and Osiris, the Syrian Mithra, the Greek Dionysus, and many others. The list is very long. Although the lore has Jesus being born around this time, the artificial assignment of Jesus' birth to this time of year was obviously predictable. It was not actually formalized until at least the fourth century CE. In many ways, Jesus was just another "borrow" from history. I'll not go into all "his" other fabled qualities, which were plagiarized from other religious traditions as well.

Of course, the hotly debated question is, Did Jesus actually exist? Despite theologians' confidence, the clear answer is "no." Here's a brief review of the matter. Let's start with the notion that one does not have to prove non-existence. I'll get into more of this kind of logic in Chapter 8. For now, if this matter is subject to any argument, or any sort of overt discussion among humans, the burden of proof is on the myth believers.

Yes, it is a myth. As we've noted, there have been prophets, including ones with supernatural powers, in many cultures and for as long as religious cultures have been around—for many thousands of years before the alleged appearance of the so-called Jesus of Nazareth. We would need some reasonable evidence to establish that this figure is distinguished as the one out of that very large group that actually existed. In fact, there is actually no evidence that such a person ever lived. As Bertrand Russell noted a century ago, "Historically it is quite doubtful whether Christ ever existed at all." (435), and, since then, no convincing evidence has appeared.

There are countless historical inaccuracies that render the Jesus myth as such—too many to be detailed here.[7] But let's consider some of the critical issues. One that's particularly enigmatic is Jesus' resurrection. There is no documentation or archeological evidence for this, of course, and there is no report of any observation before or after he was supposedly crucified. (If in fact, he existed as a mortal being, as some theologians have suggested.) In fact, there is also no actual physical or literary evidence from contemporaneous sources that he did exist. The historicity of the putative Jesus comes up negative. There are many writings, statues, coins, and other evidence of important persons living during the early first century CE, but nothing at all about a Jesus character. There are dozens of first-century writers—historians, naturalists, philosophers, literary figures, and others—whose voluminous works have survived with no mention of this character at all.

All records usually cited by theologians are from later periods. It is interesting that, despite numerous other documents of the time written by historians and other scholars, there is no contemporary record of the Jesus character who allegedly performed feats that would have drawn incredible notice. One would think that performing miracles would garner some attention, with information spreading to wider parts of the culture where such things could be written down. That such never happened obviously casts serious doubt about them and the Jesus character itself.

To be specific, there is a great wealth of surviving literature from the early first century in that part of the world, including the writings of the Roman naturalist Pliny the Elder (CE 23–79, who died in the eruption of Vesuvius of 79 CE), the historians Plutarch (born Greek, Lucius Plutarchus as a Roman, living ~ 45–127 CE) and the Roman Livy (Titus Livius ~ 59 BCE–17 CE), just to name a few. Historians of the time leave no written trace of the supposed Jesus of Nazareth, and there are no other contemporaneous written records of his existence. As we now know, the Jesus character is not mentioned in any of the Dead Sea Scrolls, dated approximately from the 3rd century BCE into the 2nd century CE. (To illustrate just how desperate religious apologists are, some supposed Dead Sea Scrolls housed at the Museum of the Bible in Washington, D.C., have now been determined to be modern forgeries. (58)) No early-first-century historians or Roman records mention this individual, despite numerous contemporary references to Jews at that time. In this context, University of London professor G. A. Wells provides extensive analysis and conclusion that this issue as a myth. In one of the seemingly last historical straws, he notes:

> There is widespread belief that Jewish and pagan testimony establishes Jesus' existence beyond all reasonable doubt, and makes any further discussion of the matter unnecessary. The principal Jewish witness is the historian Josephus. . . . He makes no mention of Christians, who presumably existed only as an obscure sect during the period about which he was writing (which extended up to AD 73), and *not* as followers of a historical Jesus of renown. His works do indeed contain two passages about Jesus the Christ. But the longer of these *has been shown fairly conclusively to be wholly a Christian interpolation.* It is a glowing description which no orthodox

> Jew could have written. If Josephus had really believed what he is here represented as saying, he would not have restricted his remarks to a paragraph of ten lines. Furthermore, the passage occurs in a context concerning the misfortunes of the Jews, with which it has no connection, except from the standpoint of a Christian, who would naturally regard Jesus' condemnation by Pilate at the demand of the Jewish leaders as the very worst misfortune ever to have befallen the Jews. (536)

Quite obviously, Josephus' words were altered by someone else, likely one or more Christian fabulists. This is how myths survive. Something is said or written which may later be distorted or edited in some way; someone else embellishes and/or further distorts the story; the misinformation is spread; a legend without any basis becomes a shared meme in some group or subculture; and someone else comes along and writes it down (or edits something else) as if it had any reality.

The fact that some books written by mortals mention a character called Jesus doesn't mean there is any reality to it. Ancient history is replete with such inconsistencies. There is actually no substantive writing about a Jesus character until appearance of the New Testament Gospels, and these were *not* written by contemporaries of the fabled character. The gospels are most likely anonymous and were initially written, by all reasonable reckoning, between approximately 70 and 115 CE at the very earliest (I'll provide some additional detail on the date problem below). The lifespan of individuals at that time would be extremely unlikely to enable any personal or second-hand knowledge of such a person. Further, the writers were obviously well-educated folks. However, those living at the time of the mythical Jesus character were particularly uneducated and illiterate folk, who likely spoke Aramaic, didn't write, and could not have laid down information which may have been used by Biblical scribes who lived many decades later. The oral lore that scribes used is incredibly doubtful, and there's no reason to believe it wasn't just all made up. (Such religious fabrications do happen, as is more obvious to modern people; for example, about the ludicrous fantasy of how Joseph Smith founded the Mormon religion. That any more credence is given by some to the Jesus character is only a function of elapsed time.)

Fast forwarding to modern times, we haven't heard much lately about the *Shroud of Turin*, which was long claimed to be the cloth which wrapped the body of Jesus of Nazareth. This is because we now know that it was clearly a fraud, a creation by someone done between approximately 1260 and 1390 CE, as demonstrated independently by radiocarbon dating at three different universities. (113) Radiocarbon dating is not perfect, and there are some known errors of measurement. However, done properly, it is accurate to within a few decades, which is why you're seeing the range quoted. Nevertheless, it is clear that the shroud is at least a millennium more recent than the time in which this Jesus is alleged to have existed.

The notion of semi-mortal messiahs or prophets, of course, is just another set of myths, which differ from culture to culture. Again, one is not any more valid than any other. The 17th-century intellectual, Native American Kandiaronk, put it best:

> If it were possible that God had lowered his standards sufficiently to come down to earth, he would have done it in full view of everyone, descending in triumph, with pomp and majesty, and most publicly. . . . He would have gone from nation to nation performing mighty miracles, thus giving everyone the same laws. Then we would all have had exactly the same religion, uniformly spread and equally known throughout the four corners of the world, proving to our descendants, from then till ten thousand years into the future, the truth of this religion. Instead, there are five or six hundred religions, each distinct from the other. (189)

With the various messiah myths, the best characterization is that they are "conspiracy theories"; and the Jesus one is not an exception, following many other characters who were "sons of god" or others that had both deistic and mortal characters (I'll talk more about conspiracy theories in Chapter 11). Whether the notion was invented anew or borrowed from other, older religions is immaterial (though the latter is likely). It is a meme which has been fabricated and enhanced over the centuries and has become an incredibly strongly held myth, particularly in the West. Consider this—If a Jesus appeared today (as promised) and claimed to be the son of God, what would people think? Perhaps some would believe him. He might even perform what *seem* like "miracles." (I'll discuss the nonsense of miracles below.) New conspiracy theories might even evolve. However, he'd generally be viewed as a megalomaniac or a schizophrenic. If such a person, self-identifying the way legend depicts, existed two millennia ago, people at the time would likely have come to similar inferences.

A final word on this meme. Many "liberal" theologians on this topic entertain the notion that Jesus actually existed but only as a fully mortal character, without any divine origin or quality. He was supposedly a prophet and a charismatic leader of a band of Jews, and he promulgated guidance and modeled moral behavior and other matters. Perhaps he was so revered that a legend of divine inspiration and, later, divine quality began to emerge, along with a few fabricated miracles. Of course, if that were the case, some evidence would have appeared prior to the very-much-later drafts of the Gospels. We can only conclude that the idea that there was a mortal Jesus is also just a fiction. It's a notion that keeps the debate going; and it apparently serves as some "cover" for liberal theologians. But that's all.

Whatever is God Anyway?

The basis for all theistic religion is the existence of some sort of god (or gods). What do we mean when we say "god?" This differs, of course, across cultures, religions, and, most strikingly, across time. There are likely as many definitions of "god" as there are cultures—or people. And what character or quality these may embody vary, from the animistic to the "nature's god" idea (see below), to an impersonal deism. But these are not what is considered as God by the three Abrahamic religions. In general, these embody a personal God, one which provides guidance and exerts control over the world and its humans. He also promotes great fear and threat of punishment, the possibility of eternal salvation; and He is one who must be worshiped.

I've gone through some of the problems with this God, the absurdity of His putative powers, etc. (Below I'll detail some of His other "mistakes.") It was Nietzsche who famously posed the question, "Is man a blunder of God? Or is God a blunder of man?" (373) If God created man, He did a pretty poor job, as there have been many large course corrections along the way. And if man created God, he obviously still can't agree on who or what He is. There have been gods created for every religion and culture, and, for some religions, many gods throughout history. God is a human-fabricated idea. We invented it. Biblical text proposes that God created man in His own image. (Genesis 1:27) Of course, this is just the lore as handed down by men who wrote Bibles. It is clear that we invented God in our own image—not the reverse—and that there wasn't a lot of serious thought behind the creation. It has always been interesting that God was rendered male, not surprising, of course, since the Biblical writers were male.[8] There have been recent calls for God to be female, but these haven't made a great impact on Abrahamic faiths so far. We might also question why God even needs to have a gender at all. It is a pretty shallow and unthinking God who would choose one gender, or any gender, for Itself over any other.

And different cultures created different characters or images of a god. It's not clear why there needed to be some personification of a god in the first place. The ancient Greek philosopher Xenophanes proposed in the 6th century BCE that, if horses and cows had hands, they would draw gods conforming to their own images. Horses might draw gods that looked like horses, and cows would draw cow gods. As I noted in Chapter 5, Baron de Montesquieu, the eighteenth-century French political philosopher, offered, "If triangles had a god, he would have three sides." And as we also noted, if an advanced AI system had a god, it might be some sort of "supercomputer."

Of course, God is apparently whatever you say He/She/It is. Conceptions of a singular god are more common today than in the Bronze Age, though the "nature" of this is somewhat different in Christian, Jewish, Islamic, and certain other faiths. There is the classic Judaic view, the God Yahweh, whose role was written as a vengeful, vindictive entity whose control of human thought and action was based on fear and threat. "Thou shalt not" was the predominant theme. The Christians, of course, extended the roles to reward and punishment, the options of Heaven and Hell. The Islamic God, Allah, is not so different from these.

Nevertheless, slightly different conceptions of God in such religions have been a catastrophic problem for our species (I'll look at this more below). Every religion claims to be the "one true religion," having the one true god. That's a bit of a conundrum on this planet to say the least, as Kandiaronk eloquently said. But how can this be? Obviously, it's nonsense. As Dennett notes, "Whatever religion is yours, there are more people in the world who don't share it than who do, and it falls to you—to all of us, really—to explain why so many have gotten it wrong, and to explain how those who know (if there are any) have managed to get it right." (123) The "one true religion" appears to be pretty variable—across geographies, across cultures, and across eons of time. It grows out of the heterogeneity of cultures around the world. Why is Christianity a minority religion in the Far East? Why is Catholicism dominant in South America? Why is Hinduism a minor religion (statistically) in America, although it is growing? There are exceptions, of course, but you get the point.

This is always a difficult concept for religious apologists. Christian Evangelicals, for example, continue to proselytize their religions, each of which is the "one true religion." Besides being a logical pain for any theologian, such concepts have always been the breeding ground of discrimination, hatred, violence, and war (below). And we can find this pattern even down to the various sects of the major religions. Many Christians, for example, do not consider Mormonism "real" Christianity (obviously not focusing on the fact that the legends of Mormonism are, in fact, no crazier than those of Catholicism or any Protestant denomination). The twentieth-century violence between Protestants and Catholics in Northern Ireland is hard to ignore. And the sects of Islam also suffer from this problem. Great animosity and violence continue between followers of Sunni and Shia Islam in the Middle East. We invented the idea of God and continue to define the myths and traditions associated with He-She-It-Whatever in ways that "suit" us—for better, and usually for worse.

Nature's God

Many people do not conceive of a personal, supernatural God. They use the term in various other ways. There is the philosophy of *pantheism*, suggesting that God *is* the universe or synonymous with it. As an attempt to define God, of course, this falls short. Merely relabeling the universe provides no clarification.

Many great thinkers have been accused of believing in God. Even if that were true, it is not evidence that God actually exists. But we should also point out that the views of most of these thinkers have been misinterpreted. For example, many consider "god" to be an embodiment of the natural world, not a supernatural being. That view was essentially espoused by Spinoza. He viewed God as the embodiment of nature, an infinite nature. However, in that context, he also saw God as a "substance," a substance which had caused itself. Despite the attempt to "depersonalize" the idea of God, this position has proven logically untenable (as I'll address further in Chapter 8).

Many cite the example of Einstein believing in God, perhaps to bolster their insecure faith. He repeatedly tried to correct that view. His view was somewhat similar to that of Spinoza (without the baggage of the "cause" argument). To correct the record, he did not believe in God. To those who try to interpret his view of "God" to be anything but wonder at the laws of nature, Einstein wrote the following in 1929:

> It was, of course, a lie that you read about my religious convictions, a lie which is being systematically repeated. I do not believe in a personal God and I have never denied this but have expressed it clearly. If something is in me which can be called religious, then it is the unbounded admiration for the structure of the world so far as our science can reveal it. (144)

He further clarified, "I believe in Spinoza's God, who reveals himself in the lawful harmony of all that exists, but not in a God who concerns himself with the fate and the doings of mankind." (235) To be clear, he used the concept of a god purely as a metaphor, essentially endorsing an awe about the laws of nature. Nevertheless, he was often accused of being "religious." In fact, he was raised in a Jewish household but did not actually practice the religion. His answer was, "To sense that behind anything that can be experienced there is something that our minds cannot grasp, whose beauty and sublimity reaches us only indirectly: this is religiousness. In this sense, and in this sense only, I am a devoutly religious man." (235) This is similar to the concept of "Nature's God," which has been used by many throughout history. It was used by Thomas Jefferson in the Declaration of Independence, (which I'll address below). But none of the above are the fabricated personal God or gods of the Abrahamic religions.

Afterlife

Probably the most significant factor in the rise of religion is humans coming to grips with the reality of death and the possibility of some sort of afterlife. The superstition of an afterlife arises from the powerful instinct for survival along with the human capacity to imagine. To imagine that life goes on, even when the body fails to move, is a fully understandable superstition; and it has been the central organizing myth for religions through the ages. Even Sigmund Freud astutely observed that religion clearly came about as a way to "survive death." (174)[9]

Obviously, evolution of the afterlife idea is specifically related to the survival instinct, a more powerful force in human behavior than just about anything else. It is fundamental to our nature, and we strive to avoid death at nearly all costs. The instinct for survival exists in all living things, something which is almost entirely driven by our genes. We can even trace it back through evolutionary history to more primitive species, to early life forms on this planet, and possibly even back to genes themselves. As Dawkins notes, "The replicators [e.g., DNA or RNA molecules] that survived were the ones that built survival machines for themselves to live in [organisms]." (111) It is only metaphorically that we can refer to genes surviving, of course, though there may in fact have been self-replicating RNA-like molecules early in the evolution of life on Earth. But the principle is undeniable—the power of the push for survival is a fundamental property of life, from the most basic living entities to supposedly advanced humans.

Fear of Death

As the survival instinct is natural, so is the fear of death natural. But death is inevitable.[10] Death is the end of life. But the human effort to deny this is powerful. "Fear of death" and instinct for survival are obviously related: one a subjective response to something largely hard-wired, the other purely genetic.[11] Given the cognitive sophistication of humans, that there have been superstitious inventions of an afterlife, heaven, reincarnation, etc., is not surprising and is easily understood in the context of the behavioral evolution of our species. Our need for survival, so to speak, has led us to all manner of ideas for immortality or avoidance of material death; the most popular of these is the religious notion of "life *after* death." That afterlife, in some form and fantasy, is a fundamental principal of Christianity.

The notion of an afterlife exists primarily, and perhaps only, to make people feel better—to reduce their fear and anxiety. There is clearly aid and comfort in the notion that "life" will not end, and providing this idea has been a powerful comfort for dying humans, albeit a clearly invented superstition. Christians and others do try to alleviate moribund suffering with the notion that relief will soon follow. There are even formal rites to support the effort, and such rites are even more potent when delivered by a member of the clergy. One cannot argue with the notion of relieving suffering, and this is the continuing basis for promoting an afterlife; but we must again acknowledge that, in this case, it is based on a superstitious fantasy.

Primitive Conceptions

We cannot communicate with other creatures about their impulses, of course; but all other living things have the same genetically based survival instinct that we do. They "fight" for survival with extreme ferocity. And many social animals tend to provide assistance to ill or injured members to evade or forestall their death, "realizing" that each member provides a positive ingredient in group survival. The idea of individuals dying is not easy to accept, and many animals do engage in death rituals, as we discussed. Does the animal believe that it has some sort of a heaven in its future? Does it believe that it has an immortal soul? These are nonsensical questions, of course, since we do not understand the thinking processes of other species, including even the ones with advanced nervous systems. But you can see the motivation for the evolution of such an idea. Accepting death of a member of the group can be difficult for many species.

We might ask where a conception of an afterlife might have arisen in the evolution of the animal kingdom. The concept of life after death certainly began to develop long before modern *Homo sapiens*. Primitive humans often engaged in ritual burial of their dead and included various implements and even food, as they believed that such were necessary for their deceased ancestors in their life "elsewhere." Constructing an alternative to death was quite understandable. There "needed" to be some sort of life after death. It was present in some of our other cousin *Homo* species who also engaged in death rituals. How about the earlier *Homo erectus* or *Homo habilis*? There's no evidence here; we just don't know.

Of course, the nature of an afterlife could not be easily envisioned by primitive humans (any more than it is today). But it must have been governed by some supernatural entity, much as the sun, moon, and stars were. It was understandable therefore that what lay above—the sun and the bright lights in the night sky and the seemingly endless expanse overhead—should be looked to for the source of the apparently supernatural power, the gods. The mysteries of the world must have come from above, and that is where the afterlife must be. Of course, some Western religious traditions continue to look to the heavens, so to speak, when referring to an afterlife. In general, theistic religions have always looked upward for a god or gods and referred to Heaven as "up" there somewhere. Whether or not religious apologists offer this only metaphorically, it is easy to see how the tradition came into being.

Early Civilizations and Cultures

Although primitive humans conceived of some sort of "afterlife," it was not until the advent of some modern religions that such fantasies took on more palpable qualities—Heaven, Hell, the Mormon's planet Kolob, etc. Shermer describes the varieties of afterlife concepts humans have invoked to reduce the natural anxiety of death, and these include some ethereal immortality, various utopian heavens, or reincarnation. But he also notes that our drive for immortality also leads us to the idea that an afterlife is based on our human "pattern of information" that could be transferred to another medium and that could be stored for a time "approaching infinity." (472) The commonality, if you will, is that some sort of soul, spirit, etc., gets "uploaded" to some kind of alternate, ethereal existence. An afterlife, after all, would not be valued if it didn't preserve our own personas.

Afterlife was different, of course, in many ancient cultures—in ancient Mesopotamia, in Egypt, later in Hinduism and Buddhism, and much later in Christianity. All human cultures—including ones older than Christianity—had taken up this idea. There is evidence that the ancient Egyptians, as long as 4,000 years ago, may have had elaborate death rituals which affirmed it. The inside of temples, tombs, and coffins were often adorned with representations of stars in the sky; and souls of the deceased were directed to various stars under the assumption that one's soul could be reborn there. Pharaohs and other esteemed people in the society were directed to more fixed stars. But lesser folks were directed to stars that dipped below the horizon at certain times of the year. The star charts on the inside of coffins are believed to provide this mapping of where people were sent after their deaths. (510)

The prevalent view in ancient Mesopotamia was that when humans die, their spirits must go to a dark underworld. This was frightening, of course. The effort to avoid such death in ancient Mesopotamia is well described in the major poem, *The Epic of Gilgamesh*, written probably about 2,100 BCE and often described as the "earliest classic of world literature." It tells the story of Gilgamesh, who was the mythical king of the city of Uruk about 4,700 years ago. (519) He was a demigod, ⅓ human and ⅔ divine. He was a great warrior king, the most capable man in the world. He could defeat anyone in battle, and he resolved that he would never die. Nevertheless, his dearest friend died, and Gilgamesh became terrified that he too would die. He would somehow find a way to defeat death, and he set out on a quest to discover if there was a way to avoid it. Historian Harari notes, "He found out that he could not reverse the decree of the gods and would later die himself. When the gods created man, Gilgamesh had learned, they set death as man's inevitable destiny, and man must learn to live with it." (203)

This was generally unacceptable in some cultures. As time went on, all cultures developed the superstition of an afterlife. Many modern religions have sustained themselves largely by propagating the myth that there is continued life after death. It is a way to promulgate an organized religion and to promote various kinds of behavior—whether it's praying, engaging in various superstitious rituals, going to church, or killing others in the name of some god—all in the service of potential immortality. The fantasy of an afterlife is the pivot around which religion survives and does its "dirty work." As Harris puts it, "A single proposition—*you will not die* —once believed, determines a response to life that would be otherwise unthinkable." (206)

Heavens and Hells

If there were to be an afterlife, humans needed to ascribe some character to it, or at least a name that could take on some palpable quality. Long before the Christian conception of such things, other ancient mythologies had construed such death outcomes. A couple of examples: Without getting into details, both Greek and Roman mythology had conceptions of a positive or rewarding afterlife. A few select mortals who were favored by the gods could be sent to the Elysium (Roman) or Elysian Fields (Greek). If you weren't favored, or had committed some offense, you might be sent to a punishing afterlife, the Underworld, under the control of Hades (Greek) or Orcus (Roman). The various gods, of course, had control of who went where. And, sometimes, if you displeased the gods, you could be sent to the underworld by the gods directly, even before your otherwise natural death (We'll look at a classic example of this in Chapter 9.).

In modern times, the Christian mythical conception of these became known as Heaven and Hell. Without getting into debate about the origins, some conceptions of a positive and negative afterlife likely arose in ancient Jewish tradition, which also added the Devil to the mix. (147) Of course, the Biblical legend of Jesus gave the matter a push, and a formal Heaven and Hell developed into mythical places of eternal life. Just as the ancients had laid out before, where you went depended on how some supernatural being thought you lived your life on Earth. Indoctrination with the two possible afterlives has been the near universal contingency for the conformance to moral codes in the Christian tradition. The concepts of heaven and hell, while originally very evocative of reward and punishment, eventually came to signify anything "good" or "evil." And they became synonymous with eternal reward and the mortal fear of eternal pain.

History has given us a number of examples of this model. Of course, there is no validity or coherence to any of them. Consider an *equally valid* one, from a parody in a modern comedic film titled *The Invention of Lying* (Warner Bros. Pictures, 2009). In the plot, a world is depicted where everyone speaks only the truth. People have no concept of lying or deception. The protagonist's mother is about to die, and she is suffering and terrified about the nothingness of death. Her son is bereft at the situation and learns to lie for his mother's comfort and his own selfish purposes, reducing his own anxiety. He invents the idea of a joyous afterlife for her; she is reassured, and she dies happy. Word spreads, and he becomes famous, promulgating the afterlife. He boasts, "Today I stumbled upon something that no man has ever stumbled upon before. They'll write about me in history books for generations to come. And yet, moments ago, it was unfathomable not only to myself but to mankind as a whole. It's hard to describe but it was as easy as . . . how do I explain this, I said something that wasn't't!" He even begins spreading word that he talks to a "Man in The Sky." Of course, it is no more unlikely as a parody than the Christian myth of exactly the same plot.

What about Hell? Hell is the ultimate punishment. If you're not going to go to Heaven, there must be some other kind of afterlife for you. And that is Hell. The threat of such punishment is necessary to maintain the religiosity of some. Of course, some religions avoid the problem entirely by assuming some type of reincarnation. Threats of suffering an eternity of "fire and brimstone" have been a staple of Christianity for a couple of millennia, and children have been frightened with this from an early age. As Andrew Seidel observes, "Churches have since abandoned burning and stoning those accused of witchcraft. But Christian churches today still terrify parishioners and children with the prospect of being burned alive for eternity: hell." (457)

It's always been a place you don't want to go, and yes, it's always been a place. Nevertheless, it appears that there is still some controversy, for example in the Catholic Church, as to whether Hell actually exists. The last Pope (Francis) declared that there is no Hell, per se. That obviously lit up a controversy. In 2018 the Pope had apparently denied the existence of Hell in an interview with an Italian journalist. The *National Catholic Reporter* followed up by discounting the report, but did not actually deny his comments outright, only suggesting that we should refer to other, previous Papal comments. (424) Nevertheless, the Pope had also indicated in 2015 that, "If God forgives everyone, why does hell exist?" We can take a bit of comfort in the fact that when an "afterlife" is the subject of semantic and political confusion within a religion, the whole concept, of course, is nonsense.

Most Christians are unconcerned about the heat death of the universe or that our own planet will become uninhabitable in a relatively short time. Of course, the ultimate Armageddon for humans is predicted to occur well before that. (Many proposed apocalypse dates have come and gone.) Many believe they will be saved into some sort of nonmaterial afterlife. In the end, of course, if you believe in an afterlife of some sort, which is it? It depends on what god you believe in. Are there different afterlives for Christians, Muslims, Hindus, etc.? If you profess that there is but one God, then he must be worshiped in the way you've learned, e.g., Christianity. But that's a pretty bigoted deity. For example, some are "saved" and others not, based on their own "ignorance." (The theological escape hatches offered here are just embarrassing.) I'd suppose that missionaries and other evangelicals are racing to enlighten the rest of the population before the apocalypse leaves some behind.

Pascal's Wager

Of course, most people, even though they may espouse the notion of an eternal soul going "up" to Heaven, are really not entirely sure. They believe in the idea, but doubts do creep in. This leaves an interesting dilemma. Should I believe or not?

Of course, the endorsement of a heaven and fear of a hell are natural outgrowths of death anxiety. This anxiety can be lifelong; it can come about from some specific religious upbringing or conversion; or it may be acute in terms of any anticipation of death—from a sudden accident, or impending catastrophe, or a major or terminal illness. (75) The aphorism, "There are no atheists in foxholes," is often quite apt. "Now's the time to get onboard." Many start to believe when the time is near. As opposed to so-called knowledgeable philosophers and theologians, comedian George Carlin probably said it best, "I was thinking about how people seem to read the Bible a lot more as they get older, and then it dawned on me—they're cramming for their final exam." "Last minute" conversion is often termed Pascal's Wager, the argument posed by Pascal. If you "bet" that God exists and pray accordingly, there is nothing to lose; and there is a notable risk in assuming that He does not. Therefore, your path to salvation, if it exists, can be assured. The problems of logic aside, devout Christians might opine that a lifelong history of sin and failure to worship might not necessarily get you into Heaven anyway. After all, heavenly rewards for good behavior occurring only at the last minute wouldn't exactly counteract an "immorally" lived life.

At bottom, of course, it makes no difference. Afterlife fantasies, of course, are "just" another set of superstitiously conditioned behaviors. But they are the glue which has held theistic religion together for centuries. All things die, including human beings. While the fear is rational, its coping strategy of an afterlife is often not. And if it is, I refer you back to the above-mentioned film. On the other hand, some people, at some point, espouse that they do *not* fear death, or at least say that they don't. Mark Twain's reflection on this is worth noting, "I do not fear death. I had been dead for billions and billions of years before I was born, and had not suffered the slightest inconvenience from it." But we cannot discount the notion that the fear is still present in most of us.

Miracles and Superstitions

Afterlife superstitions have always been critical in maintaining religion. They are the hallmark of religion. However, they are merely superstitious memes passed down with cultural evolution. Besides the afterlife and gods themselves, there are many other superstitions which are notable memes of religion. We've seen how superstitions likely developed in early humans. Such fantasies included unseen forces, likely leading to the advent of animistic spirits. Although such entities are less frequently worshiped today (there are exceptions, of course), religious adherents still believe in a variety of superstitious memes—called miracles. Jesus walked on water and turned water into wine. Something created a snake out of Moses' rod, and he parted a huge body of water. There's a long list of Biblical miracles. Miracles are noted not only in the Judeo-Christian Bibles. They appear in Hinduism, Islam, and other religions, even in the ancient Chinese philosophy of Daoism. Though the Buddha eschewed the idea of miracles, legend has it that he also walked on water.

Superstitions

Superstitions were a part of early human cultural and religious development, as I described in Chapter 4. Humans continue to believe in all sorts of things—without any rationale, of course. As Carl Sagan observed about superstition, "It is merely belief without evidence." (440) Beliefs have merely been handed down over the generations, though there's no good reason for them. Early humans and Bronze Age people developed many superstitions which persist to this day. Don't walk under a ladder is one of the oldest superstitions. That would create the image of a triangle, a shape sacred to ancient Egyptians (similar to the pyramids). Doing so might bring misfortune. Don't spill salt. If you do, throw some of it over your left shoulder. Salt was once a valuable commodity, sometimes a form of currency. To spill it was considered wasteful, and spilling it could bring bad luck. But if you throw some salt over the left shoulder, that keeps the devil or bad luck off your back and behind you. Cover your mouth when you yawn. Otherwise, the Devil may enter, or perhaps your soul will escape. Carrying a rabbit's foot probably won't bring you good luck, but the ancient Celts thought that rabbits somehow could communicate with the gods. Knock on wood. In Pagan times, spirits were believed to live in trees. If you hope for good fortune, symbolically knocking on wood may expunge the evil spirits that lurk there and interfere with your hopes. Crossing your fingers for good luck is likely an emulation of the shape of the Christian cross. The origin of saying "God bless you" for sneezes is a bit unclear; but during the 17th century with the Black Plague raging, sneezing was an indication that you were near the end, and Pope Gregory I actually issued an edict that persons who showed such symptoms were to be blessed in hopes that the prayer would ward off death.

Superstitions arise merely from the accidental association of events (I'll discuss this more in Chapter 10). One of the oldest superstitions is praying for rain. If you pray for rain, you might just get it. And even though it doesn't work all the time, it "works" sometimes, and some humans continue to do it. However, most superstitions tend to fade away if not reinforced by some consequent event. Thus, they need to be maintained by various cultural memes. We do continue to knock on wood and avoid ladders. And we engage in various religious superstitions—prayer, worship, many religious rites—the list is endless, even down to acknowledging a sneeze with "bless you."

Miracles

The most fervently held superstitions of theistic religions are "miracles." What is a miracle? An act of God? Just a legend? Or something else? Miracle is likely from the Latin *miraculum,* or "an object of wonder." It is used when we wonder about something. We talk about the miracle of birth, or we're emotionally moved by a miraculous sunrise. We also admire and wonder about some things because we don't understand them. We call things miracles, almost euphemistically, when things happen that are beyond our awareness or knowledge. Sometimes we label things miracles just because we're in awe. "His recovery from cancer was a miracle." "It was a miracle that they won the game." Some will recognize the "Miracle on Ice" from the 1980 Winter Olympics.

The traditional use of the term, of course, comes from the idea that the causes of something are unknown—and cannot be known with ordinary knowledge. Needing an explanation, we fill in the gap with God. We can't go back in history to view someone walking on water or somebody rising from the dead. But there are some contemporary illustrations of historical events that are cited as miracles. Religious healing of the sick is a prime example: the "miracle" of "faith healing" and similar endeavors. Some people *have* recovered from illness "in association with" such rituals, though the performance of such is usually a hoax. The performance of exorcisms to cast out demons has largely died out, though it still occurs from time to time in some cultures, and some people have benefitted in the short term associated with such rituals. That does not mean that the possessing demons in exorcism or the presumed powers of faith healing actually existed and that miracles were performed. Such cases always derive from hucksterism, some quasi-hypnotic effects, misdiagnosis of an illness, powerful placebo effects, or hysterical illness. We no longer engage in the procedure of trepanation, the practice of cutting the skull to release evil demons or spirits. You might think that this procedure was a death sentence and was truly ancient, but the Incas were carrying out this "surgical procedure"—without anesthesia, of course—until at least the fifteenth century. (290) It sometimes worked; survival rates were as high as four out of five, actually higher than for similar procedures performed on American Civil War soldiers for bullet wounds to the head. Neither prayer nor any other "channeling" of supernatural power can be seen as operative—they all involve a number of variables, including predominantly superstition.

No actual miracle has ever been observed. It is as though things just "magically" happen, which is inconsistent with our human nature to imbue everything with a cause. Significant events without explanations make us uncomfortable. Evolution has dictated that. When we don't understand, we attribute the cause to a mysterious higher power. We often just don't have reasonable explanations for some things. Sometimes we do, and the "miracle" loses its luster. For example, take one of the most famous stories of folklore, the mythical Moses allegedly parting the Red Sea. Biblical authors could not have conceived how this could have happened, and the lore was that of a miracle. The Red Sea problem has been addressed by many, and reasonable meteorologic-hydrodynamic explanations have now been offered for the possible land bridge of exposed mud flats in the area, which would have been easy to traverse. (139) The ancients could not have conceived of this mundane explanation.

Let's look at the concept of *magic*. When you see a magic show, you may be impressed, but you know that there is a rational explanation. When a young child sees magic, however, he cannot explain it. It is magic! For any significant events that we cannot explain, magic comes into play. That "gap" can be uncomfortable, and magical explanations start to appear. These are superstitions—they provide nominal "gap" explanations, explanations which are shared within the culture. Supernatural forces and miracles develop in this way. If we look back in history, the phenomenon becomes clear. What we see as understandable today would have been a miracle to our ancestors.

The principle is well captured in science fiction writer Arthur C. Clarke's third law of technology. "Any sufficiently advanced technology is indistinguishable from magic." What would *Homo erectus* think about men who could fly in machines? What would Neanderthals think about curing an otherwise-fatal injury with surgery and anesthesia? What would Neolithic Cro-Magnons think about cell phone video communications? These are miracles by any definition to the observing hominins. And you can see why a supernatural power might be invoked at the time.

We can't go back and observe such things, of course. But let's take a modern example. During the Second World War a number of indigenous South Sea Island tribes gave rise to what became known as "cargo cults." The term has been used to describe how superstitions, including god-like ones, can arise out of cultural exposures, e.g., the Taro Cult of Papua New Guinea. In general, these cargo cults evolved because of exposure to the activities of modern societies conducting the war. Such cults made inferences, more accurately described as superstitions, to imbue deistic qualities to the events and people they saw. (163, 560) Here's how it worked: A number of South Pacific Island tribal people during the War would see airplanes (both American and Japanese) conduct airdrops or land with various materials that were strange—"otherworldly" items such as advanced weapons, canned food, radios and antennas, bottled drinks, clothing, cigarettes, and other items. These things had never been seen before. It was magical, and the primitives felt there must be some supernatural connection. Though such cults, in effect, had been around since the 1800s, the intensity of exposure to those flying foreigners led to religious awe and worship. This superstition became tribal religious practice in many parts of the South Pacific. It seemed as though the aircraft and visitors were from "another world." After the war and the disappearance of foreign visitors, islanders began imitating various rituals to "cause" planes to land or items to be dropped. They constructed things like wooden mockups of radios, headphones, and antennas. They made runways with fires along the sides to entice planes to land, and they made effigies of aircraft. As we know, effigies and hope for return of an overlord are common in religious practice today. The natives expected "otherworldly" supernatural gifts, merchandise from aliens. This is reminiscent of Michael Shermer's update of Arthur C. Clarke's law, which seems to hold in many cultures, "Any sufficiently advanced extraterrestrial intelligence is indistinguishable from God." (467) Most contemporary gods, in fact, are extraterrestrials.

Conclusion

As cultures developed, invoking supernatural explanations for unexplainable events (miracles) evolved as cultural memes, and that sort of behavior has persisted, even when empirical explanations became widely supported and available. We do understand significant weather changes—typhoons, hurricanes, etc., but some continue to invoke God as the responsible party. When "miracles," i.e., unexplained phenomena, happen, humans still invoke the supernatural. It's easier than trying to understand any complex phenomenon. Superstitious religious behavior can overcome critical thinking, especially if you "believe in God." We must be careful not to "think too hard" about things we don't understand or reason problems out. Demonology as explanation for disease persists, even though modern "germ theory" and rational medical thinking have been around for a while now. Those would interfere with God's miracles. As Martin Luther said, "Reason is the Devil's harlot."

The supposed miracles, like ones in the Bible, are often cited to support the existence of God. Unfortunately for religious apologists, no historical or contemporary "miracle" has ever been confirmed. To assume a divine intervention is merely irrational. People cling to such memes only to support their ideas about some god. It's a vacuous exercise, of course, but theologians can't seem to fully give up the notion. Religion professor Elaine Pagels concludes about miracles, "We can take them seriously without taking everything literally." (288) I never know what those kinds of comments actually mean. It's an oft-heard position about many religious fables among academics, clergy, and the flock, reflecting no more than their own ambivalence about and discomfort with the whole topic, especially the Jesus fables.

Today, even most Christians do not consider unexplained events as miracles, although the appearance of Christ's image in pieces of toast still captivates a number of suggestible individuals. Such things are today's superstitions. They extend from our informal language to elaborate religious rituals, and we invoke God's name wherever possible in such circumstances. We even use miraculous language to describe almost anything of wonder or things that are merely difficult to explain. It is part of our lexicon. "I hope to God he comes back from the war; it'd be a miracle." There's the "miraculous" recovery from terminal illness, such as cancer, because "God cured me." We cite it even in mundane language. It's probably no accident that we've referred to one of the most famous football plays of all time in 1972 as the "Immaculate Reception." We use the language of a God and miracles to describe very unusual things. But language is communication, not empirical truth.

Worship and Prayer

It's not possible to look at the nonsense of theistic religion without considering the idea of worship. Why would humans actually worship an idea? The answer, of course, lies in the evolution of superstitions. For primitive *Homo sapiens*, great powers that couldn't be seen came to be viewed as controlling all the important aspects of nature—the sun, moon, rain, prey animals, illness, death. Wishing and hoping for salutary events sometimes "worked." We hoped the sun would come back higher in the sky, and it gradually did, coming back into its zenith at the summer solstice. Not surprisingly, sun worship is likely the oldest of the major superstitions. We wished for rain to replenish the ponds where we drank. We desperately wished for the prey animals to come back to graze the plains, and they did. Hope and wishes enabled anticipation, and we looked to the sky where the powers resided to satisfy our eagerness. And as more sophisticated language began to develop, our anticipations took on verbal pleadings. There were undoubtedly gestural and other nonverbal behaviors associated with these as well. Many cultures still do use elaborate nonverbal activities in their acknowledgment of or appeals to the gods (some chimpanzees do this too, as above), not the least of which is assuming a submissive position in the "presence" of the more powerful authority.

Worship

As the gods granted humans' supplications, superstitious as they were, more and more reverence developed. And the gods took on particular personages, as I've discussed. It is perfectly natural to respect and revere anyone who has authority over you and gives you something. And a supreme authority would demand supreme reverence, what we today call worship.

Worship takes on many forms. One of the universal ones appears to be submission. Submission to the higher authority or his/her dicta is necessary for a religion to survive. We take efforts to submit to the will of a god—submitting your own desires to the dogma of the religion. Even bowing one's head, in prayer or otherwise, is a symbol of submission. For the Muslim, we note that the word Islam, in fact, actually means submission. Submission is necessary and effective for religious leadership to work its influence, and leadership by a god is the ultimate form of this. Where gods are not worshiped, religion dies.

Prayer

Prayer is a form of submission and pleading. However, it's also easy to envision how cultural development in primitive humans led to overt or private verbal requests for supernatural powers to intervene in human life. When the ordinary course of events doesn't result in good outcomes, you need to call in a "higher authority." One of the most ingrained ways of doing this is the religious superstition of prayer. Prayer is a behavioral meme of most religions. But it did not come into being as such. It likely evolved culturally over the centuries, developing postures, specific language, overt or silent talking, singing, etc.

And what is asked for can vary widely. People pray to a god for all sorts of things—for safety, strength, moral courage, and so forth. Praying for others, and particularly recovery from illness, is especially popular. (I'll describe some of the classic science related to people using intercessory prayer in Chapter 10.). They also characteristically pray for things they believe may not come about without some supernatural help. They often ask for miracles—recovery from terminal cancer, world peace, to be taken to an afterlife, etc. They would like a little violation of the laws of nature to render outcomes that might not otherwise happen. Ambrose Bierce, the nineteenth-century author, had the best definition of prayer I've seen: To pray is "To ask that the laws of the universe be annulled on behalf of a single petitioner confessedly unworthy."

Sacrifice

Of course, if desired events—even prayed-for events—don't come as expected or wished for, there must be some reason for this. Why would the gods not answer prayers? Remember, they had human personas, and when such beings don't respond the way you want them to, a logical inference is that they must be upset or angry with you. And so they must be mollified or appeased. There is a long history of sacrificing to appease gods. From the earliest religious sects, the idea of appeasing presumably angry gods by sacrifice has been a fundamental part of human culture, from animals to the barbaric, sacrificial murder of children. Such behavior may have stopped in the first century BCE in ancient Greece; but there is evidence that the Mayans were still religiously sacrificing children in the first century CE (29); children were sacrificed by the Incas as late as the 15th century CE (142); and the practice likely continued until the 16th century CE among the Aztecs. (181) As sacrifice continued and supernatural power became monotheistic, we should ask, "Why would God be angry?" "Why would a supernatural entity have human emotions?" "Why would It be appeased by the murder of children and other humans?" Other than the idea that we've created Him in our own image, this is the one of the better tests of omnipotence. I still marvel at the notion of an omnipotent being getting upset.

We could acknowledge that human or animal sacrifice is rarer now than it used to be. However, modern Abrahamic religions still do use some sort of ritual sacrifice as part of their culture. Some modern religions employ symbolic sacrifice, each year giving up food or valued things for a period of time. It is unclear if such appeases whatever god is prayed to. In the end, religion demands ritual propitiation of gods—by worship, prayer, and sacrifice. I'd submit that it is a very insecure deity who needs validation by the subjects He has created. These human activities are merely ritualistic behaviors. They do not in any way support the notion of a god. And when prayers are "answered," of course, that only reinforces the superstitions which led to the irrational belief in a supernatural being in the first place. Throughout history, humans have perpetuated the god memes with their own sacrificial rituals—often the most barbaric we can contemplate.

What is Religion?

We've seen that theistic religion evolved naturally and is based on some unwarranted assumptions, including various superstitions. But let's take a step back and look at what religion, particularly organized religion, really is. There is no coherent and consistently accepted definition of a religion. Is it some oral lore that is passed down? Is it a set of traditions and principles that clergy tell us it is? Is it a genetic trait passed down from your parents? We are left to conclude that there is no such thing as a coherent idea about religion—it is only a label inconsistently used by various sects to describe a seemingly infinite variety of social behaviors.

Do adherents follow what parts of a religion they "believe in" and eschew or disagree with other parts? Obviously. The source of any such tension or conflict often comes down to what is laid down in a religion's so-called sacred scriptures. Is it the scripture as originally written, a current version, or some historically intermediate text? There is considerable debate about this, particularly in the Abrahamic religions. You can find scriptural support for almost anything you do or want to do. Some followers of particular religions assert that certain activities are or are not consistent with their religion. Others profess, and offer quotation, to the "religious validity" of just the opposite. All religious sacred writings, most notably the Quran and the Old and New Testaments, have a wealth of self-contradictory provisions (I'll review some of these below). Are sacred writings merely allegorical guidance, or are they to be taken literally? If *all* specifications in religious sacred writings were followed today, nothing but chaos would be evident. For the faithful, there are three basic ways to avoid this. There is the "pick and choose" method of viewing the validity of any passages in scripture. You can view your scriptures as purely allegorical, which pretty much supports anything you believe and anything you do. And you can join particular sects that believe some things more than others—Protestant as opposed to Catholic, Reform vs. Orthodox Jew, etc. This relegates the notion of "extremism" to arbitrary interpretation; and that view becomes relevant only to the culture in which it occurs, not a coherent interpretation of the scripture. Religious behavior can, and is, maintained by whatever passages of text one chooses to follow, or presumably follows.

It would be impossible to define religion in a way that captures all theistic faiths and sects. Perhaps religion is actually what people do in the name of some superstition. In fact, religions are made-up ways of controlling people by organizing around a set of myths. Do you believe in God? Which one? If we look at modern religions, one thing is obvious. Religion is merely, and cruelly, a method of control—built on fear and punishment. It requires rules, rules which are promulgated and enforced by authorities. *At bottom, religion is merely ritualistic group behavior based on mindless obeisance to supernatural superstitions.* But there are a few dynamics which seem universal.

Escape from Suffering

The path to an afterlife is only one hallmark of religion, albeit an extremely potent one. As we noted, religion's use of this meme is something that maintains its existence. To suffer assignment to the wrong afterlife is the ultimate punishment. But mundane human suffering is universal. The need to avoid or reduce this suffering is also a potent force in maintaining religion.

In *The Future of an Illusion,* Freud argued that religion evolved, in part, based on humans' fervent desire to avoid death, but also mortal suffering.[12] He observed that the initial development of gods was associated with our desire to explain the unexplainable and that later civilization was built up on the back of religious behavior; but as science came to explicate more and more of the world, the value of religion became merely superstitious and a way to relief from mortal suffering. It's worth quoting here:

> Another procedure goes forward more energetically and more thoroughly, seeing reality as the sole enemy, as the source of all suffering, with which it is impossible to live, and with which one must accordingly break off all relations if one wants to be happy in some sense. The hermit turns his back on the world and wants to have nothing to do with it. But one can do more; one can desire to recreate it, to build up another world in its place, in which the most unbearable features are eliminated and replaced by others conforming with one's own wishes. Whoever, in desperate outrage, takes this path to happiness will as a rule accomplish nothing; reality is too strong for him. He becomes a madman who, in carrying out his delusion, usually finds no helpers. Yet it is claimed that each of us, in some respect, behaves like a paranoiac, corrects some intolerable aspect of the world by means of a wish-construction, and carries this delusion into reality. Of special significance is the case in which a considerable number of

> people, through a delusional reorganization of reality, collectively try to gain for themselves an assurance of happiness and a protection against suffering. We must also characterize the religions of humankind as *mass delusions* of this kind. The delusion, of course, is never recognized as such by anyone who shares it. (174)[13]

Many would argue that this is exactly what religion does. It offers comfort and aid in coping with life. The hopeful escape from suffering is the basis for a great deal of religious faith, and such is often maintained by various religious traditions and rituals (The psychological reader will recognize this as no more than negative reinforcement, as I'll discuss in Chapter 10). Of course, there are other, secular ways of reducing suffering, and humans have been engaged in those for as long as we've been around. But the failure to reduce suffering adequately (and this is always a unique, individual determination) makes one vulnerable to "belief" and the various social influences on religious behavior. It is the reason that people become *dependent* on religion. It is the way out of suffering, or at least a way to reduce it. Religion becomes the cultural organization of this dependency. I'm reluctant to quote the infamous Karl Marx's notorious aphorism that religion is the opiate of the masses, but it's almost unavoidable. The actual verbiage is, "Religion is the sigh of the oppressed creature, the heart of a heartless world, and the soul of soulless conditions. It is the opium of the people." (328) The German scholar of religious history Walter Burkert further defined this "Opium thesis. Religious ideas and practices are accused of fulfilling human wishes in a fantastic, unrealistic, and possibly detrimental way, just as drugs do when they provide the illusion of happiness while misusing and overriding the normal cerebral functions." (67) Religion is clearly a set of institutions organized around the idea of mitigating human suffering.

Authority

Theistic religions are also based and maintained on the notion of authority. Of course, the ultimate authority is considered by many to be God. In the Abrahamic tradition, He is considered the "father," the highest authority. As Sigmund Freud astutely observed, "At bottom God is nothing more than an exalted father." (173) There are many varieties of this, but they all share that same impulse and effect—humans look to "father figures" for guidance and control (whether this is always wise is a whole other matter).

All organized religion is based on authority, via both scripture and clergy. If a God is not telling you what to do, a designated authority will suffice. The emergence of leaders in any community is necessary for its survival as such, and religious leaders have fulfilled that role throughout history.

Authorities are group members with exceptional knowledge and power. Early in the evolution of religion, authorities were shamans and other tribal leaders. In more modern times, God proxies are clergy, elders, and certain others in whom we vest the authority of telling us what we "ought" to believe and do.

We respect members of the clergy as leaders. Even though much of the information they promulgate is pure superstition, we see them as devout leaders. Adherents to theistic religion do look to more knowledgeable cohorts for guidance. Such clergy are in an enhanced position to exert authority based on religious dogma. They may, and often do, promulgate advice which is consistent with cultural norms and expectations, but this is characteristically done by citing some Biblical prescription or in the name of Jesus or other religious figure. Thus, the superstitions become incredibly reinforced.

Authoritarian control continues to be a driving force in religious belief, even though the basis for such authority is fantastical. But we should realize that it doesn't take much to *act* like an authority on almost any subject. Many people will flock to hear motivational speakers (both good and not so good), peddlers of get-rich-quick schemes, and various other snake oil salesmen. All you have to do is be personally charismatic, promise fulfillment of a desperate need, and offer a convincing presentation. The billions of dollars spent on advertising, and the "belief" and subsequent material consumption of millions based on such communications, is more than enough evidence. We certainly see this in charismatic clergy, and it's raised to impressive levels in televangelists and other hucksters. Without compelling authorities, in your neighborhood or leading your faith from afar, organized religion would die.

Cultural Survival

A theistic religion cannot survive on its own. It requires a degree of cultural coherence to be maintained. If religion is purely individual and not openly shared by others, it will disappear. One can argue that, even under the hypothetical circumstance where every individual has his/her own view of religion, what some call "spirituality," organized religion would be in serious trouble. Belief isn't often enough to keep a religion alive. Over the long term, other methods—primarily social—are required to ensure its survival.

First of all, religious leaders have to teach and coach others. It reduces their own insecurity about beliefs. We all know that teaching is a great way to reinforce your own learning and behavior. Proselytizing a religion is self-reinforcing. Religious fanatics also regularly attempt to impose their beliefs on others. In Christian culture, evangelism is often seen as some sort of selfless desire to spread the word of a Christian God. However, selling their God is purely a way to control others, and evangelists clearly have a need to do this. Again, it is just as likely that the process of doing so is a way for the faithful to reduce their own doubts or insecurities and reinforce their own beliefs. As faith declines in the developed, industrialized world, evangelicals continue their missions to Africa, for example, to convert Muslims and others to Christianity. Then there's Islam (which I'll talk about more below). Social maintenance of the Islamic faith is a prime method of sustaining this religion as well, and an elaborate system of rules is promulgated. Of course, the penalty for nonbelievers and for apostates is death, and the Islamic scriptures actually do prescribe that such people should be killed. The Catholic Church has gotten over this method of social control, but it's crystal clear in its history as well (more on both of these below).

Natural-law Religions

There are a few cultural trends which you may or may not consider religions. These are what Harari describes as "natural-law religions." (203) In the first few hundred years CE, there appeared Jainism and Buddhism in India, Daoism and Confucianism in China, and Stoicism, Cynicism and Epicureanism in the Mediterranean basin. These religions tend to eschew the idea of gods, though supernatural aspects do appear here and there in these traditions.

Let's take Buddhism as a putatively non-deistic movement. The central personage in Buddhism is the Buddha or Siddhartha Gautama, and he doesn't have supernatural powers. The central idea that he promulgated is that humans have "suffering secondary to craving"; and the answer to this is to "accept things as they are." The strategy of Buddhism, for example, is then largely to eschew the striving which is putatively associated with suffering. To the degree this is successful—via meditative activity and/or change in overt behavior—one sees less reliance on, or complete avoidance of, any god idea. But it's not that simple. Gods do come into play. "Buddhism does not deny the existence of gods—they are described as powerful beings who can bring rains and victories—but they have no influence on the law that suffering arises from craving." Most Buddhists don't attain nirvana. "So they continued to worship various gods, such as the Hindu gods in India, the Bon gods in Tibet, and the Shinto gods in Japan." (203)

Some argue that Buddhism is actually not a religion. However, it does profess a number of ideas that require some sort of supernatural force, e.g., reincarnation. There are no known scientific principles—or common sense—that would allow some part or character you possess to "come back to life," even as some sort of other organism. But Buddhists do believe in reincarnation. It's like endorsing a supernatural force without overtly ascribing it to a personal god. Buddhism is a cultural phenomenon, but it does border on and certainly emulates the qualities of a religion.

Modern Religions

We've seen some of the history and superstitious aspects of religion.

One can appreciate the natural evolution of historical religions—from animism to primitive superstitions to "divine" kings to organized religious groups. But let's examine a few which have sprung into existence in recent times. I use the term "modern" rather advisably—only to suggest something of relatively recent vintage. The purpose is merely to illustrate the ludicrousness of organized religion. The absurdities should suggest that the premises of older theistic religions aren't any more or less coherent or believable. All religions are invented. Some are just more obviously or recently invented. Judaism, Christianity, and Islam are a bit older, but let's take just a couple of well-known, more "modern" examples.

Mormonism

Mormonism was essentially founded by an early 19th-century charlatan named Joseph Smith. Smith had actually been convicted of hucksterism. He claimed to have "found" a couple of golden tablets, which he orally "translated" to others recording his words, which became the *Book of Mormon*. The tablets, of course, never existed, and he promulgated the notion that for anyone else to see them would mean their death. He is the prophet of Mormonism or the Church of Latter-Day Saints.

If that wasn't enough, some incredibly fantastical premises were outlined. For example, the Mormon God was actually born—of human parents—on some planet similar to Earth. He was a mortal, but for some reason he eventually ascended to also become a god and somehow ended up living on a star or planet named Kolob (a fictitious body noted in the Book of Abraham, though the Bible, of course, is unclear just what that body is). Mormons join their god on Kolob when they die.

Although any "official" statements from the Church of LDS may offer softer interpretations, this is what many Mormons do accept as true. It's truly remarkable that anyone believes these things. However, such *is* the origin story of this religion. Mark Twain once lampooned the Book of Mormon as "chloroform in print." Too easy a pun, since it actually contains the "Book of Ether" (once used as an anesthetic). In any case, the origin story of this religion is impossible for any rational human to believe, including by most who identify as Mormons. Some don't believe it, but that would be merely hypocritical. It's just another example of a religion not endorsing its own literal heritage. Nevertheless, the religion continues to flourish.

Scientology

Scientology was founded in the early 1950s (when some of the readers of this book were alive) by a fellow named L. Ron Hubbard, who became essentially its prophet. He was a prolific novelist, writing over 200 books of fiction (mostly fantasy and science fiction), but for some reason he decided to start a religion.

Scientology is based on admittedly created myth, and Hubbard likely used various ideas from his science fiction novels in creating the religion. Here are some of its bizarre notions: Millions of years ago, there existed a galactic confederation of planets, and their leader, Xenu, transported billions of his followers (using spaceships) to a planet called Teegeeack, now known as Earth. He placed them around volcanoes and killed them with hydrogen bombs. Then their souls hung around there to inhabit the bodies of humans of the future. Those humans were considered spiritual beings, called thetans, essentially souls merely residing in physical bodies. They were immortal and had countless past lives, where they had lived in various extraterrestrial civilizations. Humans were supposed to have infinite capabilities, including the possibility that we could learn to control matter/energy and space and time. (That physics potentially allows for such is not what was generally meant; we certainly can't do this with our minds.)

The "souls" are also the source of human great spiritual angst and unhappiness, conditions which can only be relieved with special Scientology techniques. One example—there is an "auditing" procedure for spiritual confusion or anxiety using a machine called an "electropsychometer." It is a device which supposedly reveals a person's spiritual anxiety.[14] Then there is essentially brainwashing and often restraining a person from leaving a physical facility where the indoctrination is occurring, a clear reflection of how bizarre—and controlling—Scientology ideas are.

If this sounds truly mind-boggling to you, you have a grasp on reality. The science fictional and delusional quality of these beliefs is clear. There is, in fact, evidence that Hubbard had bouts of hallucinations and delusions, and suffered from paranoid schizophrenia, which certainly is consistent with the above components and tenets of Scientology; and it may have contributed to his founding of the religion. (396) This movement seems like a hoax, and it is. But it has tax-exempt status as a religion in the United States and a few other countries. The bottom line is that it is a deliberately invented religion—*as all are*. He may have had other motivations as well. As Hubbard once said, "The only way to make real money is to start a religion." He died a very rich man in 1986.

Conclusion

On the face, these "faiths" seem patently absurd. Some consider them cults that follow truly bizarre dogmas, as opposed to "true" religions, though any distinctions really don't seem evident. But any skepticism—or the outright absurdity—of these faiths arises because of their youth, only because they are of recent vintage. The only reason we know about the putative prophets Joseph Smith and L. Ron Hubbard is because they're reasonably contemporary figures and their histories can be more reasonably exposed. People tend to more quickly discount their accounts than more ancient philosophies. More ancient figures are not as easily debunked, though there's no evidence for their existence as well. We don't seem to apply the same skepticism to older religions. Who could imagine with a straight face that someone could walk on water or turn it into wine or raise someone from the dead or that a Moses character came down a mountain with some stone tablets provided by a deity? If such stories appeared in today's news, would you believe them? Religion has always required ignorance, and today, at least, the willful variety.

Scriptures

We entertained the question of "What is a religion?" So how do we know the real dictates of a religion? How do followers know how to behave and what to worship? The tenets of religion, of course, are just memes passed down through the centuries. The only consensus answer often comes down to the notion that it is based on some historical documents, also known as "holy scriptures." These are the actual bases of the various religions, which are supposedly written—or totally inspired by—some deity. However, there are serious problems with this, as we'll see. I'm not launching a particularly new critique by any means, of course, but it's important to highlight here.

All the well-known sacred writings—Old and New Testaments, the Quran, the Buddhist Tipitaka, the Mahabharata and other sacred Hindu writings—were orally handed down by humans for hundreds, if not thousands, of years. The problems of Biblical texts, in particular, have been extensively written about, and I have nothing special to add here. Nevertheless, I would be remiss not to point out some of the glaring nonsense, since religious adherents do often actually cite Bible passages as justifications for their beliefs.

The folly of theistic religions is easily displayed by illustrating the absurdity and self-contradictory language of nearly all scriptures, most notably those of the Abrahamic religions—Judaism, Christianity, and Islam. (It is unclear who wrote the text of the Quran. Different sects of Islam have different views on this, and I'll not devote any further comment to this here. I'll talk about what the Quran mandates of Muslims later.) But this is not enough to dissuade followers, who are often comfortable with the texts. Adherents are content to suggest that the word of God must be "religiously" followed—which is literally impossible, of course. Quoting Biblical phrases is almost an obsession with many.

Whose Words?

Of course, it is a bizarre exercise to consider what scripture may be the One True Word of God or gods. Debates on this obviously go nowhere since there is no evidence to support any one fantasy over any other. Judeo-Christian religions profess that they are based on scripture from divine inspiration. They are works which would not be taken seriously if not believed to have been inspired by a deity or actually "written" by one. Bibles are the disjointed works of mortal men. (There's no evidence that women made any significant contribution; that the Book of Hebrews was written by a woman is still in some dispute.) Psychologist Steven Pinker summarizes it well: "Historical scholarship has amply demonstrated that holy scriptures are all-too-human products of their historical eras, including internal contradictions, factual errors, plagiarism from neighboring civilizations, and scientific absurdities." (405)

God's "words" have always been passed down orally from the personal perspective of various, less educated folk, and they later—much later—began to be reduced to writings. And the editing, of course, has a very long history. They underwent considerable revisions, even after they were written down and even after mass printings. Even today, translations and meanings of words and phrases are open to interpretation—sometimes violently so. One can only imagine how these legends and moral prescriptions changed with oral and edited transmissions over the generations. That they would have been unrecognizable to an original fabulist is not out of the range of possibility.

Is the Bible the inspired word of God or something else? Those who believe the former, of course, have great difficulty—essentially an impossibility—in explaining the contradictions and inconsistencies (see below). There are so many that we must conclude that Bible verses were written by many different men, or by a few men who kept changing their minds, or—if you believe in the supposed "word of God"—by some supernatural being who couldn't get His mind right (a pretty sound argument against omnipotence). The issue was probably best summarized by Thomas Jefferson. In a letter to ambassador William Short in 1820, Jefferson wrote, "I find many passages [in the Bible] . . . of so much ignorance, so much absurdity, so much untruth, charlatanism, and imposture, as to pronounce it impossible that such contradictions should have proceeded from the same being." (247)

The general consensus, from various methods, is that the Bible is in no way a reliable historical document. (469) There is no evidence that the Old or New Testaments were written contemporaneously to events they describe or that any events alluded to actually happened. And there have been many versions and edits throughout history. And, of course, there are many different "versions" of the "Bible" in modern times. I leave this to the historians, but as an example, one of the more blatant pieces of editing is worth noting. It occurred in the early 19th century in something called a "Slave Bible," used to introduce Negro slaves in the British West Indies to Christianity. (264) To prevent any possible slave "uprising," it omitted any reference to the supposed exodus of Israelites from slavery in Egypt. How obvious is that?

It is clear that, whatever literary and allegorical value it might have, the Bible is not the work of a deity, but something developed and revised over the centuries for whatever purposes the authors had in mind. The

Bible is merely a poorly written authoritarian document, with one imaginary creature as its author.

Old Testament

The Old Testament is a set of various unlikely and nonsensical writings, which were constructed by various men over many centuries. Many classical writings from various civilizations were passed down in most cases almost *in toto*. But clearly the Biblical texts were passed down orally and written down long after the named prophets would have existed. The Pentateuch of the Old Testament (the first five books—Genesis, Exodus, Leviticus, Numbers, and Deuteronomy) were passed down as lore over at least a millennium, perhaps up to the second century BCE. There are no clear dates as to exactly when some of these folk tales were first actually written down, with the most likely range of most of the texts being approximately 500-600 BCE. (116)

We know, however, that much of the texts of the fictional tales undoubtedly were either wholly fabricated at the time of writing and/or went through enough revisions such that it would be impossible to recognize where they'd come from. The Bible describes many events which could not have occurred at the time period alluded to. Let's take one of the more well-known fabulist errors. The Old Testament describes the use of camels multiple times, including stories of the mythical characters Abraham, Isaac, and Jacob using camels as pack animals. (Genesis 24:11) (The theological assertions that these were actual humans, of course, is not supported by any actual evidence.) However, it has come to light that this is most likely untrue. Israeli archaeologists have discovered that camels were not domesticated in that part of the world until many centuries after the events described in Genesis and "not earlier than the last third of the 10th century BCE." (443), generally agreeing with other studies of the same problem. The Bible texts describe domesticated camels supposedly hundreds of years before that. Obviously, this was just a nice story but had no factual basis.

The Old Testament is filled with other truly bizarre stories. There's the preposterous story of Adam and Eve. Are humans really the descendants of Adam and Eve, as the Bible asserts? Remarkably, an incredible number of clergy still believe and preach this nonsense. However, some have come around to reality. For instance, now even the well-known Christian apologist and philosopher William Craig admits that humans evolved from older species and that there could not have been "a pair of sole genetic progenitors for humans within roughly the past 500,000 to 700,000 years"; and he now asserts that the relevant Genesis verse should not be taken literally (448), which is probably true for everything else in the Old Testament. (Of course, he also doesn't seem to know that humans in the form of *Homo sapiens* actually arose likely a few hundred thousand years after the interval he quoted. If he meant our hominin ancestors, he's off by several million.)

Then there's the fable of the Jews and their exodus from slavery in Egypt. "I am the Lord your God, who brought you out of the land of Egypt, out of the house of slavery." (Exodus 20:1-3) There is no archeological evidence for the Jewish enslavement in Egypt or their migration across the desert. (116) There is absolutely no evidence that a Jewish exodus out of Egypt ever occurred. There is even no reasonable historical evidence that the Israelites were ever *in* Egypt or that they wandered the Sinai. The American Rabbi Sherman T. Wine summarized the matter bluntly:

> The Jews did not emerge as a nation under the leadership of Moses. They were never rescued from slavery in Egypt. They never stopped at Sinai. . . . It is highly unlikely that a large number of foreign slaves would have led a successful public rebellion against an Egyptian monarchy at the height of Egyptian power and then would have proceeded to invade a territory under Egyptian control. It appears that the Exodus story was fashioned by the devotees of a religious cult that did not exist at the time of the emergence of Israel. The Exodus story was the creation of later prophets and priests. The general consensus of modern scientific biblical scholars is that the biblical Exodus never took place. Even a prominent Conservative rabbi in Los Angeles, David Wolpe, was willing to confront his congregation with that possibility! An exodus of two million former slaves-600,000 adult males (Numbers 2:2.4)—from the most powerful nation in the Near East—accompanied by colossal catastrophes—with no reference to it in Egyptian texts or in the propaganda of Egypt's powerful enemies—is the stuff of nationalist propaganda, not truth. (552)[15]

A Hateful and Loathsome God

It's unclear why the Old Testament was written down anyway. But we can reasonably assume that it was a codification of lore intended to serve as some sort of moral guidance for a culture, as had numerous "holy" scriptures that preceded it. Of course, what we might currently consider "moral" is at odds with Biblical guidance—an understatement if there ever was one (We'll look at the concept of morality, per se, in Chapter 9.). The Bible is filled with endorsements of and prescriptions for slavery, racism, bigotry, massacres, incest, child sacrifice, and other abhorrent practices.

The fairy tales in the Pentateuch are replete with terror from the cruel and vindictive God Yahweh figure. Just a sampling. As the fable goes, He was pretty upset with the Egyptians for not freeing the mythical slaves. In addition to all sorts of plagues and turning water into blood, He mandated that, "Every firstborn son in the land of Egypt shall die, from the firstborn of Pharaoh who sits on the throne to the firstborn of the servant girl," (Exodus 11:5) a particularly poignant cruelty. Commanding the Israelites to commit genocide (of the Canaanites and others) is a pretty malicious order. I guess He couldn't do it himself, though these kinds of things are part of His history.

Of course, there also needed to be some method of enforcement for human noncompliance with prescribed behavior. It relied heavily on the threat of punishment from a divine being. And He prescribed the death penalty for all sorts of mundane human behavior. A partial list—male homosexuality, working on the Sabbath, cursing your parents, adultery, a son disobeying his father, blasphemy, bearing false witness, premarital sex, worshiping other gods, and many others. Even engaging in prohibited thinking can be a "sin," again what more modern literature has called a thoughtcrime, e.g., the sins of coveting all sorts of things of your neighbor. (Exodus 20:17) He doesn't prescribe death outright for such, but we do find it elsewhere, "For the wages of sin is death." (Romans 6:23)

The worship problem is especially troubling—"Thou shalt have no other gods before me." (Exodus 20:3) Of course, what is not often recognized is that this God is essentially admitting that other gods exist, something anathema to Judeo-Christian theology. And if other objects or idols (e.g., a golden calf) were ever worshipped, such mundane objects would not be a threat to an omniscient, omnipotent being. He rather pathologically tested humans' faith. God ordered Abraham (another mythical figure) to incinerate his son Isaac and then excused him from the task at the last minute. Then there's God's test for Job in which He allowed Satan to kill all his sheep and cattle and children and servants. Job remained faithful to God, who subsequently rewarded him with more children and a lot of money. A God who needs to resort to such cruelty to assess "faith" is a good deal less than omnipotent, wouldn't you say?

And we might ask why an omniscient god requires worship in the first place. There must be a level of insecurity, narcissism, or pathological need for control in such a requirement. He comes off as a bully, a jealous, insecure figure who has to intimidate others for His own needs. The bizarre cruelty of Yahweh was best summarized by Richard Dawkins, "The God of the Old Testament is arguably the most unpleasant character in all of fiction; jealous and proud of it; a petty, unjust, unforgiving control freak; a vindictive, bloodthirsty ethnic cleanser; a misogynistic, homophobic, racist, infanticidal, genocidal, filicidal, pestilential, megalomaniacal, sadomasochistic, capriciously malevolent bully." (113) Just to be clear, despite such, countless millions of people around the world still believe in Him!

New Testament

The New Testament was an obvious attempt to supplement or replace the punishing code of a cruel and vindictive God with a more compassionate and hopeful one. That certainly may have been on the minds of the men who wrote it. Two problems. First, it's clear that just because something is written down doesn't mean that it's real. Second, of course, is that we have no idea who wrote these things.

I'll not get into too much detail on this subject, but the origin of the Gospels (Matthew, Mark, Luke, and John) remains obscure. Let's start with this. It is generally acknowledged that the writings were not done by any eyewitnesses to the events described, and no contemporary evidence exists for the events described in the Gospels. Only lore may have been passed down by generations, and, of course, lore is lore, something that undergoes continuous revisions. And it's equally likely that they were all simply fabricated from scratch. Nobody knows. Dawkins notes that, "No serious scholar today thinks the gospels were written by eyewitnesses, and all agree that even Mark, the oldest of the four gospels, was written about 35 or 40 years after the death of Jesus." (116) This assumes, of course, that such an individual actually existed, as I've discussed. Mainstream scholarship puts the dates even later. The renown Biblical scholar Bart Ehrman puts the writing of the four Gospels no earlier than approximately 70-100 CE. (148) Others put it at a later and longer interval, e.g., over the first couple of centuries CE. (167)

And things continue to change. There were numerous "inventions" that followed. For example, the idea of a Trinity didn't appear in any bibles until the 11th century. In addition, no matter what had been laid down before, it was the humans who populated the Roman Catholicism's Ecumenical Councils of the Middle Ages, which decided what should actually be considered as dogma of the Church. Of course, the Reformation led to even more significant changes. The only conclusion from all this is that there is not—and never will be—any consensus on the supposed word of God.

God's Mistakes and Confusion

To wit, the Bible contains some of the world's most truly bizarre literature (actually, if some or all the words were literally handed down, any god so responsible would need some lessons in grammar, consistency, and logic). There are literally thousands of fabricated, bizarre, and impossible dicta and legends in the Bible. I'll note here just a few of the more glaring examples to illustrate the absurdity of these works. By the time the New or Old Testament Bibles were written, they had passed as hearsay from generation to generation many times, undergoing whatever changes authors seemed to like at the time. It is no wonder that there are more contradictions than consistencies in both the Old and New Testaments.

As most scholars know, the huge array of blunders, inconsistencies, and self-contradictions in the Bible makes it an incredibly flawed, hypocritical, and all too human-authored document. I'll not detail them all, but one cannot be impressed by the following lowlights. The inconsistencies and contradictions in the Bible have been well documented, but it *is* instructive to take a look at some of the more blatant ones.

An Ambivalent and Confused God

There are just too many to detail here. God keeps changing His mind, a rather bizarre behavior for an omnipotent being, or He is just confused about what *He* actually believes.

> He was confused about the fate of the Earth. " . . . the earth abideth for ever." (Ecclesiastes 1:4) " . . . the elements shall melt with fervent heat, the earth also and the works that are therein shall be burned up." (2 Peter 3:10)
>
> It is unclear whether He wants to remain anonymous. "No man hath seen God at any time. . . " (John 1:18) " . . . I have seen God face to face, and my life is preserved." (Genesis 32:30)
>
> One of the more famous ambivalences. " . . . thou shalt give life for life, Eye for eye, tooth for tooth, hand for hand, foot for foot, burning for burning, wound for wound, stripe for stripe." (Exodus 21:23-25) " . . . ye resist not evil: but whosoever shall smite thee on the right cheek, turn to him the other also." (Matthew 5:39)
>
> Of particular and confusing cruelty. "The son shall not bear the iniquity of the father . . . " (Ezekiel 18:20) "I the Lord thy God am a jealous God, visiting the iniquity of the fathers upon the children unto the third and fourth generation . . . " (Exodus 20:5)
>
> How are we to be saved? "For by grace are ye saved through faith . . . not of works." (Ephesians 2:8, 9) "Ye see then how that by works a man is justified, and not by faith only." (James 2:24) And also. " . . . he that goeth down to the grave shall come up no more." (Job 7:9) " . . . the hour is coming, in which all that are in the graves shall hear his voice, and shall come forth . . . " (John 5:28-29)
>
> After separating day and night, He went on to create the sun, a truly bizarre sequence. He also "saw that the light was good." (Genesis 1:3-5) I find it curious that He would have to observe his creation of light to see that it was "good." Didn't He know?

The list is endless, but you get the idea. Either mortals disagreed with each other over time, or an omniscient being could never make up His mind. The Bible documents how God also made an awful lot of other mistakes, which is somewhat curious, to say the least, for an omnipotent being. To highlight just a couple. He messed up so badly that he had to wipe things out and start over. Jesus worked on the Sabbath but somehow escaped His punishment. He forgot to use "intelligent design" for millions of creatures which appeared over the countless millennia; and why did He have to extinguish 99.9% of Earth's species (so far)? I guess they were mistakes. It does seem indicated to question *His* intelligence. The Bible is filled with such nonsense.

What to Believe?

It is difficult to understand how people still continue to "believe" in Biblical texts; and, if so, which to believe and how. All sacred writings in the Abrahamic religions are no more than fables, of course, lore amplified by hearsay upon hearsay over the centuries. Yet religious apologists continue to believe that that the Bible is the word of God. The bizarre and self-contradictory specifications in the Bible make this a ludicrous document and an impossible set of choices. The notion of using the Bible as if it were the word of God is just to *delude* oneself, as Freud observed. To believe these things, in addition to the power of superstitions and doing what we're taught, an incredible set of rationalizations must appear.

Religious apologists are fond of quoting Bible passages. In fact, they're often preoccupied with such activity. Leave aside for the moment that religious followers *never* follow scripture reliably in their behavior. Given the Biblical hypocrisies, that would be impossible. Worshiping a book is less a method of guiding their own behavior than engaging in self-satisfaction and criticizing the lifestyles of others.

Should Bible verses be taken literally or only as metaphor? One merely chooses what one likes to rationalize behavior. No one follows the letter of scripture—which would be empirically impossible. Religion then becomes merely a type of self-identification in which one professes which words of a deity one favors. One refers to only what one wants at the time, ignoring the fact that the Bible is replete with inconsistencies and hypocrisies. Despite this "cherry-picking," Christian apologists continue to assert that the Bible *is* the word of God.

Further, there's the oft-quoted notion that we ordinary humans may not be capable of really understanding God's words. Which leaves two options: For some texts, we have clergy to interpret them for us. For others, we may be comfortable with the bizarre "God works in mysterious ways" argument. Many theologians deal with the problems of inconsistency, hypocrisy, cruelty, etc., in the Bible by considering scriptures to be inspirational and only allegorical. Nevertheless, without the threat of eternal punishment or reward/salvation, it is no more than a collection of fables. But the bottom line is that guidance by the Bible (or any religious scripture for that matter) has no force other than the perceived attitude of the people who worship a god. Many Christians do consider the Bible as allegory, but still do not give up the faith presumably underlying it. Literal interpretation has been surrendered, but the author God has not.

Theists reliably discount the contradictions in the Bible as unimportant. The worth of the Bible is based on some other perspective—one that is actually impossible to comprehend. Whether it's allegory, or inspirational, or just enables people to feel better when reading something—it's just dishonest. The Bible is fictional literature, period. If you want great fiction, you might as well read Shakespeare or Ernest Hemingway or A. A. Milne. And we know works by these authors *are* fiction. To imply anything else about the Bible is just deceitful. The only reason that clergy and parishioners continue to read and cite the Bible is because they believe it is "nonfiction." If it were ever discovered who actually wrote this stuff, the entire edifice of Christianity would crumble—though many desperate people would still "believe" in it.

Bottom line. It is clear that the Bible (both Old and New Testaments) was written by humans—for their own purposes. In the end, the Judeo-Christian Bibles are religious texts formulated over the centuries—by poor recollections of lore, by many, many authors, by random editing, by essentially plagiarizing other religions and cultures, and by forgeries. They represent conspiracy theories about history—not history itself—and do nothing to support the idea of God. As Albert Einstein once suggested, "The word God is for me nothing more than the expression and product of human weakness, the Bible a collection of honorable, but still purely primitive, legends which are nevertheless pretty childish." (151) As Harris puts it, "It is time we admitted, from kings and presidents on down, that there is no evidence that any of our books was authored by the Creator of the universe." (206) Enough said on that.

A last, interesting word on this. If you still don't know what to believe, now we have artificial intelligence to the rescue. Various chatbots, based on large language models, have been launched for various faiths, including not just the Judeo-Christian Bibles, but the Quran, the works of Confucius, and others. (561) They answer questions and provide advice on any personal or spiritual matter, based on interpretations of huge amounts of historical, current, and various translated materials and scripture. You can find out what God really thinks and what to do about almost anything in your life. As an extreme example, one AI system based on the Hindu Bhagavad Gita actually condones murder if it's done for the "right" personal reasons. Nevertheless, there is absolutely no reason to doubt the advice of these systems any more than one's own interpretation or that of some priest or imam. Yes, advanced AI systems do hallucinate—they make things up, but so do so-called "learned" clergy, who interpret for their flocks what God really means.

Business of Religion

Why does organized religion endure? We'll talk more about religious behaviors, per se, in Chapter 10. But one thing is clear. Organized religion provides social and other functions, which may account for some of their persistence, though that doesn't make them unique as cultural institutions. Whatever their origins, whatever their stated goals, their principal mission is to maintain themselves. As organizations, they are self-serving. In terms of organizational behavior, as Skinner puts it, "The laws of . . . religions are maintained primarily for the sake of the institutions." (486) That is for a specific purpose—to survive. A religion is a social organism, and like all organisms, its activities are directed toward its own survival. And, in most Western cultures, the principal mechanism for this is financial.

To endure, all organized religions have become, above all, businesses. More than anything else, large or small, they are businesses. There is a need to organize, keep leaders in power, grow and maintain capital structures, and in this day and age, maintain some broader types of communications. Even small denominations and churches work on financial bases. Pastors must be supported, capital investments always seem to be needed, and the plate is always passed. All in order to maintain and grow the flock. People tend to think that the classic Abrahamic religions, and other theistic cults, are principally faiths which, purely out of necessity, have to raise some money. But the reverse is more likely the case—they are business organizations enabled by promoting superstitions, for their own benefit. The Mormon Church, in fact, promotes the idea that God himself has mandated that the flock contribute ten percent of their income to the Church. It is hard not to laugh at the notion that a supreme being is focused on income percentages. The notion that such must be done in order to service God, of course, is merely a marketing ploy. If money is not raised, the business dies out, and all that'd be left are the strange individuals who we see running around spouting bizarre ideas.

We see this most acutely in the accumulated wealth of the various religious organizations. The net worth of some of the major religions around the world is staggering. It is impossible to specifically calculate the net worth of these groups, but here are some data that have achieved reasonable consensus among researchers.

By any measure, the Catholic Church is one of the wealthiest organizations in the world—of any type of business or organization. Though there are various estimates, the Holy See and the Vatican have a net worth of probably at least $30 billion, including incalculable investments in all sorts of industries around the world, as well as innumerable holdings in gold, artworks, real estate, and other tangible assets. (55) In addition, though it is realistically impossible to calculate, we might include the holdings of the various dioceses and the other Catholic-owned organizations around the world. These are worth many billions more; some are in the billions each. The total value of Catholic holdings, *in toto*, is most likely in the trillions of dollars. We know that a fundamental character of organized religion is hypocrisy. This takes the cake. Despite the putative vow of poverty (most priests do not take it), it is likely that the Catholic Church is one of the richest private entities in the world.[16]

In the Middle Ages, the old Anglican Church (now primarily of England, the Episcopalians, and certain others) made the mistake of splitting off from the Catholic Church and rejecting the Pope, and it made them poorer. The Church of England, for example, is probably worth *only* around $10-12 billion now.

The point is best illustrated by what we think of as some "modern" religions, those founded in the last couple of centuries. The charlatan Joseph Smith founded Mormonism only about two centuries ago. Now, by most estimates, the Church of Latter-Day Saints has a net worth well over $200 billion. It's unlikely to shed any of this value since the Mormon requirement of tithing persists. The ritual is strong enough that it's a formal expectation to maintain good standing in the religion, and records are even kept in this regard.

The Seventh-day Adventist Church was founded only about a century and a half ago. It is a very small sect with very conservative leanings. But it holds about $15 billion in assets. Poor Scientology may be worth only about $2 billion now. A great deal of the movement's wealth was lost to founder L. Ron Hubbard, as an individual, perhaps more than half a billion dollars.

Dogma in Christian religions, in particular, continues to change and, in recent times, somewhat rapidly. And religious leaders have conformed, albeit a little more slowly, to the behavior of their flocks. This satisfies the principal motivation—to maintain attendance on Sundays, or other days in other faiths. It's not as if God adapts His message to the tenor of the times. It's that dogma and teaching methods need to flex to the needs of the parishioners (I'll discuss this more below). Otherwise, quite simply, money is lost. Flexibility in the clergy is necessary to maintain the business. One sect which seems to change more slowly is the Catholic Church; some minor changes have appeared, but obviously, it has such huge reserves that is really doesn't have to be as flexible.

The purpose of large-scale organized religions is profit, growth, and the lifestyle of its leaders. You may not view your local church that way, but it is fundamental to any organized religion. As religious organizations become larger, the self-maintaining and self-aggrandizing become more evident. Preachers begin to claim they have been in communication with God. At least, they get paid for actually knowing and communicating to the flock what God's words mean. And our Western culture has grown to include movements which are unapologetic in this regard. The United States, in particular, is replete with mega-churches and entrepreneurial evangelists who profess some special theological understanding and skills. So-called televangelists reap huge profits in their ministries, and of course, it's often unclear how much is diverted into untaxed personal benefits. I would refer you back to journalist H. L. Menken's astute view of this, "Deep within the heart of every evangelist lies the wreck of a car salesman." The effort is to continue to grow the flock, with its clear self-serving financial implications. Many have condemned L. Ron Hubbard, who was a bit too honest about his financial motivation. Yet, millions follow—and donate to—Christian televangelists and megachurches. Although we might think of televangelists and similar individuals in this monetary regard, it is certainly not limited to them. Every time you put a dollar in the plate, you're reinforcing the same thing.

Organized religion is a business, and its consumers have been buying the product for centuries. And religions are able to survive and flourish, with the tax burden borne by their customers, other businesses, and other citizens.[17] Bottom line—the notion that soliciting money and holding vast sums of wealth is necessary to preach superstitious fantasies and proselytize religious dogma is the height of organized religion's abuse of power and disrespect for its adherents. Nevertheless, it *is* what keeps religion going, and the parishioners are the all-too-willing, and sometimes unwitting, customers.

Religious Identity

Despite the well-known problems of scripture, hypocrisies, and other problems of religion, most people continue to identify themselves with some religious label. That has been changing, of course, and now, for example, over a quarter of U.S. adults have no particular religious self-identity (We'll look at a variety of statistics in Chapter 11.).

In certain contexts, people often ask, "What are you?" with the implicit probing of what formal religion one carries around with them—as though there must always be a specific answer. The question is rarely intended to elicit an occupational response. If you answer, "I'm a teacher," the questioner is generally taken aback. If she really wants to know, there is persistence, "No, I mean what religion are you?" One might say, "I'm Catholic," or "I'm Jewish." But if you're uncomfortable answering because you don't really believe much or any of the religious teachings of your childhood anymore, you might answer, "I was raised . . . " Of course, this is a vague response, particularly for those who may not be practitioners or who see themselves as agnostic. And if you decline to answer, this is pretty much a communication that you're an atheist or you're afraid to say because of the particular social context. But you must have an answer since, most of the time, when such questions are asked, the questioner really wants to know. If you were born to parents of a given religious identity, then you must be *that*, as if religion is a genetically determined trait. Of course, a more reasonable question might be, "What mythology do you subscribe to?"

Since religious faith has historically been a way for groups to discriminate against each other, i.e., tribalism, what persists is a variety of social contingencies designed to elicit a religious identity and behavior—which vary from culture to culture, of course. In many cultures, it is simply unacceptable not to have a particular religious identity. The consequences of this go from social discomfort to natural segregation (e.g., Islamic neighborhoods in Paris), to overt discrimination, to being a pariah (history of Jews excluded from country clubs), to directed violence, to outright homicide and war (see below). Questioning others' religious identification is merely a way to further promote discrimination. When the question is clearly intended to elicit a religious response, not providing a good one can be a problem. And it is particularly often so for children. It is unfortunate that the "what are you" question isn't perceived more generally, for example, as "Who are you?" We can only hope that over time it is.

War, Religion and Hatred

There is no question about religion as a major force of hatred, violence, and war in our species. Unfortunately, one cannot discuss god memes and religion with touching on this subject. I'll not attempt any sort of comprehensive review, but here's a very brief look at the matter.

We've seen that "faith" and "belief" are matters of superstition. Ideas themselves, of course, are valid to the individual holding them. But superstitious memes become a problem when they're held and shared by numbers of people. When such communal belief is strongly held and when different beliefs held by others are devalued, tensions begin to develop. Such is the basis for discrimination, leading to social tensions, hostility, violence, and war. Private beliefs do not usually constitute a threat to others (though there are obvious exceptions). It is when such beliefs are externalized into social expectations and rituals—organized religion—that problems develop. We know that religions have been maintained, in some part, by the need to maintain a basis for exclusion when the there is no obvious visual difference among people. This enables the closer coherence of a group or tribe. But the net effect, of course, is that the religion is maintained, in part, by its discrimination against others, which is almost always a prelude to violence at some point.

Religious freedom is formally acknowledged in many modern countries. Nevertheless, there persists an enduring belief on the basis of one's own religion that it is, in fact, the one and only true religion and offers prayers to the one true god. This has obviously reinforced the notion of separateness and tension between different faiths. This undercurrent bubbles up from time to time in hatred, terrorism, and wars. A country's endorsement of religious freedom is not a reliable bulwark against such tragedies.

People have been attacked for their religious beliefs throughout history. Religion is associated with the full range of discrimination, hatred, and violence we've seen through the centuries. Most people are familiar with the medieval Crusades, suppression of women under Islam, and pogroms against Jews. In America and elsewhere there has been a rise not only in antisemitism but also in anti-Muslim violence (which I'll review below). We needn't review all the tragic history of religious violence. It is well known, and I'll not attempt any exhaustive review here. Just some lowlights, only a few egregious examples. I'll briefly address Christianity and Islam, traditions followed by more than half the world's population. Nothing is implied about all the other religions' violence and war that I've not addressed here. There are excellent sources detailing much of this. (206, 211)

Are Religions Violent?

The answer, in a word, is yes. By their nature, they are violent. The concept of God was invented. It became reified, a pivotal part of culture, which then led to ostracism and eventual punishment of any doubters and certainly apostates. One can see how this situation evolved into outright hostility toward non-believers and the history of conflicts and wars to enforce belief, or at least conversion in lieu of violent assault and death. To infer otherwise is to ignore history.

Many people espouse that their religion is a peaceful one. In current discourse, it's often considered inappropriate to criticize a religion as being violent or even hostile towards others. It's politically or socially frowned upon—at least in the West. But there's no such thing as a theistic religion which is reliably moderate or liberal. All religions are, by their nature, hateful and violent.[18] It only depends on what time period and geography you're looking at it. If there are moderating changes in a religion, they come not from the religion itself but from the social and cultural conditions in which it exists, and not from any historical text or God changing His mind. Changes can be very significant over time, which we'll review below. However, organized religions do tend to be intolerant of those who practice differently, and at some points in their history they become violently so. At any time in history, any religion will be, or has been, essentially fundamentalist, eschewing all other views, gods, religious practices, tolerance of others, and acceptance of their existence. As Bertrand Russell astutely observed,

> You find this curious fact, that the more intense has been the religion of any period and the more profound has been the dogmatic belief, the greater has been the cruelty and the worse has been the state of affairs. In the so-called ages of faith, when men really did believe the Christian religion in all its completeness, there was the Inquisition, with all its tortures; there were millions of unfortunate women burned as witches; and there was every kind of cruelty practiced upon all sorts of people in the name of religion. (435)

Violence and war are clearly prescribed by religious texts, and the list of God's instructed killings is pretty long. Of course, the idea of mass killing goes all the way back to the Old Testament. Lest we not forget, in Deuteronomy 20:17 God commanded the Israelites to commit genocide of the Canaanites and others (The actual historical record is ambiguous, but the point is made.) Some claim that their religion is inherently peaceful, based on scripture, and that violence is merely a corruption of their faith. Many Christians espouse the idea that Islam is a violent religion, whereas Christianity is a religion of peace. Again, it only depends on where and when you're looking at it, which I'll review below. Reliance on scriptures —and they are variable and "convenient"—offers no clarity. We saw how inconsistent and hypocritical scriptures are in this regard. As Dawkins notes, "How can there be a perversion of faith, if faith, lacking objective justification, doesn't have any demonstrable standard to pervert?" (113) The changing text and interpretations of the Quran and Bibles over time, to fit the needs of followers, is just dishonest; and any sort of war or peace can be rationalized by "cherry-picking" and quoting particular verses from those scriptures.

Though there are isolated examples, it is truly rare that a religion is nonviolent over a long history. For an isolated example, some cite the Amish, who've practiced a form of Protestantism since the 18th century. However, they do discriminate against members who are noncompliant with the faith. In one instance, an Amish separatist group in Ohio repeatedly attacked other members of the faith in 2011, forcibly cutting off their beards. The attackers were often relatives of the victims. For some time now, in Myanmar (and in Sri Lanka), even supposedly non-violent Buddhist monks have perpetrated oppression, hate and violence, and killings against the minority Muslims. (511) No matter the difference in what people believe, hatred and violence in the name of religion is part of human culture, as Harari observes:

> A religion that recognizes the legitimacy of other faiths implies either that its god is not the supreme power of the universe, or that it received from God just part of the universal truth. . . . monotheists . . . have been compelled to discredit all other religions. Over the last two millennia, monotheists repeatedly tried to strengthen their hand by violently exterminating all competition. (203)

Is War Natural?

Let's take a step back. No one doubts the long history of human wars. Humans have fought over territory and resources since we organized ourselves into tribes. But is war in our nature? There are continuing debates about how "natural" war is or isn't. Whether tribal conflicts and war, per se, are fully instinctive or largely "voluntary" has been debated for a long time. To address this, we first need to look at other species, including our own ancestors.

We have to acknowledge that humans are clearly not the only species that goes to war. Many other species fight battles, and some animals often go to "war." Our nearest living relatives, chimpanzees, are notorious for this. Over 40 years ago, Jane Goodall was able to observe chimpanzees engaged in violent tribal conflicts. (186) They not only fight with other groups, but they actually use a number of tactical maneuvers in their battles. Certain other, "lower" species also engage in group-based violence and wars to gain or hold territory and resources. The list of warring species is seemingly endless. But well-known species which engage in battles are lions, meerkats, wild dogs, and wolves. Even some insects do this. Ants and termites, for example, fight fierce battles with other groups, and ants fight wars against termites (who always lose). Despite what we might wish for, war between groups of animals has a long history. Whatever humans are, we *are* animals, and many animal species go to war. That "war is natural" is an inescapable conclusion. Whether such tendencies can be controlled is a whole other matter.

Human War

Let's back up to consider some history, some very ancient history. Human "wars" have probably been waged for as long as humans have existed. There is debate, since definitions vary, over whether early humans engaged in what we actually consider "war," or at least some crude version of it. It is likely that Paleolithic humans did engage in violent acts. When groups coalesced and became organized in the late Paleolithic, and when resources may have been in contention, the tribal competitive was assured. This likely led to violent conflicts between groups, something that might reasonably be called war. (267, 298) Our cousin species, the Neanderthals, likely engaged in intraspecies tribal conflicts (though the archeological record is admittedly less than fully conclusive). *Homo sapiens* probably did engage in battles with Neanderthals. (318) As we noted in Chapter 4, the factors leading to extinction of the Neanderthals are still in some doubt or may have been multifactorial. *Homo sapiens* likely outcompeted, had sex with, migrated differently, and adapted better over time than Neanderthals; but we fought battles with them too.

Battles among groups of *Homo sapiens* certainly occurred in the Paleolithic era, and there were many "wars" conducted from the Neolithic up through the Bronze Age. The specific factors are not fully known. However, the idea of "others" is perfectly natural, leading humans to exhibit a propensity for being suspicious of and discriminating against "other" individuals and groups, just as other social animals do. As tribes of humans grew, alien groups would certainly have been seen as threats. Gods and religion became unifying principles in this era, and we know that different cultures created different gods. There were different powers, values, and formal guidance inferred from these gods in different parts of the world and in different cultures. As the world became "smaller," these cultures clashed. It cannot be ruled out that this was at least associated with religious symbolism and identification. Religion creates a powerful force of cohesion, and it enabled the symbols or "rallying cries" for violent aggression against those who believed differently.

Violence and war have been perpetrated in the name of religion throughout human history. Only a small percentage of these might be seen as "caused" by religion, per se. Wars are generally fought for territory, resources, and power, but being "fought in the name of religion" enables cultural coherence and solidarity in the warring population, and thus popular support for war. In the last few millennia, violence and war have been increasingly conducted in the *name* of religion, whether it was for religious purposes, or in the name of religion for other reasons. Nevertheless, religious significance in many wars cannot be denied. Violence has been initiated by many religions, and many religions have been the victims of violence and war. For those who doubt or discount the relevant history, we'll look at *only* a few of the lowlights of the two dominant religions in the world today.

Christian

We've noted some of the horrific and violent guidance in the Bible. But reference to the religious significance of violence and war alludes to some very old conflicts. There was the Battle of Megiddo (in what is now Israel), occurring in the 15th century BCE. The Egyptian Pharaoh Thutmose III defeated Canaanites and expanded the Egyptian empire. However, it also became pivotal in Christian mythology. The Biblical reference to Megiddo (Revelation 16:16) notes that it is supposedly the site where the "final battle" will occur, between "good" and "evil," bringing on Armageddon or the "end of the world." It hasn't happened, but violent efforts attempting this have occurred ever since.

Is Christianity a "peaceful" religion, as its adherents claim? In modern times, Christianity was one of the first religious offenders. The well-known Crusades began with one launched by Pope Urban II in the late tenth century. These medieval Crusades lasted two hundred years. Don't gloss over that. Catholicism was killing people for two hundred years. The purpose was to purge the "Holy Lands" of the "infidels," mostly Muslims and Jews. It is one of the most tragic examples of religious violence in all of history. One to two million people died as a result of these actions by the Catholic Church. Similarly, the 15th-16th century Spanish Inquisition was basically an attempt to wipe out all heretical beliefs. Likely hundreds of thousands were executed in the name of Catholicism, in Spain but also in North and South America. Though it was prosecuted in the name of religion, the net effect was also to expand and consolidate the power of the Spanish Kingdom. As Dennett notes, "It is always worth reminding ourselves that not so very long ago people were banished, tortured, and even executed for heresy and apostasy in the most 'civilized' corners of Christian Europe." (123)

Of course, we can't ignore the prime example from recent history, the Nazi Holocaust. World War II was not specifically motivated by religion. In fact, there were numerous secular causes that were primary in this regard. The condition of Germany in the post-World War I decades, the worldwide Depression, Hitler's expansionist efforts and the rise of fascism cannot be denied. However, Hitler did blame the Jews for much of Germany's post-war conditions. His solution was to expel or kill them—the "Final Solution." Exterminating the Jews was a priority, and millions were killed, one of history's most extreme examples of violence against a religion. Further, lest you think the Catholic Church, for example, was not supportive of the Nazi agenda, hundreds of priests actually joined Nazi troops on the front lines to offer spiritual guidance. On Hitler's 50th birthday in 1939, Nazi flags actually flew on many churches, and congregational prayers were offered for protection of the "Führer and the Reich." Further, as Harris points out, "The truly sinister complicity of the church came in its willingness to open its genealogical records to the Nazis and thereby enable them to trace the extent of a person's Jewish ancestry." (206) It took until 1998 for the Church to actually apologize for its "inaction" during the Nazi regime, though the apology did not address, for example, that it celebrated Hitler's birthday from the pulpit of its churches nor Hitler's use of Christian dogma to promulgate his "rationale" for Jewish genocide. In fact, more recent evidence, gleaned from Vatican archives, reveals that Pope Pius XII had a back channel communicating with Hitler, that he deliberately remained silent about the genocide, and that he was afraid of offending Hitler and Mussolini (who had helped him rise to power, including giving sovereignty to the Vatican). Further, the Pope had held secret thousands of letters from Jews begging for relief and citing the shame of the Pope's silence at the time (those letters were released only in 2022). (269) It would be naïve to infer that the Church did not have a role.

And there is other modern history of religiously inspired violence and genocides. And Christian-inspired violence continues to occur on a smaller scale. Even without war, there is a substantial history of Christian-motivated violence in America, particularly on abortion clinics, and some abortion doctors and clinic staff have been killed in the name of some fabricated Christian principle. (There are good sources with fully comprehensive reviews of Christian-motivated violence in America. (257)) Around the world there have been numerous shootings and arson attacks on synagogues and other venues. Witness the religiously motivated mass killing of Jews in a Pittsburgh synagogue in 2018 and at a Hanukkah celebration in Australia in 2025. Some attribute these kinds of acts merely to individuals. However, that they are committed in the name of a Christian God should not be forgotten. To deny Christian violence is to deny both history and current affairs.

Of course, Christians have been the *victims* of violence over the ages as well. There is the Armenian genocide during and after WWI, in which likely about a million died at the hands of the Ottoman Empire, led mostly by Muslim Turks. This included thousands of Armenian children; many were essentially kidnapped and forced to convert to Islam. More recently, many Christian churches have been attacked and destroyed, often in India, Pakistan, and China in the last few years. In 2014, we saw "ISIS" fighters (known at that time as the Islamic State) conduct killings, rape, and a near-genocide in Iraq of Christians and Yazidis. Yazidism is an ancient religion, possibly derived from Zoroastrianism and Islam, but also Christianity, with elements taken from each. This derivative quality has often led the Yazidis to be referred to as a sect, noting, of course, that when you look at history, all religions are in fact sects. It's clear that ISIS did not ask about derivative beliefs before conducting the violence. They merely killed, raped, and kidnapped thousands of Yazidis.

Of course, Christians have perpetrated violence on other Christians. The 20th-century Irish conflict between Protestants and Catholics in Northern Ireland is a prime example. There is the notion that Catholicism and Protestantism are different religions. We might call them sects. They might not agree, but again, all religions are sects. Nevertheless, from the perspective of a naïve outsider, they do have nearly indistinguishable belief systems. Yet, violence in the name of religion was the source of innumerable killings there.

Islamic

Despite religious identification and belief in a deity declining in the United States and Western Europe, there is one religion which is growing—Islam. In fact, Islam is the fastest-growing religion in the world, now about a quarter or the world's population. Although this is largely a function of demographics (age, birth rates, etc.), the word is also spreading—throughout Europe and the Western World—via migration and information technology. (315) Most predominantly Islamic countries around the world continue to subscribe to varying degrees of enforcement of the religion under Sharia, God's divine law according to the Quran, and writings of the prophet Muhammad (the Hadith and Sunna). The law varies from essentially from very minimal use in some countries, such as Turkey and the Caucuses, up to nearly complete subscription in some places in the Middle East (noting that reliable information cannot even be obtained in many of these countries).

Is Islam a "peaceful" religion, as many of its adherents claim? Nothing could be further from the truth. Let's first look at the dogma here. The Quran, despite various interpretations, does urge Muslims to conduct jihad numerous times. It prescribes jihad until the whole world is either converted to Islam or killed. I won't get into literal scripture here, which may have been altered many times, as I've said; there are various conflicting verses in the Quran and the Hadith, and various interpretations lead to different conclusions. But I would point out one phrase that seems to summarize the intent of the prescribed Islamic jihad, "Know that Paradise is under the shades of swords." (Bukhari 4.52.73, 72)) As Sam Harris points out about Islam being violent, "It is widely claimed by 'moderate' Muslims that the Quran mandates nothing of the kind and that Islam is a 'religion of peace.' But one need only read the Quran itself to see that this is untrue." He makes a substantive case, with several examples, that Islam *is* a violent religion, as the Quran, and especially the hadith, so dictate. "Anyone who can read passages [from the Quran] and still not see a link between Muslim faith and Muslim violence should probably see a neurologist." (206)

We could again ask ourselves whether a religion is what its scriptures mandate, or what people do in the name of it, and/or what its "priests" suggest. On all three points, Islam can only be viewed as a stimulus for violence. Many *wish* for Islam to *be* a peaceful religion, and many people who self-identify as Muslims, of course, do behave peacefully. But this is because they have all manner of reasons to do so, not because Islamic doctrine suggests or mandates it. Dawkins points out that Islam has been a "peaceful" religion only for about 13 years in its history. (113) Countless have been killed in the name of Islam over the centuries.

Let's take *only* a few isolated examples. Muslims have killed innumerable people throughout the Middle East and thousands more around the world. There is a long history of violence and wars between Muslims and Jews in the Middle East, beginning with the U.N. recognition of the state of Israel in 1948. Since then, several wars and numerous violent skirmishes have been launched by Islamist organizations or nations over the so-called Holy Land, the geography now widely recognized as Israel, though clearly not by all. There have been numerous terrorist killings and military operations launched by these Islamic groups in the area over the years, killing countless thousands. And the war launched on Israel by the Islamic Hamas in 2023 has killed thousands on both sides.

Most people know about the violent attack by Arab Muslims on America on 9/11/2001. The notion has been floated that the attack was for political or socioeconomic reasons, but many middle-class fundamentalist Muslims were among the perpetrators. Islam was clearly a factor here, and their suicide mission is more than evidence of it. Witness also the violence that followed publication of cartoons depicting the prophet Muhammad in Denmark in 2006, or the mass murder of personnel at the French satirical publication *Charlie Hebdo* in 2015 for the same reason.

Sharia law also demands a long list of what we would generally consider rather bizarre and violent sanctions. A small list includes amputation of the hands for theft, the death penalty for leaving the faith, marrying a non-Muslim, a woman's adultery, and so-called honor killings for other offenses such as pre-marital sex or becoming the victim of a rape.

Beginning in the late twentieth century, the Taliban in Afghanistan carried out incredibly violent acts against non-believers, and penalties for various "offenses" by women have been severe, often to the point of beheading.

A couple of other recent examples: The recent Syrian civil war, which began in 2011, was stimulated, in large part, by an extreme drought in the area brought on by climate change (265), the worst on the instrumental record, leading to population unrest, hunger, and migration. But the governmental response included violent suppression of minority Muslim sects. The government, headed by a leader of a particular Shia sect and supported by a predominantly Shia country (Iran), portrayed the opposition as Sunni extremists. Even though drought was the igniter, religion eventually took over and *was* critically important, and thousands died.

The Hadith essentially prescribes female genital mutilation, as conservative Islamists also continue to assert. One could quibble with the language in scripture, but millions of girls around the world have been subjected to this mutilation in the name of Islam. Believe it or not, even in the United States female genital mutilation was not a crime until 1996 (18 U.S.C. 7 §116); in only 35 states is the procedure a felony for a practitioner who performs it, and mandatory reporting of the procedure by others is required in only 10 states. (5)

And, of course, Muslims have been *victims* of hatred, violence, and wars throughout modern history. The medieval Crusades are a prime example. But anti-Muslim hatred and violence continue. Recent history has coined the term "Islamophobia," but this doesn't really capture the idea (phobias are fears, but they don't generally lead to hostility). Some lowlights: Mosques have been attacked and destroyed around the world. And witness the 2019 terrorist killings of scores of Muslim worshipers in mosques in New Zealand by a right-wing extremist group.

There have also been numerous attempts to suppress people of Muslim faith. Let's take a prime modern example. The official Chinese government position on religion is atheism, but it does officially recognize a number of religions, including Islam. Nevertheless, there is an ongoing effort to isolate and suppress it. As is becoming more widely known, China has essentially imprisoned as many as one million Uyghur Muslims in internment camps in Xinjiang Province in China. China has called them "re-education camps" to "round up everyone who should be rounded up" for having "been infected with the virus of Islamic radicalism." (422) That there have been some terrorist attacks by Islamic followers in northwest China is undeniable. However, it is difficult to see how this justified the incarceration of this population. China has even begun forced medical procedures to block Muslim women from having any more children. If that's not genocide, what is? Despite official government proclamations on the matter, the "reeducation" camps have not been closed. (65, 422), *and* the government has now placed cameras in public spaces in that area of China, with AI-assisted facial recognition, to track the movements of Uyghurs (and others in the country, I should point out). The U.S. has condemned these efforts as "genocide" (558), and a few other nations have subsequently come on board.

But Muslims of different sects have also been the targets, e.g., Sunnis of Shia, and vice versa. There has been some level of this conflict since the religion was reportedly "founded" in the early seventh century CE. They've been killing each other ever since on the basis of a particular difference of opinion about the religion.[19] At least half a million Muslims were killed in the long Iran-Iraq war in the 1980s. Violence between Sunni and Shia Muslims continues to this day, e.g., the current Shia insurrection in Yemen.

Non-Religious War?

It's often claimed that "secular" war has killed as many or more people than religious ones, and that's probably true. Of course, many large-scale violent conflicts do not necessarily involve religious connotations, e.g., the Mexican-American War, the Vietnam War, many wars of independence, various civil wars, etc. Only a small minority of wars carry an outright religious mission or stated inspiration. However, statistics is not the point here. Although only some wars are fought in the *name* of religion, they have troublesome hints of it in a very large percentage of the violent clashes between cultures and nations.

Of course, one would be foolish to assume that religion has been the only source of violence and war, though it often does play a significant role, albeit sometimes not well-recognized—godless violence is not always entirely secular. National leaders often need to assume rather supernatural status, in effect, to carry out subjugation of a population and to conduct violent conflicts. The divine right of kings persists for such. It endures in some countries which might not be considered theocracies. For example, Kim Il Sung (1912-1994) became the self-appointed head of state of the Democratic People's Republic of Korea (North Korea) in 1948. Shortly after WWII, he led a violent invasion of the South, leading to the Korean War. He was the self-appointed "Eternal President" and was worshiped as though he were a god; at least he was believed to have divine origins. His successor, son Kim Jong-Il, was considered the son of a god or "Sun of the Nation." Some divine quality has been a way to maintain power, since gods are considered infallible and thus must be followed. As George Orwell once wrote, "A totalitarian state is in effect a theocracy, and its ruling caste, in order to keep its position, has to be thought of as infallible." (385)

No matter the operational factors for violence and war, religion is often the motivating or contributing force, and the assumption of religious powers or rights enhances the effort. It's more than clear that the American Civil War was fought over slavery. But it's worth noting that slavery was blessed by many Christian churches prior to and during the Civil War. Over 90% of the South belonged to Baptist, Methodist, and Presbyterian denominations at the time, and there was a split in each of these, with some congregations being able to easily cite the numerous Bible passages endorsing the idea of slavery—and they did. Religion was clearly a factor in that war. (347) While we rightly don't consider her an authority, Margaret Mitchell commented on this phenomenon in "Gone with the Wind," and with considerable insight. Rhett Butler dryly asks, "All wars are sacred," he said. "To those who have to fight them. If the people who started wars didn't make them sacred, who would be foolish enough to fight?"

Conclusion

It's an especially cruel deity that would violently pit one "true" religion or sect against another. Modern history has seen violent conflicts between sects of the seemingly same religious persuasion. Catholics and Protestants haven't always gotten along. Violent conflict between Sunni and Shia Muslims rears its ugly head from time to time. Muslims and Jews have gone to war many times.

There has always been hate and violence between humans of different religious persuasions, and there will likely always be. You cannot extinguish religious behavior you disagree with by punishing or killing it. The Crusades, the Chinese suppression of the Uyghurs, the many pogroms against the Jews, and many other movements have demonstrated this pretty well. But such efforts do continue. As beliefs persist and intensify, religion becomes the motivator and the fuse for all manner of hostility. *That is what religion gives us. It survives, in at least major part, on the blood of others.* To infer otherwise is to deny reality.

On Social Good

The argument that organized religion provides social comfort and welfare to others is not doubted. However, this is not an argument for the existence of a god or the many religions around the world. Theistic religions have always been organized for fealty to superstitions, discrimination, and violence toward others. However, we often hear the argument that "religion makes people feel better." This is offered almost as a rationale for its existence. A religious community does provide social support. It can serve a social purpose—socialization, sharing of values and ethos, and help in times of need. There is no doubt that it can be a source of hope for certain individuals who need it, and it does help to relieve suffering for some people. Parishioners do assert that their religion "makes them feel better."

Of course, it is obvious that all sorts of human social interactions do this. We might again ask, "Is religion actually a comfort for suffering?" In fact, most "comfort" reported by religious people is associated with social support and with rituals of a religion—communal services, prayer, Bible readings, sharing such things with others, etc.; and any relief that follows becomes superstitiously associated with some "higher power." Thus, "feeling better," while obviously valued, also reinforces the superstitious beliefs. It is not God that makes people feel better. It is rituals, memes, practitioners, communications, and the culture surrounding the religion.

However, relief of suffering or just "feeling better" does not require superstitious beliefs. There are innumerable ways of achieving such, from all manner of normal human interactions, to informal groups, to secular organizations, to professional help. In fact, many clergy often counsel their congregations in ways other than "spiritual guidance." There has even developed a profession of counseling for religious adherents, which makes more use of accepted psychological techniques than theistic inspiration. Finally, we know that, in general, "mental well-being" (and various other so-called health measures, as they are conceived by laypeople) is no different among atheists or Christians or those who practice certain Eastern religions. (497) Theistic religion is not necessary to achieve social support or relief of suffering.

Altruism

Religious organizations also provide social goods in terms of food, shelter, money, volunteer services, and other charitable efforts. We can only applaud such. However, these in no way support the idea of the religion they're associated with. There are also innumerable secular organizations addressing the full panoply of human needs across the globe. Such good works often are accomplished by secular people and organizations. Informal groups, online organizations, secular charities, governments, and NGOs around the world provide a level of relief of human suffering and socioeconomic hardship that seriously dwarf the efforts of organized religions. Moreover, many religious people participate in those secular groups.

We might describe such activities as altruism. And what does altruism actually mean? It is technically behaving in ways that sacrifice something of individual value to enhance the health, survival, or various needs of others. It is natural for people to act this way, religious or not. It is certainly clear that many religious persons, in the absence of religious auspices, are involved in such altruism. There are numerous secular organizations and millions of humans who also show such altruism, directly or through charitable giving. The idea that membership in a religious group might increase an altruistic desire or actually help others is a fatuous proposition. In fact, currently "Atheists and agnostics tend to participate in civic life at rates matching or exceeding religiously affiliated people." (491) Religious adherence is in no way an argument for the maintenance of organized religion or any supernatural being around which it seems to be organized. And we know that some religiously affiliated charities often come with overt or covert efforts at recruitment or conversion. To work in some charitable organizations, in fact, there is actually requirement for an overt or implicit affirmation of the religion. One might rightly find abhorrent the idea of reinforcing superstitious beliefs in the context of our natural human tendencies to help others.

Human beings tend to help others, especially those in need, but we are not the only species to do so. We are not the only species that can act altruistically. Such is likely fundamental to many species of social mammals. It is a critical set of behaviors that has contributed to survival in humans and other primates. Many animals provide social empathy and altruistic behavior (I'll cover this in more detail in Chapter 9), and they don't require a god. The altruistic value that many assign to their organized religions is just another excuse for maintaining their superstitious beliefs.

Conclusion

The argument for theistic religion based on "social goods" is vacuous. Social good is done by people, not ideas, and channeling a fantastical god for this purpose is not necessary. One does not need an organized religion to promote and support positive humanistic values. Invoking of religion in meeting social needs only serves to reinforce and maintain the superstitions. And those superstitions continue to enable continued hatred, violence, and war—things most religions would declare they do not support. But as we've seen, there is some debate about this, and history isn't very agreeable with that view.

Government and Theocracy

The integration of religion and social power is as old as supernatural superstition itself. Early leaders of tribes sired their inheritors (and in more primitive primate cultures, alpha males passed on the legacy of dominance), which led to natural lines of succession in early humans. Such royalty eventually emerged in various forms and is still with us today. The divine right of kings is the legacy; and subjects rarely challenge this—though there are notable modern exceptions.

Since the beginning of human cultures, there have always been attempts to validate gods and religions by associating them with the extant mortal power in a society. Throughout history, theistic religion and government have been tightly fused together. Governments, and particularly their leaders, have drawn authority from the superstitious power of gods. From ancient Egypt to Babylonia to the Islamic Caliphates to the fusion of civil government with the Catholic Church in the Middle Ages, leaders have always drawn on the divine to promulgate and enforce their ideas.

Governments have continued to draw authority from associating themselves with divine qualities or inspiration. And organized religion has always used government and politics to validate itself and to reinforce its survival. Religions tend to fade if they're not reinforced, and the best way to do that is with the power of the state. The last few decades have seen some waning of participation in organized religion in many parts of the world (which I'll review in Chapter 11). In lieu of violence, this has led to increasing efforts to imbue governments with religious authority and use governments to sanction religious practices.

When the fusion is nearly complete and religion essentially controls public policy, you have a theocracy. Today, many countries around the world are essentially or officially theocracies. In the last few centuries, we have seen this throughout the Middle East and certain other Muslim countries which endorse the idea of Sharia law. Principal among these are the ones with Islamic "governments." I'll not address this here but refer you to one of the better volumes dealing with it, Sam Harris' *The End of Faith.* (206)

Let's pause for a moment on this concept. In the West, we do tend to reject the idea of theocracy (though some forces always attempt to move us in that direction). Westerners tend to assign a pretty loathsome quality to some contemporary theocracies, and the reasons for this are quite understandable. But we should note where such trends come from. There is a long history of this type of political power, from primitive shamans to Pharaohs to the divine right of kings in general. Where different parts of a culture have different gods, conflicts arise. Behavior surrounding hatred and violence requires a lot of energy, a problem which is reduced if most everyone believes the same way. This is enabled by an authoritarian political system based on a unified religiosity. Human beings—like many other social mammals—favor an authoritarian system which *imposes* and enforces particular kinds of belief on a population. That *is* our human history. The trends toward more democratic political systems are only very recent. Actual democracy lived in in ancient Greece for only a few hundred years, and it died in Rome in the first century BCE. Subsequently, the power of kings wasn't actually challenged somewhat until the English Magna Carta of 1215; and true democracy, independent of a royal, didn't really begin to take hold in Europe until the late 17th century. The overwhelming course of human history is dominated by monarchs who promoted a particular religious belief or by actual theocrats. The fusion of religion and political systems is a potent factor in what has kept theistic religion alive for thousands of years. We shouldn't ignore this. Theocracy is the typical form for human societies. It is—so far—the default state of our history. Anything else is an exception.

The United States, A Modern Example

Let's take a prime example, the United States, where religion is attempting to increase its power in the culture. This is an interesting modern case, so let's devote a bit of space to this. There is a never-ending movement in the U.S. to inject Christianity into the government. I'll not review the full current and recent history of the issues here, but it is consistent with the age-old attitude that those who believe in a god do try to have their beliefs enforced by the reigning political authority.

One prime illustration is the so-called Christian Nationalist movement. Christian theists and countless evangelicals have long been attempting to create a theocracy in the United States. We often see Christianity as regularly attempting its influence and self-validation through both the American federal and state governments. In America, the movement attempts to insert Christian "values" into law, sometimes surreptitiously, sometimes quite overtly (I'll not deal here with all the controversial issues and whether or how much success has actually been achieved in this regard.). They reliably condemn theocracies that are based on religions like Islam. Yet they continue to assert that the United States is a Christian nation, despite the fact that the religion itself is fading. (Chapter 11)

One of its primary contentions, however, is that the United States was *founded* as a "Christian nation." Nothing could be further from the truth. Fortunately, all evidence from the historical record overwhelmingly contradicts this assertion.[20] For now, let's consider some of the historical facts about the founding of the United States and the role of religion. First, contrary to what Christian nationalists promote, it's likely that the so-called Founding Fathers were not serious Christians or "believers," despite any public pronouncements. Belief in a personal god was rare in that group, and they are often misquoted in this regard. They were deists or even atheists, at least likely doubters, a position which was not socially acceptable at the time and would be seriously criticized or worse if publicly proclaimed. Most American Founding Fathers, in fact, were deists and did not believe in a personal god or use such a meme in founding the United States. (457) For example, one of the most potent assertions of the secular founding of the United States was offered by John Adams in 1787:

> The United States of America have exhibited, perhaps, the first example of governments erected on the simple principles of nature; and if men are now sufficiently enlightened to disabuse themselves of artifice, imposture, hypocrisy, and superstition, they will consider this event as an era in their history. Although the detail of the formation of the American governments is at present little known or regarded either in Europe or in America, it may hereafter become an object of curiosity. It will never be pretended that any persons employed in that service had interviews with the gods, or were in any degree under the influence of heaven, more than those at work upon ships or houses, or laboring in merchandise or agriculture; it will forever be acknowledged that these governments were contrived merely by the use of reason and the senses. (3)

Our Founding Documents

Christians often mention the God reference in the Declaration of Independence. However, the author, Thomas Jefferson, was most likely a deist, and the reference to "Nature's God" in the Declaration was clearly intended by Jefferson to be a deistic reference at most. He was likely referring to the "laws of nature," pulling from Spinoza's conception (and presaging Einstein's use of the idea). There is no evidence that he was implying a personal god or certainly any God of the Judeo-Christian tradition.

There is no reference whatsoever in the United States Constitution to God, Jesus, or any other god figure or religion. In fact, as we all know, the First Amendment specifically mandates the separation of church and state. Christian nationalists try to forget this or sometimes, remarkably, try to suggest that the words don't mean what they say. It's one of the most classic examples of willful ignorance. Actually, in laying the ground leading to the First Amendment to the U.S. Constitution, Thomas Jefferson, in the "Virginia Statute for Religious Freedom" of 1786, specified that "no man should be compelled to support any religious worship, place, or ministry whatsoever." (42) In promoting the bill, he specifically averred that it was "meant to comprehend, within the mantle of its protection, the Jew and the Gentile, the Christian and Mahometan, the Hindoo, and the infidel of every denomination." (339) It became the stimulus and basis for our First Amendment, and it was no accident that it came first, a bane of modern Christian religionists.

The Constitution further specifically attempts to keep religion out of governmental authorities. In Article VI it states, "The Senators and Representatives before mentioned, and the Members of the several State Legislatures, and all executive and judicial Officers, both of the United States and of the several States, shall be bound by Oath or Affirmation, to support this Constitution; but *no religious Test shall ever be required as a Qualification to any Office or public Trust under the United States*." [emphasis added] It was more than apparent that the founding documents of the United States specifically eschewed, and actually sought to block, the intrusion of religion into governmental affairs.

Treaty of Tripoli

If the Declaration and the Constitution weren't enough to lay down our secular roots, that idea was communicated to the rest of the world many times. The world was only slowly becoming aware of our Declaration and Constitution. However, one of the efforts that reinforced our secularism was evidenced in what's commonly known as the *Treaty of Tripoli*. There had been many years of attacks on merchant ships by Barbary pirates in the Mediterranean, sponsored by North African states, and we fought battles over this. The pirates may well have been motivated by the tenets of Islam, according to various accounts. In fact, the Tripoli ambassador told Thomas Jefferson that the rationale for the piracy could be found in the Quran.

In attempts to appease Islamic leaders, the nascent American government took specific pains to eschew any religious affiliation, and

Jefferson announced that he was providing assurance to the Muslim states that we were *not* a Christian nation. The *Treaty* was actually our first one, a treaty of friendship with other governments. It was intended to assure shipping protections and avoid conflicts based on religion with the Muslim lands. President John Adams had been initially hesitant to approve it, but it was unanimously ratified by the Senate and signed by President Adams in 1796. To wit:

> As the government of the United States of America is not in any sense founded on the Christian religion; and it has in itself no character of enmity against the laws, religion, or tranquility of Mussulmen; and, as the said States never have entered into any war, or act of hostility against any Mahometan nation, it is declared by the parties, that no pretext, arising from religious opinions, shall ever produce an interruption of the harmony existing between the two countries. (21)

The *Treaty* became a foundational affirmation of the United States as a secular nation.

The Distortions Continue

Nevertheless, distortions of history by Christian politicians and others persist to the present day. One recent United States Attorney General, William Barr, argued in 2019 that "In the Framers' view, free government was only suitable and sustainable for a religious people . . . Our Constitution was made only for a moral and religious people." (30) He forgot to read the First Amendment, as well as Jefferson's affirmation for equal rights of people with no "denomination."

In 2018, there was Attorney General Jeff Sessions, who attempted to fuse Christianity with the US Government. He promoted an immigration policy of the government separating parents and children at the U.S.-Mexico border; that is, governmental kidnapping. (Many families have still not been reunited, and the government had no plan at the time to do so. That situation may still persist. It is nearly impossible to get any honest information about this.) Sessions then attempted to justify the practice by citing a Biblical phrase, "I would cite you to the Apostle Paul and his clear and wise command in Romans 13 to obey the laws of the government because God has ordained them for the purpose of order." Of course, he cherry-picked even this, and Biblical scholars Margaret Aymer and Laura Nasrallah point out the several passages which directly contradict Sessions' assertions. (22) (Of course, this is not surprising since the number of self-contradictory passages in the Bible is extensive.)

There were even attempts to overtly revive the divine right of kings. A person running for President in 2016 was revered by many as "chosen by God" for the office. And as Seidel observed, "Evangelicals were simply seeing in Trump a character they'd been taught to revere. As if to prove a point, conservative TV commentator Ann Coulter called Trump her 'Emperor God.'" (457) Attempts to construe our founding as a Christian nation and build a theocracy never seem to end, and that person has been elected to the presidency twice.

Nevertheless, the United States was quite obviously not founded as a Christian nation, despite attempts to portray it as such and despite ongoing attempts to inject Christianity into the government. The efforts to enhance the influence and power of Christianity in America are based on historical distortions, lies, and attempts to proselytize Christian myths through governmental authority. In fact, the founding fathers specifically eschewed religion as a guiding principle, as Jefferson always did. He put it best in an April 11, 1823, letter to John Adams:

> The day will come when the mystical generation of Jesus by the Supreme Being in the womb of a virgin, will be classed with the fable of the generation of Minerva in the brain of Jupiter. . . . But we may hope that the dawn of reason and freedom of thought in these United States will do away with all this artificial scaffolding. (249)

Conclusion

We still have work to do to actualize Jefferson's vision and avoid a movement toward government-religion fusion. Many countries around the world are still dominated by royals in one form or another, whether dictators, kings (occasionally queens), or other autocratic designations. Even countries with putative democratic governments continue to employ the leadership of royalty in some capacities. In the Kingdom of Jordan, the King has essentially full executive power despite there being a parliament. In Great Britain, the Royals still have various roles as the head of the government—such as signing off on new laws, and opening up or dissolving Parliament, and appointing certain officers—though in reality these are mostly ceremonial. Such anachronisms persist as symbols of the Divine Right of Kings, a homage to the theistic character of governments.

Religionists do continue to try using government to bolster their claims and increase the flock. Of course, this is merely a reflection of the weakness of their beliefs. Thomas Jefferson had it right when he asserted that "religion will decline if not supported", but his most poignant comment on the matter might be as follows, "It did not speak well of the power of God . . . if He needed a human government to prop Him up." (339)

Culture and Religion

I earlier posed the question, "What is a religion?" Is it what literal scripture says it is, what clergy and others espouse, how people self-identify, or what people actually do in the name of their god? No matter the answer(s), religion is a cultural phenomenon, steeped in all sorts of supernatural memes. It is part of the culture of all modern human societies. And it is perfectly natural, albeit with superstitious memes, as I've shown. Religions are cultural traditions. However, as we noted, humans are not the only animals to display cultural traditions. Do such animals ascribe their cultural traditions to a "higher power?" We have no way of knowing, of course, and most people would scoff at that idea. However, it seems just as ludicrous to ascribe religious behavior to a supernatural influence when the specifics of such behavior have evolved culturally in various ways over thousands of generations. Religion is a set of cultural practices, nothing more.

Religion Follows Culture, Not the Reverse

As a cultural phenomenon, religion is subject to change with the societies it inhabits. In fact, culture often has a more significant influence on religion than the reverse. Religious dogma or "law," like the concept of *law* in general, is only effective to the degree that most people are willing to follow it "voluntarily." That is, we behave consistent with cultural norms, including religious ones, because we are conditioned to do so, not because to do otherwise would lead to consequences in some hereafter. The notion that we behave as we do because a religion so prescribes it is just pure nonsense. Quite the opposite is true. We only label our behavior as consistent with moral or religious codes because that is what we've been told or conditioned to espouse.

As people become more and more aware of this phenomenon, and as competing incentives take hold, religion's grip on us sometimes weakens. Taking the view of only modern history, we can confidently say that, in general, culture shapes religious behavior more than religious dogma shapes culture. In that sense, religion becomes more and more democratic; a rather heretical statement, of course, for any religion, but in most parts of the world—though clearly not all—it is holding true. When religion manifests more reasonableness because of social influences, e.g., becoming more liberal/less fundamental, it gives up some dogma. When it does, it reveals itself as subscribing less to its putative theological bases, but rather as an institution serving its own needs for control of the flock and its financial resources. The hypocrisy becomes ever more manifest.

The most important trend in theistic religions has been its weakening or softening. What is preached from the pulpit and how people subscribe to sanctioned or prohibited behavior has changed rather dramatically over the last few centuries. Since the advent of the Abrahamic religions with Judaism, probably around the fourth century BCE (give or take), strict adherence to religious law has weakened in Western cultures to the point that it is barely recognizable in ordinary daily behavior. Only a few sects are identifiable by their garb; most people in the United States and Western

Europe do not attend religious ceremonies; and very conservative sects of religions tend to isolate themselves—likely as backlash to cultural influence on their religions, though they often term it *contamination*. Modern societies no longer sacrifice goats or children—unless, of course, you possibly consider the death of children due to the denial of life-saving medical care for some vaguely defined religious reason to be such a sacrifice (see also below).

Theistic religion, far from being an unerring work of a god, is constantly changing. Of course, this often leads to more liberal interpretations of scriptures and a relaxation of required religious behaviors. Clerics and other leaders feel free to promulgate "God's word" in whatever way they want to cherry-pick and interpret it for the congregations they lead. Thus, religions adapt dogma to their host cultures, not vice versa. To adapt is to survive. In essence, for theistic religion to maintain itself and grow—there's no better way to say it than this—they "make it up" as they go along.

History is the evidence. The 16th-century Reformation was associated with some notable changes in Christian religious traditions. But it was the Enlightenment and the growing spread of secular education to the masses in the 18th century, for some, and the 19th century, for others, that tempered some rigid Christian doctrines. People in large numbers began to openly question formal religious dogma, and secular culture, pragmatism, and even a growing respect for science gained more and more influence in people's lives. Traditional Christianity began to lose some of its appeal.

Modern

In recent years, many organized religions have adapted to modern culture, though some more quickly than others. Let's take a couple of modern examples. How many self-described Jews eat pork? Believe it or not, most Jews work on the Sabbath if they need to, and you'd be hard-pressed to find some who avoid TV or cars on Saturdays. Do they allow their children to play video games or baseball on the Sabbath? (By the way, the legend that the great Hall of Fame Pitcher Sandy Koufax didn't play on the Sabbath is untrue, although he did sit out on certain Jewish holidays.)

Catholicism arose, in effect, over the first two centuries CE. It has been one of the slowest mythologies to adapt to culture, but it too has never been immune. The message of the Reformation of the 16th century was a clear message to Catholics in this regard. In more modern times, it has gone through a number of revisions, and the behavior of self-identified Catholics is testament. Most Catholics rarely follow dictates of the Church "religiously."

Catholics are no longer burdened by having to eat fish on Fridays. How many Catholics use pharmaceutical birth control? Catholics do have sexual relations for purposes other than procreation. Self-identified Catholic women do have abortions. Remarried Catholics can now receive communion. Some Catholic dogma has been slower to extinguish or mold to social pressures or than others. It was not until 2006 that Pope Francis urged compassion for gay and lesbian humans, obviously spurred by the exploding social movement of the times. He also decreed that same-sex couples can be blessed, and transgender individuals and children of same-sex couples can be baptized. In a more recent and dramatic change, he was "considering" allowing priests to marry. He called the matter a "temporary prescription" (though he may have meant "proscription"). Where is the Bible on all this stuff? Depends on who you "listen to." All this is heretical to conservative Catholics, but adaptation does occur. As I mentioned, it only happens more slowly with Catholicism because, in a word, it's richer any other faith.

Islam, as we know, has been slow to come around, especially in most of the Middle East. That women have been able to shed the hijab in public and drive in Saudi Arabia in the last few years is probably not great cultural progress, but it is change. In Iran wearing a hijab in public is still compulsory. In Afghanistan, education is denied for girls after the age of six. Some cultures have gone another way, showing poor adaptation. In France, for example, it is illegal to wear the accoutrements of any religion in many public places. The notion that actual suppression of religious views is of any practical, moral, or long-term effect isn't realized there. This is politics. However, liberal Muslims in the West generally endorse values of women's rights, respect for LGBTQ individuals, interfaith marriage, and many other values of the cultures they live in. The vast majority of American Muslims endorse the idea of democracy, something about as far from Sharia law as you can get. (351)

Demographic

Religions also adapt by increasingly moving toward certain populations and geographies and as functions of socioeconomic or other demographic characteristics. It is a way to maintain membership (and money) while resisting some of the cultural pressures they get exposed to. First, critical thinking and intellectual sophistication tend to weaken blind faith in the supernatural (though there are clearly exceptions, which we'll address later). Many people of all stripes still believe in God. But therefore, organized theistic religion still skews toward the less-educated, though the relationship is more complicated than a simple trend. Second, religions tend to be driven by socioeconomic factors. In fact, the origins of religion tend to focus on the poor and those who may follow more "blindly." The "flock" has been an apt metaphor. It is used many times in the Bible to suggest that people are simple, naïve creatures who are in need of leadership and herding. As Christopher Hitchens observed, "Almost all religions from Buddhism to Islam feature a humble prophet or prince who comes to identify with the poor . . . It is hardly a surprise if religions choose to address themselves first to the majority who are poor and bewildered and uneducated." (221) And unfortunately, we cannot ignore "race." Some cultural movements toward religious worship were enabled by slavery and its modifications in the United States. A very unhappy historical example shouldn't be ignored. Religion was adapted to slavery in many parts of the American South and became a reinforcer of religion among black slaves. For example, some Lutherans who settled in Georgia in the eighteenth century were opposed to slavery. However, their position softened when there was a shortage of white laborers, and retaining slaves gave them the opportunity for some evangelizing. Lutherans and Baptists began using slaves, and the slaves were excused from their chores on Sundays to attend church. They were also excused from work if there were religious services or activities on weekdays. In fact, several bills were enacted in some of the colonies to formalize this strategy. (251) The incentives were obviously potent. Race became an adaptation to help various sects of Christianity grow in the early American Colonies.

Then there are women, who are half the human population. The case of how religion treats women requires some special comment here. Historically, organized religion's behavior toward women has been characterized by oppression and subjugation (and often much worse, as we've seen). Some faiths in some areas continue that trend. Even without violence, almost all theistic religions have historically provided for the dominance of males over females. This is no way to grow the flock. It has been slow, but religions are now adapting. Modern religions have increasingly recognized the equality of women and welcomed women into greater participation in religious activities. Many faiths now even welcome women into the clergy.

When Religion Doesn't Adapt

Nevertheless, when organized religion doesn't adapt to modernizing culture, major problems arise, such as the hatred and violence we've noted. But there are numerous other mundane, though fairly problematic, examples. Let's look at what happens when religion doesn't adapt well.

One prime example is in the area of medical care, especially in the United States. It's well known that Jehovah's Witnesses often refuse blood transfusions, believing that such would deny them eternal salvation. Countless people have died from the refusal of medical care, including some children, with parents making or supporting such decisions. We should note that it is difficult to obtain reliable statistics on this issue, though the list of credible anecdotes is impressive. There are currently a number of religious sects who refuse medical care for their gravely ill children. In fact, there have actually been a few studies on this, some showing that as many as 90% of children lost to illness would have survived if care had not been blocked for religious "reasons." (16) Parents even sometimes refuse to provide routine vaccinations, endangering not only their children but many others in their communities. There was a recent Muslim edict against polio vaccination in parts of the Middle East, primarily in Afghanistan and Pakistan, with the potential to stimulate an epidemic of a disease that was essentially eradicated in the rest of the world (vaccinations were apparently resumed in 2021). (518) In the United States in recent years there have been several preventable measles outbreaks due to parents eschewing vaccinations for "religious" reasons. I still can't find God's proscription of this in any scriptures, people's bizarre interpretations notwithstanding.

Let's revisit the worldwide pandemic of the COVID-19 virus in the early 2020s that we noted in Chapter 2. We saw hundreds of thousands (perhaps millions) of people needlessly die as the result of not getting incredibly safe and effective COVID-19 vaccinations. Some of these deaths were due to some generic antivaccination movements. There is a long history of such movements, beginning with Edward Jenner's vaccine for smallpox in 1796, as we noted in Chapter 2. But an antivaccination position has gained traction in the last few decades, sometimes for seemingly political reasons that are hard to reasonably fathom. However, many were due to vague religious objections. A number of Christian clerics even preached against getting the vaccinations (though they were not the only ones), resulting in the death of many parishioners and a few pastors themselves. Many religionists who refused vaccinations against this highly contagious virus, even sickened and killed many others they exposed themselves to (despite knowing the risks to others. That seems to ignore the Biblical dictum of "love your neighbor as yourself." (Leviticus 19:18)) When some of those ministers, in fact, died of COVID-19, their parishioners characteristically offered that "he's in a better place now, with Jesus." Religions failed to adapt, and there were hundreds of thousands of preventable deaths. (I should note that the Pope and many other religious leaders *did* urge their followers to get vaccinated.)

An interesting example of poor adaptation comes from the tradition of Christian Science. Christian Science acolytes refuse most medical treatment. The child abuse of refusing modern medical treatment for one's child cannot be overstated. Such a fate even actually befell Einstein's grandson. One of Einstein's sons, Hans Albert (a quite educated man himself; he was an engineer) had married a Christian Scientist. In 1939 their son, Klaus, contracted diphtheria at age 6. (235) A number of diphtheria treatments had existed for many years—including a fairly effective antitoxin, and a vaccine that had been promulgated for over a decade at that time. But Klaus died without the benefit of any such care. Christian "Scientists" prefer to believe in the "God works in mysterious ways" idea rather than the germ theory of disease. Intelligence and love of one's children were no match for the Christian Science superstitions.

Scientologists are indeed a special case. For example, they tend to shun care by psychiatrists and, in fact, are often openly hostile toward the profession. Ignoring even the history of L. Ron Hubbard and Scientology as a religion, it's likely that Hubbard himself was suffering from paranoid schizophrenia (perhaps underlying many of his delusional concepts of the religion) and was less than satisfied with his many years of psychiatric treatment. It may have prompted the hostility toward the profession. However, current Scientology practitioners think psychiatrists are an evil force and blame them for all sorts of ills of the modern world, including racism, genocide, terrorism, school shootings, wars, and even the Holocaust and 9/11.

But cultural adaptation in some quarters leads to backlash in others. When threatened, especially with loss of influence and membership, religion often intensifies its efforts. A prime example is the present evangelical movement to suppress women's rights to abortion and reproductive freedom in America. It should be noted that nowhere in any Biblical text is there any prohibition of abortion, birth control, in vitro fertilization, or any other matters of reproductive health. (As with many other cultural developments, the men who wrote scriptures two or three millennia ago believed that human culture would never change.) The bizarre lexical gymnastics and citation of "cherry-picked" passages leading American religious adherents to *infer* that the Bible prohibits abortion, for example, is just another of their "interpretations." In reality, it is just another effort to control and subjugate women. Yes, organized religion has always subjugated women, and any fabrication based on use of Judeo-Christian Bibles is just another contrived method. We should note, however, that there are many countries around the world where abortion, for example, is fully legal, including many where Christian cultures are predominant. This includes, for example, highly Catholic Ireland, where the proscription was actually reversed. The failure of Christianity in America to adapt may be costing it membership.

Conclusion

How did religions keep changing? How did some change and adapt faster than others? How did all this happen? Did God keep changing His mind? Theistic religion is a cultural phenomenon, with all the various memes changing with the transformations in its host cultures. For theistic religion to survive over the long run, it must adapt to the cultures in which it exists. To fail in this regard is to lose members. Of course, some sects are slower to come around than others. And some are actively resistant. When the latter occurs, we see tension, apostasy, social conflict, political punishment, and violence.

Pathology

Let's take a step back to consider what the so-called ideas about God and religion really are. The truly bizarre nature of religious tenets and beliefs must be acknowledged. People believe all sorts of weird things. Blind faith can lead people to believe in almost anything—anything at all

—however at odds with reality. Beliefs can range from the nonsensical to the delusional, often couched in memes heard in a church.

Belief in a personal God, with all of its associated superstitions, is one of those things. Such belief, and all of its superstitious rituals, is in fact merely *pathological*. It is the basic *disease* of theistic religion. You might consider such words a bit harsh. Nevertheless, religion is merely a large set of aberrant behaviors which grow out of a disconnection from reality. It is an aberration of thought and behavior maintained by superstitions and other behavioral dynamics. Deeply religious people, whatever that means, carry a delusion that may serve a personal function, but worshiping a fantasy is *delusional* nonetheless.

In general, human beings sometimes do have bizarre ideas, along with delusional thinking and hallucinations. However, we judge the sanity of the report by the circumstances of what's reported and observed. We've reviewed just how common hallucinations are. Intense experiences or abnormal psychological states or intense beliefs sometimes enable all sorts of delusions, hallucinations, and other extremely aberrant behavior. If someone hallucinates the image of a dead spouse late at night, we might pass on this as understandable. If they have visions or hallucinations because of brain damage or the influence of drugs, that is comprehensible. If they hallucinate because of lack of sleep or vacuous environment, that's explainable.

However, there are other situations in which delusional reports are more clinically problematic. If you hear Marie Antoinette talking to you, that might be a problem. If you sincerely believe that armies of zombies are after you, we'd question your sanity. If Martians are seriously controlling your thoughts, you may be in need of hospitalization or heavy medications. Under ordinary circumstances, all these behaviors would be considered pathological. However, when similar types of behaviors occur in a religious context, they might not even raise many eyebrows. Just change the names and circumstances a bit.

We usually don't infer anything aberrant about those who merely endorse the "idea" of God. We also don't usually conclude anything pathological about people who say they talk *to* God (also known as prayer), though regularly praying to the wrong god, e.g., Odin or Zeus, would be seen as pretty odd in modern societies, even though they are as equally evidenced supernatural figures as the Judeo-Christian one. Cultural context matters.

Those who claim to have directly communicated with a supernatural being are viewed with skepticism, including even by most religious leaders, even though such is the basis of the guidance given by some clergy. How often do we hear a believer talk about a "relationship" with God? This is meaningless, of course, as it is just an undefined idea. If they report that "God has inspired" them, we brush this off as a metaphor or merely how an individual cites a religious belief as motivation. But if they report actual conversations, they are usually seen as having some sort of psychological pathology. When you verbalize that God wants you to do something, for example, we usually hear this as metaphorical. Politicians often do this. If such behavior is taken literally, it is either doubted outright or the individual is seen as somewhat sociopathic.

Intensely repeating it, however, will lead to some clinical inferences; and the degree of specificity of reports and any associated behavior leads to how we view religious behaviors. If you see the image of the Virgin Mary in your window one morning, okay. Even seeing Jesus in a piece of toast usually doesn't provoke much of a reaction from others. For those who reliably profess that God regularly speaks to them as a reason for their actions, we might use clinical words indicative of some pathology. Persons heavily enmeshed in certain religious cultures accept such reports as valid. However, others may characterize such reported experiences as "crazy." When people hear God talking to them, we may merely brush this off. But where actual aberrant behavior accompanies such assertions, we might infer something pathological, depending on the behavior. If a person believes that God watches her every move, we might dismiss the remark. It might be merely conceptual. But if she never bathes or wear clothes because God told her not to, or if God told her to kill her children (yes, that has happened), we might call that insane.

There are too many instances of people doing things in the name of God's inspiration or actual direction to detail here. One interesting example of bizarre divine inspiration is what has been called the "Jerusalem Syndrome." This is an acute psychotic state which comes about in some people who make a pilgrimage to Jerusalem, and it has even occurred in some tourists. They have been known to experience hallucinations, bizarre religious ideas, and significant delusions, sometimes including a belief that they *are* Biblical figures. Interestingly, Christian tourists generally embody New Testament figures, and Jewish tourists usually identify with Old Testament ones. The condition can appear, and often does, in people with no psychiatric history. It does seem to usually resolve spontaneously or with a minimum of treatment. Nevertheless, in the throes of the psychosis, believers do feel that they've been inspired by God and can feel that they *are* Samson or even Jesus.

Inventing explanations in the form of supernatural influences is also a related type of pathology. Laying on of hands, for example, e.g., in the supposed cure of a disease, might be explained by some combination of placebo effects, by a coincidence of patient effort and misdiagnosis, and/or by a sudden hypnotic effect in a "hysterical" sufferer. However, if the "healer" claims to be channeling God as the causal mechanism, many observers might "buy it," noting that "God did it." (Some "healers," of course, have come to admit their own charlatanism.) If a person recovers from an illness by citing that a prayer was answered; we might dismiss this, but when the recovered patient believes in the divine power as the intervention, it is a clear delusion. Some may not like the word, but that *is* pathology. (I'll talk more about intercessory prayer in Chapter 10.)

Group Delusions

Organized religions draw people into groups for mutual support, sharing, personal validation, and reinforcement of superstitions. That is one reason they endure. Belief in some specific god is a type of behavior which is responsible for the various religious traditions. It is a cultural, pathological problem. Your faith or assumption that there is a god is validated only by the opinion of others. The pathology, as Richard Dawkins' described in *The God Delusion*, gives rise to the various memes of religion shared and passed down by cultures or groups. In fairness, the concept has been around for some time, and it was well-described by Sigmund Freud in 1927 in *The Future of an Illusion,* as I noted above.

Our tendency for group delusion runs deep. We are social creatures and have a clear tendency to believe what others believe, regardless of objective or empirical truth. The growing epidemic of conspiracy theories, for example, is ample evidence. The United States government is hiding evidence of UFOs in Roswell, New Mexico. The murder of children at Sandy Hook Elementary School was a staged event. COVID-19 vaccines are used to inject a microchip into your arm so the government can keep track of you. The list is endless, and millions of humans actually believe these kinds of things. Given widespread willful ignorance and gullibility, it is incredibly easy to see, for example, how the myth of a human being crucified and arising from his tomb was shared and promulgated.

Theistic concepts, in the end, are no more than conspiracy theories. And when they are widely shared, they become group delusions. When conspiracy theories grow out of theistic ideas, they become religious cultural memes, and the notion of them as delusions tends to be overly minimized. Nevertheless, they constitute the pathology of the group that shares them. Neuroscientist Harris put it well, "Clearly there is sanity in numbers. . . . The danger of religious faith is that it allows otherwise normal human beings to reap the fruits of madness and consider them *holy*." (206) That is the sum and substance of organized religion.

Behavior

Again, what is religion? To do something religiously is to do it over and over again in a consistent manner. That's the definition of religion. Einstein, as most know, defined insanity as doing the same thing over and over again and expecting a different result. That's religion. It continues to merely enrich itself, suppress people, and engender discrimination and hatred. If that's not insanity, I don't know what is. It is a form of human *pathology*, an irrational deviance of behavior. Look at a few examples. First, a thought experiment—take the God meme out of the picture, so to speak. We see people worshiping, praying, and engaged in other traditions that characterize organized religion. We see people sitting in a building, all bowing their heads, chanting, sometimes swaying back and forth. They report hallucinations and delusions about things that are physically impossible; being intensely fearful of ideas, going home, kneeling, talking to someone who isn't there; and many other bizarre behaviors. If that's not pathological, I don't know what is. Without a God meme, some describe such group behaviors as a "cult," but of course, all religions are cults. The social acceptability of all this is the only thing that limits our impressions of outright pathology. And take it a step further. If someone gets the idea that it's okay to abuse someone or manipulate others for one's own benefit, and perhaps does it, we might often use the labels of "psychopathic" or "sociopathic" for such behavior. Theft or conspiring to steal fall into this category. However, taking money from parishioners in the name of a superstition and using it to enhance your own lifestyle generally raises few of their eyebrows. If that doesn't meet the criteria, I don't know what does. Individual pathologies, group delusion, cult behavior, and psychopathic conduct are more than enough evidence to clearly characterize theistic religion as *pathological*.

Conclusion

Religion and the bizarre notion of gods grew naturally with human evolution and the cultural development of our species. These are human behaviors, not a function of something supernatural. It grew naturally out of primitive superstitions, and humans continue to promote those superstitions to the present day.

Organized religion might be the greatest "con job" humanity has ever pulled on itself, and the leaders who promulgate it—clergy—might be the most imperious in the history of human culture, despite any veneers of humility they may manifest. For those "true believers," when you are *certain* about your god and have the "real truth," you feel you have the right to control and suppress others. Such has been the throughline from the divine right of kings to the various modern organized religions. And humankind continues to pay the price—by allowing and promoting some of the greatest of human tragedies in discrimination, hatred, war, and genocide—all curious, dreadful, and vile actions for an allegedly loving omnipotent God. We can only hope that more and more people grow tired of the "God's plan" and "mysterious ways" excuses.

Yes, a belief in a god, albeit a mythical figure, can and often does make people feel better and even reduces some serious suffering. Again, there are other ways that such things are accomplished. And we must weigh such palliation against the destructiveness of organized religion, which faith has reinforced throughout history. Further, that religious ritual may ease suffering for some is no argument for the existence of a god. It is the ritual that is comforting, not the superstition it is based on. Let me sum up with one of the best perspectives on this, from George Bernard Shaw, "The fact that a believer is happier than a skeptic is no more to the point than the fact that a drunken man is happier than a sober one."

8
PHILOSOPHY IS DEAD: BUT SOME LAST WORDS

Philosophy is dead. —Stephen Hawking

Extraordinary claims require extraordinary evidence. —Carl Sagan

And always in mind is that big question, "How come such a strange thing as existence?" And "How come the quantum?" And, in connection with the quantum. "How do we get the impression that there is one world out of the records of many observer-participators?" If these questions verge on philosophy, then perhaps we can adopt as motto, "Philosophy is too important to be left to the philosophers." —Physicist John Wheeler

Philosophy is about "big questions"—What is reality? Is there a god? How do we know? If It exists, what does It want from us? Is there a soul? What is moral? These are questions which have been the subject of the discipline of philosophy through the ages. Of course, there are no ultimate or verifiable answers to such questions. They are the sum and substance of speculation and opinion—of philosophy. This chapter is about the "philosophical" issues and arguments on the question of a god—not a small undertaking, but we'll hit some of the highlights.

Philosophy is very good at asking provocative questions, but very poor at providing answers. The answers depend only on who you ask. As Hawking observed, "Traditionally, these are questions for philosophy, but philosophy is dead. Philosophy has not kept up with modern developments in science, particularly physics." (214) Philosophy is not dead, literally, of course. Hawking was not in the business of trying to suppress dialogue about ontology, morality, etc. He was merely expressing the view that philosophy doesn't actually explain anything and that physics was a much better way of understanding reality. Moreover, as he suggests, we have no evidence—none at all—that God or any god exists or has ever existed. Carl Sagan was right—the God idea is an extraordinary one, and any evidence for it continues to be missing.

What is philosophy? The word comes from ancient languages, generally meaning a "love of wisdom." But today we consider wisdom to derive from learning about the world and having a sophisticated understanding of it. Ancient and Enlightenment periods had "natural philosophers," who might have been considered scientists of a sort. Philosophy now, as a discipline, has devolved into purely armchair theorizing (although one might argue it has always been that) because one can do it without any reliance on data or observation (noting that the word "theorizing" is obviously somewhat misused in this context).

Philosophy as a discipline fails to offer anything heuristic about the nature of reality. Philosophical ideas are only speculations and opinions which can never be tested or evaluated. Nevertheless, philosophical treatises on all manner of subjects have been developed and handed down for thousands of years, and presumably there were opinions and lore even before that, shared orally. In the main, however, the basic questions philosophy considers have not changed over the course of history—what is reality (metaphysics), how should people behave (morality and ethics), the study of knowledge (epistemology), and certain other questions about human behavior (e.g., political philosophy). Answers to such questions are merely opinions and haven't changed appreciably since people started to consider them.

We can consider that philosophy is associated with asking interesting and thought-provoking questions; but, as a discipline, it has nothing to offer in understanding ourselves or the nature of reality. Philosopher Daniel Dennett himself summed it up pretty well when he said, "I think whenever you don't know whether you're asking the right questions, you're doing philosophy. In any field that if you've got a perfectly good question, you know how to answer, and you're doing physics or chemistry or biology or history, that's not philosophy. But if you suddenly wonder, am I wasting my time on this? Are there better questions to be asking? Then you're doing philosophy." (525)

We're always asking questions about ourselves and our universe. I'll use the term "universe" here, realizing we could also take it to mean multiverse. Citing both of these would just be awkward now. We'll also equate this with the "world," which is how such ideas have been alluded to historically. But, of course, we don't know if they're the "good" questions to ask at any particular time. We don't know until we look further. When we do, that's science. Philosophy only asks questions for which there *can be no* answers—by definition. As such, philosophy is probably not a "waste of time," but it has no value in understanding our world. Perhaps we must redefine philosophy, but it would be to the chagrin of most philosophers to do this. Philosophy may be trying to redefine itself. There are works described as "experimental philosophy." These derive from asking people how they view such concepts as free will, morality, ethics, etc., and how those might change under different conditions or if the world were somehow different. But such is essentially social psychology masquerading as some sort of philosophy. And one final intent of philosophy: Philosophers and theologians have been trying to prove the existence of God for millennia. It hasn't happened, but in this chapter, we'll look at some of the critical arguments and "proofs" which have been offered.

Religion and Science

In order to consider any philosophical arguments about the god question, let's look again at comparing science and religion, which I touched on in previous chapters. Are these disciplines compatible? Are they equally valid ways of knowing about the world? Is it reasonable to hold both perspectives at once? Can we even address the god question with mundane thinking, or logic, or philosophical speculation, or scientific reasoning? We should address these questions before moving on to the theological arguments.

One might first ask why a god would give us the ability to diagnose his flawed work (perhaps a problem for an omnipotent being) and to establish scientific theories about the universe which eschew any need for his stimulus, guidance, or intervention. That seems a rather bizarre game. Again, the "god works in mysterious ways" argument to this is nonsensical. It's just a conversation ender.

Religion and science have had their conflicts. Religions always have a vested interest in maintaining their myths, and religion has traditionally been in the business of suppressing science, a pattern which persists to this day. Probably the most famous instance occurred in the well-known Galileo scenario. The Earth was seen as the center of all existence at that time. In 1633, Pope Urban VIII took exception to Galileo's work, which demonstrated the sun was the center of the solar system. Galileo pled guilty to the charge of suspicion of heresy and was sentenced to house arrest, where he remained until his death in 1642 (He was allowed, however, to continue his work, though he eventually went blind from his use of his powerful telescopes pointed at the sun). At least he wasn't burned at the stake by the Roman Inquisition, as Friar Giordano Bruno had been a few years earlier (in 1600), for also advocating the same heliocentric theory. (Interestingly, Bruno had not only promoted the idea that we weren't the center of the universe, but that there were also planetary systems surrounding other stars, an incredibly prescient perspective in the sixteenth century.)

Galileo's conviction for heresy was eventually reversed, but only when he recanted, though only to avoid more severe punishment. The Church finally essentially admitted that Galileo was right in the eighteenth century, in 1758, when it dropped the prohibition on books endorsing the heliocentric theory. But it wasn't actually until 1992 that the Catholic Church actually got around to apologizing and forgiving Galileo for developing a theory that was clearly factual, over three centuries after its discovery, and it absolved Galileo of the "sins" of his scientific discovery. It's not perfectly clear why it took so long for Galileo to achieve forgiveness; perhaps the Church had a harder time ignoring science in the twentieth century, or perhaps nobody really cared to address it. But the

Vatican withdrew its condemnation only after 13 years of "study" of the problem by the Pontifical Academy of Sciences (now there's a pretty interesting concept).[1]

The Church had relented. Today we no longer believe that the Earth is flat or that it doesn't revolve around the Sun (at least most of us). We trust science for those kinds of inferences. And we've extended our tools of observation (and mathematics) to be even better explain the phenomena we see. This has led to an increasing respect for science and a diminishing respect for religious explanations for the physical world. But the tension does remain and is growing in some quarters.

Are Religion and Science Compatible?

Some believe that one can simultaneously support both a theistic and also a scientific view of the world. Is that possible? This is a very, very old problem, and efforts to resolve the conflicts between these two human endeavors have never yielded any satisfying results. So, let's examine the problem in a little more detail.

The most important question for our purposes here is: Are religion and science incompatible? Science is concerned with objective facts. Religion is not. Religion is concerned with myth and superstition. Science is not. They are inherently incompatible. Many have tried to resolve this dilemma or construct some common ground, but it proves to be a futile exercise. Einstein, Wolfgang Pauli, Niels Bohr, and many other towering figures in the history of science concur—there is an irreconcilable conflict between science and religion. Bertrand Russell, whom I've alluded to a few times, put it best. "Science tells us what we can know, but what we can know is little, and if we forget how much we cannot know, we become insensitive to many things of great importance. Theology, on the other hand, induces a dogmatic belief that we have knowledge where in fact we have ignorance." (435)

Religious apologists often quote Albert Einstein to suggest that science and religion are, in fact, compatible. Einstein did famously say that, "Science without religion is lame; religion without science is blind." (152) However, he was undoubtedly relying on his own view of the "inspiration" and beauty of the universe in expressing his obviously poetic views on the matter. He went further to qualify his meaning to indicate, "To this there also belongs the faith in the possibility that the regulations valid for the world of existence are rational, that is, comprehensible to reason." (152) He famously said that he did not believe in a god "who concerns himself with the fates and actions of human beings." He did not believe in the compatibility of religion and science.

The conflict between science and religion is not trivial. Although religious identification has been waning somewhat (see Chapter 11), there is a troublesome trend to attempt suppression of scientific explanations for the physical world and reinstitute religious ones in the education of children. Such a paradigm, of course, is hypocritical. Rejecting evolution, for instance, while supporting the teaching of photosynthesis in biology classes makes no sense. The worship of religious dogma and rejection of science often takes on a ludicrous quality. Still, many Americans don't believe that humans descended from more primitive creatures. If you ask Americans whether they believe in the statement, "Human beings, as we know them today, developed from earlier species of animals," about half will agree. This is up only slightly from a couple of decades ago and is likely a function of increased education, but it's also a far cry from the 80% in the rest of the industrialized world. (346)

Let's revisit the all-too classic example I mentioned in Chapter 2. Many Christians assert the age of the earth to be about 6,000 years, despite the overwhelming evidence that it's approximately 4.5 billion years. There are exceptions, of course. Many years ago, I interviewed some oil field workers who were members of a fundamentalist congregation in west Texas, where the 6,000-year-old Earth was being preached by the local pastor. They were well aware that the oil came from fossils millions of years old. They merely sloughed off the bizarre inconsistency and just went about their lives. Many religious zealots don't.

The case that religion and science are inherently in conflict has been reviewed in detail by biologist Jerry Coyne in his book *Faith vs. Fact: Why Science and Religion are Incompatible*. (100) As he notes, science operates on the basis of empirical evidence, experiments, and the consensus they enable. Religion operates on the basis of faith, i.e., belief which can't be validated or even tested. We must look at science and religion as inherently incompatible.

There are other views, of course. Many have tried to resolve the differences or merely ignore the inherent conflict between the two, attempting some independent validity for both. Paleontologist and evolutionary biologist Stephen J. Gould describes a model he termed "Nonoverlapping Magisteria." (188) He asserts that science and religion could not comment on each other, but each has its own and separate domain. He had initially published the model for this in 1997, noting,

> The net of science covers the empirical universe: what is it made of (fact) and why does it work this way (theory). The net of religion extends over questions of moral meaning and value. These two magisteria do not overlap, nor do they encompass all inquiry (consider, for starters, the magisterium of art and the meaning of beauty). To cite the arch clichés, we [science] get the age of rocks, and religion retains the rock of ages; we study how the heavens go, and they determine how to go to heaven. (187)

This is a popular conception. We don't mix science and religion—or we ought not to—an increasing problem in children's education. Nevertheless, Gould's language, although valid on its face, misses an important point. Religion obviously cannot inform science; but science *does* have a great deal to say about religion. As I've reviewed, this includes various subjects, including description of how religion and religious behavior came into being and are maintained. I showed how religion likely arose in humans, and I'll detail how religiosity is maintained in Chapter 10. These are scientific pursuits *about* religion. Of course, nothing about science can be explained by religion.

Comparison

Religion, especially in its various fundamentalist forms, is opposed to science; and those who practice such are opposed to the findings of science. Theistic religion and science are inherently and specifically incompatible. It is not possible to reconcile these two things, even though some people have attempted to do this throughout history (You'll recall, many of the scientists of the Enlightenment were religiously trained, and many were members of a clergy.). It takes such a level of cognitive gymnastics to do this that the tenets of theistic religion are destroyed. Or as a result, science and the findings of various sciences tend to be rejected outright by religious people.

Science and religion are inherently and specifically in conflict. Some have actually tried to resolve this, but such efforts are transparently forced and absurd. One might say that biological evolution, for example, one of the most well-evidenced theories in all of scientific inquiry, is God's will. You might believe that DNA mutations, the biochemical underpinning of evolutionary change, for example, are caused by certain knowable events, but to invoke a supernatural force for such strains the mind. One could offer that the Biblical six-day creation of the world is only a metaphor, and that gravity might have been instrumental in the process, but you might still believe that God was the responsible party. We often see this kind of reasoning. (One continues to wonder why a god would take billions of years to accrete the earth when such a supernatural power could do it rather quickly.) Science stands on its own: the knowledge which it gives us does not depend on there being a supernatural individual to enable it. The "gap," in fact, continues to close; but science and religion just continue to be incompatible ways of thinking.

Further, the notion of gods, or the supernatural in general, are "non-falsifiable" ideas—by definition. Unlike science, which draws conclusions from facts, theistic religions and the notion of God—in whatever form—are superstitions in which "facts" are invented to fit the conclusion. *Religion does not draw conclusions from facts; it starts with the conclusions and offers fabricated explanations (albeit poor ones) to fit them.* If you can't explain evil, for example, you might say that "God works in mysterious ways." If you experience difficult situations, hardships, tragedies, etc., you might say that "God is testing me or testing my faith." If you cannot explain something, you might offer that "it is God's plan." As we saw in the last chapter, one might continue to wonder about those dying of curable diseases who refuse effective medical care that science has given us. We saw this repeatedly in the recent COVID-19 epidemic. People refuse vaccination and treatment, including lifesaving treatment, often invoking "God's will" as the answer. The explanation is always God. You can't disprove that; it's just another sad conversation ender.

Science, as opposed to religion, works on the basis of "disprovability." That is, scientific facts and theories are always subject to being falsified. The idea of a god cannot be falsified or disproved, whether it is the god Yahweh, Christ, or any of the many Greek and Roman gods.

Hypocrisy

Theists claim that the hand of God continues to exert influence on everything (why else would one pray?), despite endorsing scientific concepts and their resultant technologies. Most of these folks have no problem accepting and benefiting from the products of science, be they medicine, chemistry, engineering, computing, etc. They merely reject science out of hand (i.e., with no knowledge whatsoever) when it comes to fields such as cosmology and much scientific knowledge which affects their own lives—psychology, climate science, etc.

Religious views of reality all depend on a supernatural force or being to initiate or direct events. Science makes no such assumption. Attempts to resolve some coherence between science and religion merely ignore this. Of course, there are those who reject scientific explanations altogether or only employ them selectively, which is bizarre. Some people believe that bacteria can make you sick, but they don't believe that antibiotics can help you get well; it would only be the hand of God that could make you recover (often with drastic consequences, as we've seen). Some do believe that the Earth was created 6,000 years ago, as I've said, but they have no problem driving cars powered by fuels scientifically distilled from fossils millions of years old.

Scientists

Can you be religious, i.e., a theist, and still endorse the principles of science? This was done, as I noted, throughout much of history. Not logically, of course; and *scientists* who believe in a god—and there are small and dwindling numbers of them (see Chapter 10)—admit that their beliefs are outside the bounds of logic. For some scientists, an ingrained (or learned) respect for science renders the idea of God unpalatable, though, curiously, some professional scientists are in fact "believers." And what about the doubting scientist? If God enabled humans, at least some humans, to believe in what derives from evidence, i.e., science, why would He give them the capacities to doubt or disprove His existence? Scientists would seem to suffer from St. Augustine's problem of the "disease of curiosity," or Martin Luther's caution that "Reason is the Devil's harlot." William James again summarizes the matter well:

> Science . . . has ended [the debate or comparison] by utterly repudiating the personal point of view. She catalogues her elements and records her laws indifferent as to what purpose may be shown forth by them, and constructs her theories quite careless of their bearing on human anxieties and fates. Though the scientist may individually nourish a religion, and be a theist in his irresponsible hours, the days are over when it could be said that for Science herself the heavens declare the glory of God and the firmament showeth his handiwork. Our solar system, with its harmonies, is seen now as but one passing case of a certain sort of moving equilibrium in the heavens, realized by a local accident in an appalling wilderness of worlds where no life can exist. In a span of time which as a cosmic interval will count but as an hour, it will have ceased to be. (241)

Authority

Now let's consider the philosophical arguments for the existence of God. This would be a monumental undertaking, and countless books have been written about this, but I'll hit just the highlights. Let's dispense with the easy ones first.

First, there is the argument from authority. People tend to believe all sorts of things that they cannot observe—because they are told to do so or they obtain some authoritative information about it. This is a complicated subject when it comes to human behavior, and I'll not address it fully here but return to it in Chapter 10, on learning and conditioning. For now, I'll observe that there is a kind of cultural hegemony that pervades a population on the God issue. In every culture there are clergy and scriptures telling people what to believe. The Bible told me. The Bible proves the existence of God. We've seen how ludicrous that argument is in Chapter 7. All scriptures were written by mortal humans, with their own flaws, biases, and agendas. Bibles and other holy documents are some of the most often cited sources of "evidence" for God. But as we've seen, these documents are merely human creations.

The priest told me—or rabbi, pastor, etc.—what is true. Or the Pope, often considered *the* authority by Catholics. Again, an authority figure, but not an authority on anything. My parents told me. Many people believe only what authorities tell them. And the authorities that children tend to grow up with in religious households are parents, Bibles, and clergy. These are pivotal enemies of truth about how the world works and generally lead to a distrust of science and critical thinking. There is no cogent argument from authority, whether it's a person or a "sacred" book.

Asserting something doesn't make it so. How many times have we heard, "The end of the world is coming" or "God will take care of you." Faith, if you will, is left with no legs to stand on.

Design

We covered some of the nonsense of the "argument from design" or Creationism in Chapter 4, but it deserves a bit more scrutiny here. First, there is the "God of the gaps" argument offered as "evidence" that God exists. The argument from design, of course, is based largely on the notion that science has not explained everything that calls out for explanation, particularly all of life and its evolution to present forms. Theists assert that evolution itself and the complexity of life that we see haven't been explained to their satisfaction. As science proceeds, the "god of the gaps" becomes less and less relevant, of course, to the point now of being a totally specious argument.

Most conservative religious sects outright deny the "theorem of evolution," as Dawkins puts it. Creationist perspectives are ludicrous, of course, and have been denounced not only by science but by more liberal religious groups. There is considerable effort is some quarters to meld the notion of a supernatural, designing God with the theory of evolution, which is inherently an incompatible notion. Such an interaction strains the mind.

Nevertheless, attempts continue. In 2014, Pope Francis, for example, said, "The evolution in nature is not opposed to the notion of Creation, because evolution presupposes the creation of beings that evolve . . . He created beings and left them to develop according to the internal laws that He gave each one, so that they would develop, and reach their fullness." (555) What? Of course, the notion that God "created beings" is specifically incompatible with scientific fact, since "beings" clearly evolved from more primitive chemistry. And of course, He made many, many mistakes, as we saw in Chapter 4. He even admitted His mistakes, and He had to start over. He had to save a few from the Flood with which He wiped out most living things, obviously not reaching "their fullness." If God did all this, He certainly was pretty incompetent at it. As Russell noted,

> When you come to look into this argument from design, it is a most astonishing thing that people can believe that this world, with all the things that are in it, with all its defects, should be the best that omnipotence and omniscience have been able to produce in millions of years. (435)

Physicist Victor Stenger has eloquently addressed this issue in another way, referring especially to the logic of the argument:

> If God exists, he must appear somewhere within the gaps or errors of scientific models. Indeed, the 'God of the gaps' has

> long been a common argument for God. Science does not explain everything, so there is always room for other explanations and the believer is easily convinced that the explanation is God. However, the God of the gaps argument by itself fails, at least as a scientific argument, unless the phenomenon in question is not only currently scientifically inexplicable but *can be shown to forever defy natural description.*[emphasis added] God can only show up by proving to be necessary, with science equally proven to be incapable of providing a plausible account of the phenomenon based on natural or material processes alone. (506)

These arguments do relate to the theists' conceptions of "intelligent design" and their perception of "irreducible complexity." Much of life, especially human beings, is far too complex for science to ever understand. As we've seen, this argument is repeatedly shown as bunk. Theists also think that life itself, and the rest of the universe, is too complicated a problem for science, but God knows since He created it. We've addressed the issue of life and the so-called "fine-tuning" of the universe previously. Obviously, we don't know the entire history of the universe. There is no reason to believe that "life" must be a function of the precisely "finely tuned" values that we currently observe. Theologians are fond of citing the assumption that the values of the fundamental constants in our universe are extremely improbable. Of course, this is also an assumption, and "improbable" is just a reflection of our own ignorance. In fact, the probable notion of a multiverse, with an infinite number of realities, renders the current dynamic quite probable, even guaranteed. The theologian offers the criticism that no multiverse idea has, as yet, been proven; and I emphasize *as yet*. But you also see the hypocrisy—no proof of a god either.

The intelligent design and "gap" notions always fail miserably. We need to acknowledge that there will always be things we do not know, and that is a hard pill for religious apologists to swallow. Finally, we should realize that the whole concept of "design" is a human inference, i.e., the organization of matter to enable a specific outcome or purpose. It is just as valid as Voltaire's sarcastic assertion on the matter that, "Obviously the nose was designed to be such as to fit spectacles." (435) The attribution of "design" to some supernatural being is but one more theological fantasy.

First Cause

Let's tackle the one argument favored by most theologians—that the universe itself was brought into being by God. This is probably the most often used of the philosophical arguments which are invoked for the putative existence of a supernatural god, an ultimate gap "explanation," if you will.

I briefly alluded to this in relation to physics and cosmology in Chapter

3. But let's look a little more closely at this supposed philosophical argument, generally known as the "first cause" or "cosmological argument." It asserts that everything that exists has a cause. That is, every event is contingent on some other event; thus, it's also known as the "argument from contingency."

Though it has been put forth many times throughout history, it was the thirteenth-century Dominican Catholic priest, St. Thomas Aquinas, who most famously offered it as the so-called proof of the existence of God. As a putative logical argument, it goes something like this:

All things existing have a cause
The world exists and thus must have a cause
Therefore, the cause is God—the only "thing" that is "uncaused"

Of course, as a "logical" argument, this falls flat on its face. It's the most tortured, fantastical syllogism ever proposed, unless of course by "God" you mean "I have absolutely no idea." The "first cause" argument is illogical. It merely assumes the conclusion. The same type of argument for God was proposed a few hundred years later by Leibniz. He argued in his "principle of sufficient reason" that "No fact can be real or existent, no statement true, unless there be a sufficient reason why it is so and not otherwise." (301) Since the world existed, he assumed that God created it, just like Aquinas did, reprising the contingency argument, "For one Being, dominating the universe, not only rules the world but he creates and fashions it, is superior to the world, and, so to speak, extramundane, and by this very fact is the ultimate reason of things." (301) Such has been the characteristic argument of theists throughout history.

However, if we objectively look at the broader context, there are only three possible scenarios to the argument *for* God. (1) God "preceded" the universe, (2) both God and the universe are eternal, or (3) God is a "part" of the universe or a "character" of or in the universe. I'll consider each of these in what follows.

Was There a Cause?

The basis for Aquinas's and Leibniz's argument is that since God exists (an assumption), and since nothing can exist without a cause, then God is the cause of Himself. This has been an acceptable argument to theologians, but, as you can see, it strains all logic, credulity, or linguistic sense.

There is another central question vis-à-vis Aquinas' argument: If there was no beginning, if the universe is eternal (going forward and backward), the cosmological God argument falls apart. As we reviewed in Chapter 3, there need not have been a beginning to our universe. It is only our human need for everything to have a specific cause that leads us there. Human common sense is ill-suited to provide a reasonable perspective on this problem. Our experience, which is tantamount to Newtonian mechanics, suggests that each observable event has a prior cause. We cannot mentally conceive of an occurrence which was "uncaused," if you will. Further, we showed in Chapter 3 that the simple "cause and effect" reasoning is not necessarily valid when it comes to the ultimate nature of our reality or even our universe.

Again, the argument offered by physicist Krauss comes into play. "What is the difference between arguing in favor of an eternally existing creator versus an eternally existing universe without one?" (284) Many have tried to refute this logic—unsuccessfully. Philosopher Peter Williams openly criticized Krauss' argument from eternity. (547) He argued that:

> 1) Everything that exists has an explanation of its existence, either in the necessity of its own nature or in an external cause.
> 2) The universe exists.
> 3) Therefore, the universe has an explanation of its existence.
> 4) If the universe has an explanation of its existence, that explanation is God.
> 5) Therefore, the explanation of the universe's existence is God.

Same problem. A rejection of the cosmological argument, including Williams' first premise, is offered by physicist Paul Davies. Theologians offer that,

> Every thing and every event in the physical universe must depend for its explanation on something outside itself. When a phenomenon is explained, it is explained in terms of something else. But if that phenomenon is all of existence—the entire physical universe—then clearly there is nothing physical outside the universe (by definition) to explain it. So any explanation must be in terms of something non-physical and supernatural. That something is God. The universe is the way it is because God has chosen it to be that way. Science, which by definition deals only with the physical universe, might successfully explain one thing in terms of another, and that in terms of another and so on, but the totality of physical things demands an explanation from without. This line of reasoning takes as its basis the assertion that all physical things are contingent upon something else. . . . In a sense that contingency argument falls a victim of its own success, for suppose we enlarge the definition of 'universe' to include God. [And theists do assert that God *is* everywhere in the universe.] What, then is the explanation for the total system of God plus the physical universe of space, time and matter? In short, what explains God? The theologian answers: 'God

> is a necessary being, without need of explanation; God contains within himself the explanation of his own existence.' But does this mean anything? And if it does, why can't we use the same argument to explain the universe: The universe is necessary, it contains within itself the reason for its own existence? (108)

The 'first cause' argument is based on "commonsense" reasoning, meaning what is common to our human experience. This is the reason religious acolytes actually make the cosmological argument. Aquinas succumbed to that temptation. As we've discussed, human beings (and other animals) do have a natural tendency to ascribe external causes or prior agency to observed realities. But it doesn't work here, as Davies lays out. Despite the "appearance" of a syllogism, Aquinas' notion that God is uncaused is merely something we're asked to "believe in." The First Cause/cosmological argument has become the last refuge of the theologian. Many continue to believe it as espoused by clergy and other authorities, many of whom have convinced (or nearly convinced) themselves; but it continues to be literally nonsense. They are just acutely and psychologically uncomfortable with their own position or the troublesome idea that they actually have no proof of God.

Nothingness and Origins

Given that the cosmological argument is vacuous, let's consider the second scenario, the possibility that both God and the universe are eternal, going "backward" and "forward." (Again, let's for the moment ignore the problem of "time" as a mundane concept.) If that is the case, thus, neither one can be the creator of the other. However, it brings up the related and equally important issue long debated by philosophers, "Why is there any reality at all?" For our purposes here, we might all agree that the universe does exist (maybe not all of us), which brings us back to the Leibniz question, as we discussed in Chapter 3, "Why is there something rather than nothing?" (301)

The question might be seen as essentially meaningless since there obviously *is* something. We might just consider that it's absurd to pose such questions, though many philosophers will disagree. Formulating an absurd question doesn't suggest that there can be any meaningful answers, per Dennett's definition of philosophy. As I noted in Chapter 3, we might pose any number of questions—merely since the human brain is capable of doing so—that are impossible to answer by their very nature. How big is infinity (noting that infinity is not, in fact, a number)? What is existence? What is the objective definition of beauty? It makes little sense to pose questions which, by definition, cannot be answered—other than for entertainment. Then there are matters which merely *are* and about which questions are meaningless. Why does 2+2 equal 4? Why do a triangle's angles add up to 180 degrees? Such are givens. Then there are really absurd questions. "Is it farther to Toledo or by bus?" "What does infinity smell like" "How many angels can dance on the head of a pin?" "What is the meaning of the number 42?" (Absurdly, I'll come back to that one in Chapter 11.) Just because a question is posed doesn't mean that an answer can be generated. A question correctly formulated grammatically doesn't necessarily suggest that it can be answered in any meaningful way. Thus, ultimate questions such as 'first cause' and Leibniz's wonderings are only postulated in order to invent artificially constructed answers—like a god.

Nevertheless, if taken at face value, Leibniz' question may deserve some further comment. Consider—if we have an infinite and eternal universe, why isn't it "empty?" The obverse to Leibniz' question might be just as logically proposed, "Why isn't there nothing rather than something?" We do have evidence that, in fact, "*something*" can be a default state, if you will; our reality *is* here. It exists, as Samuel Johnson so cleverly demonstrated. We showed (Chapter 3) that there may be many, possibly an infinite number, of universes. According to inflationary cosmology, there are universes which *are* "nothing." They may exist briefly or for some time and disappear, but they are "empty," as far as our common understanding is concerned. Again, as Greene noted, "As far as we can tell, nothingness is a perfectly logical possibility and so must be included in a multiverse that embraces all universes. . . . the answer to Leibniz, then, is that in the Ultimate Multiverse there is no imbalance between something and nothing that calls out for explanation. Universes of both types are part of this multiverse." (193) There can be both something and nothing.

As our universe *does* exist, and if it is something rather than nothing, the reasonable question is, "Where *did* everything come from? Was there a creation of a beginning of any kind? Perhaps not. In Chapter 3, we've shown there is no need to assume that there was a "beginning"; or perhaps the universe arose "out of nothing." Those explanations may make you uncomfortable, but they are nonetheless coherent views offered by modern cosmology. Again, as Hawking asserts, nothingness is a perfectly reasonable assumption. "According to M-theory, ours is not the only universe. Instead, M-theory predicts that a great many universes were created out of nothing. Their creation does not require the intervention of some supernatural being or god." (214) As difficult as that is to believe, it is an equally—and perhaps more—reasonable and parsimonious explanation for our existence than some supernatural being.

The theologian comes back—it was God who created the universe "out of nothing" or "out of nothingness." We covered some of this in Chapter 3, but there are some other perspectives on this. Physicist Alfred Bahr, for example, puts it very succinctly:

> In order for a god to exist, the nothingness would have to be space and time. Only in space-time is existence possible. But

> space-time always appears together with matter. That means that a god could exist only inside the universe, provided he existed. But then this god could not be a 'creator god' but has been created at the Big Bang together with the universe - as a perishable creature. (26)

I should point out again that the human mind is incapable of defining nothingness in any affirmative way. We conceive of the idea only as the "absence" of anything. The only actual referent we have is presumably "empty space," though we know that this must be defined as spacetime—something—and it is not really empty. At bottom, the notion of "nothingness" is in no way useful in looking at the God concept, including the notion of God Himself creating the universe "out of nothing." Religious adherents, of course, would disagree with Bahr, and argue that God is something *beyond* the laws of nature. Of course, this renders Her immune from any argument or rational discussion.

The notion that science may explain the nature of the universe/multiverse but that it cannot explain where all of reality came from is not evidence—or in any way a valid argument—for the existence of God. It is merely another "gap" fantasy.

Not So Omnipotent

In order to consider the third possibility, that God is a "part" of the universe or a "character" in the universe, let's back up a bit. We must consider whether the theists' God we're considering in this argument is omnipotent, omniscient, both, or neither?

Omniscience

We've covered some of this before in Chapter 6, but let's revisit the issue of free will. People believe in free will. This despite the accepted notion that we live in a determined universe. If we do not, all bets are off, i.e., the universe as we know it could not exist. The other possibility offered by theists is that the laws of nature do operate, but not for our free will. Of course, no one has ever offered an explanation of how that could possibly work.

The theological assumption is that we do have free will, and it was provided to us by God. However, an omniscient God knows exactly what we think, feel, and do, even beforehand. Thus, free will disappears. This again is the age-old paradox of free will or "theological fatalism." It has never been addressed by theologians in any reasonable, rational, or comprehensible way. God even knows whether or not humans will believe in Him, worship Him, follow His laws, etc. And, of course, hold them responsible for doing so or not. It turns out to be the most absurd of theological reasonings.

Then again, consider some other classic questions, "Can God change his mind? Can he change his future mind? Does He even have a mind (a purely human character)? Such seem like bizarre questions, but we are regularly treated to theological references to "the mind of God." If He is omniscient, none of this makes any sense. To attribute omniscience to God, in effect, renders His existence a ludicrous idea.

Omnipotence

And what about omnipotence? The God Christians and others believe in is incredibly fallible, despite their endorsement of Him as omnipotent. We noted some of the "mistakes" He made in Chapter 7, rather a problem for an omnipotent being. But let's just hit a few of the other lowlights. It's unclear why a supernatural God would do the things He has done. He creates unimaginable suffering and premature death, and always has. In fact, this is a principal reason many theists and other worshipers have left their faiths behind. Only a cruel and sadistic god would give cancer to children, burn people alive, wipe out cities with earthquakes, and spread plagues around the world. What would a being gain from such cruelty? Being supposedly omnipotent, why would He do such things if it wasn't important to Him? It can only be considered the ultimate embodiment of so-called "schadenfreude," taking pleasure in others' misfortune, pain, and suffering.

Then there are a number of odd classical paradoxes which can't be resolved. For example, one goes, "Can God create a stone so heavy that he cannot lift it?" Of course, an omnipotent being *must* be able to do this, and therefore It is *not* omnipotent. Read that again. Such logical paradoxes are rejected out of hand by theologians, without explanation. Thus, it brings into question the whole idea of omnipotence. God *must* be fallible. The problem has generated a panoply of mental gymnastics on the part of theologians; but the problem remains—either God is omnipotent, or He is not. You cannot use logic to support a position and then abandon it when it seems inconvenient.

Further, why would an omnipotent God create mortals? Why didn't he just create immortals? Unless, of course, He just wanted or needed adulation and worship. And, yes, this *is* a reflection of a pretty insecure God, something notably inconsistent with omnipotence.

More generally, if there were a preexisting omnipotent God who created the universe, why did He wait billions of years to "create" a hospitable Earth, and a few billion more to fabricate some "special" creatures, people who would scientifically appreciate how they themselves came about? And why are there "plans" to extinguish them? The universe, and all its creatures, may die a heat death or some other terminal fate. Perhaps the universe will experience a Big Bounce and start over in some way. If there are multiple universes (in whatever form), why is it that in this one, the balance of the energy density of matter and space has allowed life to evolve? And has that scenario occurred in other universes? In other forms? If not, why bother to have them? If we're so special, it seems as though a presumably omnipotent God has merely experimented with cosmology in order to "get it right," though this assumes we're the "right" one. But, of course, trial and error would seem pretty inconsistent with being omnipotent. In any case, given what current cosmological theories tell us, the notion of an omnipotent supernatural power is just absurd.

The only possible conclusion is that there is no such thing as an omniscient and omnipotent being. The recourse of "mysterious ways" and not being able to know "the mind of god" thus only serve the psychological comfort of the theologian, pastor, and parishioner.

Infinite Regress

A principal argument of theologians for God is that we know in our hearts that there just *must* have been a Creator. However, from what we've reviewed, this type of absolute declaration can only be seen as a wish. If God is not omnipotent and omniscient, if He is fallible, and if He does not precede the universe, then we have a problem. If God is "part" of the universe or a "character" of the universe, as physicist Davies suggested of theological arguments, or if God is not omnipotent, then you're left with the question of who or what created God. As Krauss asked, "Who created the creator?" (284)

Nearly all religions have an allegory of how the world came into being. There is no reasonable reason to favor one over any other. As we discussed in Chapter 3, there is the Hindu lore about the world, or "turtles all the way down." Bertrand Russell recounts the story of an Indian espousing the Hindu perspective. "The Hindu's view [is] that the world rested upon an elephant and the elephant rested upon a tortoise; and when they said, 'How about the tortoise?' The Indian said, 'Suppose we change the subject.'" (435) There is no answer to such legends, as there is similarly no answer to the Christian one.

If our universe is not eternal going back in "time," then that begs the question, "Who or what created the universe we know?" Was it some unknown force, a deistic entity, or a being of some sort? Einstein famously asked his assistant, Ernst Straus, "Did God have any choice when he created the universe?" Perhaps this was just sarcastic, but he obviously was not referring to a persona of any sort. And why would there need to be such? Can there not be a force without a persona (the deists' perspective)? In conclusion, the only reasonable position on the God and causation argument is that it is meaningless. Enough said on that.

More Philosophies

Acknowledging the so-called "death" of philosophy, as Hawking offered, let's still pause to consider the other arguments made in this field for the existence of God. There are a couple of more philosophical arguments, handed down by many through history, which have been taken seriously but are fairly transparent on their faces.

Ontological

First, there is the theologians' so-called "argument from ontology." That is, the attempt to "prove" the existence of God by reason alone. There have been many attempts at this, but the one most famously cited comes from St. Anselm, the Archbishop of Canterbury, in the 11th century. He proposed that God exists because we humans can conceive of Him. If He exists in our minds, He must also exist in reality. If we can conceive of God, there should be such a being, that is, in some form in both the mind and in reality. If we can think about it, it must be true in reality. If the logic escapes you here, you're thinking clearly. Obviously, we humans can conceive of many things which don't actually exist, from pigs flying, to being reincarnated, to a giant elephant and tortoises holding up the world. Paradoxically, Anselm also suggested that God is "a being greater than anything which can be conceived." (I'm paraphrasing a bit, but you get the idea.) Therefore, He must exist. Though such philosophical musings have been taken seriously over the centuries, it is easy to see that they are just idle ruminations without any discernable value.

Teleological

Then there is the truly favorite theological argument, the one from *purpose*. That is, the world seems to have some order to it, which is seen to imply some sort of purpose. This is an equivocation, of course. There is order in the world for which it is difficult to envision what might be called purpose—from quantum phenomena to planets without life to birds who can't fly. The oft-cited 'purpose' here ends up with the argument that the "end" purpose is us. Of course, this is merely tantamount to the notion of intelligent design. We assume that if something has a purpose, it must have been designed or imagined in some way. Where we humans are concerned, I can only describe this as "a failure of imagination."

There are two major flaws with this reasoning. First, the notion that the purpose of the universe is us is just unmitigated arrogance. The oft-cited corollary is that we exist because we're "special" to God. The notion that *Homo sapiens* is special, belies all logic and history. Although we tend to think of ourselves as unique, nothing could be further from the truth. First of all, we are of extremely recent vintage. We could map the entirety of biological life the same way. *Homo neanderthalensis* probably thought they were special, and I wouldn't put it past chimpanzees and squirrels to think the same. And currently, we are all different, from different cultures and different religions. Many cultures and religions do consider that they are more "special" than others. And we think that we're special, as more "intelligent" than other creatures. The verdict on that is clearly still out. We exist on a spatial and temporal continuum of behavior, intelligence, and other qualities, the long-term survival value of which are yet to be determined.

Theologians also tend to support their contention with the "fine-tuning" argument. They offer that there must be a god who created the universe since the fundamental constants of nature are perfectly fine-tuned for life and for us. This is the old strong anthropic principle, as I addressed in Chapter 3. Using the "fine-tuned" metaphor seems to suggest that someone did it, but metaphors are not evidence. The fundamental values are what they are. As we discussed, the various multiverse theories would explain those values. If they are different in other universes, life may not have arisen there. And if they are the same—or eerily similar—then life as we know it is certainly possible. We have no way of knowing, at this time, what other universes may share the values of our own universe and therefore may show our type of biological life. It is likely that reality is composed of an infinity of universes (of whatever types), and that no creation has been necessary. And there could be other possible explanations which we are not yet able to investigate. The "gap" of explaining God with the values of our fundamental constants, like all other gaps, is no comfort. One cannot imply some sort of purpose in this context, and "fine-tuning" is merely a linguistic effort to disguise another human fabrication of such a being.

Of course, the contrary argument that "God works in mysterious ways, "that there is a purpose, but we cannot know about it", is meaningless. One may argue that it is comforting for the believer to hold that view, but this is purely psychological, and the comfort wears out as tragedy and suffering mount. It is absurd to consider that there could be any actual argument for that. If God had a purpose, it's unclear what that could possibly be. All the arguments are old ones, but I think the issue is again best summarized by Bertrand Russell (I've quoted him a number of times, but it's hard to ignore when the matters are so well laid out).

> When I come to my own beliefs, I find myself quite unable to discern any purpose in the universe, and still more unable to wish to discern one. . . . I will not assert dogmatically that there is no cosmic purpose, but I will say that there is no shred of evidence in favor of there being one. . . . I will say further that, if there be a purpose and if this purpose is that of an Omnipotent Creator, then that Creator, so far from being loving and kind, as we are told, must be of a degree of wickedness scarcely conceivable. A man who commits a murder is considered to be a bad man. An Omnipotent Deity, if there be one, murders everybody. A man who willingly afflicted another with cancer would be considered a fiend. But the Creator, if He exists, afflicts many thousands every year with this dreadful disease. (436)

If God's purpose was us, it's unclear why, as we've reviewed before, He waited so long to bring us about and why has He been so cruel—one might say "evil"—and why does He plan to do away with us? The teleological argument is just fantasy.

Arguments: A Final Word

There are numerous philosophical arguments for the existence of a supernatural god, but they all boil down to the assumption that existence, whatever that is, must have an original or First cause, as Aquinas epitomized. Of course, Aquinas chose the Biblical God Yahweh, if you will, as the Creator. He eschewed the Egyptian god Atum, the lore which preceded the Biblical God by at least a millennium, as it wasn't in one of his favorite books. Atum was the creator of the world . . . and of himself. Sound familiar? And Atum was essentially bisexual, something that would have been anathema to Aquinas. And so he offered what he considered "proof."

Proof

Having dispensed with the vague philosophical arguments, we should consider what we actually mean by "proof." Many philosophers, most notably Aquinas, claimed to have "proven" God, and supposedly logical arguments are put forth to support this. Whether it is attempted or not, let's examine whether or not a God *could* be proved. We saw in Chapter 2 that there is no manner in which a "logical" proof can be constructed for this. All syllogisms are tautologies. One cannot logically prove the existence of God. When theologians resort to deductive logic, they can only tie themselves into intellectual knots.

We do not actually "prove" things in the real world. We just go by the best evidence available. Science does not *prove* things either. It uses observation and what's called "inductive logic" to show what the most likely state of reality is. And the better the observations, the better the descriptions of reality. (As I've noted, the only reasoning that can legitimately be called proofs are in mathematics.) For example, we might observe a few trees, a few hundred trees, or a few million trees, and then we might conclude that "trees are green." That's our best inference. However, we know that there are trees whose leaves are not green (there are actually quite a few), and in the fall many trees show a whole variety of beautiful colors. Inductive logic is the basis for science, not philosophy, and it is always imperfect. Theologians do not use inductive logic, since there is no evidence from which to make an inference.

Disproof

Here's the last refuge of the theologian. "You can't prove that there's no God." That is true. However, technically it is not possible to prove a negative, especially an idea.[2] We can gather evidence to suggest that some things are empirically *not true*. Green leaves cannot walk or run. Human beings can't fly. But you'll notice that, depending on how such statements are construed, there might be exceptions to such assertions.

There are many things in human experience and construction that cannot be disproven. I cannot prove that fairies don't exist, or angels, or Bugs Bunny. The human mind can invent all sorts of ideas about beings that are admittedly unreal—Santa Clause, Tooth Fairy, gremlins. Despite their imaginary nature, it's interesting that gremlins were only invented in the twentieth century, and popularized in a children's book by Roald Dahl. (104) It is the story a British World War II fighter pilot, who looked out on the wing of his plane. He saw a little man, about six inches tall, and he was drilling a hole in the plane's wing. The concept was appropriated by the British RAF to explain mechanical problems of aircraft. It seems to have become a hypothetical "god of the gaps" for any unexplained electronic or mechanical problem, but the term has also evolved to suggest an imagined creature responsible for any sort of trouble. We allow and even reinforce the *reality* of such imaginary beings to our children. We then allow them to reject most of them as they age, but continue to hang on to the fantasy of God. There is no difference in the "reality" of these ideas—they are all human inventions—and cannot be *disproven*.

One cannot disprove the existence of God. The purpose here is merely to explain that it is inconsistent with any logic and reality. In the end, it is not necessary to prove or disprove something that is a human fabrication. We can invent all sorts of things and ask for disproof. You cannot disprove many entities that humans believe in—the ideas of fairies, unicorns, or the centaurs (as the ancient Greeks clearly felt were real). The breath of the human imagination is virtually limitless. These are products of human fantasy. We do not reject these ideas merely because they are fanciful or haven't been observed, but because they are so unlikely as to be merely ludicrous.

The "disproof" request is a totally specious argument. But theologians go around using the "non-disproved" argument all the time. "We can't prove God exists, but you can't disprove it either." It seems compelling on its face, but the notion of comparing "proof" and "disproof" is a *false equivalency.* The burden is on the fabricator. That such arguments persist is only an attempt to feel comfortable about one's own beliefs.

Evidence

It is clear that we invented the idea of gods throughout history—for the reasons I've noted. But that is not an argument for their existence. We would like some kind of evidence. The one thing that most of us rely on is evidence. It's not just for science; we all require evidence in conducting our daily lives. We require evidence that the stove heats up when you turn the dial, that you get the correct change in the store, that the money deposited in the bank is still there, that "you love me." When the evidence for these things is lacking, you take action, sometimes drastically.

It's only for fantasies like God that we sometimes make an exception —no evidence necessary. As we've seen, the Christian Bible, or any other scriptures, is not anything we'd be able to call evidence; and as we've reviewed, and no "miracle" has ever been validated. When Bertrand Russell was once asked what he would say to God about his disbelief if upon his death he encountered Him, he noted that God had not provided him with enough evidence. This seems a bit absurd; but if an omnipotent God wanted to actually enable some reliable faith among humans, He could have certainly been expected to have done a better job of providing some evidence.

Unless there is some sort of evidence that God exists, the effort to assert this becomes essentially meaningless. The god hypothesis *is* testable. We just need some data. Physicist Stenger asserts, "If we have no evidence or other reason for believing in God, then we can be pretty sure that God does not exist." (506) Let me close this section on a favorite story by Carl Sagan, which bears on the matter of proof, disproof, and evidence. It's worth recounting here. He hypothesized that a dragon lived in his garage. (439)

> "A fire-breathing dragon lives in my garage." Suppose . . . I seriously make such an assertion to you. Surely, you'd want to check it out, see for yourself. There have been innumerable stories of dragons over the centuries, but no real evidence. What an opportunity!
> "Show me," you say. I lead you to my garage. You look inside and see a ladder, empty paint cans, an old tricycle—but no dragon.
> "Where's the dragon?" you ask.
> "Oh, she's right here," I reply, waving vaguely. "I neglected to mention that she's an invisible dragon."
> You propose spreading flour on the floor of the garage to capture the dragon's footprints.
> "Good idea," I say, "but this dragon floats in the air."
> Then you'll use an infrared sensor to detect the invisible fire.
> "Good idea, but the invisible fire is also heatless."
> You'll spray-paint the dragon and make her visible.
> "Good idea, but she's an incorporeal dragon and the paint won't stick."
> And so on. I counter every physical test you propose with a special explanation of why it won't work.
> Now, what's the difference between an invisible, incorporeal, floating dragon who spits heatless fire and no dragon at all? If there's no way to disprove my contention, no conceivable

> experiment that would count against it, what does it mean to say that my dragon exists? Your inability to invalidate my hypothesis is not at all the same thing as proving it true. Claims that cannot be tested, assertions immune to disproof are veridically worthless, whatever value they may have in inspiring us or in exciting our sense of wonder. What I'm asking you to do comes down to believing, in the absence of evidence, on my say-so.

The burden of proof is on the religious acolyte, not the doubter.

Conclusion

As we've seen, so-called "proofs" for the existence of God come down to no more than an assumption. But in the case of supernatural beings, we might as well call them daydreams or fantasies. There is no more reason to believe in the Christian God than Carl Sagan's dragon. Or the Egyptian god Atum. Or Mother Goose. No special proof is required to debunk a human-created fantasy. It is sufficient to show how such ideas have come about and the evolutionary, cultural, and behavioral principles which have maintained such memes.

The philosophical arguments are merely weak metaphysical rationales for the human behavior of needing to invoke a supernatural being to explain things we can't explain with mundane concepts. Having dismissed all the metaphysical and "logical" arguments for the "unexplainable," the theists final word is that we mortal humans cannot understand the problem, again usually offering "God works in mysterious ways" or "God's plan." These assertions, of course, are literally nonsense. Why would God need a plan? The latter is just a restatement of our admitted ignorance about a fantasy. Religious apologists are just acutely uncomfortable with "not knowing." That is a psychological problem, not a cogent philosophical argument.

What is God?

Having dispensed with the philosophical arguments for God, let's turn to some other theistic and "philosophical" issues. To do this, let's first revisit the question of "what is God," which we briefly considered in Chapters 1 and 7.

You might believe in a personal god or perhaps one that is just a "force" without any other qualities. The latter is essentially deism. That is, you might believe a god who's responsible for the universe coming into being, but that we're now "on our own." As Carl Sagan once observed, "If there is a real God of the Western sort . . . God has been evolving toward what the French call *'un roi fainéant,'* a do-nothing king, who gets the universe going and then retires or goes somewhere else." (440) Where, of course, isn't clear. Nevertheless, if you take such an "impersonal" view of "god," you have innumerable logical and philosophical problems. If some random god created the universe, i.e., got it going, what did it then do if it had no further interest in what became of the creation? Does it still exist, does the "power" still exist, or did it extinguish itself? The more you drill down on this idea, the more untenable it gets. And, as we've seen, there is no reason to believe in a "creation," by either a personal god or an impersonal force. In most cases, taking a deistic view is merely an agnostic one. That is, you assume that something or someone created our reality, but you don't know if it has some sort of persona or not (I'll address agnosticism more below). Nevertheless, this type of belief—an impersonal deism—is, in fact, what has been evolving in many modern cultures, in some more slowly than others. Then again, why "believe" in such an idea anyway. There is nothing to be lost or gained from this view. The universe merely exists. End of story.

You might believe in a personal God, one that takes interest in the affairs of humans. But which one? If you take such a view, you're left with belief or faith in something with no evidence, no logic, and no consistency across cultures. Or perhaps even species. Who's to say that Cro-Magnons or Neanderthals didn't have gods? The identity and qualities of such a god depend on where you were born, what fantasy your parents had, and/or what religion dominates your culture. There have been countless fabrications of supernatural beings with an interest in human affairs that we've seen throughout human history.

I still find it interesting, though clearly not compelling, that God is continually anthropomorphized. God was clearly created in man's own image—as opposed to the Biblical assertion of the reverse. We did make him male, which was the dominant cultural value in more primitive eras. Many suggestions of female or neutral gender have appeared in the last couple of decades, though such haven't really taken hold. (Recall that the Egyptians got around this by giving Atum both so-called male and female qualities.) It is curious why a presumptive supernatural force had to be given a particular persona at all. What is the personage you envision? Or do you just endorse an idea? To imbue character, will, creativity, compassion, forgiveness, cruelty, etc., is only to make It human, which is a good distance from supernatural.

As individuals—or groups—we have very different ideas about God. In an interesting series of experiments, over 500 self-identified American Christians were asked to choose pictures of faces which represented God, and faces were chosen for a variety of characteristics by the subjects. (237) It turns out that for American Christians the God image was a loving being, white, and "young." However, the chosen images were fairly correlated with each subject's age and race. They were also unsurprisingly a function of social, political, and a few psychological variables. Although almost all subjects still considered God to be a man, conservatives tended to envisage God as more powerful, masculine, and white; and liberals viewed God as more dark, complicated, and loving. Apparently, God's persona is mostly a function of who's imagining It. Ideas about God do directly grow out of our own limited personal history and experience.

Remember Xenophanes' notion that horses and cows would conceive of gods in their own images; Baron de Montesquieu offered that triangles would certainly have gods with three sides; and as I noted in Chapter 7, perhaps a conscious AI will regard its god to be some sort of supercomputer. Such seemingly bizarre notions are no more evidence of the nature of a god than the Judeo-Christian one. That some conceive of their God in a certain way is meaningless to the question of whether there is any such existence. The theologians' answer is often that God—the Judeo-Christian God—does exist but that "He" is merely "perceived" differently by different people and cultures, is an embarrassingly weak argument. It belies human history, in which various cultures have each affirmatively endorsed their own "one true god" over all the others—often violently, needless to say.

Spirituality

Let's return to the philosophy that "truth" is what people experience. In this vein, people report a variety of experiences they consider "spiritual," whatever that means. This is an idea whose meaning defies definition; and the variety of interpretations and descriptions of this is nearly as great as its adherents. (496) Nevertheless, adherents of spirituality often endorse the notion that some sort of nonmaterial existence or force exists in the universe and that we, as humans, can access this. It is sometimes labeled as "spirituality." It varies considerably from culture to culture; but various personal experiences of it are often reported, quite often with very positive connotation. I'll not attempt to discern what people mean by being "spiritual," and of course, there is no definition of "spirituality" which effectively communicates from one person to another.

Some associate spirituality with the natural world, being one with nature or the rest of the universe. For some, we often hear, "I don't believe in God, but I am spiritual." And some people go further. They sometimes associate these experiences with a nebulous force or with an undefined supernatural being. Some people use it to signify, "It is how I get closer to God." They report spiritual experiences that, at least to them, "prove" the existence of God. "I know there is a god, I can feel His presence." It is the notion which also tends to consider spirituality, or being spiritual, as the belief and personal subjective experience that there is some supernatural power "out there" somewhere. It might be guiding or overseeing or suffusing the universe; and although it might be an "impersonal" force, it might still deserve respect, reverence, and attempts to "commune" with it. Of course, this only comes close to religious prayer without giving the force a name.

Some people refer to themselves as "spiritual," having experiences often described as merely states of extreme calm or pleasure—without any specific meaning or referent. Such spirituality is viewed as experiences that are more meaningful about the world than ordinary perceptions and actions. In such conception, spirituality is a personal experience and satisfaction with the qualities of the universe that bring you awe, wonder, beauty, and inspiration. This is essentially consistent with the Spinozian or Einsteinian view, the principal difference being any attempt to "commune" with it. In the end the notion of being spiritual without the idea of a god or supernatural power is no more than a subjective state that you find in some way soothing in the face of the knowledge that the universe is "merely" a physical system fully obeying the laws of nature.

Whether or not spiritual views carry a supernatural overtone, such experiences do boil down to mundane human behaviors. In addition to some philosophical stance, such often include variously labeled meditations. They involve subjective states which are pleasing to the practitioner and are labeled with such terms as oneness, being, calm, connectedness, being in the present, etc. The practice of meditation, for instance, is often described as involving unusual, profound, or "strange" experiences. The particular meditation label is pretty much irrelevant here, whether we call it mindfulness, Zen, etc. Religious people who pray also report such experiences. These "feelings" occur in the context of being awake and eliminating attentional focus on external events or memories. There is considerable evidence that brain activity is significantly modified in many such meditative states, which is altogether not surprising. (59, 107) This is indeed an unusual dynamic in human behavior. We have evolved to pay attention to external things, as they are important for survival; and this is the way we behave almost all the time. When such a pattern is disrupted (other than sleep), we do feel strange and sometimes attribute other "meaning" to such experiences. Whether such activities are valuable in general and/or helpful to particular individuals is not in question here. Clearly, they are in many instances of positively reported "spiritual" experiences. However, what they do not do, despite some proclamations, is provide any validity to some notion of a higher power or god.

Spiritual experiences, despite proclamations to the contrary, are unrelated to the issue of any supernatural being. In the most practical sense, the notion of spirituality is irrelevant to the problem of God. It is only a word sometimes used to soften the dialectic about this issue. If you report your spirituality as bringing you closer to what you *consider* to be God, the conversation is effectively over.

Get Closer to God: Drugs

I talked above and in Chapter 5 about "religious experiences," which many take as "evidence" of God. People do sometimes report intense reveries, with a powerful experience of God. If this does not happen spontaneously, with various mediations, or other focus, it may happen with some type of biochemical assistance. Some people do report appearance or an enhancement of their "spiritual" or religious experiences with some type of chemical augmentation; and sometimes this enables a reported perception of the supernatural. It becomes essentially a philosophical stance for some. "He exists. I can feel Him."

Of course, the use of various psychoactive substances or actual psychotropic drugs (both substances that directly alter chemistry of the brain) can create unusual and bizarre experiences. They act on the brain such that "reality" is experienced differently from ordinary conditions or that new realities are created. It is this change which leads some to perceive unworldly or other experiences that are reported as unnatural and sometimes supernatural. Psychotropic substances, in fact, can lead to the distinct belief that there are forces, even personages, that are greater than the mundane world, even of a god; and such substances have often been used throughout modern human history for exactly that purpose.

Putative "spiritual" experiences have been associated with religious behavior from the time of the earliest human civilizations. And chemical alteration of the brain most likely contributed to that. Psychotropic substances have been used in religious activities for thousands of years. In fact, biologist Henry Herman cogently argues how drug use which alters perception and experience is likely to have been pivotal in the development of supernatural religions. He offers that prophets and other individuals in Neolithic Mesopotamia likely used opiates and other substances which enabled the perception of supernatural beings. (220) In fact, use of substances derived from plants may have been common in early humans' religious experiences. Torrey notes that even "beer and wine, which are well known to stimulate 'the mystical faculties of human nature,' may have been used to assist in communicating with such spirits. If this is true, then it is possible beer and wine played a significant role in the early development of religious ideas." (519) Alcohol was probably used in some form at least 10,000 years ago in the Fertile Crescent; and actual recipes for beer have been discovered dating back around 5,000 years BCE, along with some religious connotations. They included "instruction on how to curry favor from goddesses who governed the production and enjoyment of the brew." (350) And now we have a recent discovery of a "brewery" dating back 13,000 years ago near Haifa, Israel, suggesting beer was fabricated at least that long ago. (317) Evidence for winemaking, also known for its psychedelic properties, also goes back at least 5,000 years BCE. (337)

One interesting example of psychotropic-enhanced religiosity is a drink known as "Soma," which was presumably used by Hindu gods and the more mortal priests, during religious rituals. (67) The psychoactive effect of the drink is not specifically known; but it was alleged to confer immortality. It must have been fairly powerful; and it apparently did contain various chemicals with psychoactive properties.[3] Historical references abound with the name. We see in Aldous Huxley's *Brave New World* that people regularly used a drug called "Soma" to increase their happiness (I find it not coincidental that Huxley selected the name). And a very popular current medication developed in the United States is called "Soma," which is marketed as a muscle relaxant. In point of fact, the drug has psychotropic effects, acting like a sedative or tranquilizer. Soma does appear to be a part of human history, including instances of mind alteration, sometimes associated with something spiritual or religious.

A whole range of other substances have been used by early civilizations. Cannabis and opium may have been used as far back as the late Paleolithic or early Neolithic period, often in religious rituals. (446) And we know that the Aztecs and Mayans used beans of the cacao tree, which they considered a "food of the gods." Although technically not hallucinogenic, they were used to enhance mood; and they were used in various spiritual ceremonies. Drugs do seem to have been used in all sorts of religious ceremonies all over the world.

We know that ingestion of psychoactive substances is natural. Humans have been doing this for a very long time. And what about our ancestral cousins? For example, there is some evidence that even Neanderthals may have engaged in religious psychoactive drug use, with evidence from a burial site at the Shanidar Cave in northern Iraq. (341) They seem to have used a plant that contains a substance known to produce euphoria and effects like amphetamines. As the plant was found coincident with a burial site, it's been hypothesized that such substance may have been associated with a religious rite.

It's not surprising that many other animals ingest psychoactive substances. Most people are familiar with catnip, which seems to make cats go "crazy"; and it does contain a chemical which may be hallucinogenic. However, an enormous number of animals have been reported to use some sort of psychoactive substance. (476) Principal among those observed are birds, dolphins, and some species of deer, who actually eat psychedelic mushrooms. Cows, as well as other ungulates such as sheep and horses, will sometimes seek out a plant called "locoweed." This intoxicating plant acts as a tranquilizer, putting animals into a stupor of calmness. There have been numerous reports of animals also consuming fermenting foods and appearing to achieve altered behavior/experience, e.g., certain tree shrews and bats. However, the lore that many animals actually seek out rotting/fermenting fruit might be overstated. There is no specific evidence that the oft-cited elephants, for example, and others actively seek out food whose sugars have been fermenting. They likely merely eat what is available; and some food stored or deliberately or accidentally hidden obviously rots. Of course, we don't know what altered states of perception these animals experience or whether they consider them "spiritual" or reflective of some "higher power." If only we could ask them. Nevertheless, it does appear that ingesting experience-altering substances has a natural history.

Perhaps one of the most powerful effects of today's psychoactive drugs we might again refer to as "the relief of suffering." Many psychoactive substances can temporarily reduce distress which might be otherwise experienced. When paired with the mild euphoric effects directly produced by a drug, such usage often develops a profound persistence. Other than some direct drug tolerance and addiction, the euphoria and mitigation of suffering are often why people continue to use such substances. But some also ascribe "spiritual" experiences to using them.

Drug use and religion have been closely tied over many millennia. It is clear that mind-altering substances, used throughout history, tend to avoid or escape, albeit temporarily, noxious or unpleasant circumstances; and "spiritual" connotations emerge. The pairing was obviously a potent force in the persistence of early religious behavior; and some people continue to attribute the unusual power of the drugs to something supernatural. After all, such experiences never seem to arise in otherwise mundane life. But all this in no way affirms the existence of any supernatural force or being. It doesn't make such supernatural images real —only in the chemically altered brains of the user. "Getting close to God" via chemistry is merely another fantasy.

Belief and Faith

Many humans believe in a supernatural entity or god, and they often place great faith in It. But one may ask the obvious questions. *Why* do you "believe?" We'll deal with that question empirically in Chapter 10. For now, let's examine the concept of belief itself.

The answer to the question is generally muddled or some sort of circular reasoning. I believe because I *know* He exists. Because I have faith. Because He must exist. Because the Bible, His word, tells me so. In short, the philosophy of the believer is that he knows because he knows. All the nonsensical arguments about origin stories, miracles, answered prayers, etc., fall by the wayside; and all the rational arguments for a god fail, as we've reviewed. But people espouse a belief in some supernatural god anyway. All attempts to rationally address or discuss these concepts are doomed to failure. Either there are unwarranted assumptions, or one cannot share the referent of the believer. She just *knows*.

Of course, belief in a monotheistic, omnipotent God is no more defensible than paganism or other unverifiable "entities" which have appeared through history. It's just what you believe. There are those who believe because they have a sincerely held view of God. There are those who believe because they merely assume God exists. And there are those who believe because it is socially convenient to do so.

What Is Belief?

Let's examine what we mean by "belief." There is a difference in the ordinary use of the term "belief" and its use in a theological context. Belief is a word we often use to convey something subjective and often mundane. Some "beliefs" are rational, others not, as commonly considered. We consider them rational when we can see, or reasonably find out, the current or historical antecedents of that belief. I believe it will snow today. I believe John will be home for dinner. I believe you. I believe the weeds are growing. I believe the sun will come up tomorrow. These are examples of belief that's rational, based on empirical knowledge or experience. When they are not evident or otherwise empirical, we consider beliefs irrational. Belief in angels, fairies, leprechauns, heavens, and supreme beings are examples.

What Is Faith?

Closely related to belief is the idea of *faith.* For most religious acolytes, faith appears to be belief at an even more emotional level. The "faithful" argument for God—and this is paramount in the Christian and other traditions—is that "He" exists; and we know it is so because we experience it in our hearts. Most would accept "hearts" as metaphor for mind or brain. Nevertheless, the problem with this logic is clear, as we've discussed.

Where belief is irrational, we do tend to call that faith—the notion that despite the absence of any evidence, one continues to believe. The Bible is not evidence. Authority is not evidence. Your private spiritual experience is not evidence. Belief without reasonable observations or other evidence is hard to justify. Thus far, any evidence for God is clearly lacking. And so, when emotions about the matter are strong, *faith* is the only alternative.

Of course, the principal argument of religious apologists for the existence of God is that they believe in Him, and no evidence is required. They merely have "faith." The faithful merely assume that God exists. There is no actual knowledge or evidence required. Faith is an assumption or hope. Harris notes how Hebrews 11:1 defines faith, "the assurance of things hoped for, the conviction of things not seen." (206) In fact, the so-called "leap of faith" is quite aptly named. It takes a pretty good leap to cross the gap which the God hypothesis apparently fills. Mark Twain went a bit further than Carl Sagan. He noted, "Faith is believing what you know ain't so." The notion that faith in God is an argument for Its existence is absurd on its face. Faith is often just a talking point which tends to avoid harder questions or coming to grips with life in a different way.

Belief as Convenience

Most people who say they believe in God, don't actually mean that they "know," even psychologically, that there is a God. Often religious behavior is merely a matter of habit (see Chapter 10). Socially, it can be a so-called "path of least resistance." We just continue the rituals of attending services and other behaviors as habit, and it is *expected* that we profess a belief and faith. To continue the rituals and deny faith would be an intolerable hypocrisy for many (though some do it, for other reasons). They know that no god probably exists, but it would be inconsistent to behave in religious ways and denounce the God idea. When you're raised to engage in religious rituals and strongly indoctrinated to the various tenets of a religion, and when you continue to engage in church attendance, offer that you pray (though you might not), and quote religiously inspired moral codes, it'd be tough to do anything but espouse a belief in God. And of course, often this tends to handle any cognitive dissonance about the matter.

Social convenience is often the overriding factor. There is social reinforcement for espousing belief; and there is often social stigma if you are perceived as an agnostic or a non-believer (For the latter, the consequences can be drastic, even to the point of death in some faiths). It's what Dennett has described as a "belief in belief." (123) One espouses the belief without necessarily or privately being convinced, or at least seriously doubting, that God actually exists. It's the right thing to do. It is difficult—actually impossible—to know what is in a person's heart, i.e., brain. What we know is what people verbalize or in some other way communicate—what they say and do. It is useful to say you believe. It is often awkward to say you don't. If you're going to Church with the family, it'd be pretty uncomfortable to say anything other than you believe. And it is sometimes even more awkward to say you don't know. A classic story in this regard comes from the experience of the 18th-century philosopher David Hume. He reportedly slipped and fell into a bog and was unable to extricate himself. A local woman happened by and recognized him, and she noted that he was "Hume the atheist." On those grounds, she was reluctant to assist him. But the story takes an interesting twist:

> Hume implored, "But my good woman, does not your religion as a Christian teach you to do good, even to your enemies?" "That may well be," she replied. "But ye shallna get out o' that, till ye become a Christian yourself, and repeat the Lord's Prayer and the Belief!" What did Hume the Atheist do? Did he stand his ground and ridicule the superstitious beliefs of this woman? Of course not. To the surprise of the fishwife, he readily complied, and true to her word she helped him out of the bog. Hume retold this story frequently, always commending the fishwife as the most acute theologian he had ever encountered. (237)

Doubters

We might conclude that there was no belief in Hume's heart. But this is only because the example is rather extreme. It is reasonable to suspect that a significant ambivalence exists in many who profess belief in a supernatural god. Some continue to openly endorse a belief while also privately having serious reservations. One questions one's faith as doubts creep in. Lacking evidence always creates doubt, with the belief needing to be maintained only by the dynamics of any superstitious ritual or social convention. And for a critical-thinking adult (even many children) it is perfectly understandable that doubts about God creep in. Then there is evidence that reinforces doubt. Why did God do that? Why aren't my prayers answered? How could God let that happen?

Nevertheless, ambivalence tends to linger, often intensely, in many people. On the one hand, what they know to be true is what has been empirically found, either in their own lives or what they trust from science. On the other hand, they would very much like there to *be* a god, who could eternally save them, remembering the idea that our survival instinct is one of the most powerful. Ambivalence is quite natural here.

One "answer," of course, turns out to be Pascal's Wager, as I noted in chapter 7. You might as well believe since this increases your chances of salvation if there is a god; and belief is of no penalty if there isn't. It becomes more of an issue the closer you are to death. One worships in hopes of avoiding death of one's soul. There is nothing to lose by assuming God exists. This is why old age, serious illness, or even ordinary mortal threats lead people to wager that God exists. But it's not a position of much thought or intellectual integrity. And for some people, holding the ambivalence about God reliably over time can be associated with anxiety or fear, sometimes acutely so. And again, placing the wager, even if only privately, is the cognitive dissonance that reduces such distress.

Belief as Exposure

What people "believe in" is specifically and purely a function of the times and cultures and circumstances in which they find themselves. If you are not exposed to ideas about God—or any gods—you're not likely to believe.

What about those humans who've never been exposed to the dogmas of Christianity, for example—primitive tribes or isolated civilizations? What about the Jews and others who lived before the alleged Jesus figure? Did God do such a poor job of earning worship from thinking humans that he had to provide a symbolic offspring to motivate the doubting masses? And was that the best demonstration that an omnipotent being who created the world could muster? Surely he could have gotten better attention with some more obvious miracle. And as I'll ask again, why did He take so long to do it? We humans had been around for about 300 millennia before that. Did Neanderthals or possibly *Homo erectus*, species with language, believe? They likely had no exposure that would have enabled faith in a monotheistic god. Theologians have "answers" to such arguments, but they are all attempts to read God's mind or other fabrications. Having dispensed with the idea of free will, it is unclear why God gave some the ability and interest to believe and others not. If you're not exposed to any evidence for God—including others talking about It—you won't believe.

Conclusion

We have many beliefs, and sometimes they pan out. I think it'll rain tomorrow. I believe I will be promoted. I believe the car will last longer. We generalize about our beliefs about the future. However, when it comes to belief in any particular meme—and religious acolytes will take the greatest umbrage at this—*it's just a matter of convenience*. Is it okay to believe in Jesus, or in the angel Gabriel, Zeus, Atum, or no god at all. It all depends on what place and in what time of history you live. None have said it better than Robert Frost:

> *For, dear me, why abandon a belief*
> *Merely because it ceases to be true.*
> *Cling to it long enough, and not a doubt*
> *It will turn true again, for so it goes.*
> *Most of the change we think we see in life*
> *Is due to truths being in and out of favour.*
> (The Black Cottage)

Belief in gods is like that. You might believe, doubt, lose faith, and believe again. Or you might believe in the Judeo-Christian God, lose faith and believe in Krishna, become atheist, try Scientology, and come back to Yahweh. What comes into favor is a function of many things. In the end, belief and faith are merely a set of learned memes (Chapter 10), not something derived out of some divine inspiration. We might think of belief as a choice. But as we've discussed, choices, like all behavior, are determined by their behavioral history. Some people believe in a god. Some do not. Some change their minds—in either direction. Belief is a fragile thing. It needs to be heavily reinforced to survive. What were cultural memes 4,000 years ago seem to have fallen "out of favor." But perhaps some of them will come back.

Agnosticism

You either believe, or you do not. However, some people take a specific philosophical position that we do not know, and cannot ever know, whether there is a god. This is being an agnostic, a term likely coined by the English biologist T. H. Huxley in the mid-nineteenth century. At bottom, it is the position that human knowledge is limited and therefore we can never affirm or deny the existence of God.

But it is a problematic philosophical position. As we've reviewed, the arguments for and against God's existence constitute a "false equivalency." Further, since the God meme comes from no more than a human fantasy—much like fairies or gremlins—there is no need to plead ignorance on the issue. It isn't a coherent philosophy. I could say I'm agnostic about Zeus, or Apollo, or Ra, or Vishnu, or Santa Claus, etc. But most people today would laugh at that position. However, it's no more absurd than being agnostic about a God called Yahweh. There's no more evidence or logic bearing on one vs. any of the other thousands of gods which have "appeared" throughout history. If there's some reason for believing in a Judeo-Christian God as opposed to those others, I've not heard it.

As far as our daily lives are concerned, it matters not how one comes to endorse an agnostic view. Many people just don't prioritize this issue at all and don't want to be bothered contemplating it. Yet, we as humans are conditioned to make inferences on these kinds of questions—the need to explain things we can't otherwise observe. In such context, as in much of life, we still do tend to make important decisions on emotional, as opposed to rational, bases, as we've reviewed. (259) Agnosticism is merely the emotional escape hatch regarding any serious attention one might pay to this uncertainty; and cognitive dissonance on the idea—the position that we can never know—does reduce any discomfort about the matter for many. For many critical-thinking people, both atheism and theism are objectionable positions; and thus, the idea of agnosticism or "I don't know" is the path of least resistance for them. It makes no sense, but it does seem to "work" for those who espouse it. Some self-described agnostics maintain their position is that they've not seen evidence for God. On the other hand, they don't know where everything came from, and they're willing to entertain the idea that a being created things. Thus, agnosticism is merely a way to feel more comfortable about the issue.

Many people are atheists "at heart," even though they may self-label as agnostics. There is an unwillingness to openly acknowledge that blind faith is an illogical and superstitious belief, and there are in some places negative social contingencies to openly acknowledge that one holds an atheist view. Many atheists do not profess their atheism publicly, a behavior which is not well-reinforced in modern American culture and may actually be punished by others (though this is changing, as we'll see in chapter 11). Being agnostic is more socially acceptable.

Most people deal with the issue in a more mundane or practical way. For some, it just doesn't matter. Some deal with the philosophical issue in whatever way they approach life otherwise. Voltaire famously addressed the problem with his own unique satirical sense, reportedly asking that his own tomb be constructed with half inside the church and half outside. Robert Ingersoll, called by some as "The Great Agnostic," was a 19th-century lawyer, son of a Presbyterian minister, and a pioneer in what has come to be known as the "freethought" movement. In actuality, he did not believe there was any difference between atheism and agnosticism. To him, it didn't matter. He felt that one could not "know" anyway, and people should direct their emotional and practical efforts toward other humans. He was an advocate for many secular social causes, including rights for women and African Americans, reducing cruelty to animals, and prison reform. For him, and like many of us today, as Hawking noted, philosophy is less important than how we treat each other.

Conclusion

In the end, philosophy has nothing to say about the existence of supernatural beings. There is no metaphysical or other philosophical position which can enlighten the idea of whether God—any god—exists. You can even ask, if you believe in God, why does God exist? Obviously, these are absurd questions, which, by definition, can't be answered. Again, just because a question is proposed doesn't mean that it is a good question. Philosophical questions beg other questions, which can be addressed only by still other questions, until one finally reaches the need for empirical answers. The bottom line is that there is no good answer to the god question and that to ask it is merely a polemical exercise.

Nevertheless, humans do want answers, however great the questions. We have a mostly innate need to explain the world around us. Humans have evolved to seek explanations of events—even seemingly incomprehensible events—and be unsatisfied with lack of explanations and understanding, even where such is impossible. Sometimes there aren't any good answers; so we invent the explanations. Any number of unanswerable, seemingly ultimate questions can be posed, as Dennett allowed as the substance of philosophy. The need to fabricate an answer is powerful, and so it is satisfied by the irrationality and pathology of religion.

That's where theology comes in. Where there aren't any explanations, the "gap" must be filled. In fact, we've invented "gap gods" many times in many different ways throughout history. Theology suffers from the same limitations as other human explanatory behavior. It merely invents God and God's behavior, sometimes relying on fabrications in the ancient writings of other humans. Where no commonsense explanations can be found, theology resorts to strange inferences about God, often excusing His behavior. God "works in mysterious ways." "It is God's will." "We can't know God's mind" Let me reiterate. Assuming God has a "mind" is probably the height of pathological anthropomorphizing. Why would God need a mind?

The classic philosophical "arguments" for the existence of a god are many, which we've reviewed. All the "rational" arguments have been well debunked. They are a meaningless set of lexical exercises. And the ontological and teleological exercises go nowhere. The appeals to people's desires, social support, intellectual avoidance, and issues of morality and salvation are just philosophical mumbo-jumbo. As Bertrand Russell cogently notes,

> The arguments that are used for the existence of God change their character as time goes on. They were at first hard

> intellectual arguments embodying certain quite definite fallacies. As we come to modern times, they become less respectable intellectually and more and more affected by a kind of moralizing vagueness. (435)

If philosophy isn't as dead, as Hawking offered, at least we know that theology is philosophy run amuck. It hasn't changed appreciably in thousands of years. Only the names and fantasies have shown some variation. Theologians, like many people, have difficulty overcoming difficult and often troublesome questions. Whether they're mundane or cosmological questions, to put it bluntly, there's often great difficulty in saying, "I don't know." (Merely saying this is not agnosticism. It is just an honest response, and the matter may or may not be pursued in some way.) Overcoming the explanatory "need" about such things, based on millions of years of evolution, is something that can be learned. If it makes you a bit uncomfortable to "not know," there are two options—adopt an artificial superstitious belief or entertain your natural curiosity about other aspects of life.

In conclusion, the philosophical notion of an omniscient and omnipotent god is literally *nonsense*. There is no *sense* in which it is a defensible idea—no common sense, no scientific sense, no logical sense, no sense based on reason, no sense based on purpose. In point of fact, the notion of a god existence actually diminishes the value we may place in our existence—here and now.

9

GOD, PURPOSE, AND THE FICTION OF MORALITY

Religion is an insult to human dignity. With or without it you would have good people doing good things and evil people doing evil things. But for good people to do evil things, that takes religion. —Steven Weinberg[1]

Morality is doing right, no matter what you're told. Religion is doing what you're told, no matter what is right. —H. L. Mencken[2]

For the belief in a single truth and in being the possessor thereof is the root cause of all evil in the world. —Max Born[3]

This chapter is about morality, what it is, and where it comes from. Theologians offer the notion of morality as pivotal to the existence of God. The classic conception is that God has handed down His moral code to us so that we can behave in certain ways and to derive meaning and purpose to our lives. You must behave according to God's laws to be fulfilled and to obtain the primary purpose of life—an afterlife. Without God, the "logic" goes, we would just not know how to behave toward one another. It is a ludicrous assumption, of course, but it continues to be offered as a pivotal idea about God.

Do we need a god to be moral? Do we need divine inspiration to decide what to do? Do we need the threat of eternal damnation to behave "morally" toward others? Do we need a god to have purpose and meaning in our lives? The answer to all these is unquestionably "no." But organized religions continue to tell us that it is "yes." In this chapter, we'll see that "no" is a much better answer and that our "morality" is actually *not* a function of any religious teachings but rather determined by other factors. No god is necessary in this realm.

I will review some of the ideas about morality and explicate how these are historically associated with various authoritarian dicta. But we'll see that they are more basically reflective, in general, of the kinds of social behavior which tend to enable species survival and cultural development. They are natural phenomena, having evolved biologically and culturally, not from the guidance of some supernatural fantasy.

But let's first consider the question, "What is morality?" Right and wrong? Good and evil? What constitutes moral behavior or actions? These are difficult questions; and there are certainly no universal answers (though some have offered them throughout history, as I'll detail). The only answers typically given refer to some code of conduct which is endorsed by the culture in which it is found. But such codes vary considerably from culture to culture and across time. We end up labeling moral or immoral behavior only as a convenience when we approve or disapprove of some behavior, usually to enable some sort of enforcement. There is no inherent meaning of the words, save any formalized dogma. Nevertheless, we'll take a look at how moral notions have come about and what this has to do with theistic religion. In this vein, it is again helpful to remember that we've dispensed with the notion of free will, which would make the whole notion of morality irrelevant, in any case. My comments here are only to expose that the idea of morality offers no solace to the theological arguments about God.

We'll see that so-called moral values which are beneficial to a culture do not depend on religious attributions. They are merely certified, albeit superstitiously, by religious texts, prophets, and the interpretations of clergy and the rest of us. A secular view of morality is actually different from a religious one only in how it is labeled or viewed, not in its behavioral operation, as we'll see. And the evolution of moral behavior predates the advent of any concept of God, supernatural influence, or religious teachings. Much of it, in fact, predates the evolution of humans or even our early ancestors.

What is Morality?

Morality, per se, is actually a fiction. Morality and values don't exist, per se, in the ether. These are concepts we use to label sets of behaviors and to promulgate behavioral rules. At bottom, morality is a summary term for the way people behave toward others.[4] Morality, per se, is really *a set of behaviors which can be construed as beneficial to a group in which it occurs*. We consider behavior which benefits others as moral and that which harms others as immoral.

A little bit about terms. I use the term "morality" to describe a general concept about what any particular culture considers normative and helpful. That is, moral behavior is what most people in a society emit based on the accepted standards of that group. Those standards are based on what a group or culture actually *values*. Where such values are formalized, we refer to them as *moral codes*, which we'll discuss. Where they refer to acceptable behavior on the part of certain defined groups or occupations, we refer to them as *ethics*. Finally, where there are formal sanctions by a culture applicable to certain behaviors, we refer to them as *laws*. I'll often use these concepts almost interchangeably depending on the generality of what we're discussing. But we can think of "morality" as the most fundamental of the concepts.

Values: Right and Wrong

Traditionally, we consider morality to involve discriminating behavior, which we label as right vs. wrong, good vs. bad, or sometimes good vs. evil. Morality is based on what we value, and such has been variously debated and written down as "moral codes" throughout history.

Good and bad, right and wrong, are labels for ideas about behavior. From an empirical point of view, there is no such thing as "good" and "bad," evil, etc. Morality doesn't exist in this way. It is a set of abstractions about a set of behaviors with no other inherent secular meanings of their own. Unless one subscribes to a religious view of morality, the notion of morality and values may be speculated about (and many philosophers have done so); but from an objective standpoint they have no meaning other than as generalizations about behaviors that are commonly known within a culture to operate in certain ways (more about this below). The notion that "morality" can be understood purely with psychological or philosophical concepts is absurd to any serious scientist or critical thinker. The abstract notions of "right" and "wrong" or "good" and "evil" are purely philosophical ideas, which ignore the realities of natural law and the laws of behavior. They are fictions.

Divine Inspiration?

Where does our morality come from? Theists, of course, cite God. In fact, that most people behave "morally" toward one another is offered as an argument *for* the existence of God. So let's give some attention to this reasoning here. Human cultures have always relied on the idea of moral codes being handed down by the supernatural, and often enforced by their self-proclaimed mortal monarchs, who often claim divine inspiration. Of course, this suggests that humans cannot be left to their own devices and, in some traditions, that the threat of divine punishment is necessary to assure moral order. Nearly all cultures lay the responsibility for their view of morality on a god, the "one true god." Unfortunately, His or Her identity varies a great deal.

During the Enlightenment, the Church began to lose some of its potency in the lives of ordinary citizens. Nietzsche had a poignant philosophical view on this. Since all religious moral philosophy is based on the idea that we mortal humans cannot know "right" from "wrong" without guidance from a higher authority, we "needed" God as the beacon and perhaps the enforcer. The philosophy was most famously promulgated in Nietzsche's "God is Dead" aphorism. The actual quote was, "God is dead! God remains dead! And we have killed him!" (374) What he meant has been debated by philosophers *ad nauseam*; but of course, it is likely that Nietzsche did not mean this literally (some devout Christians might disagree). There is reasonable consensus that he was suggesting that life has no inherent meaning, and without the idea of God we are left awash in terms of values. What man had done, with the advent of enlightened thinking and modern science, was to abandon the idea of God in favor of wanting to better appreciate the world as it is. Most consider that Nietzsche was an atheist for most of his life. However, with the decline of Christianity and the whole idea of God being progressively rejected, he felt that there was a risk of nihilism, chaos, and despair. Similarly, a century earlier Voltaire had somewhat poetically offered, "If God did not exist, it would be necessary to invent him." Both were commenting on the problems of morality and social order, though Voltaire may have been merely affirming the need for humans to believe in *something* in order to function well. One also wonders, given his style and various writings, whether Voltaire was actually being sarcastic in this pronouncement (There is some ambiguity concerning Voltaire's private beliefs. He may have been a deist or enamored with the views of Spinoza. During the Enlightenment, it was still socially unacceptable to espouse a denial of God.) Nevertheless, such types of moral philosophies have been embodied over the millennia —we cannot function without guidance from the supernatural.

Good and Evil

At the zenith of many theistic perspectives on morality are the concepts of "good" and "evil." Of course, these are merely labels for right and wrong, usually at the extremes. When right and wrong are elevated to such by religious dogma, pragmatic and philosophical problems begin to emerge. Without theistic religion, there would be no concept of "evil," though we've also certainly come recently to use the term in an ordinary sense.

The "good and evil" problem depends on some bizarre religious tenets. It goes something like this. If there is a singular God, what we call monotheism, then the world operates according to some given set of His laws or rules that enable order in the world. However, God has also given humans free will to choose "good" or "evil." If you are allowed to choose evil, for example, then there is obviously free will. Otherwise, there is no explanation for either good or evil. As monotheists assume free will, they also have a good deal of trouble explaining why there is so much suffering in the world, much of it provided by humans' "evil" deeds, while an omnipotent God stands by and lets it happen. Further, why would an omnipotent God bother to create beings who would make "evil" choices, only to provide them with eternal punishment? The idea of evil makes no sense in a monotheistic view of the world. Then there is the so-called dualistic view of the supernatural, with an Evil power, variously named, in constant battle with a benevolent God. If that's the case, "Evil" also has control of events in the world. Two sets of laws would certainly result in a fully chaotic disorder of the world and humanity. If Evil exists, we could rightly ask who actually enforces the order we see. As Harari notes, "Monotheism explains order, but is mystified by evil. Dualism explains evil, but is puzzled by order." (203) It's difficult to conceive of a thinking human who's not acutely bewildered by this.

Of course, what you consider "good" or "evil" depends on your religion—what god and scriptures you believe in. Let's take just one major example, the notion of an "afterlife." If you believe this, then your moral inclinations are based on what will get you there. The many extant religions have somewhat different moral codes, with often very different ideas about good and evil. An afterlife for Muslims, for example, is based on very different behaviors—and a different sort of afterlife—than the one for Christians. And of course, they've killed each other over the centuries based on such differences.

This has all been the bane of theologians for millennia, and no convincing—or even halfway reasonable—explanations have ever emerged. Theistic religion is helpless in explaining the idea of good and evil, the basis for its so-called morality—one more nail in the coffin of the God idea. Religious or divine inspiration of a moral code is essentially another superstition. We are left only with a confusing and often nonsensical impression of any divinely inspired morality. The obvious conclusion is that there is no such thing. As Albert Einstein put it, in letter to a Baptist pastor in 1953, "I do not believe in the immortality of the individual, and I consider ethics to be an exclusively human concern with no superhuman authority behind it." (144)

Organized religion, in fact, has been the opponent of moral progress —at least the kind of moral progress most of us value today. As Bertrand Russell put it,

> You find as you look around the world that every single bit of progress in humane feeling, every improvement in the criminal law, every step toward the diminution of war, every step toward better treatment of the colored races, or every mitigation of slavery, every moral progress that there has been in the world, has been consistently opposed by the organized churches of the world. I say quite deliberately that the Christian religion, as organized in its churches, has been and still is the principal enemy of moral progress in the world. (435)

Moral Codes

There is not just one moral code or set of morals, a guide sent down by one supreme being. They differ widely over the geography of the planet, over different cultures, and over vast time periods. Why? We'll deal more with this below, but obviously such codes are a function of changes in mundane human behavior and culture. If you don't believe that, the only other alternatives are that there are either many gods, or one god keeps changing His mind.

Yet, there is some commonality of moral guidelines promulgated by nearly all religions. There is nothing really unique in moral principles that have been ascribed to a Judeo-Christian God, and it doesn't require religious guidance to follow. We've generally agreed on them. That there are some commonalities merely suggests only that various human cultures have found similar notions to be of survival value. Human tribes obviously discovered that "thou shalt not kill" was critical to survival of a group, and such long predated the Ten Commandments. The same can be said for theft and several other non-helpful behaviors. The codes of behavior which are reinforced by most (though clearly not all) modern cultures have many similarities. It makes little sense to kill others, to steal, to disobey one's elders, etc. These are things that enable cultural cohesion, survival, and growth. We take them as appropriate ordinary principles, having nothing to do with any scriptures. Let's take a prime example, the so-called Golden Rule, "Do unto others as you would have them do unto you," often associated as a Christian axiom, with Jesus' fabled Sermon on the Mount proclamation. (Matthew 7:12) However, nearly all moral codes have such a tradition. It was noted in ancient Egypt in the second century BCE and even with Confucius a few hundred years before that. The Golden Rule exists in nearly every society and also appears, in one form or another, in many religions, including Hinduism, Taoism, Buddhism, Zoroastrianism, and others. (457)

Human behavior is judged as moral to the degree that it conforms to the expectations of the culture in which it exists. But moral codes generally evolved with reliance on hierarchies for enforcement, including reference to or appeals to a deity, or enforcement by figures claiming divine right; and they eventually evolved into more formal codes of behavior, what we call *laws*. (127) If there is general agreement in a culture about what is moral, much of that landscape is encoded into law, the set of principles that most people follow. We should note that "law," per se, whether written or not, is only a specification of the set of behaviors that most people do agree to abide by voluntarily. When they don't, law itself becomes impotent, e.g., for private sexual behavior, for use of certain substances, for language, etc. In such cases, laws begin to change—often with some difficulty—but as noncompliance with any law increases, formal laws begin to weaken and fall. Morality is always changing, and law is always trying to catch up.

The great legal codes of history—the Code of Hammurabi, the Magna Carta, the United States Constitution, its Bill of Rights—were based on what had been learned over the centuries about what would "work" for the group where it existed. And of course, even those have changed over time. Yet there is always resistance to change from those who hold to inflexible moral codes, particularly those based on theistic religion. And again, gods do seem to be a roadblock to moral progress.

Historical

Morality, per se, developed in ancient times through the leadership of divinely inspired rulers or "governments," such as they were in those times. Those in power, the lawgivers, promulgated codes of conduct which had been handed down by the gods. Thus, the rules of conduct carried divine force.

Moral codes significantly well predate the Abrahamic religions. The Code of Hammurabi was one of the earliest formal moral codes to be laid down in a civilized culture.[5] Hammurabi, king of Babylon, handed down his divinely inspired Code in approximately 1750 BCE in ancient Mesopotamia. As Harari notes, "The gods Anu, Enlil, and Marduk—the leading deities of the Mesopotamian pantheon—appointed Hammurabi to make justice prevail in the land, to abolish the wicked and the evil, to prevent the strong from oppressing the weak." (203) It is the first known example of a formal *legal* code. There were 282 specific laws governing nearly all aspects of civil life. It specified some interesting and progressive ideas, such as a minimum wage, bearing false witness, and the presumption of innocence. Of course, it also established or "encoded" the notion of castes, with superiors, commoners, and slaves, as well as a hierarchy of men and the subjugation of women. Quite notoriously, it promulgated some very draconian punishments[6], most notably punishing offenses by cutting off various body parts, including breasts, hands, eyes, and tongues. You could get your hands cut off for hitting your father. Thievery, breaking and entering, kidnapping, and many other offenses were often punishable by death.

Hammurabi's Code, and some of its unfortunate laws, have persisted to modern times. For example, it specified that if anyone should find runaway male or female slaves in the open country and bring them to their masters, the master of the slaves shall pay him two shekels of silver. Financial considerations about such things were still evident in America over two thousand years later. The U.S. Constitution, in the Fugitive Slave clause, and the Fugitive Slave Acts of 1793 and 1850, emphasized mandates and penalties for aiding runaway slaves; and many states added their own statutes, e.g., Texas in 1841. Hammurabi's legacy persists to this day. A marble portrait adorns the exterior of the U.S. Supreme Court building, along with other historical figures. Much of the Code of Hammurabi also lives on in the codes of the Abrahamic religions, Judaism, Christianity, and Islam. Let's take just a few other examples.

Islam

Without taking a comprehensive view, we note that the supposedly divinely inspired Quran also specifies draconian punishment for all sorts of crimes. The death penalty is warranted, not just for murder, but also for adultery and for rejecting the teachings of Islam. And death can be accomplished via stoning or cutting off some limbs. And the subjugation of women continues. Punishments can be applied for a woman not concealing her body or face. They may be stoned and even beheaded for behavior for which most of the rest of the world find acceptable. In conservative Islamic cultures, the treatment of women is consistent with the "morality" of the Quran and, of course, largely inconsistent with that in most modern, Western cultures. As we all know, the only explanation is that it derives from worship of some theistic fantasy—Allah.

Judeo-Christian

Judeo-Christian "morality" is little different, but it also bears great similarity to Hammurabi's Code. The Bible's prescriptions are seen as divinely inspired instructions - only the names of the gods have changed. In Hammurabi, "If a man put out the eye of another man, his eye shall be put out." The Bible, "Thou shalt give life for life, eye for eye, tooth for tooth, hand for hand, foot for foot. burning for burning, wound for wound, stripe for stripe." (Exodus 21:23-25) (We should note that translations of all these texts do vary somewhat.) There are many other examples not worth mentioning here. Yet, it is clear that the Bible is not exempt from prescriptions for punishment by deadly violence. The Old Testament requires the death penalty for such things as incest (Leviticus 18:6-18, 20:11-12,14,17,19-21), blasphemy (Leviticus 24:14,16, 23), and many other offenses. And similar to Islam, the punishment for adultery—death to both the adulterer and adulteress. (Leviticus 20:10)

The Judeo-Christian Bible itself is filled with endorsements and prescriptions for slavery, racism, bigotry, massacres, incest, and child sacrifice; and there are prescribed punishments for many sins, even for thoughtcrimes. One must wonder how these things would have been divinely inspired, as this would certainly render God an acutely immoral character. And the guidance has continued. Slavery, in particular, is variously endorsed in the Old and New Testaments; and until very recently Christian churches and slaveholders did actually use the Bible to justify slavery in this country.

Probably the most widely cited moral code is the Old Testament's Ten Commandments.[7] Why do we have ten? Is it because we have ten fingers? Maybe. The ancient Egyptians had a base 10 numbering system. Other systems developed, e.g., ancient Babylonians had a base 60 numbering system; much later the Mayans used a base 20 system. But it was 10 that stuck. It seemed like a convenient number. But the notion of 10 commandments, of course, makes no inherent sense. Some religious scholars do point out that many other "commandments" are scattered through the Old Testament, but the lore of "ten" has persisted. Surely there must be more to morality than this.

Legend gives the story that God putatively gave Moses the Ten Commandments, but he dropped and broke them when he saw the Golden Calf his followers had constructed as an idol; and he had to go back for another "copy." No wonder that God elevated to number one the notion of not having other "strange gods before me," after seeing the Golden Calf.

Of course, this is consistent only with a pretty insecure and jealous God, as He Himself confesses. "I the Lord thy God am a jealous God." (Exodus 20:5) It is hard to conceive of an omnipotent being with the human foibles of jealousy and insecurity. But then again, we know that He was conceived in *our* image, not vice versa. Perhaps the original copy of the Commandments had a more pragmatic set of moral codes. Perhaps there were some good dictums in those—don't abuse your children, don't forge documents (like a set of commandments), don't subjugate women, don't keep and trade slaves, or don't commit genocide (which He himself had ordered of the Canaanites).

God's "choice" of commandments here has as much validity as anything else humans have constructed. Parody is sometimes as convincing as reality. An equally valid Commandments "theory" is portrayed in comedian Mel Brooks' movie *History of the World, Part I* (Brooksfilms/20th Century Fox, 1981). In this version, Moses descends Mt. Sinai with *three* tablets—15 commandments, but he drops and breaks one of them when he sees his followers worshiping the golden calf. What happened to the other five? Maybe there were some good ones in those five too. We can laugh, but Mel Brooks' version *is* as valid as any fantastical scripture written by earlier authors.

Conclusion

Throughout history, men in leadership roles have attempted to provide potency to their views of morality, codes of conduct for behavior, through ascription to divine inspiration. Of course, this has reinforced the idea of the existence of gods; but whether any particular moral codes survive is an empirical question. Absolute moral memes, such as religious dogma, are either rewordings of common sense, don't work, or don't survive in and of themselves. They are memes passed along by hosts, often in sacred texts. For instance, the meme "thou shall not kill" has survived pretty well, but this is because humans in many cultures promote it. Others, like Jews not working on the Sabbath or Mormons not consuming a "hot drink," are only followed by very small populations, usually by some fundamentally religious people. But survival of even these memes is subject to the behavior of their "hosts." They would not survive if not promulgated by the individuals who carried the memes in their behavior. And as conditions change, they themselves are changed. In fact, there are pork-eaters among people who also do consider themselves Jews or Muslims, and many Muslims drink alcohol. Morality, like religion, is a function of its environment, not absolute prescription.

There is no reliable vector from religious or other formal moral codes directly to our behavior. Indoctrination is usually insufficient; and threat and enforcement are generally necessary, not only by religious leaders but by the legal codes of the culture in which they live. However, the religious threat of punishment as a motivator for human actions is a poor method of behavior control, which is why many cultures try to fuse legal and religious proscriptions, i.e., theocracy. Without such, many moral memes begin to evaporate.

Moral codes are presumably consistent with social behavior, which would enable better survival of cultures that host them. Whether ascribing them to divine dogma will "work" or not is still an open question. Religiously inspired moral codes have been around for only a few millennia. We are only beginning to see cultural evolution doing its work. Islamic prohibition on the education of girls is now enforced in only a few parts of the globe. Most Jews do not walk to the supermarket on Saturdays. And countless numbers of Mormons have had premarital sex and/or don't give the required tithes to their church. As religiosity declines and fear of divine punishment is replaced by more functional social expectations, religious moral codes may continue to lose potency. It will be the functionality of agreed upon codes that may characterize our species and cultures, as it has for the social organisms which have preceded us. No god would have a voice in this evolution. History certainly thinks it won't; but many things go in cycles.

Morality in Animals

Let's back up in history a bit. Humans are not the only animals to exhibit so-called moral behavior. Most of what we consider "morality" has evolved from earlier species, just as *we* ourselves have. We've invented some other aspects; but much of the commonalities of moral behavior which have survived (though clearly not all), have arisen because they have been functional for the animals that preceded us.

People, like other social animals, discriminate between what is good for those close to them, and for the group, and what isn't. There is considerable evidence that this is instinctual, with many social animals naturally constraining their selfish instincts in favor of what benefits their group in various situations. (286) Is this a moral principle? Let's consider a fundamental one we've already noted, "Do unto others as you would have them do unto you." One human or other animal behaves toward another with the expectation that such will be reciprocated. Sharing food, picking insects off each other, screeching predator alerts, assisting with child rearing, and many other behaviors. Such moral behavior occurs in nearly all higher primates—monkeys and apes, including us.

Then there's related idea of cooperation. In fact, some types of cooperative behavior are crucial to the existence of all social animals—zebras, orcas, elephants, and many others. I speak here mostly of social mammals, though there are relevant others in some lower taxonomic classes, e.g., many birds. There are still other species, of course, who combine themselves into groups, e.g., ants, bees, schools of fish. But we are concerned here only with social mammals, from whom we evolved. Non-social mammals don't display any behaviors we might consider either moral or immoral, e.g., leopards, shrews, aardvarks. It appears fundamental that when animals developed groupings for survival, cooperative behavior became a necessity.

Cooperation does appear necessary to maintain group cohesion, which is necessary for survival in social mammals. The renowned primatologist Franz de Wall has shown that cooperation, including punishment for noncooperators, is characteristic of most primate species; and our unique abilities in these areas are likely the underpinnings of our social and biological evolution. (130) He provides a strong case that morality is an evolutionary product, not something that has come about because of religion. (129) Observations and experiments suggest the presence of cooperation, earned favors, empathy, adoption, and many other moral behaviors. This occurs in many primates, especially chimpanzees, for example. They expect that if food is shared or other favors are done that the kindness will be returned at some later point. If it isn't, the offender will be punished. Reciprocity is expected, and it is punished if not given. (130) Failure to reciprocate would be immoral. Take another example. Capuchin monkeys, like many monkeys, will share food with others in the group. However, if one tribe member seems to be holding an unfair share of food, it will be punished for such hoarding. (302) This appears to be not a function of a punisher not having the food, nor of emotional arousal, but rather a deliberate social sanction for the hoarders' behavior. To hoard would not be beneficial to the group, i.e., not moral, and that kind of behavior is therefore punished.

A significant extension of cooperation, we know as altruism, i.e., doing something for others even without the possibility of imminent reciprocation. Altruism does tend to enhance the survival of a group. However, it seems to occur as food is plentiful, and it wanes when resources are scarcer. Altruism is rather fundamental to many primates, including us. We share our resources without the expectation of reciprocation to the degree that resources are available. At base, it may be instinctual; and it may have arisen as a generalization of maternal behavior. Mothers must do all sorts of things without an expectation of imminent reward. As de Waal notes, "Maternal care is the prototypical form of altruism, the template for the rest." (129) And no moral code or divine inspiration is necessary.

Lest you think that only primates or "big-brained" mammals have moral tendencies, consider an example in another species—dogs. Wild dogs do have a panoply of social behavior one might describe as moral, e.g., sharing of food and assisting in the raising of others' pups. But they can also exhibit some fairly complex morality; and we can demonstrate this experimentally. In one very interesting study, for example, domestic dogs were asked, in effect, to guide some of their behavior based on a "moral evaluation" of what humans were doing. (87) In the experiment, over 50 domestic dogs were asked to view their owners struggling to open a container. They viewed "actors" in the experiment who either helped an owner, did nothing (the control group), or actors who turned away, refusing to help (a non-helper group). When food was subsequently offered to the dogs, they usually chose it from the helpers or the control actors, but they avoided accepting food from the non-helpers. Apparently, dogs do have the capacity to favor cooperative behavior and avoid or shun those who are not cooperative. Of course, it's unclear if this is driven by some innate propensity or the dogs may have learned such tendency "at home"; but they clearly have the capacity for such moral actions.

Morality, so to speak, is cooperative and altruistic behavior. Its opposite, of course, is immoral behavior, as we humans often label it. To wit, even many animals violate what we might consider moral behavior. They can go to great lengths to cheat, lie, and steal. We know that squirrels, for example, will engage in mock burying of nuts to deceive others. Birds, in particular, have well-developed repertoires of selfish and deceitful behavior. Many bird species will do what cuckoos are good at. They place their eggs in the nests of other birds to incubate, thus reducing their own energy expenditure and promoting survival of their young. "Crying wolf" is a widespread strategy in the animal kingdom. For example, the drongos, a group of birds native to Africa, Asia, and Australia, will emit false predator alarm calls, which help them steal food from other birds or other animals in the neighborhood. The brown thornbill of Australia can even mimic the hawk warning calls of various other birds to dissuade other species from preying on their nests. There are many instances of deceit in the animal kingdom, designed to foster individual survival, without regard for others. Many animals will steal food from each other. The strongest lions in a pride usually push out weaker ones from a kill, taking the "lion's share," which disfavors others from sating their hunger (even though such behavior does tend also to strengthen the pride's gene pool). There is always a tension between what behavior facilitates individual survival versus survival of the group—which is one reason morality is always changing.

Finally, a number of extant primate species, especially apes, exhibit various moral behaviors; and these are often expressed as "emotions" which are consistent with group cohesion—such as empathy, love, or sympathy. Chimpanzees, in particular, console each other in times of pain, injury, or illness. It is as if they "understand" the role of empathy in their relationships with others in the group. They "do unto others." Experimentally, some apes can even communicate some moral concepts with sign language. (128) Morality is fundamental in social animals. We did not invent the idea of moral behavior; we only labeled it as such. It evolved naturally, not deistically.

Evolution of Morality in Humans

And what about our more "modern" ancestors? Humans displayed practices of group cooperation, work ethic, sanctions against theft, etc., during the agricultural revolution and certainly before that. Morality also predates any code of conduct which may have been promulgated by cultural or civic leaders, such as Egyptian pharaohs or Hammurabi.

We can reasonably trace "group survival" behavior back to our ancestors, other living primates, and extant primitive hunter-gatherer societies. We don't know that much about the relevant details in our prehuman ancestors. However, cooperative behavior in the *Homo* genus was probably first evident in *H. heidelbergensis* at least 400,000 years ago, with collaborative efforts in hunting and gathering. Anthropologist Christopher Boehm explains that such moral behavior developed on the basis of interactions of evolutionary genetics and human culture. (53) He variously illustrates the evolution of cooperation, e.g., the sharing of meat from kills with members of the group, at least a quarter million years ago. Environmental pressures forced human hunter-gatherer societies to cooperate, and he argues that there was already a tendency to engage in sharing and punishing "free riders." (53) We don't know if *Homo erectus* had moral codes, per se. How about *Homo habilis*? How about Neanderthals or other *Homo* species or even back to *Australopithecus*? Nevertheless, some cooperative behavior did exist in those species. We often fail to appreciate the continuum of evolution and its behavioral manifestations. We've added reason to the equation, of course (something that also evolved biologically), but morality as an avatar of group survival cannot be denied. We are not special in this regard; and no god was necessary for those other species to act morally.

Primates do use enforcement mechanisms to maintain order. Alpha male baboons attack rogue troop members. Other primates will shun noncooperators. Fast forward to humans during the earliest civilizations. Alphas and other leaders needed a better way to enforce moral codes. Advancing language and culture enabled humans to employ more symbolic methods of control. That is where gods came along. Eventually, leaders were able to invoke the threat of divine punishment as a method of enforcing moral codes. (430) As societies grew larger there were more internal pressures, and as a result, stricter moral codes arose, which depended on supernatural beings for their enforcement. Evolutionary psychologist John Teehan makes the further argument that moral codes, and religions themselves, developed because social groups grew larger; and the morality which governed kinships was no longer adequate for more complex societies. (515) Large groups need a more unified source of enforcement; and it took more of a divine threat.

The enforcement of cooperative behavior is enhanced when a god figure is imbued with the ability to punish noncooperators, by divinely inspired leaders. We still see this today in religious sects, with God in control of punishment for immoral behavior, including the threat of ultimate punishment. But such may have a very long history. Hints are provided in primitive societies. For example, there is a current tribe of hunter-gatherers, the Mentawai people, native to the island of Siberut in the Indian Ocean. They worship a water spirit they call Sikameinan. In a recent survey, it was found that they believe Sikameinan punishes people who do not appropriately share meat. If they are "attacked" by Sikameinan, they must pay the local shaman to get a healing ritual. (479)

As human groups grew even larger, some people were more compliant with moral norms than others. Subgroups with somewhat different moral tendencies began to develop, eventually separating themselves behaviorally and/or physically, what we now label as "in" and "out" groups. In fact, the "us" vs. "them" notion evolved naturally because groups were too large and complex to be governed by one set of morals and one divinely inspired leader. Social cohesion was facilitated in smaller groups, and discrimination and violence toward their out-groups arose. Such dynamics continue to pervade cultures today. They spread through our social behavior, financial interests, and language, e.g., from those who speak "our" language to "others" who speak something else. Some people consider themselves Americans, for example, and everyone else is "other." There are Muslims and "others." There is a long and varied tradition to such thinking and behavior. An interesting direct illustration is the Yupik people who inhabit parts of Alaska and northern Siberia. Their name actually means "real people." The name of the Neur people in the area of South Sudan in Africa, means "original people." These are vestiges of the "in" and "out" group dynamic which shaped moral culture in early humans. We see each other, by nature, as different—different mores, different languages, different customs, different gods. We see it all around us.

Just as other animals, on an individual basis, human self-serving behavior can persist at the expense of others. We've generally adopted mores which favor survival of our cultural groups over some of the more immediate impulses we have as individuals. However, selfish or "immoral" behavior does persist. Both of these types of behavior have naturally evolved and exist throughout much of the animal kingdom, including in humans. Humans engage in all sorts of "immoral" behavior for individual benefit—stealing, cheating, lying, etc.—which does often compromise others and impairs group cohesion. But in such instances, this doesn't represent a failure of theistic religion and individual faith, as many have argued, e.g., the often-heard comment, "Our problems are because people are moving away from religion." Rather, such usually occurs for known biological and behavioral reasons. If "moral" behavior (cooperation and punishment of cheats, free riders, transgressors) had been the only social behavior subject purely to genetic evolutionary mechanisms, then it's hard to understand how we currently have so much "immoral" behavior—in all cultures—which is inconsistent with the theistic view. If "immoral" animals (those showing more of the antisocial or selfish behavior) were allowed to reproduce but at a lesser rate than others, it would have disappeared. However, the genetic drive for individual survival is just as strong as the one for primate social cooperativeness. Both individual self-satisfaction and group-friendly behaviors are natural tendencies of social creatures; and the supremacy of one over the other does tend to vary with circumstances and over time and with different cultures.

Morality, writ large, as humans conceive it, is no more—or less—than what arises from our biological heritage—and the memes we use to describe it. In sum, it is clear that most of what we consider "moral" behavior has evolved naturally from lower animals and our *Homo* ancestors. As the great biologist Edward O. Wilson notes, "Scientists and humanists should consider together the possibility that the time has come for ethics to be removed temporarily from the hands of the philosophers and biologicized." (530) It requires no god.

Theories of Morality

Morality, per se, is not a divinely inspired code; it evolved naturally. But as complex human culture developed, other explanations for morality have emerged. If we back away from some sort of divine inspiration for morality, and we've seen innumerable reasons to do this, there are other perspectives. Psychologists, sociologists, and philosophers have put forth a number of "theories" of morality. I'll address here just a couple of the more popular ones. We'll see that although they do not depend on the supernatural, there are fatal problems with each of them. I use the term "theory" advisedly. What we're discussing here are not theories in any hard scientific sense. They're merely philosophies, ideas about what morality "is" and sometimes what it should be.

Morality, by most definitions, does appear to have evolved naturally, in both lower animals and in our human ancestors. There is an ongoing debate (as there must be in science) on whether human "morality" is purely natural and thus arose solely from our genetic history; or, despite seeming morality in other animals, we can view "it" as amplified or even purely a product of human culture, which depends on our advanced intelligence. First of all, let's clear something up. "Advanced intelligence" is something a misnomer, as we dealt with in Chapter 4. But it is itself not something one can separate from biological evolution. Brain capacity itself evolved because of its advantage in problem solving. That there may have been a reciprocal relationship between brain evolution and cultural development—such as tool making or even moral behavior—is not the point. Changes in brain capacity and function could not have evolved without natural selection. To what degree brain capacity influenced culture and morality, or vice versa, is an ongoing debate, which have been reviewed in detail by many, which I'll not get into here (286, 438); but we can rule out intelligence and brain capacity as the primary source of moral evolution in humans.

Much of current thinking is based on the following. Early humans learned to cooperate, the fundamental tenet of moral behavior, and it grew out of two factors—the "theory of mind," which I've previously discussed, and a so-called "intentionality system," which looks at human will as the agency for our behavior and the possibility of punishment. (252) That is, to know what others presumably consider and what they "intend" is the basis for how we relate to each other and conceive of some behavior as moral or immoral. Whether we consciously intend to behave in some ways is theorized to be the basis for our moral or immoral actions, not just our overt actions. It is the basis for how we hold people "responsible" for their behavior (as I noted in Chapter 6; and more on this below). This is the so-called biologic-cultural approach. Though it might make some "intuitive sense," it has been criticized for not addressing various "psychological" issues. For example, some have argued that human morality must consider such psychological "states" as conscience, shame, self-control, and guilt. (286, 287) But of course, these are immeasurable ideas in animals; we know nothing about such hypothetical constructs in our *Homo* ancestors; and their contribution in early modern humans is vague at best. We do tend to "elevate" notions of morality to some sort of intention or other cognition. However, any evidence for the historical sources of these remains elusive. Morality on the bases of psychological states is hard to explain.

Universal Morality

Then there's the philosophy that there is but one set of principles that guide human moral actions, what philosophers call *deontology*. It is most famously associated with the 18th Century German philosopher Immanuel Kant and his "categorical imperative." It is the philosophy that mores or obligations do not depend on circumstances. That is, there is but one set of "rules" that govern morality; and humans employ "rationality" to infer from these rules what is right and wrong in any particular situation. But the overriding set of rules persists as guidance for all of us. Where these rules reside, of course, remains obscure.

Of course, Kant's views, or any sort of deontology, rely completely on the classic framework of "is" vs. "ought." That is, humans do behave in various ways with respect to others—what "is." And this is variously in conflict with a universal morality, the "ought." Needless to say, this perspective assumes (1) human free will, and (2) a universality of what is "good" and what is not. Leaving the free will notion aside, the second premise is also a major problem. How we define "good" and "not good" is entirely dependent on context, as history more than demonstrates, despite what Kant or any moral or religious canon is quoted. Doctrines one might cite as putative support for moral prescriptions, of course, are not evidence that such an all-encompassing philosophy exists naturally in humans. Teehan again describes this well, "The notion that ethical truths are 'out there' waiting to be discovered is itself the remnant of a pre-scientific mode of thought. It stems back to a time when not only ethics, but science itself was under the magisterium of religion." (516)

In fact, it is impossible to conceive of a universal morality which does not depend on some extra-human, i.e., supernatural guidance. Where else would something like that come from? If it is purely instinctual, the notion of "choice" and morality itself evaporates. Kant certainly did not use Biblical references to support the notion (Though he did propose the concept of a universal morality, Kant himself was likely an atheist.). Such views are patently inconsistent with the extreme range of human individual and cultural behavior.

Naturalism

Returning to a critical question, is morality natural, or is it contrived by humans? We've reviewed the obvious evolution of moral behavior; but there is the more extreme philosophy which considers all morality as "purely natural." One "theory" in this vein has been dubbed "moral naturalism." That is, *if some behavior is natural, it must be "good."* If cooperation is natural, it must be "good." Same for altruism and empathy. So far, so good. What about violent physical punishment, something which is natural for many of the social mammals?

There is obviously a problem with naturalism as applied to morality, i.e., deriving one's values from what naturally exists. The problem has come to be known as the "naturalistic fallacy."[8] Not everything that is natural is functional for human cultures; as most of our "moral" behavior, at least in Western cultures, has evolved largely because it has been useful for group survival. As several have pointed out, just because something is natural, doesn't render it "good." Harris argues that "To say that something is 'natural,' or that it has conferred an adaptive advantage upon our species, is not to say that it is 'good' in the required sense of contributing to human happiness in the present." (206)

Let's take a couple of examples. First, war. There is continuing debate about whether war is natural, i.e., ingrained in our species. It clearly is. We looked at this in Chapter 7, but a brief recap. Many animals "go to war," generally over territory. And this includes our primate ancestors and *Homo* cousins. We've also gone to war against other species of our own genus. To suggest that war isn't natural for human beings ignores both evolutionary and modern historical precedent. This is a different question from the one it is sometimes confused with—can human instincts for war be controlled? Though war *is* natural, can we say that "war is good?" Probably not. Would it be "good" for our species in the long run? That's not answerable. All we know is that, by most contemporary reasonable consensus (though clearly not all), war, no matter how natural, isn't considered morally "good" (though we know that's not a universal view). Naturalism fails miserably on this count.

On a more individual basis, though I'm chagrined to bring it up, we might consider the often-cited idea of rape. It is noxious to consider this at all, whether we call it rape, sexual assault, or forced mating behavior (a biological concept, which doesn't necessarily eschew its obvious violence), but it illustrates the point. Many social animals engage in this behavior, e.g., chimpanzees, dolphins, some monkeys, orangutans, some ducks and geese, and even some insects. (102) To look at rape as "good" strains the mind. Naturalism does seem to fail here.

Merely because a behavior, such as rape, has occurred in other species does not imply that it has any relationship to morality. Whether or not such behavior may confer an evolutionary advantage is unclear; and it's not clear how much such behavior is purely instinctual. Some complex behaviors are learned, including by other species with large brains, such as those cited. Any evolutionary advantage of most learned behavior or individual traits is not possible to evaluate at this time. Such behavior does not usually leave a fossilized record. (We can often infer gross behaviors, such as what an animal might eat or whether there was a hominid brain large or enough to solve problems, or what artifacts were fabricated, from the fossil record; but specific behavioral responses don't leave a good trail of evidence). For example, take eye color. It is quite variable in humans; but we don't know if it has or will influence mating behavior and thus an evolutionary selection for color. So far, this merely represents a genetic drift, without obvious potency, though perfect prediction would be folly. Skin pigmentation is another, though one can depressingly envision this being a factor in the current Anthropocene Epoch.[9] The behavior of rape may have been reinforced in the above organisms, and it is clearly reinforcing in human rapists.[10] But it is folly to speculate on a survival advantage for such behavior. We in the West have countless good reasons for eschewing and punishing a rapist, but we'd be remiss not to point out that some religiously based cultures do not agree. Islam, for example, often blames female victims for being raped. That it may be partly natural, being evidenced in species that preceded us, is a difficult question; but to infer that it is "moral" would be unacceptable to most of us.

Infanticide is another, which is often seen in other creatures. It happens in lions, chimpanzees, some monkeys, dolphins, bears, several rodent species, baboons, and several others. One functionality of this, in some species, is to bring about estrus in the females, allowing the offending males to pass on their own genes, as opposed to some other impregnator. It is natural. Nevertheless, we are seriously taken aback when viewing a lion killing a cub. It seems morally offensive. We universally assume a moral violation here for people, although at times it has been practiced by humans. Most recently, the Chinese government's "one-child" policy led to infanticide of an unknown number of babies (somewhat more females than males, as females were seen as less valuable in the culture at that time). The policy was relaxed only in 2016. Despite any natural history in the animal kingdom, most would not consider the practice "moral."

Who Is It that Acts Morally?

Which brings me to the question of why we label things in such a manner. We've seen that many animals act in ways that we would label as moral—sharing, altruism, empathy, even self-sacrifice. Nevertheless, such actions are seen as purely genetic, having nothing to do with some conscious intention. We tend to think of human moral actions as guided by a conscious, freely intending actor; whereas, when other animals do such things, it must be for some innate reason. Of course, this assumes the myths of human free will and conscious motivations. And such inferences of human morality ignore the fact that millions of years of evolutionary history of primates acting in such ways is what brought us into existence in the first place.

In the end, the matter comes down to a matter of labeling, of whether one considers an agent a "moral" being. It was Charles Darwin who was one of the first to note the evolution of "morality" in humans and other animals, and he gave numerous examples of this. For him, morality was tantamount to certain social behavior, what we now think of as "prosocial" behavior; and it often came down to matters of how we conceive of such things.

> As we cannot distinguish between motives, we rank all actions of a certain class as moral, if performed by a moral being. A moral being is one who is capable of comparing his past and future actions or motives, and of approving or disapproving of them. We have no reason to suppose that any of the low animals have this capacity; therefore, when a Newfoundland dog drags a child out of the water, or a monkey faces danger to rescue its comrade, or takes charge of an orphan monkey, we do not call its conduct moral. But in the case of man, who alone can with certainty be ranked as a moral being, actions of a certain class are called moral, whether performed deliberately, after a struggle with opposing motives, or impulsively through instinct, or from the effects of slowly gained habit. (106)

In the end, the actions of humans as moral or immoral come down to a matter of labeling. Such gives us nothing of better understanding; and again, it certainly doesn't support the idea of any sort of divine inspiration or code of conduct handed down by God.

Behavioral

Morality, per se, is only a concept, an idea, not an explanation for anything. We use this term as a shorthand to describe rules of behavior which are necessary for members of a group to follow for the benefit or survival value of any particular tribe, society, culture, etc. These "rules" come out of long histories of agreed upon behavior, that is, the norm of behavior endorsed or followed by the vast majority of a group (Though it deals with behavior, this is not tantamount to the philosophical approach known as "consequentialism." That philosophy views morality based solely on a value judgment of the consequences of people's behavior. Consequentialism looks at behavioral outcomes as "right" or "wrong." The behavioral view is that behavior, moral or otherwise, is purely a function of learning or conditioning.). As language came to be an important method of group cohesion, these became set down more formally, eventually leading to the various documents and scriptures we know. However, whether and to what degree they are "followed" depends on each individual.

"Morality" is basically a set of expectations or dicta specifying what people "should" or "ought" to do. If it exists as anything more than an idea, there are three important considerations in this regard. First, this implies that there is a very good consensus about what "should" be done and not done—either universally, within a putative religion, or within any culture. This is far from certain, as we've seen.

Secondly, any morality assumes that there is actual freedom of choice in what one does and doesn't do. This bears on the question of free will, which we dispensed with in Chapter 6. Since free will is basically an illusion, the concept of morality itself loses most of its meaning. When we speak of morality or ethics, we usually mean that what we "choose" to do is either right or wrong (or some variant of these). If free choice is not empirically relevant, then the ideas of right and wrong only signify the possibility of reward or punishment for various actions.

Third, there is the matter of what we might call "practical freedom of choice," in which external contingencies confer everyday restrictions on an individual's actions (so-called "degrees of freedom"). For example, there is the moral dictum of "thou shall not steal." But it doesn't really function that way. Few modern humans steal food from the grocery store. However, this is due to the fact that most don't need to. But when resources are in short supply, e.g., when you have no money or there is little supply, we see minor thefts or actual looting. The moral code is impotent in the face of real-world contingencies. We also share our food, money, and other resources when they are plentiful. We give money and goods to charities and sometimes donate our time and labor. But when those are in short supply, e.g., when we just don't have the time or money, we tend not to. Altruism is mostly a function of real-world contingencies, just like it is in other animals, not subscription to some moral code.

With these caveats in mind, let's examine the notion of morality empirically. Morality, ethics, and values can be very amorphous concepts. We can ignore the subtle differences in these and focus here on the behavior of individuals and groups where we use these kinds of labels. It is from this perspective that a more basic analysis is possible. We've entertained some "theories" of morality; but what it comes down to is how people learn to behave in the ways that they do. We'll get into more detail on this in the next chapter. For now, the most economic approach to the issue of morality is specifically a focus on the evolution of the behavior of individuals.

We learn to behave consistent with, or inconsistent with, or with some variant of what most people in a culture consider moral. We do have some innate tendency to cooperate with others and to do things which enhance the cohesion of groups or cultures to which we belong. We also have innate tendencies to foster our own self-interest, to act in ways that selfishly benefit us, sometimes at the expense of others. How this tension plays out for each human being is a function of how we develop along the way. In the end, how we conform to any formal or informal moral standard is specifically due to what we're conditioned to do. It is the external contingencies which shape and direct our "moral" behavior. Let's take a great example from *The Adventures of Huckleberry Finn.* On contemplating the "moral" decision of whether to identify Jim as a runaway slave, Huck Finn says,

> They went off and I got aboard the raft, feeling bad and low, because I knowed very well I had done wrong, and I see it warn't no use for me to try to learn to do right; a body that don't get started right when he's little ain't got no show—when the pinch comes there ain't nothing to back him up and keep him to his work, and so he gets beat. Then I thought a minute, and says to myself, hold on; s'pose you'd a done right and give Jim up, would you felt better than what you do now? No, says I, I'd feel bad—I'd feel just the same way I do now. Well, then, says I, what's the use you learning to do right when it's troublesome to do right and ain't no trouble to do wrong, and the wages is just the same? I was stuck. I couldn't answer that. So I reckoned I wouldn't bother no more about it, but after this always do whichever come handiest at the time.

Morality always depends on the conditions present. We've already covered the notions of the morality of slavery and the runaway slave issue; but the bottom line—context is paramount, including your time in history.

We are taught to behave in certain ways. We are reinforced for doing so. We have models to show us what "moral" behavior is. And we have authorities in our lives to enforce compliance with any code of conduct—beginning with our parents and later various authority figures, from teachers to clergy and others. All moral codes are prescriptive, and they are promulgated by authorities. However, there is no evidence that they have primary control of people's behavior. We behave as we are conditioned to from an early age, as Huck clearly knew. We might then tend to attach dicta from a formal moral code to our actions—and the actions of others; but it is not the code that generates and maintains our behavior. It's not the Bible or any other dogma. It is what we've learned and the conditions present.

Whether we behave morally or immorally is a function of what we learn and often a matter of circumstances. People can be conditioned to behave in rather horrible ways in even a very short time. It's very easy to shape. Let's take the example of the infamous "Stanford Prison Experiment," which illustrates the matter (There are various summaries of the experiment in unpublished papers. (333) Zimbardo himself did not publish the study in any peer-reviewed journal; and we should note that the methodology has often been questioned. (307)). People (college students) volunteered to act as prisoners or guards in a psychology experiment. It was merely a simulation. Subjects were randomly chosen to act as prisoners or guards. Those who were to be assigned as criminals were actually "picked up" and booked at a local police station and then confined to a "prison" in the basement of a college building. Prison life was simulated as carefully as possible, with procedures, uniforms, physical confinement, and other parameters being instructed for both groups. Interesting events developed. A rebellion by prisoners ensued, which was quelled by the student guards using fire extinguishers. Over a short time, prisoners started acting like prisoners, with emotional and depressed symptoms, and guards started acting like guards, perpetrating rather "evil" behavior toward the prisoners. Finally, the experiment had to be terminated early, after only six days, because the student guards became acutely humiliating, vengeful, and sadistic. They had "learned" the evil behavior in only a short time. Don't dismiss this as only "acting" or "faking." Interviews with guards and prisoners decades later revealed that they often genuinely felt the emotions exhibited during the experiment.

This sort of consequence has sometimes been referred to as the "The Lucifer Effect"—turning good people into evil, as Steven Weinberg so aptly noted about religion. The shaping of vengeful and abusive attitudes, which many people come to develop, at least covertly, is easy. We hold prisoners or criminals with disrespect, contempt, and desire for vengeance. They are seen as immoral, not something that has come about for reasons we don't well understand—or at least that we're not willing to take the time and effort to understand. They are seen merely as immoral people.

Moral behavior, like most of what we do, is under the control of its consequences (see Chapter 10). This is the actual effective dynamic of what we consider moral behavior, as opposed to the dogma espoused by religious institutions. One example. Much religious dogma is maintained, in large part, by punishment, the threat of punishment, and fear—especially the fear of the eternal punishment. For "immoral behavior" there is denial of a joyous afterlife or often burning in hell, regularly promulgated by Christian clergy. Albeit a superstition, it is a potent motivator for many, being reinforced by parents and others in a group or community. It is a very widespread social dynamic. In one major survey, it was found that the threat of supernatural punishment, in terms of a personal God's disapproval and the notion of an afterlife, was identified as an important factor in maintaining moral norms and punishing free riders or those who commit moral transgressions. This was supported in a global sample of 87 countries and even controlling for the level of religious participation, country of origin, denomination, and education. (17) People felt that such was a culturally important factor. Of course, why individual people identify an afterlife as a factor in their own behavior is more about their childhood and social learning, not the actual threat of fire and brimstone. Nevertheless, they identified such threat as being important for their culture and its norms.

In general, the contingencies for morality have classically involved primarily fear, particularly in Christian sects. Fear is a subjective state associated with conditioned escape or avoidance responses and their associated emotional experiences—and it is often associated with the biologic "fight or flight" response I discussed. As such, it is a powerful motivator. It signals punishment, the basis for much of Christian fear-motivated morality. However, it is not a very effective method of shaping and maintaining behavior, which is why religious tenets have to be continually reinforced. That is why religious people keep going to church and/or reading their Bibles. Fear of consequences must be continually reinforced, and when couched in Christian canons, its effectiveness is often poor. In some studies, people of Christian faiths actually ascribe fewer consequences—in this life or an afterlife—to their actions than do people of so-called karmic religions (e.g., Buddhists and Taoists) or the non-religious. (546) We behave morally according to how we're conditioned, not authoritative interpretations of some archaic texts.

So what are we left with? Morality is only what we do, a set of behaviors that are conditioned over a lifetime, beginning mostly in early childhood. It is merely what we call "rule governed behavior," i.e., response sets or memes which may be well generalized in a culture, though formal scripts are not usually necessary. They guide behavior which is inferred by the community as beneficial to the group. In the long run (and it may be longer than we can currently envision), those that work for any defined group survive, and those that don't, don't. One can artificially ascribe them to some moral code or scripture, but it is our actual learning that is responsible for moral or immoral behavior in any relevant context. For now, let us remember that the idea of morality in no way supports an argument for God; and we do not need Voltaire's suggestion that we need to invent Him. To behave in constructive ways toward others requires no divine providence.

Moral Relativism

If Kant was wrong and there is no universal morality, where does that leave us? What is considered moral is quite variable across times and cultures and is justified by whoever is proselytizing it. But what "works" at one time and in one culture—what nurtures and sustains a culture—is often anathema to another. This is often construed as "moral relativism," the notion that moral judgments are valid only in relation to a particular culture or society—or even a smaller group—in which behavior occurs.

"Moral relativism" is a term suggesting that morals or values do not have any absolute, structural standard. There is no universal right or wrong. The moral values and memes of different cultures exist as is, and outsiders have no inherent right to impose their own values on others.

This is a truly abhorrent idea to religious adherents. It is seen with a truly dismissive attitude, as the concept of "to each his own."[11] It is a concept often employed to criticize or denigrate more liberal or scientific thinking about human behavior. But in fact, it is merely a crude description of the notion that what is considered moral in one situation, culture, era, etc., may not be in another. Many religious adherents denigrate such a perspective, as it contradicts an ultimate morality supposedly provided by God; and such malleability of values is not allowed. It ignores reality, of course, but it is a threat to God's "word." Fundamentalist Christians, in particular, are quick to cite "moral relativism" for most of the social "evils" that they see in the world. Bad behavior comes out of not following God.

History

Whether any moral codes "work" and survive is always an empirical question, one that is answered only with an arc of history. You must ask for whom, for how long, where, etc. There are some moral commonalities which have existed since civilizations began, but these are far from universal or perfect. Some Islamists, for example, have not abided by the mandates for "thou shall not kill" or "do unto others . . . " In fact, for some, quite the opposite is true in their eyes when it comes to apostates, infidels, and women. And as we've seen, Christians also have a dreadful history of violating God's "thou shall not kill" dictum.

History and context matter. What is considered moral or amoral has undergone significant change over time and differs a good deal from culture to culture, one society to another, or subgroup to subgroup. For example, murder is still considered amoral in "almost" all cultures since it does relate in some way to group survival (unless you excuse the notion of "ritual killings, which I'll return to below). Violating a scholastic dress code for an American adolescent may not relate to group survival, even though such may still be considered by many to be amoral. The same goes for same-sex relationships, pornography, or using machines on Saturdays. The value of a prescribed code of conduct is related to whether it still has a meaningful function in the culture. If it proves not, it tends to wane as a moral standard. Incest is not immoral in all cultures. Neither is taking many adolescent wives. "Cruel and unusual" punishment is considered moral by some, immoral by others. Only history can opine on what will survive as "moral" behavior.

History is a great teacher. Consider a couple of what we would now consider extreme examples. The 19th-century English novelist and iconoclast Samuel Butler cogently observed, "Morality is the custom of one's country and the current feeling of one's peers. Cannibalism is moral in a cannibal country." (68) In fact, cannibalism was not uncommon in our ancestors, and there are numerous instances of cannibalism in *Homo Sapiens*. A couple of examples. There is evidence of a massacre and cannibalism of 37 individuals from a find in England dating to the late third century BCE. (452) Remarkably, there is also strong evidence of cannibalism in the American southwest dating to as recent as the 12th century CE. (541) Does it still exist? Current unconfirmed examples often cited involve the Donner Party, the mid-19th century group stranded in the Sierra Nevada mountains, and legends of people trapped in the Andes mountains or the Arctic. These are admittedly isolated examples. Nevertheless, cannibalism is a known part of our history. Lest you think the practice is "unnatural," it is also widespread in the animal kingdom, including in creatures close to us in the evolutionary timeline, e.g., most notably in chimpanzees and bonobos. All this turns the stomach, but these things are obviously "natural" and part of our history, but is it moral? Context again is key.

"Human sacrifice" would not be considered moral today, or legal; but there is no denying the history of human sacrifice associated with early religious rites. Humans regularly killed others predicated on the presumed need to palliate the gods with human blood. As far as I know, we don't do that today, although the history is not that ancient. We might think that such barbaric religious death rituals disappeared in prehistoric times. Although human sacrifice faded out in the Old World only a few millennia ago, it continued for some time in the New World. (181) In Europe, humans were engaged in certain "ritual murders," with evidence from most recent sites dated to about 3,500 BCE. (321) In such practice, known as "incaprettamento," the victim was killed with "homicidal ligature strangulation." This involved binding the throat and ankles with a rope, followed by burial. That it was still being done in the middle to late Neolithic period seems bizarre and disgusting, to say the least; but such ritual sacrifice was a *custom* at the time.

The Aztecs sacrificed thousands of humans, for example, at the so-called sacred city of Tenochtitlan in 14th to the 16th centuries CE. There we can see evidence that throats were slit and the victims' hearts and heads were removed. (528) In a recent find, hundreds of children and baby llamas were discovered from a mass slaughter, likely by the Incas, dating only from about 1100 to 1470 CE from part of Peru's northern coast. Interestingly, the ritual slaughter may have been an attempt to "appease the rage of the gods" following a "sudden climatic event (exceptionally heavy rainfall and flooding)." (142) (One can only hope that there are better current solutions to major climatic events than barbaric appeals to supernatural gods). Aztecs were still engaged in this practice as recently as 500 years ago, with numerous finds of sacrificial children. (181) By sacrificing some to the gods, further death to a tribe could be effectively prevented, as was the obvious morality of the time.

Moral Hegemony

That one culture's morality is its own, not a universal, hasn't always prevented it from trying to impose it on others. Moral hegemony rarely creates progress. More often, it creates violence. Culture clashes, wars, and other conflicts are undeniable parts of history, based on moral values. Every culture has its own set of morals (as well as subcultures and even smaller groups). Many religious sects tend to want to impose their codes on others who may have different values. Even without war, evangelism and missionary work are the hallmarks of this approach, with the belief that their morality is better for you than the one you have. Groups that don't buy that, of course, are seen as heretical or unworthy.

In America, we would like nothing more than to have the rest of the world follow our Western values (assuming there is consensus on what those are, which there usually is not). But imposing our Western, modern values on other cultures is a tenuous enterprise. History does not provide much support for moral hegemony, especially via force. Western values do often suggest that we *should* spread them widely; and economic pressures and various religious missions have been engaged in this for a long time. And pressure, via economics or violence, is seldom effective. The political debates and methods in this regard are not the subject here. What "works" for different cultures and at different points in history is variable. Cultural evolution has a very long arc; and efforts to accelerate this by force rarely succeed, as history well demonstrates.

By acknowledging the variety of moral codes that exist, I am not advocating for a blind "moral relativism." For example, respect for cultural codes which tend to dehumanize and promote suffering cannot be reasonably justified. I personally believe that social structures which reinforce cooperation and find decreased needs for punishment are most likely to survive and flourish at this point in the history of our species, though one would be foolish to claim this with great certainty. I merely acknowledge that many moral and ethical codes exist. We can label one right and another wrong, but again this is from the standpoint we occupy now. The evolution of the various mores that permeate a culture have developed for various reasons; but they did not develop because they were "right" or "wrong." They came about because of the cultural and other factors which were characteristic of the time and place. Again, this is not "moral relativism," as the term is often maliciously used. It is merely to suggest that we acknowledge the validity of "moral" codes as compilations of human behavior, not dogmas that can be ascribed to some divine providence, even though they may be labeled as such.

Responsibility

We tend to hold people responsible for their behavior. That is, no matter the influences in our lives, we're expected to voluntarily conform to the moral standards of the day. We humans use the idea of *responsibility*, that individuals are accountable for their own behavior, especially when it comes to morality. On a practical level, this accountability considers whatever moral landscape that exists at the time. Nevertheless, many faiths consider the notion of responsibility pivotal to their religion. God holds you responsible for what you do. He allows you to act as you wish, then rewards or punishes you for your actions in relation to His moral code, sometimes for eternity.

The relation of morality to responsibility is a tricky one. We talked about this in Chapter 6, but a bit more seems necessary here. We are generally held responsible for our actions by others, and this is generally for violation of current moral codes and standards, as well as legal ones of course.

Free Will

Of course, being "responsible" for your own behavior demands the existence of free will. Whatever else you conceive about morality and responsibility, it does assume free will. As philosopher Grayling maintains, "The answer assumed by ethical debate is that free will exists. Without that assumption the entire edifice of moral discourse collapses . . . [various influences] conjointly make an individual unfree in obvious ways. But the key assumption is that people are not metaphysically unfree—that is the essential point for moral theory." (190) He goes on to briefly mention neuroscience as possibly indicating that free will is an illusion. Nevertheless, he bizarrely also goes on to assume that, "We are morally free to choose, and therefore we have the *responsibility* to choose wisely." [emphasis added] We've debunked the free will argument in Chapter 6, so I'll not give this much more ink. In whatever way morality evolves, it is not an inherent free will of humans that maintains it; nevertheless, humans continue to assume free will as the basis of doing what is "right" or "wrong."

Punishment

The enforcement of moral behavior is based on consensus and the contingencies of social rebuke, ostracism, and various punishments, depending on the moral code and how seriously it is violated. Throughout cultural evolution, various sanctions and punishments have been powerful influences on our actions. In today's cultures, they are somewhat less effective, though they're still there, which is why we have formalized systems of codes of conduct and laws with specific rules and penalties. The law in most modern culture, encodes punishment for those who do not "voluntarily" follow those codes; and we use the notion of responsibility in order to enforce them.

The nature of punishments, of course, vary greatly, depending on the culture, the governing authority, and status of the individual in any society. Humans punish for everything from blasphemy to murder, though enforcement is quite variable. In the extant Western criminal justice systems, for instance, we use incarceration to protect society from harmful acts of criminals, but we're also guided by other motivations. We employ vengeance, and its legal partner, retributive justice, to enable the harshest possible punishment. It's unclear exactly from where such urges spring, and there certainly is no convenient, universal explanation. Nevertheless, religion often provides the justification. A brief perusal of the Code of Hammurabi and the Bible exposes such morality and punishment—again the words of gods.

Exceptions

On the other hand, we don't tend to hold people "responsible" for their actions if we can find some overwhelming influence on their behavior. Basically, we cannot hold you responsible for acts if you have not "chosen" to emit them. To reemphasize the point with some examples, consider the following.

Perhaps there are exceptions to voluntary "choice." If a four-year-old child kills someone with a loaded gun, do we hold him responsible for his actions? Is a man responsible for a homicide when it is discovered that he has a large tumor in the prefrontal cortex of his brain (this is an area which is critical in suppressing impulses). We could also ask whether a person with severe intellectual impairment can be "responsible" for committing a violent crime. It actually took until 2002 for the U.S. Supreme Court to find that it was unconstitutional to execute a person with severe intellectual disability (Atkins v. Virginia, 536 U.S. 304, 2002), though individual states continue to seek ways around that ruling.

You can be exempted from responsibility if you are "insane," as we've decided our moral codes don't apply in such circumstances, as I noted in Chapter 6. When you kill your children and assert "God told me to do it," if we believe you, we provide some level of unaccountability by psychiatric diagnosis, e.g., as was adjudicated in a prototypical Texas case in 2003. A woman killed two of her sons and injured a third, essentially by stoning, and was eventually found not guilty by reason of insanity. She was confined to a mental institution. (She was released eight years later.) If we believe you, perhaps you get labeled insane. If we don't believe the accused has some major aberration in such an instance, we tend to apply the death penalty or a sentence of life imprisonment (notwithstanding that prosecution and defense mental health experts regularly disagree in such cases). I find it curious that ardent fundamentalists don't believe defendants have heard from God (only some are to be believed), even though they espouse that it's God's will in other circumstances where people die by the thousands. I cannot think of a more tragic religious hypocrisy than that.

Religious

Which brings us to the notion of moral responsibility from the religious standpoint. The fundamental tenet of Christian theology, for example, is that God rewards and punishes humans for their "good" and "evil" actions. But back up a bit. As I've said, if you assume that God is omnipotent, then there must be an assumption of His total control of all human behavior. And He is omniscient, i.e., knows what you are going to do. To reiterate, the theological notion that God imbued humans with free will—so they could behave in ways unrelated to his pre-awareness, from any standpoint, is just absurd. Thus, when there is any failure to perfectly follow God's moral code, the responsibility is His, not yours. As Einstein once observed about God, "In giving out punishments and rewards He would to a certain extent be passing judgment of Himself." (150) I'll not even comment on the mental gymnastics that it takes to get around that one.

Author Susan Jacoby cogently addresses this issue in her biography of Robert Ingersoll, who we visited in Chapter 8. He was critical of the prevailing view that punishments, even the death penalty, are the province of God and that the government could carry them out since it had always been viewed as "the deity's representative on earth." (244) Moral punishment, especially the death penalty, has always been laden with vengeance, not something merely designed to control behavior. Without drilling into the debate, we know it is outrageously expensive and has never been empirically shown to be a reliable deterrent. It is a vestige of cultural trends, for example, Biblical prescription, culminating in the "an eye for an eye" mentality and attempts to rationalize it on supposed Christian values. There is the remarkable modern example from the late Supreme Court Justice Antonin Scalia, who offered the following in a speech at the University of Chicago Divinity School in 2002 entitled "God's Justice and Ours." In arguing for the death penalty, he actually proffered, "Few doubted the morality of the death penalty in the age that believed in the divine right of kings." (447) He also offered, "In the words of a Supreme Court opinion from the 1940s, 'We are a religious people, whose institutions presuppose a Supreme Being.'" Of course, he was citing only his own view of American institutions, though widely shared. It's clearly true that many in modern cultures continue to promulgate moral, and actually legal, values based on human created Biblical and other ancient history, carrying forward the purely voluntary nature of moral responsibility, as a fictional being is supposed to have imbued in us.

Godless Morality?

Does morality come from God, or do we merely ascribe it to Her? Remembering that we live in a determined universe, in any case, let's consider the following. Here's an oft-proposed question. "How would your behavior toward others change if you no longer believed in God?" More specifically, "How would your behavior change if you *knew* there was no god?" Perhaps, if you have been a believer and it were absolutely proven to you that no god exists, how would your behavior actually change? Such are rather strange hypotheticals, but considering the answers can be very enlightening. Would you murder, or steal a car, or suddenly begin to dishonor your parents? Unlikely. Such things *are* done by people who do believe in God; and most people don't do them, whether they believe or not. On the other hand, you might still use blasphemy in angry moments—just something most of us have just learned to do.

Answers to the above questions can be found in most people who eventually come to reject religion. Their behavior characteristically does not change in any meaningful way—save sometimes going to church or engaging in certain other religious rituals. You did not learn to behave the way you do because at an early age you were taught that you must behave the way God wanted you to. You learned your morality from your parents and other respected adults in your life, long before you knew what God wanted of you, or often before you knew much about the idea of God. They provided you with guidance, modeling, and inspiration to behave in ways that they themselves valued (although threats of eternal punishment are used on some children, even before they know what that really means).

No, it is not necessary for a god to exist in order to behave in ways that others consider appropriate or moral. In fact, even a majority of Americans (56%), who tend to be more religious than those in other developed countries, believe that it is not necessary to believe in God in order to "be moral" and have "good values" (488), and this has been a consistent finding over time. An interesting example. Respect for others is embodied in one of the best-known Christian Biblical passages is "Thou shalt love thy neighbor as thyself," as we've discussed. (Leviticus 19:18) However, in an interesting study, large samples of Christians and nonbelievers were studied. It turned out that Christians view "others," i.e., non-Christians, less favorably than nonreligious people do. (498) If theological guidance were a potent factor, it should've turned out the other way and significantly so. How people treat each other is a function of social and behavioral contingencies, as Huck Finn observed, not Biblical prescriptions.

Religion is often used as a cover for moral prescriptions or a way to enforce codes of conduct for people who would otherwise not be seen as amenable to them. Of course, such use of religion to enforce morality is on shaky grounds. Take, for example, the perspective of our Founding Fathers. As Seidel points out about their views,

> The educated elite, including the founders, achieved morality independent of religion, but they failed to extend the possibility of that achievement to others. They thought religion was needed for the commoners. The enlightened

> could use reason to discover morality, so they needed no religion other than a bare deism or theism, to which many luminaries ascribed. (457)

If the God meme disappeared, how would group or cultural behavior change? Probably not very much. Different cultures and different religions would likely be affected differently. But at this point in history, for most of the developed world (and perhaps much of the more poorly developed one), behavioral norms, mores, and controls have become so well socialized and encoded that lack of "spiritual guidance," dogma, or scripture would have little effect. (I exempt the Islamic theocracies here.) People would still basically follow the Golden Rule, for example, as they have done for thousands of years, well before any religious dicta, as our ancestors did for millions of years before that. It's how we got here. Remember, morality writ large, has always been present in other social animals as well. Again, God proves irrelevant to this issue.

Values and Morality

What values do you embrace? Is premarital sex right or wrong? Is eating animals right or wrong? How about disciplining a child with physical punishment? Is monetary wealth disparity right or wrong? And how much? Is capitalism or communism right or wrong? How about the death penalty? Is same-sex relationship or marriage moral? Is it moral to medically assist suicide, also known as "death with dignity." What about burning fossil fuels? Or racial discrimination in employment, housing, or medical care? The list is endless, and we can see that people hold contradictory values on such things. Attitudes about all these examples have obviously changed a great deal in recent years, though there are still incredible conflicting emotions and behavior associated with the differences that exist.

One might ask, "If God laid down a prescription for moral behaviors, why did He promulgate so many of them and change them so significantly over time?" The answer, of course, is that the theistic view is nonsense. If you embrace values based on supernatural inspiration, of course, it depends on what god you believe in. Morality, again, is just an idea we use to represent the normative values and behaviors we expect and sanction. Although our modern cultures display more complex behaviors, we still favor actions in proportion to how it benefits the particular group we're members of.

Levels of Value

But what group? The key is how we define the groups we value. Are the values one espouses for the benefit of a nuclear family, community, society, culture, religion, nation, or other grouping? That is why values and morality are social constructs, memes which tend to benefit certain identified groups.

Nevertheless, there is always an inherent conflict among these. Family values, so to speak, can clash with culture; religious values can clash with national ones; social values can clash with family values; and so forth. Such conflicts happen all the time. How they are resolved, in the long run, determines what values survive. Then there are the values based purely on personal self-interest. We all have some of these. Where they appear in conflict with more social values, we tend to label such folks with various words—selfish or even narcissistic. Where this is extreme, we use words like antisocial and sociopathic. Nevertheless, there is always conflict between individual self-interest and social values or mores. Such is inherent in the nature of our species. Evolution has "decided" that the fitness of the individual must always be advanced; and the survival of complex individuals, like us, depends on the survival of the group. This is why what we call morality is always variable.

Objective Standard?

Can there be any objective standard of morality which effectively balances these instincts? Of course, any answer will not come from God. Others have tried to answer the question. Many have proposed a "utilitarian" approach, as suggested by philosopher John Stuart Mill and some of the other Enlightenment thinkers. That is, in the words of the 18th-century English philosopher Jeremy Bentham, "It is the greatest happiness of the greatest number that is the measure of right and wrong." Seems like a reasonable view, but its implementation, of course, boggles the mind.

Fast forward to modern times, and we note that some have suggested "science" may be able to enlighten the issue. Can science have anything to say about morality? What is the most rational set of values? That is, we can ask whether values (or the moral or ethical codes they embody) tend to function effectively for a society. Sam Harris offers that science can provide a guide to appropriate morality. Similar to Bentham, he advocates for a "moral truth" with the dynamic that, "As we come to understand how human beings can best collaborate and thrive in this world, science can help us find a path leading away from the lowest depths of misery and toward the heights of happiness for the greatest number of people." (207) In general, I would agree with this—theoretically. He does acknowledge that "there will be many peaks on the moral landscape," but a "moral truth" can still be achieved. I'd suggest a lot of valleys too. This is advocacy for a rather lofty goal, to put it mildly. Science can certainly improve quality of life in certain areas, e.g., in the sense of reducing suffering, disease, and death (though, remarkably, some would and do disagree with those as values); but the fully utilitarian goal remains well out of reach as we might conceive of it.

This might be feasible if everyone agrees on all the goals and methods. But of course, they don't. Good for some usually necessitates loss of good for others. Tribalism has always been a fundamental part of human nature and is pretty difficult to overcome. If we value the "utilitarian" approach, as Harris and Bentham advocate, we might argue for the ascendance of certain overall values—respect for science, the extinction (not punishment) of theistic religion, the education and empowerment of women, and respect for the only planet we have. It would be difficult to argue against this course of action if the value is to enable the best chance of survival of the human species and its welfare. But alas, the values of individuals, tribes, and cultures do not always coincide on such things; and survival of the species is rarely considered. We can't ignore the fact that people are different and the different values which work for each of them. Harris acknowledges that there will be "different paths" for people, but many of these will be in conflict—an understatement at least. What is "right" and "wrong" is unlikely to become universal.

Nothing is simple. Let's take a couple of mundane examples. Is it moral to eat other animals? We certainly survived as humans, in part, because of the concentrated protein that eating animals provided us. One cannot argue against this being natural. But what is "natural" is not necessarily moral, as we illustrated with the "naturalistic fallacy." Eating animals continues to be a source of such protein, and much of humanity depends on this. But is it "right" to actually raise an animal, who may in fact be a conscious being (whatever that means), specifically for the purpose of killing and eating it? There is no objective answer. It depends on who you ask. Morality is relative.

We might also observe that even the matters of suffering and death are not as clear-cut as it would seem. Death is not the kind of value to those who believe in an afterlife or reincarnation as it is to others. We humans tend to avoid suffering. But is "suffering" something that "should" be universally reduced in our moral landscape? Some assert that suffering is necessary for "meaning" in life. Friedrich Nietzsche noted, "To live is to suffer, to survive is to find some meaning in the suffering." The ancient Greek playwright Aeschylus, "Wisdom comes alone through suffering." The Roman poet Horace, "Suffering is but another name for the teaching of experience, which is the parent of instruction and the schoolmaster of life." Martin Luther King, Jr., "Human progress is neither automatic nor inevitable . . . Every step toward the goal of justice requires sacrifice, suffering, and struggle; the tireless exertions and passionate concern of dedicated individuals." The Buddha, "All I teach is suffering and the end of suffering." And of course, death is sometimes considered valuable, notably as a relief of suffering to some terminally ill patients. If there is no universal value about suffering and death, any consistency in human morality on this subject is certainly unlikely.

These are difficult questions, and no helpful deistic guidance for them exists. Even mundane questions are troublesome. Science can provide us with better opportunities to survive and flourish, but it can't reasonably tell all of us how to behave and the best way to get along with each other. If it could, life would actually be pretty boring. Many would argue that it is the diversity of individuals and behavior which makes us who we are, and science can go only so far in helping without homogenizing us. Science can contribute here in achieving personal or agreed-upon goals, but it can't offer anything about what those goals "ought" to be. There is no way to conceive of a morality which "works" for all people and all cultures.

Conclusion

The fiction of morality comes down to this. In general, we're encouraged by religious moral codes to emit "good" behavior and eschew "bad" behavior. Of course, what is good and bad depends on which code you follow. And if you do "good," you'll be rewarded; "bad" will be punished. *Isn't that Santa Claus?*

There is no ultimate morality as handed down by some god—not the Bible, not the Quran, not any formal moral code. That doesn't mean that we shouldn't *attempt* to promote values which enhance the survival, health, and happiness of the greatest number of people. But there's the rub. How much survival and happiness "should" be promoted for some as opposed to a lesser level for others or "the most people." Can we have both? What values might enable this? These are questions as old as our civilized world. Nobody knows. But as long as we rely on the God idea to formulate our values, any progress in this area will remain insurmountable.

One might suggest that our evolution to our present form, our cognitive abilities, and our domination of the Earth (if that is even a reasonable standard) was based, at least in part, on shared tribal or cultural values. It cannot be reasonably attributed to the unique ethics of any particular individual or family or certainly any religion. What does seem relevant is the shared values of a larger and larger group of people who find that behaving in certain ways tends to enable survival of that growing culture. Perhaps survival of our species may be a guide to the most effective values. Of course, survival of the species is certainly rejected by many as the pivotal moral value. We all have shorter-term goals—for our families, neighborhoods, etc. What about the planet? To remind, we're not the only creatures here; 99.9% of them preceded us; and probably an equal number will be around long after we're gone.

As Confucius wisely said, "To put the world in order, we must first put the nation in order; to put the nation in order, we must put the family in order; to put the family in order, we must cultivate our personal life; and to cultivate our personal life, we must first set our hearts right." But to do that one must give up the notion of an omnipotent god who putatively sets values and morality for us. The only rational method—one that meets Confucius' standard, and Bentham's and Harris' goal, might come back to the ancient Golden Rule. To deny the potency of this value in the evolution of our species and its ability to endure would be extremely shortsighted. And again, no god is necessary for it.

Meaning and Purpose

Without God, does life have any meaning? Is there any real purpose to life itself? One of the most truly classic examples illustrates the point. In Shakespeare's Macbeth, Seyton (Macbeth's servant) appears to tell Macbeth that the queen is dead. Shocked, Macbeth speaks numbly about life and declares,

> *Life's but a walking shadow; a poor player,*
> *That struts and frets his hour upon the stage,*
> *And then is heard no more: it is a tale*
> *Told by an idiot, full of sound and fury,*
> *Signifying nothing.*
> (*Macbeth*, Act V, Scene 5)

This might seem like a disappointing view of life, and I'll argue that perhaps it shouldn't be. It is only a view of life taken as a whole. We do tend to consider such questions at various times, but most often when we reflect on our lives, when our life is near its end, or when we see others die or are near to death. Religious acolytes supposedly consider the question more often. Gilgamesh learned the lesson, but only that the meaning of life was that it has to come to an end.

For some reason, humans have a peculiar need to ascribe some overall "meaning" to their lives. We are probably the only creature that does this sort of thing; and it is likely that humans are the only animal whose individuals are aware of their own mortality. Other creatures are unlikely to conceive of their own deaths. Survival instinct is the strongest one of all, and contemplating one's death would not seem a functional behavior in the wild. Humans are aware of their mortality and potential death since we have well-developed imaginations, consider it about ourselves, and actually communicate about it. In that sense, we humans have an acute awareness of our own mortality and thus the contemplation of "what it all means."

Religious

It is that context which formed the basis for the superstition of "life after death" and other supernatural fantasies, as we saw in Chapter 7. It is clear that our hominin cousins manifested a similar perspective. Many did possess language, and their artifacts do suggest awareness of their mortality and the notion of a hereafter. As moral guidance became more formalized, humans grew to believe that behaving in certain ways would lead to an avoidance of actual death, that their "souls" would survive in some sort of eternal state. It has come to be a fundamental axiom among religious devotees that we must act according to certain moral standards in order to enter some sort of afterlife. It is not only a Christian superstition;

many organized religions profess this kind of idea, whether it's salvation, eternal spirit, reincarnation, etc. Such would be the real "meaning" to life.

Religious adherents often claim that their religion provides meaning and purpose in their lives. But that's a bit oversimplistic. A religious individual's sense of meaning and purpose most likely comes more from the social aspects of religious behavior than from any theological or devotional aspects. Many psychological studies support this view. (49, 477) More about such behaviors in Chapter 10. For instance, in one large study, it was found that "purpose in life" was generally related to religious orientation and happiness. (477) However, it was also found that an intrinsic religious orientation was inversely related to self-described happiness, whereas other, extrinsic factors related to religion were positively associated with happiness. Self-described "happiness," whatever that is, and "purpose" are clearly more likely a function of social conditions than any devotional aspects of one's faith. Yet, the devout continue to ascribe purpose with the language of their religion.

What is "Meaning"?

If we ask the question, "Does life have meaning," we must also ask, "What do we mean by "meaning" in that context." What do we really *mean* by that? In a practical sense, if there is—or is not—"meaning" in life, what is being referred to in this context? This is a difficult question if one resorts to philosophical discourse. As we've seen, just because you can put forth a question doesn't mean that a cogent or summary answer is actually possible.

When people use that phrase, it usually connotes the idea of a "purpose" to one's life. There can be no overall answer to the question, of course, other than an individual's own perspective. If the perceived purpose is the "next" life, so be it. For others, the only way to deal with the "purpose" issue is to formulate the question differently. There is a difference between asking, "What is the meaning of your life," and the question, "What do you find meaningful *about* your life?" In effect, the latter is the purpose or meaning any individual perceives.

For most of us, the purpose of life comes down to everything that we do, that we wish for, and how we evaluate those things. It is very different from person to person, of course. When individual persons are asked the question "what is meaningful in your life?", they usually refer to other persons and behaviors that they find rewarding and also those people or activities which, if lost, would represent significant grief and disruption of their lives. Their purpose is to derive reinforcement or satisfaction from those people and actions.

The Answer in a Greek Fable

One of my favorite stories illustrative of the point is the well-known ancient Greek myth of Sisyphus. Sisyphus was the mythological first King of Ephyra (now Corinth in southern Greece). Sisyphus was a rather evil and deceitful ruler, who took pleasure in the killing of guests and travelers, which upset Zeus no end. Also, it so happened that Zeus had secretly captured the nymph Aegina; and Sisyphus betrayed Zeus in telling her father, the river god Asopus. As punishment, Zeus commanded Thanatos, God of the Underworld, to chain King Sisyphus down below in Tartarus, the dungeon of torment. But it backfired on Thanatos. Sisyphus manipulated him into divulging the secrets of the imprisonment, demonstrating how the restraining chains worked. Sisyphus escaped and chained up Thanatos. Since Thanatos, who was in charge of death, was now disabled, no humans could die. This upset many, including Ares, the God of war, since no one could die, denying Ares his satisfaction in battles. As final punishment for his trickery, Zeus sentenced Sisyphus to roll a huge boulder up a hill, which continued to roll back down, necessitating him repeating this chore over and over again—forever. Which is what he did, pursuing nothing else.

The fable has been interpreted by many, most notably and philosophically by Albert Camus in his essay *The Myth of Sisyphus.* (74) Sisyphus believed he had a choice—to accept his fate or not. He accepted it. Since he relinquished any desire for greater meaning than his mundane task, he was able to triumph. As Camus suggests of the "moral" here—is life worth living or not?

> There is but one truly serious philosophical problem, and that is suicide. Judging whether life is or is not worth living amounts to answering the fundamental question of philosophy. All the rest—whether or not the world has three dimensions, whether the mind has nine or twelve categories—comes afterwards. These are games; one must first answer. (74)

Sisyphus tells us that once one accepts the reality of the mundane as the only existence possible, one is "free" to explore other questions. Sisyphus may have initially suffered what came to be called "weltschmerz," a term coined by Jean Paul, a romantic German writer of the late 18th and early 19th centuries. It generally suggests some sort of melancholy or "world-weariness," the psychological pain and sadness that accompanies the realization that one's life is inadequate in relation to the idealized physical and social world. Sisyphus' transition to a state of ennui or boredom (pushing a boulder not that interesting), in fact, was the only meaning left for his life; and it was from such that he derived his satisfaction.

Considering that Sisyphus may have been "content" with the meaning that arose from his mundane task, one would be tempted to ask, as Peggy Lee sang, "Is that all there is?" If the answer is yes, you might be demoralized. But only because you may want more than to move rocks around. Individual desires are infinitely variable. When you sincerely ask such questions, you are generally doing nothing more than affirming that there are not enough satisfying behaviors and people in your life to give it purpose or meaning. The only answer Camus proposed is suicide; but this is only a philosophical position. We know today, and it has always been known, that there are more practical answers.

Behavioral

Let's take the "moral" of the Sisyphus story a bit further. If the meaning of "meaning" is purpose, what do we actually mean by "purpose?" What purpose do you find in your life? Why live? As it was for Sisyphus. If we look at what people say about why and when they find life "meaningful," certain patterns emerge. They may cite family, or work, or money, or golf, or even religion, or a virtually endlessness list of things people actually do—their behavior.

What is the actual meaning of "purpose?" Purpose is usually defined by the "reason" something is done. If we take a more analytical approach, we note that "purpose" is actually a word we use to describe an outcome or possibly anticipated outcome of something we do. If the outcome of any action is desirable, as defined by the individual, then the behavior which leads to achieving it is considered purposeful. And we then consider that the actions and their consequences have "meaning." To be a bit technical, B. F. Skinner notes,

> Originally a synonym of propose, the word *purpose* has caused a great deal of trouble. Like other words suggesting probable action, it seems to point to the future. The future cannot be acting now, however, and elsewhere in science purpose has given way to words referring to *past* consequences. When philosophers speak of intention, for example, they are almost always speaking of operant behavior [behavior which operates on the environment]. As the experimental analysis has shown, behavior is shaped and maintained by its consequences, but only by consequences that lie in the past. We do what we do because of what *has* happened, not what *will* happen. (487)

This rightly puts the concepts of purpose or meaning into our actual behavior. When one talks about such things, we are reflecting on the history of what has been effective or reinforcing in our lives—those behaviors which are most probable or desirable for any individual. Again, the list is endless, from laughing (or crying) at a movie, to smelling a flower or a baby's head, to a successful exam, to sex, to taking a walk, to singing, to solving an equation, to reading a sometimes-tedious book, to just relaxing. However, when we consider our lives as "meaningful" (or not), we tend to aggregate many of those behaviors into larger concepts, such as love with others, physical health, occupational success, monetary wealth, emotional comfort, etc., or even the satisfaction or pride of progeny (which I'll address further below). Such is the purpose or "meaning of life" for any given individual. The problem arises when you try to generalize such things across all people. "Meaning" for each individual is as variable as people themselves. There is no way to generalize "meaning" or purpose for everyone, i.e., the so-called "meaning of life."

And then, can one even consider the notions of meaning and purpose in life if there is no free will? We might refer to the Sisyphus model again. It is what we do that determines whether life is "meaningful." Many people espouse the idea that meaning and purpose must come from "within." Of course, this begs the question of your behavioral history and the priorities you value as the result. And although actual free will is not supportable as a philosophical or scientific notion, the illusion is not going anywhere. You as an individual can formulate ideas of purpose and meaning in your life, and you already do, without regard to the history of the universe. How formal or verbal or shared such ideas are is a function of how you operate generally; but you do show purpose whenever you cite, openly or privately, pivotal issues about your behavior.

Natural

We might also ask the question from an even larger perspective. Can the purpose of our lives yield something greater than individual meanings? Michael Shermer effectively argues that appealing to a theistic level of analysis misses the point. Morality or meaning without a god or an afterlife is clearly possible and clearly occurs in our everyday lives. All "purpose" is defined not by a supernatural being or "place" you go when you die but by the laws of nature. He goes on.

> In this sense nature is not purposeless, as so many people seem to think it is, leading them to invoke a supernatural entity outside nature to grant us purpose. No such agent is needed because purpose is built into the cosmos and the laws of nature. The purpose of stars is to convert hydrogen into helium and generate light and heat. This is their 'destiny,' their cosmic purpose. And this is true for all things in the cosmos, and here on Earth. . . . The 'purpose' of life is to survive, reproduce, and flourish, and it has been fulfilling its destiny for 3.5 billion years in an unbroken chain from the Precambrian to today and encompassing all forms of life of which we know. (472)

Many will argue with the idea that we are "just" fulfilling a destiny, as this appears to be merely a poetic view of a determined universe. But that should not in any way lessen the value and meaning we get from our lives, and it does not depend on a god. Thus, we could ask in general, "What is the purpose of life?" Perhaps this is purely individual, as I've suggested. But if we were to infer any universal purpose of life, on this planet at least, it would be to reproduce—since that is what all life on this planet has in common. It is as true of bacteria, trees, and koalas, as it is of humans (though we humans individually find other or greater purpose in each of our lives). That is, as unsatisfying as it may be to some, we might say that *life is its own purpose*. And, as you'll recall, it is actually the "purpose" of the S*elfish Gene* which is operative here. (111)

The Coda on Meaning

Is it necessary to follow some moral code or scripture to glean meaning from one's life; and is it necessary that such guidance come from a superstitious fantasy? There is a richness and beauty to human life which is rendered no less remarkable by the fact that the idea of God is merely a superstition.

The notion that life has a "meaning" outside of direct human experience actually trivializes that experience. "Meaning" is merely what makes you feel good, though perhaps "merely" is a bit understated. When you think about your life being worthwhile, perhaps that's all that matters, as Camu suggested. It may or may not have "meaning" to others. If you're lucky, it certainly does. But it certainly has no meaning to the rest of the universe. You are a product of that universe. It owes nothing to you, and you owe nothing to it. You may, in fact, feel an obligation to other people, other creatures, even your planet. Those are laudable moral standards in most cultures, helping others and future generations survive. But the universe doesn't care about "meaning." The only meaning you have is yours, and it doesn't depend on some moral code handed down by a few arrogant men.

Conclusion

Morality itself is a fiction. It is purely a set of evolved and learned behaviors, which may or may not be associated with successful functioning of particular groups. It is merely sets of actions we agree to follow and for which there are potential sanctions.

In the end, we are left with the clear notion that authority and moral codes are not handed down from some supernatural being. And moral codes differ across groups, cultures, religions, and nations. And there's the problem. Physicist Steven Weinberg had it right, at the head of this chapter, similar to the old adage. "Religion *is* the root of all evil," does ring true. This is both a causative and categorical statement. Without the artificial scaffolding of religious dogmas, our actions would not be classified as good or evil, belying the actual reality of human behavior; and religious motivations wouldn't impose real-world "evil" upon others.

The hypocrisy of religious moral codes is obvious. We can look at just a couple of examples. To reiterate, many Muslims tout Islam as a "religion of peace." However, the morality of the Islamic religion imposes incredibly draconian punishments, even death, on its followers. Catholicism, which is widespread in the modern world, advocates giving of ourselves to help those less fortunate; but it holds for itself incredible amounts of wealth that will never be spent on this supposed moral value. It is also difficult to place confidence in a supposed God-based institution that condones and conceals the rape of children by members of its clergy. In fact, one cannot escape the hypocrisy and duplicity in the moral codes of nearly all religions.

The question arises as to whether there can be a morality which is beneficial to our species as a whole. This is an important question, and unfortunately it does not lead to any great answers, and almost by definition, certainly not any consensus. But there are some things that may lead to a growing consensus and therefore more survival-enhancing morality. Such might include (and I claim no unique perspective here) population control, economic advancement of various populations, solving the problem of energy and resources, rise in respect for science, the universal education and empowerment of women, and the relative demise of theistic religion—to name but a few.

There will always be miscreants and lawbreakers. Promulgation of moral codes, no matter how universal, whether they be statutes enforced by governments or prescriptions based in religious fantasy, will not prevent putative misdeeds. Punishment has public safety and some minimal suppressive effect. However, the only reliable answer is a morality based on individuals behaving in ways that enhance wellbeing of the group, as Nature has so wisely provided for us, something that was brilliantly laid out in the American Constitution, as the goal—to "promote the general welfare."

The Future

In Dr. Martin Luther King's oration of 1965 in Selma, Alabama, he opined that, "The arc of the moral universe is long, but it bends toward justice."[12] His Christian underpinnings notwithstanding, his implication was that a more just and egalitarian culture is a force that grows and diminishes the dogma imposed by others. That does seem to be descriptive of trends in the western world at this time, with troublesome fits and starts, and certainly not for all and not for certain peoples around the world. The countervailing force, of course, are moral codes based on organized religion. History is notably erratic in terms of people's access to justice. The long view is yet to be written; the history of human civilization is little more than ten millennia. We can see King's commentary as more of a hope than a realistic description of the world.

The real test might be what values enable survival of our species, as I opined. That will be difficult, as we continue to live in tribes, with different sets of values. We continue to divide ourselves along various religious and national lines. As President Jimmy Carter lamented in his message stored aboard the Voyager 1 spacecraft, which exited our solar system back in 2012, "We human beings are still divided into nation states." One cannot predict the future, but the long, violent history of our species obviously gives us some guidance. Morality based on some god's words hasn't done us very well. If you take the long view, what "right" and "wrong" mean may only boil down to what is *effective* in enabling our survival. That is, what constellation of behavior is effective in promoting our general survival and what is not. We currently living humans will not be in a position to evaluate this. There will be no remaining issues of human morality in a few hundred million years, as the sun's expansion makes life as we know it impossible here—and maybe, for one reason or another, a lot sooner.

10
RELIGION AND HUMAN BEHAVIOR

You are what you do repeatedly. —Aristotle[1]

Circumstances rule men . . . and not men circumstances. —Herodotus[2]

What we think, or what we know, or what we believe is, in the end, of little consequence. The only consequence is what we do. —John Ruskin[3]

Do you believe in God? The question ultimately comes down to *why.* "Why do you believe?" We've seen that the notion of God or any supernatural idea is unsupportable, from various standpoints—cosmology, neuroscience, moral or other philosophy, or any type of intellectual reasoning. We've reviewed the evolution of human religious culture in previous chapters. We've journeyed through the various arguments, which actually undermine the ideas of gods or other supernatural phenomena. We've also learned that a god is not necessary for things to exist the way they do. Yet humans continue to hold dearly onto such memes—as though they represent reality. Why? And in what gods? And why do some give up their beliefs, and why do others come to believe when they had previously not?

We are left with the questions of why humans think that there *is* a god and why they engage in worship and other faith behaviors. *In order to examine why people believe in God or any supernatural superstitions, and engage in all sorts of religious behaviors—worshiping, praying, attending churches, reading books of fables, killing each other—we must look to science for some explanations.* In this chapter, we'll discuss a few of these issues. Why do people actually believe in gods? Do we inherit our faiths? Are there any psychological theories which can explain humans' belief in gods? How do we understand religious behavior from a scientific perspective? And again, when I say "why," I'm more rightly considering the question, "how does this happen," or more precisely, "what are the conditions which are responsible for such behavior?"

Human Behavior

Aristotle had it right—what we are is what we do. And what we do is what we have learned. Nearly everything we do is because we have learned it—from saying "mama" to gesturing to doing math to worshipping superstitions in the sky. By taking an empirical approach to human behavior, we'll see that belief, and all other religious behavior, is purely a function of learning, having nothing at all to do with a supernatural being. At bottom, religion is simply a set of behaviors, responses which have been specifically learned. It may seem obvious, but without some rather complex learned human behaviors, there is no religion.

To look at this, I'll address what has come to be known as a "science of human behavior." The phrase is eerily similar to the title of B. F. Skinner's classic book *Science and Human Behavior.* (482) It's no more than a coincidence, and all credit to Skinner as the founder of modern behavioral thinking, which I'll address further below. (Yes, and I'll also debunk all the baseless criticisms of behaviorism.) I'll review this briefly before proceeding to address the above questions. It comes down to the central proposition, "Why do people actually do what they do?" Seems like coming up with any answers to that is a pretty ambitious process; the matter is as old as human thinking, with great naturalists and philosophers weighing in along the way. I'll not attempt any comprehensive answer, which would be impossible, of course. Rather, I'll outline some of the well-known principles and evidence which account for why humans behave the way they do, including religious behavior.

Religious faith arises because of the conditioning (more on that word below) of belief behavior, usually beginning in childhood, though not always. How else do we explain that some people believe, some don't, and some don't care? Also, there is a voluminous number of gods that are worshiped around the world. We're left with one of three inferences. One, despite there being an incredible variety of faiths, each religious group avows that *theirs* is the "one true religion." Two, behavioral conditioning, which is always individual, leads to changes in the behavior of humans, including some sort of faith or belief behavior. Or three, that one god has merely created a patchwork of religions to see what will triumph among us humans (no matter the cost, by the way). Since the first and third of these can be rejected as absurd, we'll move on to examine how people come to believe in something where there is absolutely no evidence for it.

The central premise: People don't engage in religious behaviors <u>because</u> they believe in God or have faith. What constitutes belief, faith, and all religiosity is a set of learned behaviors under the control of our physical and social environments. It is specifically and purely learned behavior. The natural laws that account for such learning describe the ways in which our human behaviors are shaped—from the earliest age to our personal conceptions about death. In this chapter, I'll address the science of behavior which accounts for that process. It's an exhaustive topic, but I'll review just a few of the fundamentals and concentrate on how religious behavior is thus shaped or conditioned.

It is commonly conceived that many believers are such because they've been thus *indoctrinated.* That is generally not an apt description of the process as it plays out in most Western cultures, for example, and it is certainly not the whole story, as I'll discuss later. Human beings do not develop complex ideas via osmosis. They have to be learned. When people are exposed to information, they do not merely absorb it like a sponge (many metaphors to the contrary). Information that we're exposed to either changes our behavior, or it doesn't, or it makes changes to our behavior that are not immediately apparent. But actual changes do occur. Learning necessitates behavior, and vice versa. It's a lot more than indoctrination.

This chapter is about the psychology of human behavior, specifically about the scientific principles we know to be important in understanding and predicting the behavior of humans, including faith and other religious behavior. And it is confined to those principles which have stood up to scientific scrutiny. I must here distinguish this from what most people think of as human clinical psychology, based on the often-publicized practices of psychologists and other professionals who treat humans for their various ailments, personal complaints, and dysfunctions. Such activities are of great importance to our culture and society, and they often have considerable value (though the way their effectiveness is often "characterized" is generally inconsistent with scientific principles[4]). So try to suppress some possible preconceptions or biases at this point. What I address in this chapter are the basics of how human behavior actually evolves and is maintained. Finally, some empirical evidence is reviewed, showing the relationship of behavioral constructs and religiosity.

Learning and Conditioning

A good deal of what I'll discuss here will be very obvious. Most of us accept the principles of learning for our behavior. Fundamentally, to oversimplify, we tend to get rewarded for some of our behaviors that other people value, and we might get ignored or punished for those that are devalued. But there's a great deal more to it than that. The process of learning is tantamount to what we call *conditioning*. We are specifically conditioned to behave the way we do, as are other organisms.

Behavioral science is a very young one, and we are only able to predict behavior in the most general of senses. Yet the principles are sound, and we can use them to appreciate how most human actions develop. Some readers will have a bit of trouble accepting the tenets of this chapter, some a great deal. There is a clear human tendency to believe that our behavior is freely "generated" by us, or that our brains generate behavior according to various psychological "states" and motivations, and that it is largely "voluntary." But that is not behavioral science. The central premise of this chapter is that what we do is mostly a function of the learning—the conditioning—which progressively changes our behavior. People don't like the word "conditioning" very much since it tends to imply a somewhat impersonal or robotic character and insults the idea of free will (which we've dispensed with). It is also often dismissed when it comes to complex human thinking and actions. Yet it *is* the process by which we develop nearly all of our behavior. You might consider what follows to be a pretty radical view of human actions, and many psychologists and others clearly do. But it *is* a science of behavior, i.e., one which is testable—and has been tested numerous times in various ways in various populations. People also react adversely to what might be the "deliberate" applications of conditioning principles, what's commonly known as *behavior modification*, even though this is the method Nature herself uses in shaping our behavior, whether you might consider it "deliberately" planned by others or not.

This is not a course in psychology or a technical treatise on conditioning and learning, but bear with me, as I need to drill down on a variety of concepts to explicate some critical things about religious behavior. That's what all this is about. A good deal of this will be obvious. On the other hand, some who read this will not be convinced. That's okay. As we've seen, science and objective facts do not depend on your willingness to endorse them—whether the subject is human behavior or the motion of the planets orbiting the sun. We humans "obey" the laws of nature, just as much as other creatures and events in the universe. We behave and engage in religious behavior not because there is some supernatural god, but because the laws of behavior so dictate.

History

Why do people do what they do? This is a pretty fundamental question, and it has occupied the speculations of philosophers and others for many millennia. For a long time, people believed that all of what we do arises purely from our internal states of being. From ancient times, probably beginning in ancient Egypt and ancient Mesopotamia, human behavior (and its related moods and emotions) was believed to be some combination of certain internal states. This idea was formalized by many, most notably by Hippocrates (c. 400 BCE). He is often credited with proposing the idea that human behavior was specifically a function of balances in the so-called four bodily "humors"—blood, black bile, yellow bile, and phlegm. Our "temperaments" derived from some balance or imbalance of these humors, specifically to produce certain states of being —the phlegmatic (calm or stoic), sanguine (positive or optimistic), choleric (irascible or temperamental), or melancholy (sad or despondent). These emotions also corresponded to the Greeks' conceptions of the supposed basic elements of matter composing the universe, which we noted in Chapter 3—phlegmatic (water), sanguine (air), choleric (fire), and melancholy (earth). All human temperaments and behaviors were believed to reflect some combination of these "elements." Such conceptions persisted, in some rudimentary or at least symbolic forms, for hundreds of years. For example, we find reference to this even in Shakespeare where Antony's final tribute to Brutus notes, "The elements so mixed in him that nature might stand up and say to the world 'This is a man'" (*Julius Caesar*, 5.5).

It shouldn't be surprising that a good deal of this idea of "internal states" and their combinations has persisted in academic and professional thinking to this very day, although the names have changed a bit, as we'll review below. But we've also come a long way from that, and there was also a historical trend which acknowledged the environment as the generator of human behavior. The concept of human behavior being controlled by external influences has been around for a long time, as Herodotus sagaciously noted when he observed that our behavior is a function of the environments in which it occurs. It is a function of "circumstances." Hippocrates was a naturist. Herodotus was an environmentalist. And Aristotle went even a behavioral step further. He specifically noted the relationship between our behavior and its associated rewards. (18) Perhaps he was the first behaviorist.

The laws and principles governing behavior, as effective means of shaping human actions (as I'll detail below), are both obvious and have a long history of being alluded to in all cultures. This includes religious movements themselves, over thousands of years. We see this in religious traditions and scriptures, which affirm the power of reward and punishment and other principles in control of our behavior. The Bible, for instance, is full of examples. We know that the Old Testament is replete with the behavioral concept of punishment for various behaviors, as I've previously detailed. And behavioral principles are everywhere. "If a man steals an ox or a sheep and slaughters it or sells it, he must pay back five heads of cattle for the ox and four sheep for the sheep." (Exodus 23:1) Following any physical altercation, the person responsible "must pay the injured man for the loss of his time and see that he is completely healed." (Exodus 21:18-19) (Psychologists call these dynamics response cost or negative punishment, as we'll review later). The New Testament emphasizes more positive reinforcement, a much more effective method of modifying behavior, e.g., "I praise you for remembering me." (1 Corinthians 11:2) The behavioral principal of modeling is also there, e.g., "Set an example for the believers." (1 Timothy 4:12) And the emphasis on overt behavior and its consequences is everywhere. "You will know them by their fruits." (Matthew 7:20) "In the same way, faith by itself, if it is not accompanied by action, is dead." (James 2:14-26) The Bible is actually filled with references to behavior modification, which has been extensively reviewed and catalogued in the psychological literature. (66, 297)

So-called "natural-law religions" also emphasize behavioral principles. For example, psychologist William Mikulas points out that, "Buddhism and behavior modification are primarily ahistorical in their approaches. The emphasis is on what is going on in the person's life in the present, here and now" (344). The Buddha, Siddhartha Gautama (it may have been in the 6th century BCE) acknowledged the primary role of behavior when he noted, in effect reprising Aristotle, "Every morning we are born again. What we do today is what matters most." (Whether the Buddha actually used this exact language is in some dispute.) One cannot find an organized religion in which behavior is not the principal emphasis —what you actually do being paramount—despite any theological language in which it may be couched.

Nature-Nurture

Let's take a step back. To address "why" we believe in gods, let's start with a basic, age-old question. Are human actions products of our nature or nurture? The term "Nature versus Nurture" was actually coined in 1869 by the English Renaissance thinker Sir Francis Galton (interestingly, a cousin of Charles Darwin). But the nature-nurture controversy has probably been around since the times of the earliest civilizations in some form. There were interesting debates in ancient Greece between Plato, who believed in the primacy of Nature, and Aristotle, who favored Nurture as the paramount force in human actions.

When it comes to the issues of gods, are we born with religion, or at least tendencies to worship a supernatural being? As we previously noted, some have suggested that worshiping a god is in our nature. Perhaps we have even been endowed with some genetic tendency to worship some one-true God. As absurd and academic as it may seem, let's ask the question—is God in our Nature, or does "It" have to be specifically nurtured, that is, learned? I'll touch on this a bit later.

Both?

More recent enlightened views consider that human behavior is a function of both nature *and* nurture, *and* often their interactions. That is, our behavior is a function of what we inherently are, i.e., our genetic heritage, plus what we are exposed to after birth (or more specifically after conception, since there is evidence that some learning does occur in utero). Most scientists agree that it is some combination. But we know almost nothing about how that would work for the very complex behaviors that humans emit. We can learn only certain things; some of us learn earlier and/or faster than others; we seem to naturally vary in how we best learn; and we vary in our innate capacities to learn a small number of things. It would be foolish to deny that genetics plays any major role in our complex behavior.

However, most of what we animals do is learned. In fact, behavior we often think of as instinctual, i.e., genetically driven, may sometimes be a function of learning over long periods of evolution. It turns out that some expression of genes can be due to epigenetic influences (see Chapter 4, note 4). It turns out that *learning*, which is fundamental to animal survival, can be enhanced by epigenetic changes, things that influence how genes are expressed. Instincts themselves may develop because of epigenetic influences on the "behavioral plasticity" of the organism, i.e., the ability of a species to respond to the environment in certain ways. The mechanisms are complicated and won't be reviewed here, but the evidence for epigenetic effects is growing and convincing (428). If such changes increase fitness for those individuals because certain behaviors are more easily learned, they are more likely to survive and reproduce. Environments thus select increasing capacities in regard to some adaptive learning of particular behavior from generation to generation. When we see an organism rapidly displaying such adaptation, the inference, of course, is that it is likely a pure "instinct." However, it may well have developed over evolutionary time because of learning in the species. There is some evidence for this for aggression in some insects, song communications in some birds, and many other species' behaviors that we think of as purely instinctual. Learning, in fact, may be the key to complex adaptive behavior, even though it may *appear* as perfectly natural or instinctual.[5]

However, genetics or epigenetics can't explain complex human behaviors. Those are learned (nurture), not inherited (nature). Obviously, we are all born with somewhat different physical structures, capacities, appearance, etc. And those attributes often dictate some of the shape of our behaviors. But given those characteristics, most of what we do and how we do it is specifically learned. Human beings aren't fully *tabula rasa*, as Locke offered. We do have a number of "genetic tendencies." But regardless of any nature-nurture interactions, it is a given that learning is the major factor in what we actually do.

Religion and Genetics

Given how fundamental learning is to organisms' adaptation and survival, we can logically rule out the notion that the God meme is primarily genetic. Nevertheless, there have been some attempts to study the problem. Let's take a couple of examples. As we noted in Chapter 5, despite people's reported subjective awareness of "God," there is no evidence that our brains manifest some unique pattern of neural activity that gives rise to such experiences.

Various studies have attempted to look at some inherited or genetic contribution to individuals' *religiosity*. Here and in subsequent references, that term is generally construed to mean various measures of self-reported faith, as well as associated behaviors, such as attending services, reading scriptures, praying, and other related activities. If there is some biological contribution to faith, one hypothesis has been that it might be manifest in the "birth order" of children in a family. It has been suggested that such may affect an individual's tendency for religiosity. In a study of college students raised by highly religious Mormon parents, it was found that earlier-born siblings showed clearly higher levels of religiosity than later-born children from the same family. (155) There doesn't seem to be any reason for this other than time. But what happens with time? It is somewhat unclear, but this result may well be consistent with decreased parental conditioning of religious behavior, as is characteristic of other social learning by birth order in general. Again, learning within the family may be the pivotal influence, not genetics.

Some studies have employed fraternal twins to look at possible genetic tendencies for religiosity (though we know that there is imperfect correlation of genetics between such individuals). Using data from the National Survey of Midlife Development in the United States (MIDUS), it was found that some familial contributions to reported religiosity, spirituality, and conservative ideology may occur. (57) That is, such characteristics are shared by twins to a certain degree. However, we do not know whether such behaviors are merely more susceptible to certain types of conditioning in twins than they are between non-related people. Genetics can't be implicated here.

In one psychological study, a very large sample of twins was examined for religiosity, personality traits, and other individual differences. (262) It was found that religiousness was associated with multiple factors. The relationships between individual religiousness and parental religiousness were substantial, not surprisingly. However, there were also significant associations between religiousness and sex, age, and certain personality traits (for example, "agreeableness"). There were also significant associations with best friends, spouses, and others with whom there were reliable interactions—all consistent with environmental effects. And the strength of association of these things with religiousness also decreased with age, as one would expect with environmental influences. There was no convincing evidence that genetics, per se, accounted for any significant religiosity.

Consistent with various twin studies, heritability has not been shown to be a factor in other kinds of research. For example, a very large study of religiousness in adolescents was conducted with various types of families. There were adoptive families (two adolescent siblings who had been adopted and their adoptive parents), biological families (two full biological adolescent siblings and their parents), and mixed families (one adopted and one biological adolescent sibling and their parents). Without belaboring all the details, it was found that religiousness was significantly correlated with environmental effects, regardless of the family type. There was no reliable relationship to presumed heritability. (279)

It is possible that religiosity offers a protective effect for the risk of adolescent substance abuse, and we might consider that genetics comes into play (as some attribute substance abuse to genetic influences). A very large prospective study of 357 fraternal twin pairs, 6–18 years old, was conducted to assess attitudes about religion. (135) It was found, unsurprisingly, that the notion of labeling "sin" as related to drug use was significantly influenced by environmental factors, e.g., religious practices and peer religiousness. Although the subjects were twins, there was no significant effect of genetic factors, i.e., correlations between twins' responses. It might be concluded that belief and drug use are relatively incompatible, but this relationship seems to grow out of exposures, not genetic predispositions.

Additional investigation looked at another large sample of only female twins to rule out any genetic effects. (268) It was found that the risk of negative effects of stressful life events was reduced by personal religious devotion, as other studies have found. Also, current and lifetime depression and substance abuse problems were negatively associated with religiosity, not surprisingly, and the religious devotion was inferred to be "causal" in buffering against such problems (though the latter conclusion should be considered as somewhat suspect). In such cases, religiosity is likely reinforced by its tendency to reduce the effects of life stresses, without any twin-pair correlations appearing.

These kinds of studies tend to use fraternal twins. As such, studies don't seem to find a separate genetic contribution to religious behavior. However, the variance in such behaviors and attitudes shared between twins is difficult to analyze, and attributing genetic causations to correlations is highly questionable. There is no way to rule out genetic variations in arousal, attention, and other behavioral tendencies, which may make an individual more or less susceptible to various early conditioning effects. Such would potentially include susceptibility to dogma (since that often comes with the threat of punishment) or other religious learning. We just don't know. But to assume an active genetic contribution to religiosity, per se, is unwarranted by these or any other published works.

There are no reliable studies of monozygotic twins reared apart with respect to religious beliefs and behavior. This is the case for nearly all individual characteristics. Such studies are extraordinarily difficult to conduct, as adequate samples and controlling for other effects is extremely difficult. There have been a small number of studies of monozygotic twins reared apart. However, there are none which affirm any significant genetic contribution to complex behavior in adults such as religiosity. Take the infamous large study conducted by a psychoanalyst in cooperation with a New York City Jewish adoption agency, begun in the 1960s. Interpretation of the results from this work has always been difficult and confusing, but there are no results which might suggest a genetic component to religiosity. In fact, all the subjects should have self-identified as Jewish as adults, but there was no evidence of that. There were also numerous ethical problems with the research, which have been summarized in a couple of seminal papers on the matter. (395, 456) As one of the authors astutely observed, "Genes and environments are both important for human development, but their relative contribution can be estimated only at the group level. It is impossible to say which one has greater impact at the level of the single person." (395) Nevertheless, there was no evidence for any inherited religiosity.

Conclusion

The case for any genetic component for religiosity is weak. We are not born with anything in our brains that makes us believe. There is no "god gene" for our anatomy or brain function. Religious behavior and identification are not inherited.

Most of our behavior—especially that which we consider human—is learned. Nurture—language, culture, etc., are all aspects we were not born with. These behaviors, while somewhat similar across cultures around the world, do differ significantly based on the environments in which they have evolved. The coherence of these—and there is considerable similarity—arises from the evolution of human societies thousands of years ago. But there is nothing in our genetic heritage which predicts religious behavior, prayer, worship, etc. These are specifically learned behaviors. It's also apparent that religious behavior is very different in different cultures and has been throughout the history of civilization. There is some coherence among human groups for the idea of monotheism and fundamental aspects of what is considered "moral" behavior, but such developed in parallel in different societies because of early human memes and what has been more functional in those cultures—not because of genetics.

In the end, given our determined universe, it matters little how much of our behavior is genetically driven, purely learned, or some interaction of these. Both genetics and learning do play a part in our overt behavior. However, for religious behavior, evidence for any genetic driver or an innate tendency is lacking. What we do, including religious beliefs, is not divinely inspired or genetically driven; it is always learned.

Psychological Theories

If religion is not "in our nature," we must ask why most humans seem to engage in such things—believing in gods, worshipping, revering holy books, etc. More generally, do we have any reasonable theories about why humans do what they do? As we noted, ideas about why people do what they do have been around for a long time. Psychological "theories," including some sort of personality or similar ideas, have been offered throughout history. The ancient ideas of the humors gave way to demonic possession and other primitive ideas during the Middle Ages, which gave way to various other ideas, from astrology to phrenology, i.e., physically examining people's heads to learn about their personality or character, which came around in the 19th century.

There are a multitude of contemporary "theories" about human behavior. Unfortunately, they are not very good ones. That is, they are very poor at predicting behavior. Human behavior is complicated—much more complicated than current science can currently explicate to any significant degree. As Hawking noted, "We cannot solve the equations that determine our behavior." (214) Everybody thinks they know why people do what they do, including psychologists and practitioners of various other professions. But many things we think we know about human behavior we obviously do not. The field of psychology is replete with such presumptive "knowledge." But the science about psychological knowledge here is often less than stellar. For example, in a major study, it was found that only about a third of seminal studies in leading psychology journals could actually be replicated. (381)[6] Results of many psychological experiments and naturalistic studies are just not very reliable. This is likely because "psychological" concepts, as defined in various studies, are just not very good predictors of human behavior, though a lack of rigor can account for poor replicability in a few types of these studies. (418) Remember, science works by replication and consensus, which are often lacking in psychology and related fields of study. In sum, what we think we know, even with some sort of "evidence," is often grossly inflated.[7]

The best that psychological concepts have been able to do is promote general understandings of human functioning and provide general labels or descriptions for an individual's behavior. For example, Carol didn't finish her work because she's inattentive. Alice donates to charities because she's kind. Ted doesn't care about seeing us because he's antisocial. Bob keeps interrupting because he's impulsive. *Ad infinitum.* But you'll notice the problem here when it comes to any psychological concept. Carol "doesn't finish" because she tends to focus on things other than the problem, which is her history. Alice's donations *are* kind. Ted's reluctance to see us *is* antisocial behavior. Bob's impulsiveness *is* his interrupting behavior. And so forth. Psychological "explanations" often come down to tautologies. They merely insert a label between a historical set of behaviors and any particular action that is the focus of a specific observation or prediction. It is merely the very convenient way we communicate. We say that a person avoids sharing because she is selfish. We say a person looks for the best possible outcome because he is optimistic. The language reveals the meaning. To be selfish *is* to hold things to oneself. To be optimistic *is* to see the highest or best state of affairs. But when we insert a label as a hypothetical character or trait, we're merely saying that we're trying to predict behavior based on past actions. The labels themselves offer no further understanding. Such are not good theories of human behavior. They are merely aids in communication.

When many of an individual's behaviors are "collapsed" into a general description, that's often called a *personality*. In modern times, there is a huge collection of personality "theories" which suffer from the above problem. I won't spend much more time on this, but we need to describe the difficulty with such ideas. For example, if we say that a person is an "introvert," we mean that, all things being equal (and they never are, of course), he tends to stay by himself and avoids other people. But how do we know that—obviously by a long history of observing him. If we say Ted is "antisocial," it's the same issue, just different behaviors. There have been many "theories" of personality, from the psychodynamic ones of Freud and others to various humanistic and trait theories. However, they all derive from looking at a collection of behaviors and inserting a label to describe them. That does not mean that, for example, "introversion" tangibly exists—contrary to many psychological opinions. One cannot find introversion in the brain or an ego, or esteem, or guilt, or any character of "personality," per se. Such things are simply ways to efficiently communicate, merely labels—since we generally can't efficiently describe an individual's entire history and behavioral repertoire when we refer to him or her.[8]

Psychology of the Gaps

The labeling problem becomes significant when the labels take on some seemingly tangible meaning of their own, when we give them a palpable reality. Psychological theories use them to signify some "intrapsychic" reality, that is, some quality that presumably exists in a person's "mind." These include things like extraversion, paranoia, guilt, self-esteem, narcissism, superego…and thousands more. Psychological dictionaries are filled with such ideas.[9] However, as I've said, they offer no explanatory power beyond the behaviors which are historically observed.

There's an old aphorism that well captures what we're saying here, "Labeling is not explaining." It is merely filling in a "gap" in knowledge with a label that provides some communicative shorthand masquerading as explanation. Mental or psychological explanations for human behavior are just tantamount to what I call "psychology of the gaps." That is, they invent what are called "hypothetical constructs" to presumably explain human behavior when the data are incomplete or seriously wanting. Of course, they do not explain anything. Again, they are tautologies. Saying someone usually withdraws in social situations because she is introverted adds nothing to our understanding. Saying someone cares mostly about himself because he is narcissistic is not a good theory. Noting that someone is a good leader because of her strong ego adds nothing to our understanding. As we discussed in Chapter 2, such ideas are generalizations about human behavior. We might refer to them as Carroll's "poetic naturalism" (78), or Blundell's "emergent narratives" (52), or Hawking and Mlodinow's "effective theories." (214) That is, it's the way we talk about them in their own frames of reference. But they suffer from at least four fatal flaws, much like the more ancient theories. First, they invent untestable hypothetical constructs which do not actually exist as such, other than as the amalgam of the behaviors which contribute to such labels. Second, they do a very poor job of actually predicting human behavior. Third, they cannot be falsified. Fourth, there is little consensus on what these things actually mean. They often provide comfort to psychologists and others who want to give very approximate and general descriptions for our actions; that is, they are communication shorthands, but that's about it. We need to rely on more objective explanations for human behavior.

One might assume that physical sciences, for example, do the same thing. True enough, to an extent. For instance, the concept or idea of gravity has been around for thousands of years. The ancient Greeks thought that objects fall because the basic elements of reality—the four elements we described—tend to seek their "natural place," the earth. Isaac Newton had a better idea, and he developed an equation to describe it. Einstein went even further; his equations of spacetime made predictions even better. Hypothetical constructs were no longer needed, and the matter is well described by his equations. Behavioral science is not there—yet; we don't have Hawking's equations, but inventing things to fill the "gaps" hasn't gotten us very far.

Conclusion

These problems lie appropriately at the heart of why psychological theories and their application to human behavior are not terribly well respected. We look at such "theories" as ways in which we communicate about ourselves and others, but we should be aware that's all we're doing. To assume otherwise is to unreasonably reify untestable ideas.

We do not generally apply psychological theories and concepts to other organisms. We seldom offer that our cat is reclusive because it is self-absorbed or introverted, for example—though, admittedly, some people do. We don't usually say that a lizard has anxiety based on a psychological fear of a predator. Some people do describe animal behavior in such ways, but this is merely anthropomorphizing the creatures for conversation's sake (as you'll note I have occasionally done), ignoring the objective, evolutionary (and learned) explanations for the behaviors. And we generally know that is what we're doing, not supporting some "intrapsychic" theory about those animals' behavior.

Psychological "theories" don't help us when it comes to understanding religious belief or faith. In the end, we must be able to look at human behavior, including religious behavior, as due to the learning or conditioning that we're exposed to—not the kinds of so-called intrapsychic dynamics that have characterized social sciences for centuries. If we are to explain and predict religious behavior, intrapsychic "theories" will not do the job. As Skinner observed,

> Contemporary 'intrapsychic' theories of psychotherapy tell us how one feeling leads to another (how frustration breeds aggression, for example). . . . For more than twenty-five hundred years close attention has been paid to mental life, but only recently has any effort been made to study human behavior as something more than a mere by-product. . . .The mental explanation brings curiosity to an end. (484)

Science and Human Behavior

To begin, there is no point to consider a "science of human behavior" unless we hold that we live in a deterministic universe, i.e., one in which events are explained by the laws of nature. Given that (and, of course, many don't accept that), the laws of nature must apply to us humans as they do to all other entities and events. Then this begs the question—why do we do what we do, including believing in gods? Science can potentially answer such questions. There is an incredible arrogance to the notion that science is a good way—many say the best way—to understand the natural world, but that human behavior is somehow exempt from scientific inquiry and explanation. Either our behavior is explainable (at least potentially) via scientific methods, or we are somehow different from the natural world in which we live. Here we'll deal with the former, again considering the latter just another piece of fantasy.

We actually don't often think of human behavior (including religious behavior) as being a legitimate subject of scientific inquiry. A number of factors have biased us in this endeavor. There is a history of failed and extinct psychological theories. There is the notion that we act with free will. The proliferation of popular "psychology" has been a problem. We can't ignore the poor track record of scientific and clinical psychology to thus far predict complex human behavior. In recent years, there has also been a proliferation of various so-called mental health "experts" offering armchair opinions. Finally, there is the notion perpetuated by psychiatrists and others that drugs are the answer to many of the various aberrations of our behavior and our mental health. While it may be reasonably argued that some of these developments do have cultural value, they have also contributed to the popular disrespect of the field and the notion that science cannot truly explain and predict human behavior, and it is thus not worth seriously pursuing.

Nevertheless, if behavior can be scientifically explained, i.e., predicted from known—or potentially known—factors, then such behavior must be considered as within the causal order and in the logical, scientific framework, not as an untestable psychology, an alternative philosophy, or some sort of a modified free will argument.

Measurement and Complexity

Human behavior *is* often predictable. It is just very complex, and we don't always have adequate tools to measure and track events in this area. And we don't even know how all the chemistry in the universe developed into behaving humans. There are just too many variables to consider. As Mlodinow suggests, "If one understands how all that works [the fundamental particles and forces of nature], one can, in principle, at least, explain everything that happens in the world. In practice, of course, we have neither enough information about our surroundings nor a powerful enough computer to use our fundamental theories to analyze phenomena like human behavior." (350)

You'll recognize that this is essentially a restatement of Greene's calculational impasse. We just don't have the access or measurement tools or capacity to assess our own behavior at its most fundamental levels and across time. Whether we should put more effort and resources into this is an unanswerable question. The point here is that calculational impasse continues to be the problem, but that does not escape the notion that our actions *are* theoretically predictable according to the laws of physics, chemistry, biology, and—as I'll discuss—the laws of behavior.

We all have a unique history, which is also different at each moment in time. At this point in history, we cannot well explain any particular human behavior or behaviors. No neurological or physiological assay, imaging, or other test can predict a specific (non-reflexive) response. Behavior is a function of numerous variables—genetics, congenital environment, postnatal exposure to the physical and nutritional environment, the physical, behavioral, and physiological environments throughout life, and the current environment and contingencies of behavior. These events and stimuli constitute an incredibly large volume of data. Considering that there are approximately 86 billion neurons in the human brain, which may enable trillions of connections and combinations, and the momentary events in a person's life, it is obviously impossible to consider this enormous data set in predicting a specific behavioral response at any point in time and space—even if we had the necessary measurement tools and computing power.

For example, let's consider a "simple" behavioral predictive question: Will Bob ever shake hands with Ted? At first blush, there's no way to predict that. We don't even know if they will ever meet. However, if they do, if Ted is a sales target of Bob, if they are introduced with some positive language, if they are both American, if they are physically close together at a first meeting, if there is little possibility of disease transmission (such as the flu or COVID-19), we might predict that a handshake *will* occur. That's a pretty safe bet, based on understanding of the relevant environment. Nevertheless, we wouldn't be able to predict the strength of the handshake, the timing of the interaction, the nature of other nonverbal gesturing, any verbal behavior that would accompany the situation, and many, many other subtleties. We just don't have enough information. We can reasonably predict whether Alice will keep her dental appointment tomorrow, though she's cancelled before. We can reasonably predict whether Carol will go to work tomorrow, but she's sometimes taken a day off. There are no equations for such things, as Hawking noted, but it would be easy to generate at least *probability* estimates for such behaviors. We can also mathematically describe the behavior of groups of humans. For example, the so-called statistical "normal distributions" of traits or behavior almost always appear in performance on intelligence tests, birth weight, adult height, SAT scores, shoe size, and many other variables. However, the probability of obtaining particular values or behavior for any given individual is fairly gross in such realms. Predicting how a given individual will perform on an IQ test, for example, can have some small error. We would need to know a great deal more about that person to be more precise. We are limited in such efforts not by our general understanding of behavioral principles but by the calculational impasse.

Resistance

There is great resistance to a science of human behavior. We tend to accept the validity of science when it is easily understood, when it is convenient, when there's a public consensus of experts, or when we find it useful. But we tend to reject science when the matters at hand involve our own thoughts, values, and actions. We tend to resist, sometimes vehemently (e.g., when it conflicts with religious views, as we've discussed), any scientific explanations for our own behavior, even though we accept them for other areas of life and sometimes for others' behavior.

This again is eminently hypocritical unless one assumes that human beings are not governed by the laws of nature. The only way to avoid a science of human behavior is to endorse an indeterminate universe, the unfounded idea of free will (which I'll get back to later), or the actions of some supernatural being. Behaviors have causes, and the foundation of behavioral science is the enumeration of those causes. As Shermer notes, there *are* "causal pathways" to behavior arising from various historical and contemporaneous influences, e.g., "laid down by my ancestors and parents, culture and society, peer groups and friends, mentors and teachers, and historical contingencies going all the way back to birth and before." (468) Such is what I'll be addressing here.

The Right Questions

Again, the question, "Why do we do what we do?" Perhaps the better question is, "*How* does our behavior come about?" I've noted the "why" vs. "how" problem a couple of times. In this case, the appropriate question is, "What are the conditions which lead to our actions?" We can ask why the sun rises in the east, but what we really mean is "*How* does the solar system behave in ways which lead us to that phenomenon." Similarly, we can ask, "Why are you reading this book?" If you say that it is because you want to or you're curious or you're irritated, you've merely pushed the 'turtle' down the road, i.e., "What history has generated those attitudes?" That is, "What are the historical and contemporaneous factors that give rise to such behavior?" Science can answer such questions, though it presently does a rather poor job of it when it comes to humans. Nevertheless, we'll examine how this is generally relevant, including to the question of religious belief.

Do We Live in This Universe or Not?

Let's return briefly to the issue of determinism. Some have argued that a deterministic view of human behavior is not helpful. For example, physicist Mlodinow argues that a deterministic view of human behavior is actually not a tenable idea, suggesting the following:

> Determinism in human affairs fails to meet the requirements for predictability alluded to by Laplace for several reasons. First, as far as we know, society is not governed by definite and fundamental laws in the way physics is. Instead, people's behavior is not only largely unpredictable, but, as Kahneman and Tversky repeatedly showed, also often irrational. Second, even if we could uncover the laws of human affairs, as Quetelet [Adolph Quetelet, the 19th century Belgian astronomer and polymath] attempted to do, it is impossible to precisely know or control the circumstances of life. That is, like Lorenz [the Dutch physicist, Hendrik Lorenz], we cannot obtain the precise data necessary for making predictions. And third, human affairs are so complex that it is doubtful we could carry out the necessary calculations even if we understood the laws and possessed the data. As a result, determinism is a poor model for the human experience. Or as the Nobel laureate Max Born wrote, 'Chance is a more fundamental conception than causality.' (349)

Let me take this on directly. First, there *are* relevant "laws of human behavior," which I'll be addressing below. Second, contrary to Mlodinow's assertion, Kahneman's conception does not generalize the notion of irrationality. In fact, Kahneman noted that, "Irrational is a strong word, which connotes impulsivity, emotionality, and a stubborn resistance to reasonable argument. I often cringe when my work with Amos is credited with demonstrating that human choices are irrational, when in fact our research only showed that Humans are not well described the rational-agent model." (259) Rather, what is characterized as rational or irrational in everyday life is often just how we communicate about such concepts. We commonly see rationality or irrationality, for instance, as a function of certain behaviors, e.g., to what degree any behavior is in the actor's self-interest (or the interest of certain others), how acutely or rapidly it is emitted, and certain other factors. Such parameters are not unitary human actions but are subject to the well-supported laws of behavior, as I outline below. Third, that we are not able to obtain necessary data and do the needed calculations is not an argument against determinism as a model for human behavior; we continue to make progress on objective laws in this area. Fourth, the notion that Max Born's "chance" is a better explanation is out of context. In fact, Born was referring to nature in general, based on quantum theory, not specific conceptions of human behavior.[10] Finally, we're not talking about "control" here (though behavioral science has enabled significant advances in how we deal with many human problems). The problem, of course, is, "What's the alternative?" It cannot be pure chance, remembering that there is no true randomness in the universe, and, unless Mlodinow is willing to allow free will outside the laws of nature, there is no alternative to the notion that human behavior *is* as determined as is the rest of the physical world—any behavior follows from the events that have preceded it according to the laws of physics, chemistry, biology, etc. The failure of behavioral science to predict all behavior is not a valid criticism of the science, and certainly not of determinism—any more than failures of physics to find dark matter or a quantum theory of gravity due to limitations of relevant information and measurement technology.

Physicist Carroll also argues that the context for extant "theories" of human behavior is not based on a deterministic view of reality, i.e., not knowing "any way to predict what a person will do based on what we can readily observe about their current state. Whether we think of human behavior as determined depends on what we know." (78) But isn't that exactly the point? Lack of knowledge—the calculational impasse notion—says nothing about whether behavior is determined or not. If we knew the past more accurately, we would be able to predict behavior more accurately. Lack of knowledge does not bear on the question of whether or not human behavior is "determined." In addition, "thinking of human behavior as determined" is not exactly the point. That does carry with it some negative connotation, as most people would view it. Whether or not one entertains this is a function of whether we think human behavior—and its present and historical components—is consistent with natural laws or not. You can't use lack of knowledge about particular events to eschew first principles.

Conclusion

Having dispensed with the above criticisms, we do admit that human behavior is indeed complex, more complex than we can usually allow ourselves to imagine. And yes, as Hawking observed, there are no equations describing human behavior (at least yet, with some exceptions below). (214) Human beings are just incredibly complicated, principally because they have so many more behaviors, in so many different circumstances, and with so much more subtlety than other organisms. We do not have theories about such things that allow us to significantly quantify the relevant variables. But as we've seen, this really boils down to a problem of measurement and the calculational impasse problem.

Let's take examples from the world of other sciences. We have a very good empirical understanding of what causes earthquakes, but good prediction of such behavior is still lacking. We have a very good understanding of what happens when a patient suffers a heart attack, but we cannot predict when or for whom this will happen. Seismology and cardiology are respected scientific disciplines, but these are problems of not being able to measure and compute based on all relevant variables. Behavioral science suffers from similar limitations, but that does not make behavioral theories any less worthwhile, and certainly better than Born's "chance" at providing a more reasonable explanation.

We are also often unwilling to take behavioral science seriously, which harkens back to the long-held notion that we are somehow special or unique in the natural world. Yes, we can admit that people are largely what they've learned and what they do, as Aristotle seemed to suggest. But aren't people *surely much more than that*? The answer is a qualified "no." Of course, you are also your anatomy, with your own cells and those of other organisms (the microbiome), and your subjective experiences. Yet in the end, we are nothing if not behaving organisms (figuratively and literally), and addressing this scientifically is clearly possible.

Learning

We learn most of what we do in life—from saying "mama" to throwing a ball to solving equations to believing in some superstitious being. We are not born with any of these behaviors. They are learned. Let's first address the question, though it may seem obvious: What is learning? We learn by being exposed to various stimuli. But it is not merely exposure. Learning is specifically any change in an organism's behavior associated with systematic exposure to its environment. We call that *conditioning*. Some try to differentiate the ideas of learning and conditioning, implying that the former leans on some sort of consciousness or intent. That's obviously not the case. As previously mentioned, the word *conditioning* is also abhorrent to lay people, as well as many psychologists and others who purport to explain human behavior. Nevertheless, it is absolutely fundamental to what we do, including what we believe.

Nearly all organisms learn—perhaps *all*. That is, their behavior is partly or mostly a function of conditioning, the association of environment, behavior, and consequences. Learning in most organisms does not require consciousness or even a brain. Even more basically, it doesn't require what is termed "neural plasticity," the tendency of various neurons to reorganize themselves via development or exposures. Most living things do not even have nerves. Significant learning is extant in almost all social species, and numerous experiments have demonstrated that even very primitive organisms learn to modify their behavior.

Learning is fundamental to nearly all animal life on Earth. It is a method of adaptation to one's environment. We are programmed to learn, a process that is critical to evolution. Species which are more capable of learning are better suited for survival. It starts very early in an individual's life (including in utero for a few advanced creatures), and it started early in the evolution of life itself on Earth. Even the most primitive organisms learn, i.e., change their behavior via conditioning (also called "associative learning" in more primitive animals). A few interesting examples: In classic experiments, it has been shown that even such primitive organisms as a flatworm, e.g., a planarian, can learn. It doesn't have much of a "brain" (just a few thousand nerves), but it learns well. After a planarian has learned a task, such as avoiding a shock, you cut it in two, and the two halves fully retain the learning, including the one without any "brain." (335, 357) When viewed from outside, this is a pretty "clever" strategy, though we wouldn't usually describe the planarian that way. Even jellyfish are also able to learn through a process of conditioning, though they also do not have what we would call a "centralized brain." (51) There is even evidence that even some one-celled creatures interact with their environment to enable learning. For example, consider the well-known amoeba. It can learn to change its behavior in "anticipation" of changes in its environment. (397) If you can believe it, a slime mold (an aggregation of amoeba-like single-celled organisms), which has no neural or sensory organs, can adapt its behavior based on conditioning. For example, it can learn to anticipate noxious events (442), and it can "remember" where some food is located. (166)

Remarkably, even some plants can learn. For example, in a hallmark study, pea plant seedlings were taught to grow in one direction. They learned that light needed for growing was associated with a neutral experimental stimulus. (177) They were placed in a maze that split in two directions. If they grew in the direction of the stimulus (a fan), a light would come on. If they grew in a different arm of the maze, no light would be provided. The plants did learn how to obtain the needed light (No, the plants do not have some innate mechanism for associating light with the movement of air). Another interesting study used "touch-me-not" plants, aptly named since they normally fold and wilt when touched. Through repeated association, the plants were taught to "refrain" from their natural withdrawal behavior. They were frequently and gently jostled, leading to folding and wilting, but eventually the touching no longer emitted the withdrawal behaviors. There was no danger there. Associative learning was evident; the plants "remembered" their exposures and no longer withdrew from touch—even a month after the training was discontinued. (176)[11]

Conditioning appears to be one of *the* most fundamental aspects of life, from the most primitive organisms to humans. The principles of conditioning, which we'll detail further, do explain the behavior of nonhuman organisms, and they do for humans as well. The distinction is only a matter of the behaviors of which any organism is capable, based on its genetically determined morphology and capacities. That we are capable of more complex behaviors than other extant creatures in no way minimizes the potency of the same behavioral principles that influence pea plants, amoebas, and dogs.

What is Behavior?

Religion, regardless of faith or denomination, is a set of behaviors. As I've said, obviously if there were no behavior, there would be no religion. What is important is to explicate—or at least attempt to—are the conditions which lead to such behavior. But first, we must define, in some operational way, what "behavior" actually is. This is not as straightforward as most people think. Most of us think we know, but often we don't. The most inclusive definition, and probably the most useful, is: A *behavior* is anything that can be observed, e.g., by our senses directly or sometimes indirectly with some other measurement. It is some change in the organism that we can actually detect in some way.

It is the same idea as with the physical sciences. We must be able to observe any phenomenon that is the subject of study. The motion of the planets is behavior. Chemical reactions are behaviors. Even the changes in entangled particles are behaviors. We don't always "see" such things, but we have measuring devices which enhance our ability to observe. What is relevant must be amenable to observation, and there must be some observable change, some sort of motion that can be recorded.

A few examples: Walking is a behavior. So are sitting, jumping, and sleeping. Yelling is a behavior. Anger is not. The latter is a concept and not observable. Laughter is behavior. Humor is not. Same problem. Talking is behavior; thinking is not (more on this below). If we're talking about religion, praying is a behavior; faith is not. Reading a Bible is a behavior; feeling inspired is not. Singing is a behavior; hymns are not.

Overt or Covert

Behavior can be either overt or covert. Overt is what we can observe in some way. Covert behavior is not directly easily observable, but it can still be behavior. The behavior of your lungs, intestines, and even neurons constitute behavior, even though they may be more difficult to observe directly, or they may be observed using measuring devices.

We might consider heartbeats as behavior, for example. They certainly change in relation to one's environment. They can speed up or slow down with changes in the environment, relaxation, exercise, fear, etc. But such covert behavior is also subject to learning. Most reading this are familiar with the so-called "fight or flight" response, which came up in Chapter 6. This involves an increased heart rate, along with many other physiological changes that enable one to potentially deal with mortal danger. But it also appears where there is fear of far less than that. It can appear with public speaking, taking a test, or when a certain relative comes to dinner. If that happens again, especially repeatedly, you might have a "fight or flight" response even on first hearing of the relative's visit. That is conditioning. Your body learns because of the historical association. The response of heart rate becomes conditioned. And the behavior of heart rate can also be learned directly via "operant conditioning" (I'll detail what this actually means below). We've been able to demonstrate this with the technique commonly known as biofeedback. Covert behavior, such as heart rate or blood pressure, can be continually measured, and feedback is provided to the subject as the measurement changes in a prescribed direction. People can learn this quite easily; such learning has also been used to modify the behavior of brain waves, the electrical conductivity of the skin, moving blood to certain parts of the body, and many other seemingly covert behaviors. Such covert behavior is amenable to conditioning. Believe it or not, we've taught the same skills to animals, which is how the field of biofeedback got started. In the 1960s, psychologist Neal Miller began training animals to control certain bodily functions. For example, he trained rats to control their heart rates and blood pressures. That's an easy thing to do—just relax to lower the rate or jump around to raise it. But he restrained—sometime paralyzed—the animals, and they were still able to do it for the right reward (certain electrical stimulation in the brain). They were also able to learn control of other physiological behaviors, e.g., urine formation and constricting or enhancing the flow of blood to a designated ear. Such are obviously irrelevant in practicality, but they show that almost anything can be learned. Learning can affect almost any behavior, including ones generally thought of as autonomic. Almost anything that an organism can exhibit, overtly or covertly, is subject to learning, including the ones we emit. Learning is universal, if you will, and it is responsible for just about everything we do.

Subjectivity

What about behaviors we consider "private," that is, "observable" only to ourselves? What is an idea? What is feeling? What is a thought? We seem to know the answers to those questions when they're posed. But we really have no good answer. Similar to what we covered in Chapter 5, on consciousness, there are no operational definitions for such things. Attempts to define them merely result in a never-ending series of unmeasurable synonyms.

These are what we call *subjective*; they are not measurable behaviors. We can only measure the verbal and other behaviors that a subject emits. Subjective events are *not* measurable—by definition.[12] Behavior must be measurable in some way. It is not a feeling or thought or emotion. That is not to deny a person's unique experiences. But they are not measurable. That is, there is no tangible quality to "experience" that can be operationalized for measurement. Psychologists often will offer that certain subjective states *are* measurable. For example, psychological tests or clinical assessments might be used to assess some alteration of thinking or some "emotional state," for example. However, such are simply inferences based on the verbal and other behaviors of the subject; the thinking or "emotional state" are not behaviors.

Origins

Given the "reality" of subjective experiences to the person, we might ask, "What is connoted by a subjective experience?" That is, can there be some objective referent to those experiences? In fact, what modern humans generally describe as thoughts, feelings, emotions, etc., often derive directly from actions or behaviors. Subjective or cognitive processes arise because of what we do and are even often just substitute labels for what we actually do, what we have done, or what we are likely to do. In fact, what is subjective generally has its roots in objective behavior.

The words we use to describe subjective states generally do have their referents in objective, external actions; usually, this directly parallels the history of language. The language we often use derives from our behavior or actions and has only recently come to suggest some sort of subjective experiences. Let's take a couple of examples. If you "intend" to do something, a subjective experience, that connotes some probability of actually doing it. It is usually the likelihood or tendency to behave in a certain way. "Intention," in behavioral terms (from the Latin origin, a stretching or exertion), is merely the probability of being "pulled or stretched" in a certain direction. We might also say that a person is "motivated" to act in a certain way, a subjective state, but it is merely the tendency, or probability, to "move" in that manner.

Let's take a prime illustration of this principle. As Skinner points out, "*Experience* is a good example. As Raymond Williams has pointed out, the word was not used to refer to anything felt or introspectively observed until the 19th century. Before that time, it meant, quite literally, something a person had 'gone through' (from the Latin *expiriri*), or what we would now call an exposure to contingencies of reinforcement." (487)[13] And we continue to use the word for what we've actually "gone through," e.g., he had a bad experience; that skill requires a lot of experience; you should experience Paris. The notion of subjective "experience" for an individual derives from exposure to the environment and related behavior. We have just used it more often recently to more efficiently communicate something subjective.

An extension of this idea is what people label as the idea of subjective "emotions." In fact, the concept of "emotions" itself has its referent in objective behavior. It traces back to the Latin *emovere,* to move out or agitate, and the subsequent Old French term *emouvoir* or to "stir up." When we talk about emotions, we're talking about being moved or "stirred up" in some way. Take a popular example. One of the most cited examples of emotion is what we label as "anxiety." To *feel* anxious may actually come from a Latin word for squeeze or "choke" (*angere*), and we continue to say things like "I was choked up about it" or "he choked on the eighteenth hole." There is also derivation from the Latin *ānxius*, usually meaning uneasy or troubled. Its roots are in behavior. One of the common human emotions concerns being in "agony." We often use the word "agony" to reflect an experience of severe pain or personal suffering of some sort. But it generally derives from Latin and Greek roots, meaning to openly struggle with something.

We generally use words that connote subjective states with the behaviors and history of how we got there. The list of such things is seemingly endless. (487) In describing ourselves, it is just very convenient to conceive of such notions as internal subjective states rather than comprehensively describing or enumerating the external realities that lead to such. The overt reports of subjective experience often merely suggest the history of our behavior or the probability of behaving in certain ways; it is that behavior that is our focus.

Thinking

A special word about the subjective experience we call "thinking." I've used this word throughout the book, rather casually, since it's merely a helpful word in communication. Thinking can't be easily defined, or perhaps not at all. At base, thinking is not behavior, nor even covert behavior, as we usually conceive of it. It is subjective experience. We generally consider thinking to be what we "feel" inside our heads. But in Old English, it originally suggested an *action*, such as "to say to oneself" or "cause to appear to oneself." We also commonly consider "feeling" as a type of thinking. We "feel" loved, excited, irritated, and many other subjective states, but such derive from the primary meaning of "feel," an action which senses or perceives the environment. The word implies action, which we often merely assign to some internal state. You'd also be unlikely to label your experience as thinking or feeling if something hasn't "occurred" to you (from a Latin root suggesting actions, such as that something was presented, befell you, or ran into you), e.g., "It occurred to me that … " We also often use the word thinking as a shorthand for other private actions or information, e.g., as a synonym for "*holding* an idea," holding being a behavior of action. "I will hold you in my thoughts" "We hold these truths to be self-evident." On the other hand, I "reject your idea," means to throw it back at you. Such various kinds of "thinking," like most of the words we use to describe subjective experiences, have a history of describing observable events or actions.[14] We now use such words to more efficiently communicate about such things. For example, "I'll think about it." Not very efficient would be, "I'll *look* into the matter further, *entertain* the alternatives, *weigh* them against some criteria, *examine* the possible consequences, *put* them in some context, and *select* some option."

Just like the idea of consciousness, thinking as private or subjective experience, is inaccessible to scientific inquiry, even though we may "think" it is. Nevertheless, such things are the "subject" of countless "experiments" in psychology and other fields of social and neuroscience. The idea of subjective thinking (being different, of course, from the actions of neurons in the brain) is something that we cite when we are talking or otherwise communicating, including to ourselves. It is not amenable to any empirical measurement, and it is not something that can be replicated. My thoughts are not your thoughts, no matter how much we may agree on what we may be "thinking." The only outcomes of experiments purporting to measure "thinking" are the verbal reports and other objective behavior of the subjects. Thinking may be associated with the behavior of neurons, but it is not itself overt or even covert behavior. It is purely subjective. As I said in Chapter 5, identification of some groups of neurons associated with overt reports of subjective experience is beginning. At this point, we don't have the tools to measure how all those cells act and communicate with each other, though it is a valid subject of scientific inquiry. But those actions are not "thinking." Whether "thinking" can be operationalized, consider whether it can be replicated in an artificial system. If a system with artificial intelligence claims to be thinking, but we can only measure the actions of its circuits and the output, would we say—much as for consciousness—that it has subjective thoughts?

People also tend to consider "emotions" as something different than thinking. They are not. When you have some kind of emotion, it is because your brain has done something pretty complex things, just as with various other kinds of thinking. That other, more primitive parts of the brain are sometimes involved when "emotions" are reported doesn't mean they're not thoughts. Without a reported thought, it wouldn't be described as an emotion. Whether one describes an experience as a thought or emotion depends on many factors, including the language we use to communicate about it.

Given the intangible quality of thinking, we're left with the problem of whether any thoughts are meaningful in how we behave. The answer is not as simple as you might think. We've touched on this in previous chapters, but it deserves another emphasis here. There is the mistaken impression —on the part of all of us, including most psychologists—that thinking is the *cause* of behavior, especially complex behavior. There is no evidence of this, of course. All we know in this regard is that sometimes neural activity may occur "before" a behavior is emitted—though not always (see Chapter 6). What is clear is that contemporary and historical contingencies of behavior are the controlling factor in what behavior is emitted, and thinking (feeling, etc.) are covariates of this. That is, our private experiences sometimes tend to occur near the time of our actions, but they are not the *cause*. That one may seem to precede the other at times does not make it causative.

Given the calculational impasse, if we knew all the contingencies and history which led you to think about emitting a certain behavior (I decided to have a drink), we could also predict that behavior (you drank, or not). This makes the subjective thought of drinking merely a covariate of the behavior, a related subjective event, not a cause of the action. One can conceive of thoughts as what is known as "proximate cause." That is, perhaps there are preceding factors, but the final common pathway is the nearest event to the action. Of course, there are three problems with this reasoning. First, thoughts are not measurable as causes. Second, such subjective experiences only rarely precede overt action; most of our behavior actually occurs without awareness. Third, as unpalatable as it would seem, we return to the cosmological principle in Chapter 3 that the notion of specific individual causation is not really relevant. To wit, as physicist Carroll offered, "Different moments in time in the history of the universe follow each other, according to some pattern, but no one moment causes any other." (78) To imbue private thinking as the cause of behavior is scientifically unwarranted.

Conclusion

What is subjective is not measurable. What is objective and measurable are the verbal and nonverbal behaviors that people emit—however frequent or rare, however complex, however subtle. Those are the appropriate targets for scientific inquiry. Whether they conform neatly to subjective experience is not relevant. Subjective experience, however much we value it, is not empirical or measurable. Our objective here is to examine the notions of faith and religiosity. Thus, we must shun the idea of religious "experience," a subjective phenomenon, but examine the objective behaviors that fall into such categories, as I'll address below. Religious behavior is behavior, whether it's praying, reading a Bible, going to church, or saying, "God loves me." That is our focus. It is the only way to understand why people are religious.

Behaviorism

Religious behavior—all faith, belief, and ritual—is learned. It is not a choice based on the mistaken idea of "free will." It is not inspired by a god. It does not come from your genes, as we noted above. The laws of behavior, which I'll detail below, *are* what determines a belief in a god and various religious faiths, as it does other human behavior. Those laws —like the law of gravity or any other laws of nature—operate whether we like it or not. The study and application of those laws is the science of human behavior, which has become known as *Behaviorism*.

Let's be clear about one thing that's often confusing to people about this word. *Behaviorism is an approach to understanding, predicting, and sometimes modifying the behavior of individuals.* It focuses on each individual as unique. It does not purport to explain all human behavior in general (which is also why it is quite often misunderstood, when construed that way). It is only a set of unifying principles used to generate explanations and predictions of particular individuals' (and sometimes groups') behaviors. Since we are all exquisitely different, the theory only "works" (to some degree) as relevant to the details of particular humans' behavioral history and circumstances.

Much of this will be obvious to the reader, and we tend to accept the principles of learning or conditioning for many simple or mundane behaviors—from potty training to learning arithmetic to shaking hands, for example. But when it comes to behaviors that are more complex—romantic relationships, writing a book, or religious faith—we tend to reject behaviorism and resort to psychologizing and "free choices" as explanations for our behavior—though, as we've seen, those ideas actually explain nothing.

Behaviorism is directed at explicating the conditions which lead to output of behavior—the "how." The causes are the historical and current environmental conditions and contingencies. That we cannot perfectly itemize them and predict all behavior is not the point. Science is limited in this and all other areas since we do not have access to measurement of all those past and current events—and certainly not the computing power to analyze them even if we did. This again is the problem of "calculational impasse." We humans cannot conceive, in any rational way, how the myriad of historical and contemporary events that enable our behavior can be brought to bear on this problem, but we can make some clear inferences.

History

Excuse me, but take another step back. The history of behaviorism might go back to the ancient Greeks. Again, Herodotus had it right when he noted that human behavior was a function of "circumstances." That is, our behavior is a function of the environments in which it occurs. As a result, what we are is a collection of behaviors that have been learned or conditioned, as Aristotle noted over 2,500 years ago. That is exactly what we refer to as behaviorism.

The founding of an actual modern science of behavior is generally ascribed to William James, who founded the discipline of psychology at Harvard in the late 19th century. Despite his excellent work on such things as sensation, physiology, perception, and emotion, he was still unconvinced that Psychology could be a distinct scientific discipline, writing, "This is no science; it is only the hope of a science." (239) But by 1892 and the publication of his *Psychology: Briefer Course,* the field was well under way. (240)

In the early twentieth century, American psychologist Edward Thorndike was the first researcher to develop a truly systematic approach to the analysis of behavior. He found that behavior which was closely followed by "satisfaction" or reward was more likely to recur when similar situations occurred. If behavior was followed by "dissatisfaction" or some type of discomfort, the behavior would be less likely to recur. He called this "instrumental conditioning," and the principles became known as the "law of effect." Although he convincingly demonstrated this law of effect with cats and puzzle boxes, the principles quickly became generalized by others to all sorts of behaviors and creatures. Today we know the principles as *reinforcement*; and we find that they apply just as potently to humans.

Then there's the classic experiments on dogs conducted by the Russian physiologist Ivan Pavlov that we're all familiar with. To remind, Pavlov repeatedly paired a bell and the presence of food (actually a meat powder), which generates a salivary response. At some point, the bell itself, without food present, generated a salivary response. This became known as "classical conditioning," the association of a conditioned stimulus (a bell) with an unconditioned stimulus (food). Any dog owner will tell you what a powerful process such conditioning is. Then there was the famous experiment conducted in the early 20th century by John Watson, the founder of modern Behaviorism. He conditioned a young child (Little Albert, not his real name) to be afraid of a white rat. Albert had no aversion to the animal when the experiment began; but after a few pairings of the rat and a very loud noise, he became afraid, with crying and other avoidance responses, when only the rat appeared. He was conditioned to be afraid of the rat.[15]

Behaviorism as a discipline was essentially founded by Watson and described in his classic 1924 volume of the same name. (533) He described the new science of behavior as a "natural" one. For the first time, the objective science of human actions became divorced from the duality which had pervaded human philosophy and "natural philosophy" throughout history. Rejecting everything from ancient religion to Descartes to some other psychologies, Watson observed,

> There is a fearsome God, and every individual has a soul which is separate and distinct from the body. This soul is really a part of the supreme being. This concept has led to the philosophical platform called 'dualism.' All psychology except behaviorism is dualistic. That is to say we have both a mind (soul) and a body. This dogma has been present in human psychology from the earliest antiquity. No one has ever touched a soul, or has seen one in a test tube, or has in any way come into relationship with it as he has with the other objects of his daily experience. Nevertheless, to doubt its existence is to become a heretic and once might possibly even have led to the loss of one's head. Even today the man holding a public position dare not question it. . . . From the time of Wundt[16] on, consciousness becomes the keynote of psychology. It is the keynote of all psychologies today—except behaviorism. It is a plain assumption just as unprovable, just as unapproachable, as the old concept of the soul. And to the behaviorist the two terms are essentially identical (533).

Thus, the modern science of psychology was born. If we were to examine the behavior of human beings, it was objective science that would take up the endeavor. That effort was taken up by B. F. Skinner in an approach he called the "experimental analysis of behavior" that has evolved into the current behaviorist approach. Skinner's work, and behaviorism in general, has gotten a particularly bad rap—with the public and in the psychology and other professional communities. Many people know of B. F. Skinner's experiments with pigeons and the derogatorily dubbed "Skinner Box." But that's *not* what Behaviorism is all about. It is about an approach to understanding the behavior of organisms, including people, that is empirical and thus verifiable and falsifiable (I'll respond in detail to the various criticisms of behaviorism below).

Skinner's contribution was similar to other revolutionary ones in the history of science—to use specific measurements and hard data to study phenomena. Skinner was inspired early by the work of the English philosopher Sir Francis Bacon, who, like Einstein, had a distinct aversion to authority and had championed the inductive method in science. After discovering the works of Pavlov, Watson, and Thorndike, Skinner rejected the prevailing psychological "theories." He expanded the work in behavioral science, beginning with pigeons, and the approach was later expanded to other organisms. The classic volume, *The Behavior of Organisms*, detailing the core principles of behaviorism, was published in 1938. (480) In it, he developed the central principle of behaviorism that we know today: *"behavior is a function of its consequences,"* as well as the various laws that govern behavior, and he showed that what we call *"schedules of reinforcement"* have distinct and recognizable effects on behavior. We'll briefly review some of these dynamics of behavioral science below.[17]

Autonomous Man

We humans tend to think that we can act independently of all our history and circumstances, consistent with the idea of free will, as we discussed in Chapter 6. We think of ourselves as the spontaneous initiators of our own behavior, free of any real control by our environments—that our behavior (especially any behavior we consider important, that is) arises solely from within us—that it is essentially uncaused. A person so acting is referred to as the "autonomous man," a term coined by Skinner in 1971. (484) (I should clarify that the reference to "man" here should be taken purely as metaphor and was merely consistent with usage of the times.) The science of behaviorism is based, in part, on rejecting the idea of the "autonomous man." There *are* causes for behavior. We just don't always know them well, which is why we can't meet Hawking's criteria of writing equations for them (usually, but see below). But much of our behavior is, in fact, explainable by various influences. Skinner notes that "Autonomous man serves to explain only the things we are not able to explain in other ways." It is another "gap" theory. "But autonomous man is losing ground. The more we know about human behavior - and science is teaching us a lot - the less there is for autonomous man to do." (484)

The causes of behavior can be subtle, remote, historical, and complicated. We see our behavior as "autonomous" to the degree that we don't appreciate those causes, and we value a person's actions accordingly. "The amount of credit a person receives is related in a curious way to the visibility of the causes of his behavior. We withhold credit when the causes are conspicuous." This is another reason why people reject behaviorism itself. We want credit for much of our behavior, and inconspicuous factors threaten that if they become known. "A scientific conception [of human behavior] seems demeaning because nothing is eventually left for which autonomous man can take credit. . . . We recognize a person's dignity or worth when we give him credit for what he has done. The amount we give is inversely proportional to the conspicuousness of the causes of his behavior." (484) We might not give much credit to a normal adult writing a complete sentence, as the history of that skill is potentially understandable to us. However, if she produces a best-selling novel, this warrants great credit, as the particular contingencies of that behavior are inconspicuous to us.

Relatedly, we "see" the autonomous man when we hold people "responsible" for their own behavior, as we've discussed. This is a function of both the mistaken notion of free will and because it fits into some moral picture of the world. But it also assumes some internal, autonomous generation of behavior, something behavioral science rejects. Technically, "responsibility" is merely a word that describes a behavioral dynamic. Respond-ability, again if you will, is the capacity or tendency to act in certain ways where the factors promoting it are relatively remote or inconspicuous or unknown. There are reasons people act "responsibly" or "irresponsibly." We just label behavior as such when we don't know—relatively speaking—the behavioral contingencies which have led to the actions. It is a prime example of the "autonomous man."

What is Behaviorism?

Can there be a reasonable theory of human behavior—one based on solid scientific principles? If we believe that scientific methods can be used to study ourselves, the answer is clearly "yes." We do not necessarily need to rely on people's private subjective thoughts and the autonomous man to understand and predict what they will do. Which leads us to behaviorism. What is that, really? Behaviorism deals with behavior, not subjective or unmeasurable ideas. What I'll discuss in this chapter is not a global theory predicting the behavior of all people. We are all different, of course, and the notion of a summary equation for human behavior, as Hawking correctly noted, certainly cannot be conceived (I would again note here that the quest by Hawking, Einstein, and many others for a summary equation for all the physics of the universe is also still wanting). Rather, I'll review some of the principles involved with how human behavior is learned or conditioned (and later their relevance to religious behavior). This approach is alien and sometimes offensive to many, including psychologists, and is regularly ridiculed. It is virtually heretical in the eyes of some clinical psychologists and other mental health professionals. (Again, I'll answer all the criticisms below.)

Behaviorism and the "control" of behavior are concepts with an unfortunate reputation. They imply voluntary control of an individual by someone else. However, this is not really the meaning. In fact, behavioral dynamics are just as much in "control" of the behavior of those who influence others as they are of those presumably "controlled." Behaviorism merely implies a respect for the science of behavior, endorsing the idea that human behavior is a function of events in one's history and environment. The most important thing to realize is that observable behavior is the stuff of behavioral science. That does not mean that other things don't occur, from the operation of your neurons to the actions of internal organs. But to understand behavior, you must observe what a person does in contexts in which it occurs, and what consequences are generated by behaviors. It is what you can see (or hear, etc.) that is important. As the noted behaviorist Yogi Berra once said, "You can observe a lot by just watching." The laws of behavior have been developed by carefully watching what behavior occurs, in what contexts it occurs, and the consequences that follow—in many organisms, including humans.

I will not give extensive examples of behavior conditioning for any particular person. The illustrations provided here are only to elucidate the process. Every time you make generalizations about behavior from individuals or small groups of people to all humans, you are automatically wrong. Everyone is different, with different genetic and environmental histories. Although you will see some generalizations below, be careful not to overread these. They will be operative for the people being studied —not to be generalized to all humans. They are merely illustrative of how the conditioning process works for some people in some situations. Fundamentally, we are all unique products of those unique and complex histories, more complex than we can fully understand for any individual (à la Greene's calculational impasse). The constants here are merely the various behavioral principles that are in operation and how they operate on individual behavior.

Again, the core principle—complex behaviors don't come about because we will them or spontaneously initiate them, as we reviewed about Free Will in Chapter 6. They are shaped by the environment. Behaviorism is the approach to understanding and predicting behavior based on the environmental contingencies which have acted and are acting upon it. Those environments shape and "select" behavior for survival, much as biological evolution favors certain traits and selects individuals for survival—based on those traits' adaptive functioning in their environments. It stands in contrast to the history of psychological theories and the autonomous man. As Skinner lamented,

> Presumably we must look for [alternatives to mental explanations] in the external environment, but the role of the environment is by no means clear. The history of the theory of evolution illustrates the problem. Before the nineteenth century, the environment was thought of simply as a passive setting in which many different kinds of organisms were born, reproduced themselves, and died. No one saw that the environment was responsible for the fact that there *were* many different kinds (and that fact, significantly enough, was attributed to a creative Mind). The trouble was that the environment acts in an inconspicuous way: it does not push or pull, it *selects*. (484)

There is much to be said for the commonsense notion of one core principle: *The best predictor of future behavior is past behavior.* That's actually how science works.[18] Behaviorism is largely a set of refinements of this axiom. Human behavior *is* predictable based on past behavior, given similar conditions (defining conditions in the broadest possible temporal, interpersonal, and contextual terms). The refinements of this principle are what we call the *Laws of Behavior.* You might be taken aback by the notion of "Laws," but much like gravity or other physical laws, the Laws of Behavior operate whether we like it or not. I'll elaborate on the perspective of behaviorism to show how this happens, along with some examples of religious behavior. But remember that these are only examples. The environment and its history are the salient factors in shaping nearly all human social and cultural behavior, and this occurs *uniquely* for each individual.

Laws of Behavior

As with laws of nature in a determined universe, there must be some laws or principles that can be inferred which describe human behavior. And there are. And they apply not only to humans but to nearly all other organisms. Our behavior is more complex than that of other creatures. We have an incredibly large availability of responses and all the variants and subtleties of them. However, the principles are exactly the same.

What we know about human behavior is what we have reasonably learned from a century of experimentation that followed the scientific method, and this knowledge has coalesced into various "laws of behavior." They boil down to a few principles that account for nearly everything modern humans do. If you think that description is oversimplistic, it is. But it's the framework we'll discuss further. There are a large number of laws and behavioral principles, but the basic premise is that most *human behavior is a function of its consequences*. It's not that simple, of course, and there are a number of laws that contribute to this axiom and a seemingly infinite number of responses and actions that we engage in and environments where they occur. So it will be helpful for our discussion of religious behavior to directly define the major ones here.

What follows are some of the scientifically developed principles of learning or conditioning that shape our behavior. Contrary to popular belief, they do not specifically compel behavior; they select and shape behavior, just like biological evolution does. They impact us in ways that are sometimes powerful, sometimes subtle, or sometimes curious. Yet, there is no denying that the effects are real. You may resist the idea on the basis of supposedly "being controlled." Again, that is not what we're talking about. Being controlled implies that some person or entity is deliberately compelling you to do something you may not want to do (though that sometimes occurs, the principles of conditioning are responsible for relevant authorities as well). That is a mistaken view of human conditioning. The laws of behavior simply—though critically—are the principles that govern how our environments come to influence our actions to shape the behavioral repertoires that we exhibit. We *are* the sum total of what we do, as Aristotle incisively pointed out.

The principles we'll elucidate don't each occur in isolation. We are conditioned by countless external influences throughout our lives, moment by moment, all working in concert to enable the behaviors we emit. They may exert their influence suddenly, e.g., a frightened reaction to a snake may be conditioned by only one exposure, and part of that fright may last quite a long time whenever a picture of a snake is viewed. Other behaviors are influenced over extended periods of time, e.g., learning to read requires quite a long period of shaping to acquire a multitude of skills. The thousands of behaviors we emit every day, some subtle, some more obvious, have been conditioned over a lifetime. *So when I talk about specific principles, do not consider that these operate in some sort of isolation. They all work together, in a continuous fashion, for an incalculable number of behaviors throughout a lifetime. And of course, their specific dynamics are different for every individual.* How we are conditioned will likely seem obvious in many instances, but laying out the basic principles will aid understanding. I will note just a few; there are many others. I should also note that the examples given for each are only that. These are some of the important principles; I'll lay out later some of the evidence about religiosity which derive from these principles.

Conditioning

Let me note something important at the outset. It would be unethical to conduct controlled experiments on the conditioning of humans for religious behavior, even if subjects might agree to participate in such. What I'll address here are some of the principles involved and a number of practical illustrations; I'll address what we know from other kinds of experiments later. But first, let's lay out some of the basic principles of *conditioning*. Everything we learn is the result of some type of conditioning.

That is the major reason we behave the way we do and why everyone's behavior is different. Basically, we learn by what's called the *contingencies* of our behavior—that is, every behavior is contingent on what happens in its environment. This is termed *operant conditioning*, so called because it refers to any behavior which operates on the environment. Specifically, learned behavior is contingent on what follows it, i.e., *behavior is a function of its consequences*. This is the core principle of Behaviorism, accounting for nearly all of our behavior, including religious behavior.

Some other terminology: The major concept, as you've guessed, is *reinforcement.* We often think of reinforcement as a reward. Often this is true, but we use a more precise definition. *Reinforcement is anything that follows a behavior and strengthens it.* The "strength" of a behavior refers to any enhancement of an action, e.g., its frequency, intensity, shape, timing, or probability of occurring. "Weakness" of a behavior, of course, is just the opposite. Reinforcement follows the behavior it affects.[19] There are a few types, which I'll review below. But how do we know something is, in fact, a reinforcer? It may seem obvious, but it always depends on what actually happens. If you give your dog a piece of asparagus after it "sits," it may not learn that behavior. If you give her a treat, the likelihood of sitting will probably improve a bit—reinforcement. People are not dogs, but the process works the same way. Consequences of behavior are called *positive* or *negative*. "Positive" doesn't denote something necessarily wanted or favorable, though it often is. It merely means something that is presented or appears after a behavior. "Negative" doesn't denote something aversive, though it often is. It merely means something that is withdrawn or fails to appear after a behavior occurs.

The following is far from an exhaustive list, but it will give you the flavor of what I'm talking about, including some examples of religiosity. When looking at these, remember that such principles do not operate in isolation. They operate continuously, in combination, for an incalculable number, shape, strength, and frequency of behaviors, and differently across time and for every individual. As I lay out some of the principles, keep that in mind.

Respondent Conditioning

When Pavlov conditioned his dogs, he used what's called an "unconditioned reinforcer," food. Unconditioned reinforcers are things that can strengthen a behavior without having been previously learned. Universally, we might see such reinforcers for humans to be those things fundamental to life—air, water, food, etc. When these are used to shape behavior, we call it classical or respondent conditioning. When you experience a feeling of nausea after eating at a certain restaurant, you may also get some of this feeling if you return there or merely pass by it. Getting heart palpitations after a car accident, a natural response to such stress, may lead to palpitations the next time you get in your car or when you see the repair bill. Even talk of spiders can elicit a fear response if you've had an unfortunate encounter with one before. On the more pleasant side, you might get an enjoyable physical response, even a sense of smell, when you hear that a favorite meal is being prepared.

There are many potential examples of respondent conditioning in religious behavior, but let's take one of the more potent ones, the use of *music*. Music is natural. It is characteristic of all human cultures; humans began to create music at least 35,000 years ago (93), though it may have even occurred much earlier (we just don't have the additional evidence). Even animals fancy music. And many other species create music, including at least birds, whales, bats, some mice, and others. Music is both natural and compelling. Children respond positively to music at a very young age (there is even some evidence for infants' post-partum music preferences to be consistent with music heard *in utero.* (255)) During their first two years, many children are clearly captivated by music, even humming and moving to it. We've all had a pleasing emotional experience stimulated by hearing a favorite song from your youth. Most of us listen to music on our smartphones, car radios, or other devices. Music often elicits a rather visceral and pleasant emotional response, and it can be stress-reducing or comforting. We are drawn to music like few other phenomena in our lives, and it can act as a powerful reinforcer.

Respondent conditioning of religious behavior with music occurs in many ways. For example, when you begin to hear music, your physiology changes a bit, and you may feel yourself moving a little (some people quite a lot), and if it is pleasant, you almost instinctively want to hear more. Religious rites and church attendance are often associated with powerful music. And some of the most beautiful pieces of music throughout history, many would argue, are ones heard in church or that otherwise have religious significance. A few of the more potent ones are Beethoven's *Ninth Symphony*, Handel's *Messiah*, many hymns like *Amazing Grace*, or many of the other classic so-called gospels and the spirituals (which arose out of the time of African American enslavement). And then there are the Christmas carols, good listening indeed (and quite emotional for some), regardless of your religious identification or philosophy. All these musical genres (and a number of others) are compelling enough to appeal not only to the devout but to others as well.

With beautiful music, especially with lyrics including "God" and other religious verbiage, the potency of the conditioning is clearly evident. Worship becomes powerfully associated with music, and it is incredibly strengthened. This is especially true for children, who begin to hear such music in church at an early age. When this is repeatedly paired with church attendance and other religious rituals, the various memes of religion become increasingly conditioned—God, Jesus, etc. Music is associated with great ardor in the religious experience. It is no wonder that music is an integral part of most organized Christian religious activity. It works. The probability of church attendance and all the faith behaviors and rituals that go with it are strengthened because of such respondent conditioning. It is no different, in principle, than what went on with Pavlov's dogs. It is respondent conditioning.

Positive Reinforcement

This is anything which is presented or actually appears after a behavior and strengthens it.[20] We often think of this as an actual reward, like getting a paycheck at the end of the week. But it's much more basic. For example, most humans, from infancy on, are reinforced by attention, by touching, or by smiles. A behavior followed by a smile from others may tend to recur, though clearly not always. A "thank you" can be a potent reinforcer. Money is often a good reinforcer, when it is contingent on actual behavior. Personal attention, no matter how slight, can reinforce any number of behaviors. Remember that reinforcement is always individual. It all depends on what happens.

In general, reinforcers in our environments are stimuli and actions of others which have been associated with potent experiences in the past. Hugging your child after he successfully "goes to the potty" is likely to increase that behavior. Dieting is very difficult, but when feedback from peers or family is arranged to be contingent on weight loss, dieting is more likely to happen. A cookie following vegetable eating isn't a bribe; it's a likely reinforcer. There is a general axiom here, which is classically referred to as "Grandma's Rule." For example, "You have to do your homework before you get any TV." It has been more formalized and is now known as the *Premack Principle*, after the psychologist David Premack, who proposed it in 1965. (414) It is a principle of reinforcement that suggests, all things being equal, a more probable behavior can reinforce a less probable behavior. This dynamic operates to influence our behavior on an almost continuous basis. Cleaning up your room is often less probable than playing video games, so the latter can reinforce the former. What is reinforcing for an individual, of course, is generally unique; it depends on the behavioral history of that person. Attention from a peer for a work product, for example, may be reinforcing for one person and not another. Or it may reinforce that behavior when delivered by a supervisor rather than a peer. (In fact, a great deal of work has been done in such areas to improve work performance.)

When a child is praised for her Bible-reading by a parent, that can be significantly reinforcing. Reports of prayer are generally followed by attention or verbal support. Attention from the preacher is often powerfully reinforcing, particularly in behavior like confession. And then there's the church-related socialization. Going to church is generally followed by being in the reinforcing presence of others. That reinforcement for being in church is generally potent, which is why it persists. *Organized religions always involve an organization of people for just that reason.* Maintaining the religion and all of its superstitions demands that people congregate together. If everyone "practiced" their religions solely by themselves, never interacting with family or others about it, it would almost certainly disappear. The same is often true for praying, reading a Bible, crossing yourself, talking about God, and all other religious memes. Positive social reinforcement for religious behavior is what keeps it going.

Conditioned Reinforcement

All the other things in our life that operate on behavior are learned, i.e., they are conditioned reinforcers. They are reinforcers because they have been paired with unconditioned reinforcers, e.g., food, water, or the smile of an infant. Very little of what we do is maintained by unconditioned reinforcement, e.g., food and water, though these do come into play regularly. We learn what's reinforcing from pairing these things with other stimuli. Money is reinforcing for doing a good job, but just being told that can become reinforcing. Money is also a conditioned reinforcer. It has no value of its own, only what other reinforcers you can buy with it (or in some cases, accumulating a great deal of it, which is also behavior reinforced by others).

Conditioned reinforcers are those things (people, stimuli, events, etc.) that tend to increase or shape our behavior because they have been paired with unconditioned reinforcers. A prime example is "attention." Attention from others is generally reinforcing (though clearly not always; remember we are all different). As mammals, we are hard-wired to the reinforcing quality of a parent's attention. Parental reinforcement becomes conditioned to all manner of other reinforcers—verbal praise, various physical touches, even just the presence of a parent. The greatest depth and breadth or our behavior is shaped by the conditioned reinforcement of a parent. And such becomes generalized to various authority figures in our lives—babysitters, teachers, and others. Of course, for many religious apologists, it includes clergy. Just having a preacher look at you can reinforce church attendance. Even a brief glance from a parent while a communal prayer is being said is likely a potent reinforcer.

Negative Reinforcement

Negative reinforcement is essentially anything that increases the strength of behavior when it is withdrawn or avoided, usually something noxious or aversive in some way. This is often confused with punishment, but it is different. Negative reinforcement produces avoidance or escape behavior. Escape from a painful stimulus is a prime example. A hot stove is usually a negative reinforcer. We tend to avoid putting a hand on it. It is an easily learned avoidance behavior. Its potency even generalizes. Have you ever suddenly withdrawn your hand from the stove, only afterwards realizing that it wasn't on? When you get criticized, you tend to avoid the behavior which led to it, possibly doing things differently, avoiding the person who criticized you, or some other avoidance actions. Criticism is a negative reinforcer for most people. But remember, everything is individual. Sometimes certain critical comments can be positively reinforcing—they might improve some physical performance or even writing.

Food is a primary positive reinforcer and, if you're very hungry, an unconditioned reinforcer. However, eating, especially certain foods, can be maintained by negative reinforcers. That is because such behavior can sometimes reduce stress or other aversive experiences. We're all familiar with the concept of comfort food. Eating certain things or eating too much, for many people, tends to avoid or escape some of the ambient stress in one's life. This isn't rocket science. Most honest dieting people will tell you that. It is a negatively reinforcing situation, and often powerfully so, which is why problematic eating is often very difficult to overcome.

There are many religious behaviors that can be maintained by negative reinforcement. We might make a donation when the plate is passed around —to avoid disapproving looks. The social control of "sinful" acts is based largely on negative reinforcement. You must behave in certain ways; otherwise, there is harsh criticism. It might come from parents, clergy, peers, or others. The social behavior associated with organized religion is often maintained because it avoids, at least temporarily, some other troublesome aspects of life, i.e., it may be associated with reducing personal problems. Many people engage in religious behaviors under the dynamic that it makes them feel better or, in some cases, reduces some sort of "suffering." That is negative reinforcement at work. Such often provides only temporary relief, which is why religious behaviors are steadfastly repeated in the context of "suffering."

Prayer is a religious ritual which can also be negatively reinforced, often powerfully so. When a person prays, regardless of the stated receiver of the prayer (whether it's Yahweh, Jesus, Allah, or any other fantasy), that behavior tends to escape some of the stresses and aversive events or threats that can otherwise permeate one's day. The same thing happens, of course, when similar postures and behavior are engaged in meditation, for example, even without some supernatural object being invoked. In either case, such behavior is reinforced because it escapes or avoids stimuli which are more unpleasant. It is generally relaxing and stress-avoiding. Having nothing to do with a deity, it is negatively reinforced in part, though it is also positively reinforced by others who may notice or commend you for it.

Life is full of losses and stresses. When an unfortunate event occurs, we often hear, "That's God's plan," or "God works in mysterious ways." These are verbal behaviors which tend to avoid the often-negative effects we experience in the face of major losses or other tragedies in general. Thus, they are negatively reinforced. Such views or responses are negatively reinforced because they reduce the toxicity of serious problems or losses. It is not to excuse God. It is because operant principles are at work. People say such things to reduce their own discomfort and, on occasion, to avoid or escape conversations about the "God problem" in the context of all the suffering and tragedy in the world. It is noxious for many people to confront the notion of God with many of this world's horrific tragedies. Thus, vague allusions to God's mysteries provide a potent escape. It is not a coherent philosophical position. It is purely private or social behavior maintained by negative reinforcement.

Positive Punishment

A special comment on punishment is deserved. Punishment is anything that follows a behavior and decreases the likelihood of it recurring. It is the appearance or application of some stimulus following a behavior; that is *positive punishment*. Scolding, spanking, yelling, and other such consequences are examples. It can effectively and temporarily "suppress" behavior, though it is generally ineffective in the long run. Punishment loses its potency as time goes on. For example, a child who is punished for cursing may not do it for a while, but it may certainly come back. And it certainly may do so in other environments. Criminal behavior may be punished, but in the absence of other potent factors, recidivism often occurs. Punishment is a natural process, used especially in raising offspring. Physical punishment occurs in most primates as well as cetaceans (e.g., orcas and other dolphins) and some other mammals. But in order to be effective, it must come fairly soon after a response and usually be repeated often. That is why it is not terribly potent in reducing unwanted behavior. It may increase in potency, of course, when it is very reliably repeated and generally associated with some authority. But even that is a bit tenuous. Punishment can serve as a temporary suppression of behavior while an individual learns other responses that are incompatible with the punished behavior; that is, when the environment considers behavior more acceptable or deserving of reinforcement. You might stop "playing with your food" and then get reinforced for using a spoon. You might stop cursing and get reinforced for more constructive communications.

We still have the presumed threat of punishment embodied by "Thou shall not . . . " for various "sinful" behavior. Most sins are vulnerable to ordinary punishment—by parents, clergy, and others. That is how much religious behavior is controlled. The positive punishment of a child's "doubting" behavior can suppress that sort of behavior; and this is quite often characteristic of fundamentalist parents. Of course, the ultimate positive reinforcement in the Christian tradition is salvation, and the ultimate punishment is damnation, though these are merely threats or conditioned punishers, i.e., the words of scripture, friends, and clergy are repeatedly associated with this threat. "The one who sows to please his sinful nature, from that nature will reap destruction; the one who sows to please the Spirit, from the Spirit will reap eternal life" (Galatians 6: 7-8). The Biblical phraseology is tantamount to behaviorism. You can see the operant factors at work here—punishment (by threat) and both positive and negative reinforcement. The Church knows that punishment and the threat of punishment are only temporary solutions, of course. As with all punishment or the threat of punishment, it must be repeated often to be effective; and in religious traditions it is. The threat of punishment is critical to maintaining religious behavior in many sects.

Negative Punishment

Negative punishment is the removal of some stimulus, item, or action of others following a behavior. If you throw a tantrum and your toy is taken away, you may be less likely to do that again. If you drive over the speed limit, you may incur a fine, and you're somewhat less likely to do that again.

If a child expressed doubts about God or refuses to go to church, she may have her TV privileges taken away. If you're blasphemous, some possibly valued people will talk to you less. In the Catholic Church, for example, the commission of sins (arbitrary moral codes) may result in not being able to receive communion or in actual excommunication. As is well known, the list of such sins is incredibly long, including many behaviors that are rather ordinary for Catholics and others. Since they generally do not want to be deprived of communion, they tend to keep them secret (It's just annoying to have to go through absolution). This is negative punishment at work.

Modeling

Modeling is the key to most religious behavior, as with most complex human behaviors. We learn by imitating our parents. Modeling has survival value. We are hard-wired to do this, as are many mammals. Children innately follow and mimic their parents' actions. Given brain capacity at any time, they model nearly every single behavior a parent emits that they can see, hear, etc. Children seem to learn to speak rather automatically, but this is a function of modeling—they repeat almost everything they hear from a parent, which is then heavily reinforced when it's consistent with parental behavior. But modeling can and does occur throughout our lives and with many roles and behaviors that are modeled for us—from parents, teachers, peers, and others. Much of what we learn is a function of doing what we see others doing and saying; when such leads to reinforcement, even on occasion, it persists even more strongly. It's one reason the God meme rarely extinguishes.

Your parents, pastors, and others may "teach" you how to be a good Christian, Muslim, Jew, etc., particularly the verbal and textual language of the relevant faith. But you don't merely use the information to shape your behavior. It must be emitted to be learned. If you engage in modeled behavior, you might be reinforced for it. You need to mimic the language,

"God loves me," and ape the various actions, e.g., pray and attend church. There are literally thousands of such relevant behaviors which go into shaping what we call "religiosity," which I'll address more later. For now, remember that humans generally learn to be religious by modeling their parents' behavior and getting reinforced for it.

Stimulus Control

"Stimulus control" is the tendency for behavior to be reinforced, punished, or extinguished because of the presence of something or someone in the environment. Behavior is a function of the settings or environments in which it occurs. The environments in which they occur come to enable emission of those behaviors even when they're not often reinforced. Everyone knows that emitting an unwanted act in the presence of one person can lead to punishment, while doing it for another may be ignored or reinforced. Saying "I love you" to your spouse may be heavily reinforced, but doing that with a casual coworker might result in different consequences. Context is everything. That is, behavior is often under the control of people and environments in which it might be emitted.

Openly verbalizing that "Jesus loves you" is likely to get reinforced in a church building. That's less likely in math class or with an atheist spouse. Religious behavior is more likely to be emitted on Sundays for Christians and on other days for other faiths. Talking about God is more likely to be reinforced if you're in the presence of others who share similar views. Context is everything. Religious congregations and religiosity in general are built largely on such shared stimuli and environments.

Generalization

When some specific behavior is emitted, similar behavior may occur in other contexts. If a child withdraws in a social situation and this is reinforced (often negatively), she may do that in another social situation. If that pattern continues, it will generalize to most similar situations. Eventually, it may become a dominant characteristic. That is what we call *stimulus generalization*. It is not because of an "introverted personality." You may have had a bad reaction to some food with jalapeño peppers in it, and you may have had a similar reaction to unknown ingredients of another food. You may begin to generalize your behavior to fit previous learning, avoiding anything with Sriracha or Tabasco, for example. It is because learned behavior tends to generalize to different environments or situations. You might pray at home, but such behavior may generalize to a friend's house or to the beginning of a sporting event.

You might also emit behaviors which have not been conditioned individually. If you're reinforced for picking up your toys, you might also pick up your dirty socks. If you're reinforced for doing one piece of homework, you might tackle another subject. This is *response generalization*, engaging behaviors which are in some way similar to the one that's been reinforced. If Bible reading is reinforced, you're more likely to attend church. If you profess belief in an omniscient God, you're more likely to pray. As religious behavior is reinforced during childhood, it generalizes in responses and to different environments; and voila, eventually you have a general set of behaviors. Who you "are," i.e., the totality of your behavior, religious and otherwise, has been shaped by the reinforcement and stimulus and response generalization of behaviors over time.

Schedules of Reinforcement

This is a critical topic. Bear with me; the practical importance of these *will* emerge. Among the most important behavioral concepts is what's called *schedules of reinforcement*. These are a principal influence on most human behavior. Our behavior, like most other creatures, is not shaped and maintained by reinforcing every single relevant response. Initially, every response may be reinforced for a very young child, for example, when she picks up a piece of food. But continuing to reinforce every time is not efficient. Attention or reinforcement for this fades over time and becomes occasional and eventually rare. That's how nearly all human behavior is maintained.

If we were to get every particular response reinforced, what's called a *continuous schedule of reinforcement*, we'd never be able to accomplish anything, or even behave. We depend on only occasional reinforcement in order to function, what's called *intermittent schedules of reinforcement*. There are many types of intermittent schedules and combinations of schedules that maintain human behavior, including religious behavior. We'll consider only a couple here. They all produce different patterns of behavior—in mice, people, and various groups and organizations of people. Let's take just a couple of examples.

There is a "fixed ratio" schedule, in which reinforcement comes after a defined number of responses occur. It produces fairly steady states of responding. Most strength workouts in the gym function in this way. You rest after a certain number of repetitions. You might also get paid after you sell a certain number of products. Obviously, the list is essentially endless. If there is enough verbiage in saying "grace" at dinner, eating follows. Just saying "thanks" won't do it. More content is required. Fixed ratio schedules lead to very predictable behaviors, including religious ones.

A "fixed interval" schedule is one in which the reinforcement comes after a defined period of time, provided the identified behavior or behaviors have occurred. A unique feature of a fixed interval schedule is that it produces more frequent or intense responding near the end of the interval. We're quite familiar with cramming for a scheduled exam or getting work done right before a deadline. Doing the dishes can be negatively reinforced, provided it occurs before bedtime. The behavior we call procrastination is well controlled by this. A classic example of the principle is bills passed by the U.S. Congress. The number of bills (particularly spending bills) follows this behavioral principle. The number increases as the date of adjournment approaches. It always seems to happen that way. The behavior of such bodies, and all their component persons, are partly under the control of a fixed interval schedule of reinforcement. Such makes the behavior highly predictable.

Religious behavior often follows this principle. You must get your Bible reading done before bed. The probability of faith or belief tends to increase during the latter years of life and more intensely with a possible life-threatening illness, as might be predicted. The ultimate example in later years of life, of course, is a form of Pascal's Wager, as I noted. The looming end of the interval is the controlling factor. Of course, this is not religious faith as most devout practitioners conceive of it. Despite any "last-minute conversion," the behavior is actually under the control of a fixed interval schedule of reinforcement—behavior coming before the end of the interval might get reinforced. It is essentially no different than a dog's behavior pattern prior to a meal, the behavior of a congress, or studying for an exam. They are under the control of fixed interval schedules of reinforcement.

If only some responses are reinforced and the number of them is variable, behavior may be intensely conditioned. We call this a *variable ratio schedule*. This kind of schedule results in high, steady rates of responding. We might occasionally get parental praise for doing our homework before being allowed to watch TV. Variable ratio schedules can produce some highly repetitive behaviors, and some potentially problematic behaviors as well, such as casino-type gambling. In such situations you get "paid off" only some of the time, and it is unpredictable. Depending on various other circumstances (e.g., what else is reinforcing in your life), this behavior may be very strong and hard to extinguish. It is specifically because of a variable ratio schedule of reinforcement.

Superstitious as they are, prayers do not get "answered" every time. Once in a while is enough to keep the behavior going. This is a variable ratio schedule. You might pray for an athletic victory, and it might happen. And you might "thank God" for it. In fact, you regularly see athletes point to the sky and emit other behaviors to reinforce God for the victory. You might be intermittently reinforced for praying. Sometimes your prayers are "answered." I *did* have a safe airplane trip. Yes, God did give me the strength to confront my child about her behavior; alas, it may not work too often. And, of course, you'd sometimes receive reinforcement from others for your praying. One might pray for recovery from illness every few hours or days, and it often "works"; but of course, you've ignored what might be known about the typical natural history of the illness. A common cold takes a certain number of days, frequency or intensity of prayer notwithstanding. But the praying behavior has been well reinforced on a variable ratio schedule. That's one way that superstitions are maintained.

Behavior may also be under the control of a *variable interval* schedule of reinforcement. Here reinforcement is provided after a random (unpredictable) amount of time has passed in which a specific behavior has been performed, and it usually produces a relatively low, steady response rate. The classic example—and it's one where people do act like Skinner's pigeons—is repeatedly pressing an elevator button. You have no idea if and when it will work. Pop quizzes, for example, produce more regular study behavior than similar fixed exams. Some people do tend to variably check email every so often; they just don't know when some will come. Crossing yourself is partly under control of a variable interval schedule. You might do this one or more times before it's noticed, and when that might be is unpredictable, but sometime, someday, someone will notice.

Finally, there are a large number of different combinations of these, producing a myriad of definable patterns of behavior, which we'll not go into here. This is actually how much human behavior is maintained. Take one example of what we call a "conjunctive schedule" of reinforcement. It might combine fixed ratio and fixed interval contingencies, that is reinforcement coming about when successfully doing something a number of times (FR) within a specific time (FI). Sales quotas often work that way. We might get a paycheck bimonthly, but that is usually based on having worked at reasonable productivity for the past two weeks. You might do a certain amount of homework before dinner. You're likely to get reinforced if you learn enough Hebrew and prayers before your Bar Mitzvah.

No, we still cannot write specific equations for such common human behaviors. To soften Hawking's pessimism a bit, however, we *can* estimate probabilities for many of these based on past behavior. Further, various schedules of reinforcement produce characteristic graph plots or curves. The behavior of any organism, for example including the mentioned Congressional behavior, is described by the classic "scallop" shape, with a sharp upward pattern before expiration of the interval. In fact, given knowledge of just a few variables, equations can be written for behavior under the control of various schedules of reinforcement. (202, 420) When we know the behavior, the schedule, and a few relevant variables, best fit curves can be produced which exemplify the phenomenon. Of course, such would not predict future behavior with great specificity, only mathematical approximations if the behavior, conditions, and schedule are empirically known. Such would not satisfy Hawking's cynicism about equations in general for human behavior; but it does suggest that such are clearly possible *if* we know enough.

Extinction

If reinforcement is withdrawn or no longer available, the relevant behavior becomes extinct. If you got no money or social or other reward for working, you'd likely stop or look for something else. Many people who diet and don't lose weight stop trying. We do little in life that isn't reinforced in some way on some schedule, and when reinforcement is lacking, especially in the long run, the relevant behavior disappears.

However, often previously reinforced behavior does not go away easily. If a key that opened a door stops working, you'll eventually stop trying, but not without some increasing intensity and attempts. When such a previously reinforced behavior doesn't result in reinforcement, you might try "harder," but lack of success dictates that the behavior will eventually stop. You might try a different key or a different door. And if those don't work either, then you know what happens. You just give up and try to find a different way of solving the problem. If restaurant orders for a particular dish disappear, the business will likely stop making it. If asking for the garbage to be taken out stops being effective, it'll likely extinguish. Behavior which is not reinforced in any way will eventually become extinct.

Behavioral extinction also has some rather unique side effects. One of the typical ones is sometimes intense, vocal, critical, somewhat random behavior, which is often and appropriately construed as "anger." The emission of random and intense behaviors we call "anger" is a hallmark of extinction, and it occurs in nearly all mammals. Failure to take out the garbage can result in a little or a lot of anger. When the TV remote doesn't work, it's remarkable how some people curse and bang it on the table. We can say that people get angry when they don't get what they want, but this is actually simple behavioral extinction. It is an immutable Law of Behavior. It is no different than the behavior of a "cornered" animal that can't escape or can't get to the food. Behavior initially intensifies when reinforcement is not forthcoming or available.

What's interesting here is that every schedule of reinforcement has its own distinctive pattern of extinction. When any behavior has been subject to a continuous schedule of reinforcement, it is particularly vulnerable to extinction. If your phone calls to someone have been reliably answered, and then they are suddenly no longer answered over a period of time, you're likely to stop trying. If your employer stops paying you for an extended period of time, you'll probably stop working there. If any behavior is maintained by nearly continuous reinforcement, it can extinguish quickly.

However, if a behavior is maintained by occasional or intermittent reinforcement, it is harder to extinguish. That's the way life usually is. We only get reinforced once in a while (and usually subtly). If your calls are answered once in a while, you'll keep trying for a long time. You may not be regularly praised for doing homework or studying, but praise may come eventually, and the pattern may become very resistant to extinction, over even a lifetime. You might just continue to study. Much of life is that way. We only get reinforced once in a while. Intermittent reinforcement keeps behavior going. Of course, the various intermittent schedules of reinforcement and their combinations have their unique patterns of extinction, which we'll not bother further with here. Suffice it to say that these principles do operate on humans as much as on any other mammal. It's only the particular behaviors which are unique to humans.

Religious behavior, like all operant behavior, is vulnerable to extinction. If prayers aren't answered, or otherwise reinforced by other stimuli, you might pray "harder" or more often, as is typical in an extinction paradigm. That is generally what happens for many religious adherents; and clearly, it is one of the factors in extinguishing prayer behavior, and sometimes getting upset or actually being angry at God—no different than the animal being kept from its food. If reading the Bible or going to church is no longer reinforced in any way, such behaviors will tend to fade. It may take a long time, but it *will* happen. How much religious behavior in general can be emitted without any payoff, of course, is specific to each individual. But there is an additional vulnerability. Reinforcement of particular memes may fade, but other religious behavior is often maintained. For example, social reinforcement can maintain church attendance, while prayer becomes extinct. Participation in various rites may continue, while assertions about God's existence or love may fade.

The evolution of "doubts" is a function of these behavioral dynamics. If enough doubts appear, and reinforcement from others itself wanes, a "loss of faith" often appears. Extinction may occur mostly because the behavior ceases to be of use. That is, religious behavior is no longer reinforced. Religion no longer "works." Any learned behavior—and faith is such—will extinguish if no longer reinforced. It is nearly impossible to conceive of maintaining faith behavior if no reinforcement of any kind ever appears.

Extinction of faith behavior is sometimes associated with significant losses, e.g., death of a close friend, divorce, or some other personal tragedy, or one merely viewed—the death of a few hundred children in a tsunami. However, it tends not to happen all at once. Extinction paradigms, as I said, usually result in more intense behavior before any final extinction occurs. Serious losses or "God's failure" to prevent them can precipitate a "crisis of faith." The behavioral response can lead to "trying harder." That is, some people seek "comfort" with increased praying, Bible reading, church attendance, talking to clergy, etc. Religiosity becomes intensified. This is no more than the classic extinction paradigm, religiosity under the control of schedules of reinforcement.

Behaviors under the control of continuous schedules of reinforcement are easier to extinguish than others. If religious conditioning, for example, has been associated with a fairly continuous, or nearly continuous, reinforcement, it is especially vulnerable to extinction. That is why many people who've been exposed to very intense indoctrination, especially early in life, may eventually "lose faith." When reinforcement for religious behaviors later in life don't seem to be coming, especially in adolescence and early adulthood, and where other, competing behavior develops (e.g., respect for science), religious behavior often wanes to the point of serious doubts and often giving up one's beliefs entirely. That is why rigid religious conditioning is particularly vulnerable to extinction.

Interestingly, among those who reflect and promote intense indoctrination are members of the clergy. Given the principles of more continuous reinforcement and extinction, they are notably vulnerable to losing faith, and countless of them have. There is even a large international organization whose mission is to support such "fallen" preachers, called the Clergy Project.[21]

Shaping

Herodotus had it right. We are, in a sense, governed by our circumstances. This is the key principle in behaviorism: Behavior is a function of the environments in which it occurs. It is the pivotal factor in how most human behavior develops. Thus, behaviorism attempts to answer the question, "What are the conditions under which behavior is emitted?" And by this we mean the conditions which are occurring at the time and the history of environmental factors encountered by the person. We are never going to have enough data on these environments to satisfy Hawking's missing equation. But it is from the environment that we can obtain a reasonable understanding and appreciate how most behavior evolves, including religiosity.

Human behavior comes in a seemingly infinite variety of shapes, i.e., unique responses and their particular character, type, intensity, and so forth. No two handshakes are alike. No two utterances of a word are alike. No two cries of pain are alike. No two gestures of similar meaning are alike. You get the idea. The same is true for any behavior a human being emits. Every response or action has its own unique *shape*. And those shapes are determined largely by selective reinforcement in the evolving shape of any behavior. For a gross example, a child might be reinforced for eating with her hand and later being reinforced only for eating with utensils. A singing voice might be reinforced for one level of production but then reinforced only for a finer output. A tennis stroke becomes ever more exacting as the feedback or reinforcement of progressive success appears. These are called *successive approximations* for any particular observed behavior. Eventuality, any behavior becomes more and more refined because only certain aspects or levels of it generate reinforcement. That is shaping, the gradual alteration of behavior from simple or rudimentary forms to more precise or complex ones, and the subtleties come in a seemingly infinite variety and changing shapes.

Lest you think this is all oversimplistic or just wrong, let's consider a "simple" behavior—brushing your teeth. Why do you do this? Most people would say "to avoid dental problems." That's behavior largely under the control of negative reinforcement. But that's a very remote consequence. You probably learned it as a child. Your parents probably modeled and positively reinforced it. You were nagged before going to bed until you did it—more negative reinforcement. The activity was further refined by reinforcing approximations to appropriate technique, and the best ways to brush have been gradually shaped. The dentist has warned you about possible problems—variable interval negative reinforcement. Your breath and smile may be affected, remote conditioned reinforcement. And so on. Brushing teeth is seemingly a "simple" behavior, but most people can appreciate how it is built up over a lifetime via operant conditioning. Actions such as this are often called *habits*, a term we use to describe behavior which has been well conditioned. Breaking habits is sometimes difficult for that very reason.

Let's return to the comparison of human behavior working by "natural selection," analogous to the principle of biological evolution by natural selection. Take a step back. Life on Earth has the "forms" it does because of evolution by natural selection. It has been a gradual process for billions of years. In Darwinian theory, species survive and reproduce because of their interaction with the environment as what has classically been termed "fitness." This is the ability of the organism to survive, based on its characteristics. Creatures do not "decide" how to evolve. Environments act on organisms to produce changes. Some variations are "selected for" or reinforced, i.e., they tend to survive; others do not survive. Creatures whose characteristics are "selected for" by the environment are in a better position to reproduce; others are not. What we presently observe in species is the result of countless, gradual *approximations* in this process. That is the way new forms come into being. We do not see the evolution of the human eye, for example, but it was gradually selected by the environment because approximations to it were *functional* or reinforcing for certain organisms (Chapter 4). That is, the process helped the fitness of certain creatures to survive because subtle adaptive changes were reinforced—they "worked" better than their predecessors.

The shaping of all complex human behavior works the same way. Our behavior is a function of the environments in which it evolves. Behavior of a certain shape may be functional, i.e., producing some advantage or reinforcing consequence. Therefore, it has a better chance of being emitted, while behavior of a slightly different shape might not lead to reinforcement. In reality it's a lot more subtle, and powerful, than that. Behavior becomes "shaped" because the environment becomes more and more selective about what responses and shapes of those behaviors produce reinforcement. This applies to various strengths, types, and shapes of behaviors. And the timing and types of reinforcement are reserved for closer approximations to the behavior which the environment has determined to be of functional advantage.

We tend to be aware of this process only for behaviors where we can easily identify the antecedents and consequences of the action. But the environment does the job continuously and completely. Speaking in a lower voice might produce better attention than something else. Waiting longer might garner more attention than impatient behaviors. Multiply by the thousands. All the shaped behaviors have become well-conditioned over the life of the individual. All of this might seem quite obvious, and it is so because we acknowledge the influence of the environment. But the complexity of behaviors is usually controlled in subtle ways. We often don't see the gradations of behaviors; they can change slowly over time, as any evolution might dictate.

The variability of human behavior—and clearly all learned behavior —is a testament to this model of "behavior survival" enabled by the environments in which it has occurred. (Probably the most complex example of shaping might be the acquisition of language, which I'll address below.) It is specifically what makes us who we are. Faith and other religious behaviors are no exception, being a function of shaping and other aspects of operant conditioning. Faith generally begins in childhood (though not always; and I'll address more about childhood below), and is gradually shaped by parents, congregations, religious peers, and others. It is the dynamic total of environmental influences that result in the complex behaviors we see. Religious behaviors of adults, for example, such as praying or talking about God, are the result of years of shaping by environmental contingencies in early life.

Religious behavior can appear sometimes rather suddenly. What about the adult person who suddenly claims to be "born again," usually endorsing a Christian faith, for example? It doesn't seem to happen slowly, but rather abruptly in response to some event or interpersonal contact. We don't often know what the environmental trigger is, but it exists, and it becomes salient in part because of a lifetime of shaping of other behaviors which make a person vulnerable to such a change. Of course, if it lasts, it's always maintained by the contingencies we've been reviewing. Religious behavior can also be lost, of course, when behavioral extinction takes effect. Some people "give up the faith," so to speak, and become atheist or at least nonreligious. This can also be facilitated by competing ideas and behavior, e.g., respect for science, or the influence of peers or respected secular authorities. In sum, it is the behavioral contingencies that dictate the religious behavior.

Conclusion

When you put all these together—and more—we see that behavior *is* "relatively" predictable. Human behavior, including *all* religious behavior, follows these principles. Again, our ability to predict behavior is limited by the "calculational impasse," not the base validity of the approach of behaviorism.

Belief and other religious behaviors are purely a function of operant conditioning. A prime example: It is fundamental that positive reinforcement is generally more effective in managing behavior than punishment. As Skinner put it, "Religious agencies move from the threat of hellfire to an emphasis on God's love." (484) Christianity moved from the cruel and vengeful Old Testament God Yahweh to a presumably more "loving" God when the Jesus meme appeared. In fact, the modern version of "Jesus loves you" is no more than a shift in operant principles—from punishment to positive reinforcement—to shape and expand the number of believers.

Operant principles are fundamental in shaping behavior in all religions. We don't usually think of Islam, for example, in this context, but we should. In writing about the Islamic faith, Egyptian psychiatrist Dr. Mohamed El-Islam notes,

> Religion offers belonging to a group, an approved code of attitudes and behavioral norms (code of conduct) . . . Belonging to a religious denomination is socially conditioned by *operant conditioning*. The latter provides for positive reinforcement of religious belonging by social approval and social support and positive extinction of non-religious and antireligious attitudes by social disapproval and even legal disapproval in some communities. (154)

Operant factors do have extraordinary potency in control of our behavior, including all faith and other religious behavior. The shaping starts early in life and becomes more and more potent and generalized over the years. In this fashion, many of the memes of theistic religion become "over-learned." That is, they become extraordinarily resistant to extinction. They are maintained by various schedules of intermittent reinforcement, by the conditioned responses of many different peers and authorities, and by the various potent stimuli of Bibles and other scriptures. Reinforcement comes from houses of worship, prayer and other rituals in the home, and social interchanges with others. Often evangelizing behaviors can be exceptionally reinforcing. For the devout, a potent culture—the individual's environment—surrounds and supports the maintenance of faith in its members. It is the laws of behavior, not inspiration from some god. If we know your religious behavioral history, parental and other influences, and your physical options to practice it, we can reasonably predict what you'll do next Sunday. That's science, at least at a crude level. It doesn't require free will or a god.

Criticisms

There's a good deal more to say about the operant conditioning of religious behavior. But before we go further with that, let's pause to address the often-heard criticisms of Behaviorism. This approach to human behavior, as I intimated, generally has a "bad rap" with both the general public and in professional psychology and other social science disciplines. In fact, Behaviorism often appears so outlandish to many as to be summarily dismissed as the ravings of extremists. This is unfortunate, but the problems generally stem from a fundamental misunderstanding of what Behaviorism actually is.

Over the last half century, psychologists and others have largely abandoned its concepts in favor of more global and unproven theoretical orientations (As I've noted, global intrapsychic theories of human behavior are not falsifiable and actually do a poor job of predicting behavior.). There has actually been quite a backlash against behavioral approaches to human behavior. The concepts of behaviorism are generally eschewed by most psychologists and mental health professionals when it comes to appreciating human behavior. Yet, somewhat remarkably, they easily acknowledge the concepts for other animals.

There are a number of arguments put forth in this regard, and I'll address the most typical ones here. The criticisms include that behaviorism is not really a theory, that it is too simple, that it assumes human beings are a "black box" in which nothing of value occurs, that it ignores human free will, that neuroscience has better answers, and that it is dehumanizing. I'll take each of these in turn.

Theory

Behaviorism is alleged to not be a theory at all. However, it *is* a theory; and its approach is based on analyzing the contingencies of behavior which uniquely influence the actions of specific people (and all other animals for that matter). Every individual is unique, and appreciating the distinctive nature of his/her behaviors is what Behaviorism is all about. That it is not a "global" theory is a misguided critique. Behaviorism does not attempt in any way to explain human (or other animal) behavior *in toto* according to the well-known laws of behavior. Reinforcement, extinction, punishment, environmental contingencies, shaping, etc., do provide powerful influences on the nature of emitted behavior, including in humans, but they do so uniquely for each individual.

The notion that behaviorism cannot predict human behavior is shortsighted. Behavior *is* predictable based on the available data. We all know that. The same is true for theories in all physical and biological sciences, as I mentioned. Can we predict the faith behavior of an individual, the time and specifics of a prayer, etc.? No more or less than when a given leaf will fall off a tree, for example, or the time and place an earthquake will occur. The limitation is, again, the calculational impasse. In that sense, behaviorism is no less "scientific" than other empirical disciplines. Measurement tends to be more precise in the so-called hard sciences. However, prediction in the sciences occurs within the specific realms of the measurement of each. Behaviorism deals with overt behavior and its environments. Chemistry deals with the behavior of molecules. Astrophysics deals with the behavior of celestial bodies and their components. And so forth. That behaviorism deals in animal behavior makes it no less scientific and capable of theory than any other science.

Behaviorism is an approach to assessing and modifying behavior based on the known principles. In that sense, it is totally individualized to each organism or person. If we generalize across people, then behaviorism loses potency. Nevertheless, the behavioral/environmental approach has been enormously successful in predicting behavior—from the development of infants' behavior to classroom behavior to criminal behavior to productivity at work to the behavior of the U.S. Congress. When any particular individual's environmental history and current circumstances are well known, we can, in fact, predict what that person will do—at the level of data that we have. In that sense, Behaviorism is a theory with notable predictive power. We develop "theories" about the behaviors of particular individuals. Take a typical example where the theory is applied. In a clinical situation, e.g., where someone's various behaviors—or life in general—become problematic for an individual or others in one's environment, we develop hypotheses about the development of those behaviors. When done responsibly, this is accomplished by rigorous observations and detailed histories of an individual's behavior, environments, and the shapes and contingencies of those behaviors. That is, we *can* understand how some human behaviors come about, how they are maintained, how future behaviors can be predicted, and how they may—and often do—respond to some modifying influence or intervention—on an individual basis. The Laws of Behavior allow us to do that. Such cannot be used to predict human behavior in general (which is the bane and failure of much psychological theory, for instance); but reasonable "theories" of individual behavior can and are derived all the time. That is the strength in the application of behaviorism. Evidence shows that, carefully applied, behaviorism is successful in modifying or improving the behavior of humans of all stripes.

The Black Box

You may have heard that Behaviorism assumes that humans' private, subjective experiences are not worthy of attention, which led to the idea of a so-called "Black Box." Physicist Kaku criticizes it as follows. "Behaviorism is based on the idea that only the objective behavior of animals and people is worthy of study, not the subjective internal states of mind." (261) This is surprising from a physicist, who does know that one can't study things that you can't define or measure, but he takes no position on that. The "black box" is a term which was often invoked to criticize Skinner's behaviorism. Skinner did not deny that experiences are real to the persons having them, only that we did not understand the underlying neurophysiology well enough to predict human behavior, including proximate "causes" of it in the brain. Neuroscience has obviously made advances, albeit very crude ones at this point. However, proximate neurophysiologic "causes" of behavior are only that. It is from environmental learning that the vast majority of our actions are shaped. You'll notice the similarity of the "black box" idea to its use with artificial intelligence, which we covered in Chapter 5. We don't know enough about what happens in the "black box" of an advanced AI system and how it generates unique "behavior," though we're trying to study how this happens. The only things we really know are its hardware, its inputs, and some of its algorithms. This is notably similar to human behavior. Complex operations of the human brain come about in innumerable ways we don't understand. All we know is some very gross anatomy, some of the inputs, and a few of the behaviors we can measure. This is called learning, directly analogous to the use of that term in AI, but how it happens is, in fact, a "black box" for both.

Behaviorism does not object to the study of events which are more granular than observable overt human behavior, i.e., the behavior of an organism which can be "seen" only with instruments. Measurements of such behavior, e.g. heart rate or muscle tension, aided by instruments, are studied with operant techniques. Many well-accepted phenomena in physics are not directly "seen" but are "observed" by instruments or inferred by observation. We do not "see" quarks or muons, but our instruments provide the relevant evidence. Behaviorism is a scientific approach to the study of behavior, not a denial of phenomena yet to be measured. However, the study of a non-tangible "mind," which can't be observed by any potentially theoretically designed methods, is outside of reasonable scientific inquiry.

Too Simple

It's often asserted that a "simple" behavior-reinforcement scheme can't explain all behavior (It is often mislabeled, somewhat derogatorily, as a stimulus-response theory, which it is not) or that humans somehow are not subject to this dynamic, or that experimental animal models can't be translated to humans. Obviously, humans *are* subject to the laws of behavior. Also, that the behaviors and consequences are often vastly different for humans and other animals does not invalidate the principles of operant conditioning in all creatures. Biologist Robert Sapolsky, for example, makes this mistake as he takes straight aim at the behaviorist perspective. He notes, "A pigeon is a rat is a boy. Skinner preached. Soulless droid. Behaviorists were often right about behavior but wrong in really important ways, as many interesting behaviors don't follow behaviorist rules." (444) The perspective here is obviously a bit dismissive, but it is also off the mark. "Interesting behaviors," is a pretty vague notion; and complex behaviors are not easily explained—by *any* "theory" of human behavior. We can only return to Brian Greene's calculational impasse. The behaviorist approach is limited only by a lack of data on very complex behaviors. That doesn't make the theory "wrong"; no theory of human behavior can explain very complex behaviors. It's just that behaviorism does a better job than the other extant ones, especially for a given individual. Again, we simply know behavior is a function of its consequences and that the best predictor of future behavior is past behavior (details of such below). And I assume that Sapolsky was merely being a bit metaphorical in the use of "soulless" in his criticism. Biologists don't use "soul" as a way of explaining human behavior, as he well demonstrated in his insightful denunciation of free will (Chapter 6).

Sapolsky also takes particular exception to "B. F. Skinner and his followers, who dominated behavioral science with strikingly unbiological approaches for most of the last century. We clearly have left these ideas behind." (444) Despite his clearly outstanding contributions to the biology of humans and other animals, he is wrong about that, of course. In fact, most experimental work—with primates and other species—explicitly uses operant conditioning, and Sapolsky clearly knows that animal learning in natural environments is consistent with the various laws of operant conditioning. No serious scientist would disagree with that. The casual dismissal or disparagement of behavioral approaches for being "unbiological" is shortsighted. If he has a better way of explaining learning in animal behavior (remembering we're animals too), I've not seen it. Behaviorism does not deny biology. It merely focuses on what can be observed in various biological creatures.

Free Will

Again, a principal reason that many object to the notion of scientific theories of human behavior, and behaviorism in particular, is the false notion of free will. Not much more need be said about this, as I detailed in Chapter 6. However, for critics of behaviorism in particular, free will is wholly incompatible with the notion of conditioning as a major factor in human behavior. Many, including psychologists, continue to endorse the idea that we largely "choose" the behaviors we want to learn, how to learn them, when and how to manifest them, etc. They further assume that behavior can be generated essentially "out of nothing," à la the "autonomous man." There is an almost reflexive antipathy to any "theory" of human behavior that doesn't support the notion that we can have voluntary control of most of our own thoughts, feelings, and actions.[22] The false endorsement of free will is not a valid criticism of behaviorism. In psychological reasoning, the use of "free will" as an explanation for behavior is utterly bankrupt. It is a "gap" filler without any explanatory power.

Neuroscience

Much attention has been given to the brain as the "cause" of human "voluntary" behavior, that is, the antecedent events operating in our brains. In fact, many view the brain as the primary antecedent for human behavior. After all, how could the extraordinarily complex behavior of humans be explained without understanding how the brain processes experience and produces behavior? Behaviorism is often dismissed for its failure to take this into account.

The human brain is complex, and explaining how it works is one of the most difficult—if not *the* most difficult—endeavors science has ever taken on. I have no quibble with the difficulty of the problem of where the universe came from. It's just that the human brain might be the most complicated system we've ever encountered. If I had to bet, I'd wager that the universe problem is solved before the brain one.

The whole field of neuroscience is based on the idea that if we can figure out how the human brain works, then we should be able to explain complex human behavior. I addressed some of this in Chapters 5 and 6. But we can summarize the problems with this theoretical approach as follows. First, all brains are different. Although we have the same basic neuroanatomy, human brains are somewhat variable at birth. This is also likely true for all organisms which learn very complex behavior. Of course, it was variability in the brains of our ancestor hominins that enabled us to evolve into the creatures we are. Second, brains change over time with experience and the growth and death of neurons. Third, brains respond differently to the experiences they're exposed to. You must know the full panoply of environments and behavioral contingencies if you want to predict how a brain develops over time and produces behavior.

Neuroscience itself is in its infancy. We have learned a great deal about how the human brain "works" with respect to large, abstracted areas of behavior—general language capacities, gross perceptual and motor functions, and areas of the neocortex which are dominant in certain functions. But we know almost nothing about the neural bases of *specific* behaviors, particularly any complex human behavior. We can identify neurons that stimulate moving a particular body part, and we can identify if someone is thinking about a tree or a tomato, for example, provided that such have already been experimentally conditioned and correlated with measuring devices (as we saw in Chapter 5). But we are totally at a loss in using data from the brain to understand whether someone is shaking hands, or smiling, or smelling a rose, or eating an avocado, or falling in love. At this point in the history of science, no amount of data from neuroscience measurements of one's brain can predict *any* complex human behavior. Perhaps someday—or in some future universe—science will be able to do this. Presently, we cannot even envision the capacity for measurement and analysis of brain activity that will allow us (or whoever/whatever replaces us) to predict complex behavior by "looking" at the anatomy and chemistry of one's brain. At this point in history, behaviorism is a much better theory.

The other problem with the neuroscience criticism is that activity of the brain is not the fundamental cause of behavior. Again, it is what we refer to as the proximate cause. The brain does not initiate so-called "voluntary" responses without the antecedents that enable it. Such behavior comes from a lifetime of experience, the environmental contingencies, not the work of the "autonomous man." That certain "voluntary" actions may arise in the brain "before" we observe behavior (though, as we've seen, that is clearly not always the case) in no way minimizes the fact that the operant factors are, in fact, the antecedent causes to both overt behaviors and the neural activity which *may* precede them. In point of fact, most neural activity can generally be seen as covariates of behavior, not the fundamental cause.

Behaviorism does not deny the existence of brains. The failure of behaviorism to include neurological variables is not a valid criticism of the theory. It is limited by the "calculational impasse," not the lack of neuro-psychological variables in the theory. To many, studying brains directly implies a more serious scientific endeavor than looking at someone's overt behavior, but of course, looks are deceiving. Behaviorism has provided incredibly provocative measures of overt behavior in a number of fields—education, organizational behavior, child development, clinical psychology, and others. While biotechnology continues to improve, no current neurological theory of human actions can come close to matching the predictions of our actions based on behavioral science.

It's Dehumanizing

Many argue that behaviorism is "dehumanizing." The meaning of that is somewhat unclear, but there are some issues which seem relevant. Psychological theorists and practitioners sometimes offer that behaviorism is "mechanistic." After all, aren't we any more than a structure for processing stimuli and responses? Yes, we are what we do, as Aristotle noted. But that doesn't lessen the worth and value of any person. Behaviorism is merely a way of more thoroughly understanding how our actions develop.

The actual meaning of "behaviorism," and the work of Skinner and others, has been much misunderstood. No, we are not rats or pigeons or residents of "Skinner boxes." There is general acceptance of behavioral approaches in studying and modifying the behavior of other animals. However, it tends to be rejected when it comes to humans, as if we weren't animals ourselves. The laws of behavior are relevant for all creatures, and they all learn. There is nothing dehumanizing about that.

Finally, behaviorism to some means the deliberate conditioning of some person by another and is thus unacceptable—as if the well-evidenced principles of behavior do not operate for all of us, all the time. The behavior of someone who employs them to influence the behavior of others also does so because of his or her own conditioning. It applies to employers, priests, coaches, institutions, and all other humans, including psychologists and mental health professionals. If parents did not employ the various principles of behaviorism, we wouldn't be the species we are today. To ascribe some "inhumanity" or sinister motivation to the science of behaviorism is misguided. We are all human, and the study and application of behavioral principles is, in fact, human behavior, which we undertake with seriousness, empathy, and respect for others. As Skinner himself said, "I am a humanist in the sense that nothing human is alien to me." (50) He had compassion for others and actively advocated over decades for the use of his work to improve the human condition, noting how important behavioral principles often get ignored in this pursuit. (486)

Conclusion

The various criticisms of behaviorism, as we've seen, are not well-founded. Behaviorism is an approach to studying and modifying human behavior and cannot be judged outside of the universe in which it operates, observable behavior. Behaviorism is fundamentally the science of human behavior. It is difficult for many to accept that notion, but *the laws of behavior continue to operate whether we like it or not.* That it is imperfect does not make it wrong. All science is imperfect and continually being updated and revised. If it wasn't, it would be a religion. We denounce or belittle it only at the peril of retarding our understanding of ourselves—and why we are so religious.

Superstition

The entirety of organized religion and all the memes of gods and various rituals are based purely on superstition. We looked at some religious superstitions in Chapter 7. These are memes which might be considered "magical thinking." Such beliefs are generally seen as irrational, but let's take some time to examine what superstitions actually are and the behavioral dynamics which maintain them.

Adventitious Reinforcement

Superstition is merely the accidental association of behavior and reinforcement. That is, something that would otherwise be a reinforcer appears "accidentally" following a behavior, i.e., "by chance." The consequence is not contingent on the behavior. We humans exhibit a number of superstitions. Always sitting in the same seat in a restaurant is superstitious behavior. Baseball players sometimes turn their baseball caps around to create a rally. Wearing a "lucky tie" to a job interview might also qualify. Some of these things might "work" at some time, but they are most likely purely superstitious.

When there is an accidental association of behavior and reinforcement, behavior may still be shaped, and superstitions can be shaped quite easily. Let's look at a classic example, Skinner's famous experiment with pigeons. (481) He randomly reinforced pigeons with food regardless of what they were actually doing at the time. The result was a variety of bizarre behaviors—hopping from side to side, turning around, pendulum motions of the head, and a number of others. These behaviors really had nothing to do with the reward provided; they just happened "by accident" over time. If a pigeon just happened to be doing something—anything—a reinforcer was delivered, and random behaviors became shaped. That is how superstitions develop. They happen by the accidental association of behavior with reinforcement.[23]

Some animals have developed superstitions naturally, outside any laboratory, suggesting it is merely a natural, albeit accidental, process. As I discussed in Chapter 7, chimpanzee "rain dances," are an example. None of these behaviors can reasonably be seen as contingently associated with any reinforcement—they are random associations which merely persist for unknown reasons and are transmitted socially. They are superstitions. Unfortunately, we cannot ask the chimps what they "believe" about them.

Prayer

In humans, the most frequent superstitions are clearly found in religious contexts. Let's take prayer as a prime example. Praying for rain sometimes "works." Sometimes it does rain. We can see how such a superstition develops. However, asking God for guidance for your behavior is a bit different. Sometimes prayers are "answered." Praying for the "strength" to deal with certain life stresses might be followed by some "strengthening"; an illness often gets better after we pray for recovery; and other things we pray for do occasionally happen. These are superstitions, of course. They just happen accidentally. But people continue to pray, most often because such behavior is learned, usually modeled and reinforced by others. Further, prayed-for outcomes happen only rarely, which often puts prayer under the control of very lean, intermittent schedules of reinforcement. As we reviewed, this creates behavior which is very resistant to extinction. People who regularly pray tend to do so for the rest of their lives.

Since the superstition of prayer is fundamental to theistic religion, let's take a brief look at it from a behavioral perspective. Is it actually effective, or is it merely a superstition? There have actually been several scientific experiments on the use of prayer for desired outcomes for others, so-called intercessory prayer. In a classic study of this, an initial sample of around 1,800 patients in six hospitals around the country were studied when the patients were receiving a CABG procedure (coronary artery bypass graft). (47) They were randomly divided into three equal groups. In one group, patients were told they may or may not receive intercessory prayer the night before the procedure, and *the prayers were actually conducted regardless* (by various Protestant and Catholic prayer groups from around the country). In a second group, patients were told they may or may not receive prayers, and *the prayers were not conducted.* In the third group, patients were prayed for and they *were told that prayers were being conducted.* There was no actual difference in any of the major clinical outcomes for the three groups. For example, there were no differences among the groups in mortality or readmission to the hospital—not surprisingly. The only real difference was that the group that was specifically told that the prayers were being conducted had significantly more post-operative complications—infections, lung problems, etc. Intercessory prayer clearly didn't help. We can speculate on why those who were made aware of the prayers had more complications (It's tempting to do so, which I won't here.). But the bottom line is that intercessory prayer certainly didn't make any positive difference. There have been many other studies on such intercessory prayer and illness, but systematic reviews of the literature have failed to find any reliable evidence of a positive effect. (427) Many religious acolytes have speculated on why the scientific studies of prayer and illness haven't yielded good results (methodologies, type of medical problem, who's praying, etc.), but none of the criticisms have held up. The only conclusion left for many religious apologists—and this has actually been espoused—is that God doesn't want His interventions or worship scientifically studied. If that's what you're left with, the case is pretty weak.

Prayer is essentially asking God for a miracle. That is, you're asking a fantasy being to violate the laws of nature for your benefit or someone else's. Remember Ambrose Bierce's adage: to pray is "to ask that the laws of the universe be annulled on behalf of a single petitioner confessedly unworthy." We know that when prayers are "answered," any wished-for result is obviously only coincidence. It is purely and only a superstition. Praying for a miracle can be a pretty risky business in some circumstances. Superstitious reliance on God can lead to some disastrous outcomes, as we noted in Chapter 7. In sum, there is no escaping that prayer and other religious behaviors are superstitions.

Language and Communication

We can't talk about belief in God without addressing the issue of language. Without language there would be likely be no organized religion. It takes complicated communications to organize around a set of superstitions. When we say, "God bless you" or "God loves me," what does that mean? Where do such expressions come from? The answer is "simple." How we communicate, including religious concepts, is also subject to the laws of behavior. The notion of belief, per se, is not the cause of such communications. Verbalizations of belief and all other religious memes are as environmentally shaped as is the rest of our social behavior. They are a function of the language capabilities and the behavioral principles we've reviewed.

Origins

We've used some form of language for hundreds of thousands of years. We briefly reviewed the likely origin story of language in Chapter 4. Our ancestors also likely used what we call language. Neanderthals probably did, and perhaps even earlier species. As evolution produced human beings, they learned how to communicate in ever more complex ways, leading to what we call language. In fact, it may have been that communicative ability that helped us to evolve, cooperate, and survive, what some have described as our self-domestication (180). And it is what gave rise to the various god and other religious memes.

Language

Language is basically a type of communication which uses symbols —words, tones, gestures, etc., and the patterns we use to put them together —which communicates something to another person. But how do we actually acquire language? We seem to "automatically" learn how to speak, including our grammar. This is related to the long-debated notion of whether the construction of language is somehow natively "generated" in certain patterns or is more a function of guided development and environmental contingencies. Are we somehow internally programmed to communicate in a certain way, or do we learn it?

As we reviewed in Chapter 4, there is no doubt that hard-wired brain functions underlie fundamental aspects of language to some extent. Biological mechanisms were undoubtedly involved in *Homo sapiens'* evolving language capacities. That idea was refined by the American linguist and social scientist Noam Chomsky. He proposed that there is some innate universal grammar that underlies the acquisition of language, a "generative grammar." (I would point out that there are other extant theories regarding the details of language acquisition and usage, which will not be reviewed here. It is the Chomsky theory which garnered the most attention.) The theory was popular for a time; but it has now faded due to lack of empirical support, being replaced with the better-evidenced idea that children's language is developed via learning and modeling of others' behavior. (232) Despite our innate capacity for language acquisition and speech production, generative grammar and similar approaches do not explain the wide variety of language, constructions, syntax and other aspects of grammars, dialects, and nonverbal components of language throughout the world.[24]

It does now appear that language is variously acquired, not specifically through some innate universal grammar. Our language is fully acquired, and it is not automatic, even though it usually seems that way to our casual observations. It is specifically learned, including overall grammar, syntax, vocabulary, etc. Very few Indian babies grow up initially speaking English. Chinese grammar is different from English in several ways. Ancient Romans spoke with very different grammar and syntax from modern ones. Hungarian children do not learn to speak with Boston accents. Syntax in Japanese and Hindi is significantly different from English. Many languages include tonal factors, using vocal intonations for meaning (e.g., many in China, Southeast Asia, Africa, and some Native American), and others are strictly verbal. Some cultures write their language constructions from right to left. Global theories of language acquisition, including any universal grammar, do not explain these varieties and many, many more. How we communicate, including the patterns of our languages, is specifically learned behavior. Without such learning, language would never appear; the seemingly infinite variety of languages, dialects, styles, accents, word and grammatical patterns, and other characteristics of language attest to this. We are not born with such. All of this is learned (see below).

Verbal Behavior

We tend to believe that the language we use is generated from our internal thinking about things. That is, there is a common understanding that what we actually say—what we communicate—is a matter of our own voluntary initiation of speech—that we decide to say what we "want" to. That is, what we communicate might be a function of what we are "thinking." However, as we've reviewed, the brain is only the proximate cause of any behavior, which also includes communication.

Much of human communication is based on what we call "verbal behavior," as Skinner coined the term. (483) It can be specifically verbal (using words), or vocal (emitting sounds), or nonverbal (physical behaviors which have some sort of symbolic quality), or arrangement of all these things into complex communications. The bottom line is that the way we talk, the way we gesture, and the way we use language are all subject principally to the laws of behavior I've outlined. Specifically, they are operantly conditioned.

We communicate for effect. That is, the relevant dynamic of communication is the effect it has on others. That is, the interpersonal contingencies of verbal (and other communicative) behavior are what shape and maintain the behaviors. That such learning may be very complicated—even to the point of the calculational impasse—does not mean that they are not learned behaviors. That we say certain things, use particular words, and appear to "generate" language in particular ways do not imply a "free will" or conscious choices in communication. That we might refer to such communication as such is merely a substitute for not knowing the remote, subtle, and historical contingencies which have shaped such communicative behavior over time. What we say is purely a function of conditioning, whether it's "Hello," "Please pass the beans," or "Jesus loves you."

Acquisition

Consider the acquisition of language. We learn to speak from our parents. That is the overwhelming factor in language acquisition. It begins at birth and possibly before. Actually, there is likely an evolutionary advantage to an unborn baby hearing a mother's voice prior to birth (the sound being conducted through the mother's body); these maternal sounds are especially reinforcing after birth. We suspect this since other sounds may be conditioned as well. For example, children who are exposed to certain kinds of music before birth tend to show more positive responses to those genres after birth (255), and there's evidence that the prenatal brain even begins to learn processing of some word sounds. (392) Quite obviously, there would be an evolutionary advantage to a social species whose susceptibility to language conditioning could begin as early as possible, and a mother's voice is most certainly the first influence.

It seems almost remarkable that children learn to talk and that they acquire language so easily and rapidly. Babies naturally make all sorts of sounds, and as the child develops, sounds become a few phonemes. But these only become communicative as they are modeled and reinforced. Modeling one's parents is a powerful dynamic, and a young child will attempt to mimic almost anything that is uttered or articulated by a nearby adult. The tendency to do what is modeled is, if you will, hard-wired in us, but what children actually voice in terms of verbalizations is a function of the specific behavioral modeling they're exposed to; and this includes the intricate patterns of utterances that a child hears. As the brain develops, more subtlety in phonemes, syllables, words, syntax, etc., becomes possible. Rapid learning occurs for both the content and the grammar of language. What is uttered is always what the child is exposed to and the generalization of that learning. This is the way we actually learn language. It is a specific behavioral process. It is gradually shaped, again principally by the modeling and reinforcement of parents. This happens continuously with all sorts of verbal behavior, even though we're not aware of all the dynamics taking place at a particular time. Thousands of environmental contingencies—many of them remote, subtle, and with lean schedules of reinforcement—shape all sorts of verbal behavior. That is the way language and speech are acquired.

Of course, the extension of this process leads to what is actually said, how it's said, and in what circumstances. We are reinforced for some communications, extinguished for others, and shaped into various forms of our verbalizations. Whether we say "move over" or write the Declaration of Independence, such come about because of a lifetime of the operant conditioning of *verbal behavior*. Many will balk at such concepts, but remember, the complex overt behaviors we emit are specifically learned. No other explanation is possible.

Religious Talk

The shaping of religious talk is no different. How people talk about God or other religious memes is specifically learned behavior, and such talk is due to the lifelong shaping of verbal behavior in this regard, often beginning in early childhood.

When you say, "I believe in God" or when you bow your head to pray, you are doing things that you have learned. You don't "freely" do them, as we dealt with in Chapter 6. It is your conditioning that results in such behaviors. Verbal behavior (and all the nonverbal or paralinguistic responses that go along with this) is just that. If I say, "I enjoy going to church," a listener may infer that there is some meaning or validity to that. Certainly, it may be correlated with other behavior of the speaker and other factors, but it is actually only well-developed verbal behavior, usually shaped in ways that are not fully obvious. If I say, "Jesus loves you," the listener may not ascribe some meaning or validity to such a statement. However, it can be easy to see that the speaker's verbal behavior has been specifically shaped. If one says, "God told me to run for office" or "God told me not to get vaccinated," it's even easier to see the comments as verbal behavior without any other necessary referents, just social operant contingencies. We usually do not infer some underlying reality to those comments. God did not suggest that course of action. Why people say such things is purely a function of behavior which has been shaped to the present. The principles which underlie such utterances as "I love ice cream" or "Gimme the remote" are no different in this regard than saying "Mazel tov" or "Praise Allah."

We also sometimes use seemingly religious verbalizations which don't necessarily communicate anything theistic or religious. When we use the word "god," for example, we're not always talking specifically about a deity. We use the word "god" in a variety of contexts, without any seemingly religious connotation, even though the evolution or etymology of such language may be associated with some religious meaning. "There but for the grace of god go I." Many people saying this may not actually have a 'god' in mind. We often utter the word "god" when trying to communicate strong emotional content, whether great joy, great fear or dread, or other emotion. If I say, "Oh my God" or "goodness gracious," do these imply some sort of supernatural reference? Usually not. They are mere emotional verbalizations. "Thank God" is an expression of relief. "I hope to God that . . ." is an expression of intense desire. "That goddamn . . ." is a comment on something noxious or unfortunate. "God bless you" denotes offering kindness to someone. "God only knows" is a substitute for wondering or not knowing. "God forbid . . ." suggests a wish to avoid something dreadful or threatening. The OMG meme has seriously penetrated our culture. You get the idea. The list is endless. The empirical meanings of such things in discourse, of course, are mundane. Obviously, those communications don't validate one's reference to a personal deity. Those utterances are purely verbal behavior. And of course, we invoke a Devil character or concept as well. "Go to Hell," "damn you," "curses," etc. Very few people saying such things endorse the idea of a devil. We learn all this language quite independently from any religious belief. Such things are uttered by atheists, agnostics, and children, who have no idea about the ancient linguistic history of such comments and who do not necessarily imply supernatural existences.

Technically, we use language without necessarily "thinking about it." The continuum of how we use the word God or other religious memes is merely a reflection of the fact that they are well-conditioned verbal behaviors, often passed down as cultural symbols and modeled and shaped by parents and others. How we talk about some supernatural entity, or the religious memes we utter, is purely a function of the verbal behavior we learn, specifically the conditioning we're exposed to. It has nothing at all to do with any existence of such an entity. Again, you may not believe that, but the laws of nature, including the laws of behavior, continue to operate whether you like it or not— including on what we say and how we say it, with religious memes being no exception.

Child Development

A special word about religion and children. There is general agreement that religiosity is "handed down" from parents. That is mostly true. And children learn their "faiths" via the modeling and other conditioning that are associated with that. When the various types of religious conditioning happen over and over again in a child's life, when his behavior is most malleable, the probability of him endorsing specific religious doctrine is incredibly enhanced. It was the Jesuit St. Ignatius Loyola who's famously credited with the aphorism, "Give me a child until he is seven, and I will show you the man." The original version of this concept has been credibly ascribed to Aristotle. Nevertheless, it continues as a somewhat sage observation. Although religious conditioning is a lifelong process, it is obvious that children do learn about religion and are shaped into religious faith and related behaviors beginning early in life.

A couple of cautions. I'm not going to address the range of issues which affect how children develop. We're focused here primarily on behavioral factors. Child development in the larger sense, including the nature-nurture issues which may affect a child's language development, has been well summarized by others. (474) In this section we'll consider only the behavioral development which is generally relevant to children's belief in a god and other religiosity. I'll give a number of examples. But these are for illustration only. Remember, there are literally thousands of behaviors, schedules of reinforcement, and other operant factors which go into shaping belief in a god in the early years of a young child's life.

There is a paucity of good research on child development and religion. This is almost impossible to do. The ethics of our culture and sciences are inconsistent with conducting childhood experiments on such things. Nevertheless, we'll talk about some of the factors bearing on how children likely develop religious behavior. The only thing we do know is that children, like all the rest of us, do learn according to the laws of behavior we've outlined.

Evolution Has a Hand

Children tend to believe what adults, especially parents, tell them, and clearly evolution has enabled this. Survival depends on developing the behaviors which promote growth and safety, and these are learned from parents. As Dawkins notes, "Natural selection builds child brains with a tendency to believe whatever their parents and tribal elders tell them. Such trusting obedience is valuable for survival." (113) Evolution has favored parenting, especially in mammals.[26] It is critical to the development of young offspring. Mammals, especially more advanced social ones, are born without a great deal of physical capacity and a robust repertoire of adaptive behaviors. They must be taught a huge number of things in order to survive. Thus, natural selection has favored human children's ability to learn from parents. Early in life, children depend almost entirely on their parents, who provide the safety, nutrition, and learning a child needs. Parents are both unconditioned and conditioned reinforcers and incredibly powerful models. The child's learning becomes quickly and powerfully generalized to believe and mimic whatever a parent says or displays—almost anything!

Authority

The parent, in effect, has significant control over what and how young behavior develops. This is how the notion of *authority* becomes shaped for the child. Authority is the first source the child has to depend on for safety and learning about the world. Authority enables physical survival. "Don't go into the street." "You can't eat that." Those parental dictums, of course, have immediate value. But often authority trumps exploration in ways that don't have safety or "survival" value. "You must wear this . . . you can't see that . . . you can't say that . . . can you say thank you . . . you must do your math homework." Some of this provides reasonable social skills and shapes a level of cultural conformity. However, when authority is overly or punitively used, it can impair exploration, creativity, skepticism, and critical thinking (I'll define this more below). "We don't read things like that here." Where this influence develops early, it becomes rather "blind faith," i.e., following authoritative guidance without any serious questioning. This is a pivotal issue in childhood where the child sees no empirical basis for authoritative direction. Children sometimes do question authority when they can't envision an empirical or logical basis for some guidance. "What's so bad about a hot dog?" "But Susie's reading that book." Where such responses are unrelentingly punished, authority becomes potent, allowing parental belief systems to become well-conditioned very quickly.

Parental authority is also generalized to other individuals and institutions—teachers, clergy, etc., and children come to believe in the guidance associated with those authorities. When a young child hears from a parent that "God is up there watching you," it is almost always believed, even though a young child may have no understanding of what that means. And when similar verbalizations are repeatedly employed by other authorities, the initial outcome is clear. All this is terribly obvious, but sometimes we forget that religious faith *is learned*, and mostly from authorities. St. Ignatius Loyola was right.

Curiosity

Children are naturally curious, and that behavior is critical to survival —in the wild or in an "advanced" culture. They are hard-wired for exploration and understanding of the world around them. Children (like most mammals) are born with a natural curiosity about the world. This becomes manifest in exploratory behavior, and testing and learning what works and what doesn't. It quickly evolves into what we call "play," which is the way young children learn about their environments and their own bodies and behavior.[27] Thus, our young are able to achieve greater and greater mastery of their environments, and their own lives, by satisfying their curiosity about the world around them and how it works. Born of curiosity, play is active exploration of the world, and exploration leads to learning. It is what children use to survive and flourish.

Initial explorations and play progressively develop into children's curiosity about things that can't be physically explored or examined. That is the reason children ask "why"; they seek answers from the authorities who are already models and reinforcers in their lives. "Why do the leaves fall off the tree?" "What makes the car go?" "Why is the sky blue?" "Why do I have to eat that?" Sometimes parents minimize such curiosity. Some are often at a loss to explain the world, or are just too busy. Sometimes the child is not capable of grasping the reasons why something is as it is. "You'll learn about that later," or, "It just is." These types of responses can ring hollow to most children at certain ages, and some curiosity can be inadvertently extinguished.

How curiosity is handled is often pivotal when it comes to shaping religious beliefs—or their avoidance. Curiosity can be reinforced. Or conditions can exist which tend to extinguish or suppress it. These two dynamics of child development are not black and white, of course. But they do have great power, especially in early development. Where curiosity is reinforced, especially through early childhood, critical thinking and broad perspectives on life tend to grow. When a parent answers a question directly, curiosity tends to be reinforced. Even young children can understand how the planets revolve around the sun. Remember your model solar system in elementary school. Again, sometimes the parent doesn't have the time or knowledge. But reinforcing alternatives can appear. "That's a very good question." "What a wonderful question." "I don't really know, but we could find out." "The more you learn the better you'll understand; let's talk about it." "It has something to do with . . . let's think about it." "I don't know. You'll learn about that someday, but for now keep asking the question." Sometimes the child can't yet grasp an explanation. When a young child asks, "Why do the leaves fall off the tree?", the answer may be pretty technical. But reinforcing responses can be somewhat metaphorical. "The tree is going to sleep for the winter." It isn't until much later that an "energy saving" explanation can be understood. Curiosity, of course, can be fostered and reinforced by parents and others; children who grow up with a tendency to remain curious, seek out answers for themselves, and question authority are much less likely to put blind faith in some supernatural entity.

Where there is little reinforcement for curiosity during these early ages, and especially where it is often extinguished or punished, it tends to diminish greatly, and more regimented ways of thinking and behaving often emerge. The child is more likely to blindly model such parents when it comes to religious behavior. Reinforcement for mimicking religious actions and parroting religious verbalizations tend to rise, and extinction and/or punishment can occur for thinking and behavior which contradicts the parents' religious beliefs. These are incredibly powerful influences on a child's behavior, and it is how reflexive and uncritical religiosity develops.

Where curiosity is not well reinforced or often extinguished, and blind obedience to authority is favored, children have a tendency to develop faith in God when this meme is exposed to them, generally modeled by the parents. For explanations about the world, curiosity can wane, and seemingly simpler explanations gain traction. It is almost "human nature" that we often choose the path that takes the least time and energy, and religious answers are clearly that. Maintaining curiosity, seeking information, and developing answers to challenging questions are more difficult; and sometimes lead to longer and more uncomfortable conversations. "God" is often the path of least resistance for religious parents. When a child poses the question, "Why does the Earth go around the sun," the answer may be scientifically age-appropriate or it may be something like, "That's the way God made it." (Of course, we can say without too much sarcasm that God set it up that way only when Copernicus came around.)

When you don't have to look too far for answers to difficult questions, a god and artificial answers and moral codes are much easier. It is also much easier to entertain fantastical "explanations" for things than to investigate or consider them further. It takes less behavioral effort. "The Bible says," of course, is one of the easiest ways. I would observe that this answer also arises in conversations between adults, even though many of actually know of the Bible's hypocrisies.

Fantasy

Another factor which contributes to children's belief and blind faith is the notion of fantasy. Humans have the capacity to imagine, to create or believe in things that don't actually exist, and this is evident early in childhood. Imagination is a ubiquitous and fundamental characteristic of young children.

Imagining things is also involved in play. The most important things in a young child's life are *alive*. Young children generalize this idea; and it is easy to see that they imbue other objects in their environments with life. They name their dolls and stuffed animals, of course; but they do this with other objects as well—toy cars, robots of course, and almost anything that moves. Anything that can move—or can be made to move—might get a name and eventually come to life. They imbue inanimate objects with life and character. This anthropomorphism is natural in human development, and such imagination is one of the more vital and wonderful qualities of a child's life.

As they learn more from parents and others, this capacity also begins to generalize to imaginary beings—leprechauns, the Easter Bunny, the Tooth Fairy, Santa Claus, Mother Goose, elves, unicorns, mermaids—the list is endless. These are living beings to a young child, and parents clearly reinforce this idea. It is very easy to shape a child's "belief" in ideas like God and Jesus. All these are fantasies, of course, but they are *real* to a young child. As religious parents progressively ascribe how the world works to God—much like the impossible works of Santa Claus—the concept of God as a living being becomes real to a child. It is as "real" as any other fantasy. With guidance and reinforcement from the most potent influences in a young child's life, parents, it is very easy for a child to imagine that "God is up there." And the shaping of doctrinaire religious faith begins.

Culture

Young children are not naturally religious. However, they do begin at some point to develop conceptions of the God they've been conditioned to believe in. When very young children begin to hear about God, they don't distinguish Him from some other great figures, but at some point, they learn that there is something special, something of great power, associated with this figure. They also attribute qualities to God consistent with their parents' religious tradition. For example, in one cross-cultural study across several countries, it was found that an "all-powerful" God was viewed by young children as a disembodied force by "mainstream Christian" children, while young children of LDS parents saw God as an embodied individual, an actual person. (383) Further, it was interesting to note children's conceptions of the idea of God's supernatural quality. Across several religious persuasions—LDS, mainstream Christian, Catholic, and Muslim—four-year-olds on average tend to attribute some supernatural character to God. However, to a somewhat lesser extent, they also see some supernatural character in ghosts, dads, and bugs. By age six, attributing supernatural character to God is similar; and ghosts, angels, geckos, and bugs are still seen to have some of this, but some supernatural quality for dads has nearly vanished. Of course, there was a much higher level of supernatural character for God among adults in each of these religious persuasions. It is interesting to note that, among adults of these religious persuasions, ghosts still commanded a healthy amount of supernatural character, and some few continued to imbue bugs with it. Sometimes, there's just no accounting for some people's superstition and fantasy. In sum, not surprisingly, children's conception of God is a function of both age and the religious traditions of their parents.

Various studies show this is true for certain religious traditions in a number of countries where notions of a god in terms of embodiment—appearance, mental properties, and supernatural powers—are specifically dictated by the relevant culture. This is obvious. Very few "party lines" are crossed. We might ask: what is the probability of a child of Jewish parents being raised as a Catholic? What is the likelihood of Hindu parents raising a child to be a Sunni Muslim? How about Christian Science parents producing a Sikh child? The probabilities here are somewhere between zero and ludicrous. Although children do change as they grow older, and even during adulthood, it is obvious that people learn what to believe from their early exposures, which are unique to each culture and religious persuasion.

And their conditioning leads to endorsing specific dogma. Most Jewish children have a single God. Catholic children believe in three main supernatural entities (Father, Son, and Holy Ghost). Hindu children know that some supernatural spirit inhabits all living things. And for some children, the range of spirits even greater. As Kaku observes, "Kids in Japan are steeped in the Shinto religion, which believes spirits live in all things, even mechanical robots." (261) The variety of children's beliefs around the world can come only from shaping by their parents and others. One wonders what type of supernatural fantasies children of ancient Greece and Rome had. Actually, we do know some of them. They've been handed down in the extensive mythology from those cultures. And what about Neanderthal children? The capacity for imagination likely didn't start with us.

Indoctrination

The common wisdom, of course, is that it is *indoctrination* which brings children to beliefs and a particular religious faith. While the Laws of Behavior in general are the operative factors, let's look at the concept of indoctrination.

What do we mean by indoctrination? Indoctrination is the providing of information by an authority along with the message to not think critically about it, and the connotation usually includes the threat of various punishments for not following the dicta. In organized religion, it is propaganda dispensed, usually, with the threat of both mortal and eternal punishments. It is fear that is the basis of indoctrination. Punishment (and the fear it is conditioned to) tends to suppress childhood exploration and critical thinking and promote approaches to "blind" belief in whatever an authority is dispensing. Of course, since threats are punishment-based, they must continue, at least in some symbolic form, in order to be effective. That is why children are sometimes "forced" to go to church, to pray, etc. Dogma must be continually or frequently reemphasized.

Nevertheless, punishment and threats alone are rarely effective in the long run. Thus, religious behavior must also be consistently maintained by positive reinforcement—social support, feedback from parents and clergy, and the promise of eternal reward. Otherwise, religious behavior will wane. Indoctrination alone seldom works in the long run.

It is very difficult for a young child to stray from a culture of this sort or reject it outright. When religious indoctrination is a principal tool of the parents, any behavior inconsistent with this becomes unlikely. It is as if the child has "no choice" in the matter. For this reason, a number of writers have offered that religious indoctrination of young children should be thought of as "child abuse." (113, 221) We generally think of child abuse as physical abuse or neglect which demonstrably harms the child and, in some cases, psychological abuse which can obviously affect a child profoundly. But what about suppressing curiosity and consideration of *ideas* parents don't like. What does that do to a child? The point here is that we might call a child's religious "training" abusive if it involves physical or severe behavioral punishment for not acting in conformity with a religion and aping its dogma. But what about suppressing the natural curiosity and exploration of the child? There are no criminal laws about this. Yet, we can't help but take offense at punishment for disbelief, whether we call it "child abuse" or not.

Religious indoctrination has also been described as *brainwashing.* (76) Basically, brainwashing is a lay term denoting efforts to change thinking and behavior "against one's will." It usually suggests the use of physical control, threat of punishment, and deprivation reduced only by compliance with an agent's goals. Brainwashing is generally associated with some level of cruelty, and its effects are at best very temporary. When immediate threats are no longer available, behavior generally returns to baseline, though not always. Religious authorities, of course, would actively reject using the term "brainwashing" for their "religious education." Nevertheless, it may be applicable in some circumstances, and many "survivors" of Scientology training would certainly describe it that way.

Indoctrination can be overcome, of course, but it is generally not a matter of merely "deciding" to do so. Where heavily indoctrinated children eventually reject religious propaganda, other influences are obviously responsible. Rejecting various religious superstitions, and a god itself, arises as a function of other conditioning that an individual is exposed to. Sometimes "authoritative" guidance is not enough and may not be followed by a child. This is an extinction paradigm for the parents. As we know about extinction, this can result in anger and vitriol and other desperate behaviors to suppress the child's noncompliance and questioning. That there are also punishments for children's questioning is truly remarkable. It is a reflection of the fact that it is very difficult to maintain behavioral control of an idea, especially one where no earthly referent can be experienced by a young child. Since indoctrination is also usually associated with more continuous schedules of reinforcement, it is thus exceptionally vulnerable to extinction and thus also the necessity of threats. As a result, it is not unusual for children heavily indoctrinated in a faith to give it up in adolescence or early adulthood, where such control is no longer operative.

Some children raised in a culture of indoctrination to religious dogma do eventually reject it. That happens often in families where behaviors incompatible with dogma are also reinforced and become enduring abilities of the child, particularly curiosity, questioning, skepticism, critical thinking, and exploration—even taking an interest in science. Developing an interest in science is associated with an attitude of "finding out for yourself," not blindly following the word of an authority. Of course, questioning and critical thinking are incompatible with religious indoctrination. Some religious parents reinforce curiosity and critical thinking about most areas of knowledge but actively discourage or try to prevent it when it comes to religion. This is obviously a cognitive tension for a young child. When such tension persists as a child ages, it sets up one of two possible long-term outcomes. Either they grow up with this duality and rationalize away the conflicts (see below), or they eventually reject religion. Of course, this can be a problem for more fundamentalist parents. To wit, the continuing conflicts of teaching evolution in the public schools (I'll address in the next chapter). Secular education can be a major problem for such parents, who see "truth" very differently. Returning to the wisdom of Thomas Jefferson:

> The first stage of this education being the schools of the hundreds, wherein the great mass of the people will receive their instruction, the principal foundations of future order will be laid here. Instead, therefore of putting the Bible and Testament into the hands of the children, at an age when their judgments are not sufficiently matured for religious enquiries, their memories may here be stored with the most useful facts from Grecian, Roman, European and American history. The first elements of morality too may be instilled into their minds; such as, when further developed as their judgments advance in strength, may teach them how to work out their own greatest happiness. (245)

Conclusion

Although some people become "faithful" as adults, the overwhelming development of religious behavior occurs in early childhood. And it is specifically conditioned. As Bertrand Russell astutely observed, "Most people believe in God because they have been taught from early infancy to do it, and that is the main reason." (435) When you think about it, it's a bit remarkable that children must be conditioned to believe—even to the point of being promised a heaven or being threatened with a hell—in order to adopt the magical thinking of a religion. It is apparent that belief in this context is a pretty fragile thing. Thus, you must start early and continue the indoctrination.

One of the more unfortunate aspects of children raised with a more doctrinaire attitude and conditioning for a particular faith, as I touched on in Chapter 7, is that a child will develop a self-identity with a particular religion. Young children are imbued with the notion that they have an immutable religious character, almost as if it were a genetic trait. When we ask, "What *are* you," it's as though we're trying to reveal your birthright genetic religion (As I reviewed, a particular religion or religiosity is not genetic). You *are* Sunni. We are Catholics. Never forget you're a Jew. One develops a self-concept of "being" Catholic, Muslim, Jewish, etc., long before there is any understanding of what these things actually mean. Of course, this persists well into adulthood, which clearly suggests the supremacy or potency of early childhood learning over any critical thinking. The behavioral conditioning, including indoctrination, of religion in young children is the primary reason why we have such self-identities.

Some Things We Know

I've reviewed how religious behavior becomes conditioned in individuals. At this point, we might examine at least some of the research which tells us about religious behavior in the larger sense. Religion literally means to do something repeatedly, and that's what "people of faith" do. Religious behavior includes all sorts of verbal and nonverbal actions—praying, going to church, talking about a god, reading or quoting a Bible, evangelizing, observing rituals, etc. The list is a long one, and each item on it comes in a seemingly infinite variety of forms. Such responses are generally referred to as "faith-related behaviors" and in sum are often termed as "religiosity." People have different conceptions of what that means, but we can acknowledge that people are more intensely religious to the degree they emit religious or faith-related behaviors. You might consider a person intensely religious without the panoply of those kinds of behaviors. One might hold strong beliefs, hold them quite privately, and regularly pray to a god. However, such behaviors are rarer and don't constitute the religiosity that has caused so much suffering throughout history. We'll look only at a bit of the research which bears on this notion of overt religiosity.

There is a variety and large volume of studies looking at these issues. However, to reiterate, we do not have any well-controlled experiments bearing on the operant conditioning of religious behaviors or the evolution of religiosity in populations. As I pointed out with respect to children, it would be unethical to conduct various types of controlled experiments in such areas. One must rely on uncontrolled studies, natural observations, and self-reports to provide information about certain behavioral factors and religiosity. Given that caution, here I'll address some data on just a small sampling of the more well-known factors that affect people's religious behavior. For our purposes here, it would be foolish to attempt any significant review of the vast literature in this area. What I'll do is highlight just a few of the salient and large-scale studies to illustrate the principles.

Religious behavior is not randomly distributed in any large population. It is a function of numerous social and other environmental factors, which come to shape various belief and faith behaviors. In fact, there *are* some trends that illustrate a few of the factors associated with religiosity. We might ask, for example—Why is religiosity in any population unevenly distributed with respect to education, income, geography, and other demographics? If people have fairly equal access to the information, symbols, and opportunity for belief and worship, one cannot discount the role of various personal and socioeconomic factors being associated with religiosity. In addition to being positively reinforcing, we'll see that much of this is consistent with the way negative reinforcement works on religious behavior. That is, religious behavior for many does tend to reduce some aversive or noxious stimuli or experiences, or various stresses and conditions of life. A good deal of this will seem obvious, but it is important to illustrate the point. (We know that politicians and other public figures have gotten into "trouble" when observing or suggesting the relationship of religion to such things as poverty, geography, intelligence, education, and other areas. However, this is a different matter than the empirical truth of such relationships.)

Social Factors and Children

Being with others is generally positively reinforcing. Powerfully so. People learn to espouse religious ideas and engage in various memes and behaviors because they are positively reinforced. And that's where social factors come in. Socialization appears to be a major factor driving religiosity. Social contingencies, within the family and in larger groups, have significant, if not paramount, importance in the development and maintenance of religious beliefs. And it starts very early.

As an example, here's a large-scale study that illustrates the point with young children. In a large national sample, data were gathered from over 20,000 American kindergarteners and first graders, as well as their parents and teachers. (34) After controlling for many social and demographic factors, it was found that children's "pro-social" behavior was significantly associated with religiosity in the family. That is, the tendency to be helpful, sharing, and seeking friendships were associated with family religiosity. It is easy to see that religiosity paired with social behavior can reinforce each other.

In a large international study, survey data on religious beliefs were collected from over 19,000 self-identified Protestant and Catholic respondents in 15 nations. (266) The results weren't surprising. If people lived in a statistically more religious nation, they tended to manifest a more orthodox orientation. If they lived in a more secular country, a more liberal expression of their religion was evident. On the other hand, in the more secular countries, religiosity in the family was the strongest influence on children's religious beliefs. These trends were manifested regardless of national socioeconomic factors, gender, level of education, and other variables, and they held for both Catholics and Protestants. Again, we can see that the behavioral or social factors are paramount.

In the United States, a national survey followed over 20,000 students in grades 7–12 for over a decade. There were assessments of various indices of health, behavior, family and socioeconomic factors, and certain psychosocial and cognitive measures for both children and their parents. It was unsurprisingly found that participation in religious activity substantially declines in young adulthood. This downward trend is accelerated by cohabitation, nonmarital sex, and drugs or alcohol use. The trend is slowed, however, in association with marriage. Surprisingly, if college is avoided, the decline in religious behavior is steeper than if the young adult goes on to college. This is contrary to expectations, per other studies, but other factors of the "college experience" in this longitudinal study may have been responsible for the relative lack of "secularizing effect" for college students that are found in other kinds of studies. (523)

None of these results is terribly surprising, and the results are open to various interpretations. Nevertheless, they do show that social factors for children, both within the family and larger social contexts, are dramatically associated with religiosity.

Socioeconomic

It is commonly understood that socioeconomic factors are associated with religiosity. Let's look at a couple of studies in this regard. A comprehensive survey was done of 114 countries around the world. (101) The primary measure was the proportion of people who say that religion is an important part of their daily lives. Not surprisingly, individuals in higher-income countries endorsed this less than those in lower-income countries. Interestingly, the United States, as one of the richest countries, didn't follow the trend. In America, about two-thirds of those surveyed said religion is important in their daily lives, which puts us well in the bottom half of countries on this question (I'll address religiosity in the United States more in the next chapter). Actual per capita income was also negatively correlated with the question. The lower the income of countries' individuals, the more people endorsed religion as part of their daily lives. It was noted that religiosity had a stronger relationship to "emotional well-being" in poorer countries than among those in the developed world, which may account for some of these results. In any interpretation, religiosity around the world does seem to be strongly related to lower income.

In the United States, income levels do predict a variety of religious behaviors. In a major report from the Department of Health and Human Services, it was found that low-income youths and adults show higher levels of religious beliefs and endorsement of religious principles, but young people tend not to participate as much in organized religious activity. They also tend to claim a higher level of "religious experiences." Teenagers with a lower socioeconomic status (SES), in particular, are less likely than others to engage in religious activities, but they are more likely to engage in things like prayer and reading scriptures. (256) It is unclear what each of these relationships is based on, but it is obvious that SES does negatively correlate with overt religiosity in American youth.

We do know that poverty, or any other measure of SES, is directly related to religiosity; the common wisdom is that religious behavior tends to lessen the so-called "burden" of poverty. But it may not be that simple. In one very large study—several million people across 150 countries—various measures of religiosity, economic development and "well-being" were used. (48) Results indicated that, overall, the so-called "burden" of a low SES (on an individual basis of well-being) is less severe in the poorer nations than the richer ones. This was a somewhat surprising result. However, it is partly explained by the fact that developing nations tend to be more religious overall, and this pattern may be a factor in upholding behavioral norms that might ease the burden of lower SES. Religious behavior does tend to be associated with higher reported levels of well-being. It is an interesting interpretation, though it is most likely that both coping with "burdens" of poverty and national religiosity play a role in reported well-being.

In a similar vein, others have found that it may be "personal insecurity" associated with lower SES that is associated with higher religiosity. This was tested in a large, multinational, cross-sectional study. (423) Here measures of "income inequality" were used as a proxy for personal insecurity and, interestingly, religiosity was assessed as the "mean frequency of prayer" in each population. Results indicated that income inequality (the index of personal insecurity) was directly related to the prayer rate. Further, that rate was higher in proportion to other variables—notably higher violent crime, more corruption, shorter life expectancy, higher infant mortality, and higher abortion rates. It's not surprising that such factors might be associated with religiosity and personal insecurity in large data sets, but it is always enlightening that some indices of religiosity are associated with higher rates of such social problems, not lower.

It is likely that the above, and numerous other SES factors, are associated with religiosity. Of course, we can't draw any causal connections here. Nevertheless, it is hard to ignore that religious behavior is associated with SES and how people respond to their situations. Again, those factors are a function of how behavior is shaped, and, as a reminder, different for each individual.

Stress, Suffering, and Mental Health

An often-quoted benefit of being religious is that it provides comfort in one's life. This includes some relief from physical or emotional pain, trauma, loneliness, loss and grief, life-threatening disease, and various other seriously problematic experiences. Few would disagree with this. A lot has been said about religion in this context, from sociological circles to extreme political ones, as we saw with Karl Marx's infamous quote. But there is no denying that people often gravitate to religion or benefit from religion because of the comfort it may offer. The value of religion in providing comfort may have been a factor in the early cultural evolution of formal religion, but its ability to comfort in times of trauma and other significant life stresses is nearly universal. As psychiatrist Harold Koenig put it, "Exact beliefs differ between religions, but they all have one thing in common—devout adherents say that religious beliefs provide comfort and meaning when tragedy strikes." (278)

Let's review some examples which highlight this issue. Most people will endorse the notion that religion makes them "feel better," even on a daily basis and even where there is no significant stress or tragedy. If you attend a religious service, feeling better afterward is pretty typical. Over time, if that didn't happen, your attendance might drop. That's a change in behavior enabled by negative reinforcement. Here's how it works. If you feel better because of attending the service, then logically you didn't feel as well going in. The effects of some of the problems or stressors in your life have been temporarily lessened. Some aversive stimuli have been reduced, even without your awareness. That's negative reinforcement, pure and simple. If the stressors or life conditions are severe, affiliating with a religion or engaging in religious behaviors can be even more potently reinforced, and we see that in numerous observations and studies. Here are some examples:

In a wide sample of 17 developed, so-called First World countries, it was found that religiosity actually tends to be abandoned when socioeconomic conditions are sufficiently benign, as they are viewed principally as mechanisms of coping with various stresses, anxieties, and social conditions (393). In a comprehensive survey of attitudes and behavior in those countries, there was a significant positive correlation of religiosity with a wide variety of behaviors which likely present notable stress in the population. Some of the major ones were poverty index, income inequality, underage-5 mortality, births by 15–17-year-olds, STD infections, homicide, and incarceration—both of the latter contrary to the popular belief among some that religion tends to be a hedge against crime and other social aberrations. Religiosity does appear to be more prevalent is societies where there are more problematic social conditions and behavior.

American high school seniors (approximately 17,000) were examined for their level of religiosity. A variety of religions were endorsed—Catholic, Jewish, LDS, various so-called "fundamentalist" and "regular" Protestant denominations, and others. (9) Despite any differences in parental education or employment and other variables, it was found that marijuana and alcohol use was significantly less likely with increased religiosity, and this relationship held for all religions. The behavior dynamics would be difficult to tease out in this case, but one cannot rule out that some social factors associated with religion were pivotal.

Similar patterns occur in adolescents in other, non-Christian religions/ cultures. For example, a random sample of 378 high school students in Iran was surveyed for religiosity. (563) Adolescents manifest a variety of emotional and behavioral problems, most of which do not present for clinical contact. But the data show that such aberrations in the Iranian adolescents tended to be less frequent if they had a reported an emotional "attachment to god." Apparently, faith behavior may be associated with reduced psychological dysfunction in Iranians just like it may operate for Christians or Westerners. Again, this is negative reinforcement at work.

It is well-known that adverse childhood experiences can lead to adult mental health problems. Among a vast literature, this has also been studied via the comprehensive National Survey of Midlife Development in the United States (MIDUS). This is a large longitudinal survey of a representative sample of American adults assessed from 1995 to 2005. Childhood adversity was defined as any physical abuse by a mother, father, siblings, or other person, emotional abuse by any of those parties, parental divorce, and certain other less notable variables. It was found that religiosity (endorsed religious attendance, spirituality, and importance of religion in life) in adults tends to buffer the negative psychological effects of the subjects' childhood adversities, particularly on their negative affect (emotional expression). Religion tends to act as a coping resource for such individuals—to minimize the effects of previously experienced adverse childhood stress. (258) In effect, religiosity tends to be negatively reinforced because of reduction of those particular aspects of mental health difficulties.

In a very interesting study, a large sample of depressed psychiatric inpatients who had a history of childhood abuse were examined for suicidal ideation and previous suicide attempts. (125) What was found was a significant inverse correlation between suicidal ideation and religious beliefs. It is well known that childhood abuse is a risk factor for depression and suicidal behavior, and it appears that religious beliefs may well act as a coping mechanism in preventing suicidal behavior. Again, this is consistent with religiosity being a negative reinforcer by reduction of serious mental health risks.

Depression, as commonly understood, is a major problem in contemporary America. In a major review of the literature, it was concluded that religiosity tends to offer a "protective effect" on the risk of depressive symptoms or an actual diagnosis of "depressive disorder." (336) The longitudinal research in particular is consistent with the idea that religious behavior endows a protective effect against the incidence and persistence of depressive symptoms or disorders. But it is interesting to note that such associations tend not to hold for "private" religious activity, suggesting that it may certainly be the social dynamics of religiosity that are maintained by negative reinforcement, something we would expect, of course.

In sum, aversive quality of life stressors tend to be reduced by the focus on and practice of religion, be it prayer, the engagement in rituals, the sharing with others about God, or the socialization with others in religious contexts (although this socialization, of course, can and often does have reinforcing effects independent of any specific religious behaviors). Religious behavior is negatively reinforced by the mitigation of those stressors. As far as acute stress, mental health, or certain aberrant behaviors are concerned, whether other alternatives might do a better job of palliation and avoid some of the superstitious behavior characteristic of religion is an open question. But many people turn to religion as the available and well-conditioned path. Of course, providing comfort is a function of behavioral factors, specifically negative reinforcement, not the intervention of some god.

Intelligence/Education/Cognition

A very long-standing question is, "Is religiosity, at least in part, a function of a person's education or intelligence?" This is obviously a complicated and controversial topic. However, there *are* some interesting data to consider. In the United States, the more education you have, the less likely you are to be religious. In a national sample in 2017, it was found that people with higher levels of education do tend to be less religious. For example, they are less likely to have a "belief in God," to pray, or to aver that religion is important in their lives. (489) There are some exceptions in some studies, where adolescents "choosing" to eschew college, for example, have shown a religious decline coming out of adolescence. (523) However, the overall trend of religiosity and level of education being inversely related has held for a long time.

As abhorrent an idea as it is to many, we do know that intelligence and religiosity are negatively correlated. In a very large meta-analysis, it was found that this is empirically true for both the general population and for college students. (566) The correlations, while statistically significant, tend to be very small, but the reliability of this relationship holds in a large number of studies. In a very large sample of U.S. adults, it was also found that lower intelligence is most strongly associated with higher levels of fundamentalism. (308) It is also apparent that the United States is not unique in this regard. The negative relationship between intelligence and religious belief, specifically belief in a god, has been found in both the United States and Europe, and it holds for well over a hundred countries around the world. (322) In fact, over a third of the variance in religious beliefs is explained by a level of general intelligence.

There is no question that intelligence, specifically general intelligence, and religiosity are negatively correlated. The evidence on this is clear. Though innumerable exceptions abound in individuals, one cannot deny this type of finding. One can construct various possible explanations for such findings. Higher levels of intelligence are, almost by definition, associated with critical thinking and openness (to secular ideas). Perhaps this ability, and the tendency of more intelligent people is to understand the world from an informed or empirical perspective, as opposed to from religious dogma, or to think analytically as opposed to intuitively are the keys. Such are just speculations, so I'll not give much further weight to this issue. Except to say, there's something there, and it is generally a function of what we are as diverse humans, not our access to religious information or institutions.

As I intimated, there may also be some relationship of religiosity or belief to a style of thinking. That does appear to have some effect. As demonstrated in a few experiments, people who tend to use intuitive thinking, as opposed to a more reflective style of thinking, are much more prone to believe in God and to strengthen their belief in God over time. (465) This effect appears to be independent of IQ, personality characteristics, education, income, political orientation, or familial religiosity. This makes sense, of course, as people who are more reflective in their thinking are more prone to seek evidence for what they believe and to consider things more carefully, while intuition relies more on some rapid and emotional character.

What might happen to belief if certain aspects of cognition or perception were somehow different from how most people think? For example, in an interesting series of studies, psychologist Ara Norenzayan and colleagues found that adolescents diagnosed with "autism spectrum disorder" showed reduced belief in God compared with other adolescents. (379) This was the case after controlling for IQ and a number of other variables that could affect belief. The term "autism spectrum disorder" is generally applied, by professionals and others, to individuals who tend to manifest some type of ongoing social, interpersonal, emotional, or communication abnormalities.[28] The results were interpreted as indicative of "spectrum" individuals being limited in certain "social-cognitive capacities," which is often the case. That is, belief seemed to be relatively associated with impaired ability to "mentalize" about God, that is to develop a picture or character of a God. This again tends to confirm the idea that belief is related to intuitive or emotional characteristics of a person. The trend toward disbelief was stronger for males than females. This is explained partly by men having more disbelief than women, but also that the "mentalizing" ability in women is somewhat greater than in men. In sum, it is likely that the tendency to conceptualize a god and its character, in whatever form, is important in belief. The more you can conjure up a mental picture of a god or some particular characteristics (as we saw with cultural variation in children), the more likely you are to believe in It.

The bottom line here is that lower levels of education and intelligence, as well as less critical thinking, are pivotal factors in whether or not people believe. It is possible that belief may be more easily conditioned in such individuals, though we have no specific evidence of that.

Rationality and God

If intelligence is inversely related to belief in God, we might reasonably ask the question: How is it that so many very intelligent and rational people continue to believe and, in fact, persist in various overt religious actions? This is a puzzle that deserves a bit more attention. As we've seen in previous chapters, many of the great thinkers of the Enlightenment were avowed Christians; some of them were even religiously ordained. Despite their renowned scientific and logical thinking, many probably did believe in God. We actually don't know, realizing that overt "belief" would certainly have been influenced by the tenor of the times.

For them, and for contemporary humans, however, we should remember what it actually means to "believe in God." Some people are devout, "true believers." Others believe there may be a god, and they "might as well believe." Some take Pascal's Wager at some point. Finally, there are those who, as Daniel Dennet observed, merely "believe in belief." (123) That is, they merely think it is the right thing to do. In any of these cases, it is not some particular reasoning or any philosophical approach to the matter of God which is at play, even though such may be espoused. It is the behavioral history of each individual which dictates these approaches to belief.

Critical Thinking and Reason

If one tries to use reason to affirm a belief in God, that becomes a vain pursuit, as we saw in Chapter 8. Recall Martin Luther's caution, "Reason is the Devil's harlot." At least conceptually, he was right. It's clear that reason is antithetical to religious beliefs. Like all devout practitioners, he understood that actually examining one's faith is a good way to lose it. Nevertheless, some very intelligent people who believe in God will offer some sort of logical argument for why there *must be* a god. We've dealt with those already. Logical arguments are impossible to conduct about God.

Despite the vacuous arguments, some people who espouse them also engage in *critical thinking*. We engage in such for many problems in our lives. Don't get hung up on the word "thinking." As I noted, we don't have access to people's subjective experience, and we don't really know what goes on in brains during complex behavior. So what do we mean by "critical thinking?" It is the act of overtly or covertly solving problems by investigation, realistically considering and evaluating information, considering or testing alternatives, and making inferences. It's tantamount to a personal "mini-science," if you will. Many behaviors are involved, and those behaviors are the sum and substance of any critical thinking process.

People who tend to think critically are less likely to believe in God. But only less likely. Many or most of them still do. Obviously, they do not employ critical thinking when it comes to their religious beliefs. As I've suggested, that would be a dead end. Do such people go to church, honestly pray, actually believe, or "believe in belief?" The only thing we know is that their religiosity is maintained by environmental contingencies, not critical thinking. The accompanying rationalizations one hears from such people, including by clergy and theologians, are merely excuses, not in any way logical arguments. We often hear, "We can't use logic or any human capacities to understand the mind of God." Such comments are truly bizarre. They are only conversation enders or ways to make the speaker feel more comfortable about his views. Cognitive dissonance again.

Intelligence

Many highly educated and intelligent people do maintain irrational faiths. Let's look at some interesting examples. We think of scientists, for example, as very smart people, people who are well informed and who use critical thinking in their lives. They seem to have respect for the laws of nature and to rely on evidence for their decisions. Why would such a person believe in God? Yet there are still many who do. Actually working in a scientific field is not always a hedge against irrational faith.

Let's take the point with one modern, well-known example. Dr. Francis Collins, the long-time Director of the National Institutes of Health (and the deservedly famous leader of the successful Human Genome Project), is an avowed evangelical Christian. He is actually a very brilliant man, as a physician and researcher, and holds a master's degree in quantum mechanics. Dr. Collins is not bashful about his faith, actually subscribing to all the notions of Christianity, regularly quoting Bible passages, and referring at times to God as a "perfect" being. He reportedly "converted" from atheism after seeing his patients achieve some comfort and reassurance from their faith, something he was not able to put into context in any other way, and because he felt that "moral law," a guide to right and wrong, is uniquely human. (534) Does he actually believe this stuff? It seems so. Despite his offered reasons, I make no comment about the actual factors responsible for his "conversion", but the above rationalizations, of course, are just that. The reduced stress of not being able to explain the course of the diseases and the suffering of his patients was possibly a negative reinforcer for his beliefs. He obviously also fails to recognize that moral behavior actually evolved from other animals, and didn't require a Biblical myth. Intelligence had obviously not been a hedge against his superstitious beliefs, and it wasn't enough to prevent his conversion. It was a matter of the contingencies of his behavior, not a matter of some philosophical discovery of the "truth," as evangelicals sometimes term it.[29]

In fact, a number of scientists continue to believe. In a survey of scientists who are members of the American Association for the Advancement of Science, it was found that about a third of them said they believe in God. (402) However, we might also consider the most elite organization of scientists in America, the National Academy of Sciences (NAS). Scientists are admitted to NAS for "distinguished and continuing achievements in original research. Membership is a widely accepted mark of excellence in science and is considered one of the highest honors that a scientist can receive." (363) A survey of these scientists revealed that about 7% of NAS members held belief in a personal god as of 1998. (296) Although methodologies are not directly comparable, we see that disbelief in a personal god among leading scientists has been increasing precipitously over the past century. In 1914 it was around 70%; in 1933 it was around 85%; and in 1998 around 93%. Although very recent data are not available, one can reasonably suspect that this statistic may have held or increased somewhat. It is not likely that it has decreased.

Intelligent people maintain or convert to theistic religion because of the circumstances in their lives—not reason or critical thinking. It is evident that scientists, who are human beings, are subject to the same laws of behavior and conditioning as other people. They can hold incompatible cognitive views, though such is pretty rare for scientists, but those views are purely a function of learning. They are merely superstitions shaped in those people. This can only be explained by conditioning. Most scientists, and perhaps other very intelligent people, have eschewed the god idea. Nevertheless, there is some small percentage of hangers-on.

The Coda on Rationality

People tend to abandon critical thinking when it comes to belief. If they didn't, belief would essentially disappear. Most religious apologists, including very intelligent people, readily admit they have no proof or evidence for God. They merely "believe." Many intelligent people are guided by rationality and critical thinking in most of their lives, yet they continue to believe. It is merely a choice based on "faith." They fall into the trap of endorsing free will for humans and leave the rest of the natural world to the laws of physics. It exposes the fallacy of believing in a supernatural being, but to call it hypocritical misses the point. That's a term based on voluntarily holding incompatible views and is somewhat disrespectful. Rather, faith is not a matter of intelligence or rationality. It is a matter of conditioning, pure and simple.

On the other hand, there are some people who identify with a particular religion but are "critical thinkers" and have doubts about a Biblical God. There are many examples of this; for instance, many self-identified Jews, in fact, do not believe in a supreme being, in fact about a quarter of them (132), but many continue to engage in various religious rites, memes, and traditions. This sort of inconsistency can only be explained by conditioning.

Superstitions do tend to fade in the face of critical thinking. Where logical explanations are forthcoming—or can be envisioned as forthcoming or knowable via critical thinking or science—superstitious thinking and behavior tend to become extinct. The ultimate superstition, God, is left with no more reasonable influence or accountability and becomes seen as irrelevant. And so we are left with the tension between the operant factors which maintain illogical superstitions and those reinforcing other, more well-grounded behavior. It takes considerable mental gymnastics for a "rational" human being to think critically about most things and to also maintain the superstitions of God and His handiwork.

Here's the prototypical example. I've repeatedly noted often-heard phrases such as, "God loves you" or "God loves us" or "I believe in a loving God." Such is the omnipotent God that gives fatal cancers to children, commits genocide, and creates all other manner of human suffering and death. Loving? Again, the generally heard rationales, for example, are "God works in mysterious ways" or "We can't know the mind of God." Putting aside for the moment the question of why God would need a mind, such comments are clearly intended to reduce the believer's discomfort about a "loving" God. Divine cruelty is hard to explain. It's just easier to dodge the question or endorse ignorance. It takes less effort and reduces any anxiety or ambivalence about one's own beliefs. Such rationalization is the clearest example of human cognitive dissonance I can think of. Those sorts of positions are not "rationales." They are simply verbalizations negatively reinforced by the discomfort they avoid. It's a prime example illustrating that religious behavior is operantly conditioned human behavior, nothing more.

Conclusion

We must conclude that behaviorism provides a better explanation for faith and other religious behavior than anything else. Your faith or belief in a god is not a "free" choice. It is not from divine inspiration. It is not being born into one faith or another. It is not genetics. It is a matter of conditioning.

Also, that you "believe" is in no way supportive of the idea that God actually exists or that you have "chosen" to believe. There is no escaping the conclusion that faith or belief are "merely" well-learned behaviors. You might acknowledge that particular religious actions might be shaped according to the laws of behavior. However, you still feel that it is your subjective faith that is the driving force. Fortunately or unfortunately, there is no evidence for such an inference. Remember, there is no actual evidence that subjective experience is anything more than a covariate of behavior, something proximate in time, not an original cause of our behavior, religious or otherwise.

In sum, belief in God and other religious superstitions are purely a matter of conditioning, memes culturally passed down for millennia and reshaped for every individual. It matters not that "comfort" or escape from other aversive events is seemingly based on faith in some god. Gaining comfort from a superstition works pragmatically, but it is not evidence of anything more than a principle of behavior. It is "merely" conditioning. None of the fantasies matter.

Some people "choose" to believe while still maintaining that, "This is the way I was raised," a tacit acknowledgement of the fact that it *was* religious conditioning. Conditioning is such a pivotal part of our lives that it becomes our identity, as Aristotle sagaciously noted. This occurs in various realms, but it is particularly problematic when it comes to religion. Organized religions (and their congregations) also behave like conditioned organisms themselves. They learn and survive based on behavioral principles: largely money (in fact, great wealth), jobs for clergy, reinforcement by parishioners, tax exemptions, etc. There is no escaping the fact that organized theistic religion itself has survived because of its conditioning.

The denigration of behavioral science as applied to human behavior is merely a wishful denouncement of the principles and clinging to the notion of free will. There also are always objections to behaviorism based on the presumed demeaning of human behavior and experience that seem to flow from a scientific understanding of our actions—the emotions, beauty, sadness, and countless other experiences and phenomena. But this is not necessarily the case. The more we know about the dynamics of human behavior, the more we can (and do, in many instances) promote constructive behaviors in occupations, education, creativity, and other areas that we value, and we may find ways to constructively avoid or diminish behavior that is aversive or unwanted by our culture.

We live in a determined universe from which the laws of nature, including the laws of behavior, arise. You might argue that behaviorism is not a valid or reasonable way to understand human behavior. Nevertheless, the laws of behavior operate whether we like it or not. No principle of physics changes because we don't like it. Gravity is there no matter how we feel about it. The same holds for the laws of behavior as well. Behaviorism is a young science. As Skinner put it,

> What we need is a technology of behavior . . . But a behavioral technology comparable in power and precision to physical and biological technology is lacking, and those who do not find the very possibility ridiculous are more likely to be frightened by it than reassured. That is how far we are from 'understanding human issues' in the sense in which physics and biology understand their fields, and how far we are from preventing the catastrophe toward which the world seems to be inexorably moving. (484)

The good news is that we are enabled, by evolution and contemporary learning, to continually upgrade the efficiency of our actions in the direction of physical, intellectual, artistic, cooperative, and other valued objectives. As we learn more, we can use the laws of this young behavioral science to accomplish our shared goals and lifestyles. And as we understand more about human behavior, we will also be able to mitigate the unfortunate consequences of theistic religion.

11
IN THE END

Imagine there's no countries
It isn't hard to do
Nothing to kill or die for
And no religion too
Imagine all the people
Livin' life in peace...
—John Lennon (Imagine, 1971)

I've reviewed a number of topics relevant to the god problem. In the end, the critical question remains: Is there a god or not? This is a subject where nearly everyone has some kind of opinion. No one *knows* the answer, of course. No one has any more insight than anyone else—not your parent, your grocer, a cosmologist, a theologian, the Pope, or some artificial intelligence.

We can ask a question like this and never come up with satisfying answers, unless fantasy answers are what are gratifying. That is a psychological phenomenon, not reflective of some actual reality. And in the end, we know that psychological phenomena are the result of conditioning. What you believe is purely a function of what you have learned. And this includes notions of God.

To briefly review: The fabrication of and belief in gods are based on our need to *know* why things happen. Our abilities to perceive causes and to imagine have led to the creation of supernatural fantasies. That doesn't make them true. The notion of God is a human invention. It began with our primitive ancestors ascribing mystical forces to natural phenomena, followed by personification of these, followed by millennia of revised lore upon lore, leading to the god memes we have today.

We have seen that the notion of an omnipotent, omniscient god is illogical and unnecessary. As you are likely anticipating, the obvious conclusions about God or gods are that (1) the First Cause argument is a patently bankrupt one, (2) humans evolved in a manner which led to the fabrication and maintenance of god-like superstitions, (3) we live in a determined universe, and free will, the presumed gift from God, is an illusion, (4) religion itself is a set of actions, often incredibly destructive, shaped by the laws of behavior, and (5) supernatural guidance for so-called moral behavior, that which binds human groups together, is fully unnecessary and redundant to what evolution and secular aspects of culture have given us.

Let's sum it up this way. Ask yourself this question—seriously. Why would a presumably omnipotent being need the worship of some bipedal apes wandering around a small planet orbiting an ordinary star, in an arm of an ordinary spiral galaxy, in an infinite universe? If your only answer is that we can't understand the mind of God, you might rightly feel embarrassed.

Copernican Principle

We began with the *Copernican Principle* in Chapter 1: that we and our world don't occupy any privileged or *special* place in the universe. The Copernican Principle does tell us that we are not as special as we think, and in the grand scheme of things, we're probably not unique. Let's briefly review a bit of that here.

Back to the Cosmos

Our Milky Way galaxy is but one of hundreds of billions of galaxies (perhaps a lot more) in our vast universe, which may in fact be spatially infinite. Moreover, it is also conceivable that ours is but one of many universes (perhaps an infinity of them) existing in various forms, places, dimensions, and times in a vast, unimaginable cosmos. Some of them may be very different and some may be like ours. In some of them, we may have personal duplicates, with the same behaviors, cultures, and values. In some there may be very different forms of life, cultures, and values. We may not be unique at all. The future of physics may well narrow these possibilities, but our uniqueness is most likely a very tentative assumption.

We don't live in any special time either. Our universe began to take on its present form some 13.8 billion years ago. However, there is no reason to believe that it needed a God figure to "get it started." The universe may be cyclical, going into a freeze, expanding and contracting forever; it may die a "heat death"; or it may mutate into some form of a "reality" cosmologists haven't even considered yet. Whatever exists has always existed and will always exist. Only the apparent forms change. Accepting the idea of such "eternality," going both forward and backward in time, is difficult for us humans to grasp. It may be uncomfortable as an idea, but the universe owes us no solace. Nevertheless, the idea of an eternal reality is just as valid—and certainly more—than the fantasy of a God creating it.

Life

We are also not special or unique in terms of the biological world. Our planet has been around for only about the last 4.5 billion years. Life arose fairly quickly after formation of the Earth, but it took a long time for us to get here. As a species, we've been around for only about 300,000 years, which is only about 0.007% of the time elapsed since our biological "last universal common ancestor" lived on this planet. Beginning with Darwin, and after a century and a half of biological research since then, we know that we are merely a small branch of an evolutionary tree (or bush, as some call it) on this planet. We and all other extant creatures, are, in fact, the product of 4 billion years of "failure" of our various predecessors. Remember, at least 99.9% of all species that ever existed are now extinct. If this was "God's plan," it clearly wasn't a very good one.

We can look back even further. We know that we are here because we naturally evolved, but life itself, as we know it, represents but a continuum of chemical processes from those which preceded what we consider life. In fact, the chemical elements that built us have been around much longer even than the Earth. Most of your body is water, each molecule of which contains two hydrogen atoms. The nuclei of those atoms (each being one proton) formed within a couple of minutes after the Big Bang, and all the quarks that make up those protons had appeared in less time than we can actually conceive. Later, the heavier elements in your body—carbon, oxygen, iron, phosphorous, and all the others—were made by the stars. As Carl Sagan famously observed, "We are all made of star stuff." Later, all sorts of biochemical structures started forming in the universe, and it took billions of years for all that to come together to "form" human beings. That is how we got here, not via "creation" by some fantasy figure.

Also, we are only here for a unique period of time. In a billion years or so, the sun will expand and burn off the oceans (281); but biological life on Earth will not be possible even well before that. It's unlikely that human beings will be around in a few hundred million years. The most likely scenario is that if we continue to exist while there is still some sort of atmosphere and a magnetic field sufficient to protect the Earth from solar and cosmic radiation, some of us will have evolved into other species. We humans are only special or unique in a brief time and place in the great span of cosmic history.

Who Are We?

Is there something special about being human? Of course, there is. We are indeed special creatures, but only to ourselves. Neanderthals undoubtedly felt that way about themselves. Orangutans are most likely special to themselves, as are dogs. But we are not special in the eyes of the natural world; we're only some of the apex predators running around on a small planet.

We also think of ourselves as smarter than other creatures. Time will tell about that, of course. And, in any sense, we're not smarter than our ancestors of 60-100,000 years ago. The human brain and its basic "intelligence" haven't changed over the millennia. We do have more information to use, and we've invented technologies to help us (which some have reasonably opined might be making us dumber). But we have the same capacity to reason and solve problems that early humans did. Again, one of my favorite illustrations of this: If you time-traveled back to late Paleolithic times (perhaps 10-15 thousand years ago), you'd find that a human newborn then would be just as capable of doing advanced calculus and producing creative arts as we are today.

It is hard to believe, but we are not today unique as people in the long history of our species. We just have a few more tools at our disposal—for better or worse. Our advancements have been cultural—people working together to create physical, technological, and social systems. We are special only to the degree that we share information and certain cultural behavior with others, something that can be, and most often *is* done, without any reference to a mythological deity.

Truth

What is the truth? What is it, really? This question has inspired—and plagued—humankind since we existed (and quite possibly before). Without bothering to summarize the philosophies of metaphysics and epistemology espoused by humans since they began writing these things down, there are actually only three ways of looking at this problem. Truth is what you get from your own experience, or from science, or from religion.

Do we believe our own eyes, what science reveals, or what a religion tells us? Most people depend on some combination of the first two, and some people accept selected parts of what their religion tells them. Some of the latter is historically reasonable and effective behavior masquerading as divine instruction, e.g., "Do unto others as you would have them do unto you." Of course, religious acolytes believe only some of their holy books, and as we've seen, highly selected parts of them. If you believed everything the Bible tells you, you couldn't survive.

Let's first address the notion of "truth," as the word is commonly considered. "Truth" for most people is what is often called "absolute truth." The only things that can be actually proven to be true, as I've discussed, are contained within mathematics. The "absolute truths" that come from religion are just fantasies.

Common Sense and Science

The only other type of "truth" is empirical. What is true is what we experience with our own senses. When we experience things in similar ways, we call that common sense. But common sense does sometimes fail us. We can't observe everything, and we certainly don't agree on everything we see. We've also seen that common sense is sometimes a poor way to understand the world at its most basic levels. We can't use common sense to conceive of our universe being 13.8 billion years old, or light traveling at about 300,000 km/sec, or how our bodies produce carbon dioxide, or something being both a particle and a wave. We defer to science for enlightenment in such areas.

With common sense, what is "true" for one person isn't for another. That is the problem that science addresses. When you get down to it, science is no more than a systematic way of observing reality, and it is based on the consensus of observations by many people. It can seem very intricate and complicated, which is why some people are suspicious of the empirical truths that come from science. But science gives us the best approximation of what the world is really like. And of course, despite the impressive body of knowledge science has given us—from cosmology to particle physics to modern medicine and other technologies—scientific knowledge can never be absolute. It only approximates reality, albeit often to an incredibly precise degree. It does not enable absolute truth, something that's often misunderstood. If it did, it would be a religion. When people say, "Science is just what you believe in; it is your religion," they are misguided. Science gives us the best understanding about the nature of the world and of ourselves. It is not a matter of belief. It is a matter of observation and consensus. It is constantly changing, but it is the source of knowledge at any particular time that can be generally relied upon.

Religion

On the other hand, the "truth" that religion provides is very different. As we've seen, it is merely arbitrary and just made up. It is presumably absolute "truth." Of course, there is no such thing, and absolutes turn out to be only assumptions. What is "true," as exemplified in Judeo-Christian Bibles, the Quran, various Hindu texts, or any other religious scriptures, can only be taken "on faith." Religious apologists are fond of saying that life's truth is "God's word" or Jesus' putative homily, "You will know the truth, and the truth will set you free." Cities and suburbs are filled with signage such as "The Bible is true"—notwithstanding the idea that if it were "true," you probably wouldn't have to say it.

Religious truth is based only and purely on authority—on some writings of mortal men and some authorities who purport to know the meaning of what some fabricated god tells us. As we've seen, the blind following of Bibles and other scriptures as "truth" leads only to suppression of curiosity, critical thinking, and dangerous superstitions. To reiterate Einstein's perspective, "Authority gone to one's head is the greatest enemy of the truth" (149) Religious "truth" has killed more people than we can even conceive.

Conspiracy Theories

Which brings us to the crux of the matter. The truth is that theistic religion is based on a superstition. But more than that; it is a lie. It is perpetuated by those who are conditioned to believe it but, shamefully, by those who have a financial or other powerful interests in doing so. But the bottom line is—it is a lie.

Lies tend to crowd out truths in many circumstances. Lies are often simpler, more convenient, and easier to understand, and they carry a good deal of emotional tone. As we all know, truths have a difficult time making inroads. As Jonathan Swift most aptly put it,

> Besides, as the vilest Writer has his Readers, so the greatest Liar has his Believers; and it often happens, that if a Lie be believ'd only for an Hour, it has done its Work, and there is no farther occasion for it. Falsehood flies, and the Truth comes limping after it; so that when Men come to be undeceiv'd, it is too late; the Jest is over, and the Tale has had its Effect. (509)

When a lie is shared by many people, it takes on a life of its own. In recent times, we've come to know such developments as "conspiracy theories." People develop shared memes based on lies. We live in an age when memes of lies or untruths abound as conspiracy theories; and such "disinformation" memes are amplified by the rapid and widespread workings of social media and other global information mechanisms.

There are all sorts of conspiracy theories. Take just the example of the millions of people who believe that interplanetary aliens have been visiting us here on Earth. This again illustrates how, when phenomena occur that we don't understand, we tend to fill in "gaps" with imaginary ideas. This is not to say that we can definitely rule out the presence of such aliens. Anything *is* possible. But our capacity to imagine emerges ever stronger when we don't understand something—whether it's unusual lights moving across the night sky, or how we got here—and we fabricate information and spread it to satisfy ourselves.

We can subscribe to "truth" from one perspective or the other—science or religion—but not both. Science and theistic religion are specifically and absolutely incompatible. Science is based on empirical evidence. Religion is based on faith, which is belief without evidence. I can't help but quote Mark Twain again, *"It ain't what you don't know that gets you into trouble. It's what you know for sure that just ain't so."* When that surety, the lie of religion, spreads among a population, that's when the conspiracy theories of theistic religion begins to take hold, and it has been going on for thousands of years.

Theocracy and the Future

Of course, if too many people believe in such conspiracy theories, there is risk of a culture moving toward a society and a government *by* religion, the theocracies I talked about. A brief historical tour and a glance around the ancient and modern worlds is more than adequate evidence of this. The Middle East is currently awash in theocracies. Christian theocracies dominated the Middle Ages. However, current, mostly secular cultures are not necessarily immune from such a prospect.

Let's look briefly again at the United States, a country with some risk in this regard. We have a Constitution that mandates the separation of

Church and State. Nevertheless, many don't believe that, even to the point of insisting that the United States was founded as a Christian Nation, a meme I debunked in Chapter 7. Theistic religion is declining in the U.S., though only slowly (I'll review some data on this below). The side effects of behavioral extinction, and common sense, tells us that this tends to produce more intense counteractions on the part of religious evangelicals. In fact, efforts are afoot to infuse our secular institutions with Christian theology. In 1963, the U.S. Supreme Court ruled that schools could not require mandatory Bible readings and prayer. (1) Yet in recent years, there have been a number of rulings by the Supreme Court to weaken that doctrine and begin infusing Christianity back into the public square. Efforts continue in various states with executive actions and laws, such as requiring posting of the Ten Commandments and teaching based on Biblical text in public schools. Asking children—or forcing them—to engage in religious rituals is probably the most powerful way to shape a culture into an organized religion, and of course, evangelicals know that. And the effort to weaken the teaching of evolution to our children continues, along with efforts to enable "teaching" of the "creationist" alternative.

The efforts to turn the United States into a Christian theocracy continue. We can't know if this will ever fully come about, but it's a bit scary. Representative secular democracy is under threat, and religious zealots have gained more and more influence in government, legislatures, and the courts. Public school prayer, denying women's personal and biological freedoms, and many other Christian-driven trends reflect a waning of democratic forces and a rise of theistic control of our culture. Whether, in the long run, the country will persist as a mostly secular democracy or a Christian theocracy is an empirical question. The future is long.

Atheism

If you don't believe in God, what are you? There are a few alternatives. Humanist, naturalist, spiritualist, freethinker? These are certainly well-regarded identities, but they don't convey the essential message. One that does, of course, is *atheism*. To clarify, atheism is not what would be called a philosophy, a religion, or a political position. Merriam-Webster recently changed its definition of atheist to "a person who does not believe in the existence of a god or any gods." Specifically, it is merely a lack of belief in gods. Atheism does not address the question of our existence or other philosophical matters. It is merely a lack of belief in gods. (8)

Some continue to offer that atheism is a religion. Nothing could be further from the truth. That atheists "believe" there is no god is not the point. Religion is the various repeated actions in service of the superstition of a god. Atheists don't go around engaged in organized rituals of God-denialism. Atheism is a point of view, not a religion. Whether particular atheists are comfortable with their position, have no particular need to impose their perspective on others, or just have something more important to do, of course, is quite individual. In any case, one cannot consider atheism in any way a religion.

Some develop an atheist perspective in revolt against a repressive, dogmatic upbringing. Others have grown up without religious indoctrination and were shaped in the ways of critical thinking adequate to reject god notions. For others, seeing the horrors that have resulted from religious violence and war push them to atheism. Or perhaps the stimulus was God failing them when He should have "done something." For some, the vast, seemingly preventable suffering in humanity has been the tipping point. And for some, the idea of God is just personally irrelevant.

Historical

The idea of atheism has been around for a long time. The word itself comes from the Greek *atheos*—from *a theos*—meaning the absence of a god, from at least the fifth century BCE. This seemed to have worked its way into early Greek culture in competition with the prevailing superstitions, the panoply of gods that the Greeks had been worshiping for hundreds of years. As the British literary historian Tim Whitmarsh points out, it was Democritus who noted that fantasies about higher powers were developed and became lore because earlier cultures could not really understand the laws of nature. (544) There were many known atheists among the great ancient Greek thinkers—the poet Diagoras of Melos, the philosopher Empedocles of Sicily, and many others. Of course, many kept their views quiet in the face of possible punishment for impiety.

Socrates did speak publicly about gods. However, he was essentially forced to imbibe the hemlock after being formally accused of atheism, as well as entertaining the possibility of gods who were not the official ones of the state. In fact, for most of history, "atheism" was basically not believing in God the way you were supposed to. Not worshiping the "correct" god continues to plague us, and awareness of that problem may have led many to atheism as a result.

One of the more interesting atheists was the 17th-century French Catholic priest Jean Meslier. He performed his duties as a Catholic priest for over 40 years despite being a non-believer. An extensive manuscript detailing his views, kept secret during his life, was discovered in 2009. In his *Testament*, he repudiates Christianity, all other religions, and even the concepts of deism, the soul, and free will. He most notably observed, "All the laws and orders that are issued in the name and authority of God or the gods are really only human inventions." (343) He did not have a fellowship such as the current non-believers' Clergy Project available to him (as I discussed previously). He obviously found that behaving in conformance with the religious culture of his times was the advisable path, and he continued his work. Despite his position, he eschewed any elitism or monies and was a shining example of serving the needs of the poor, something many of today's Catholics and other religious acolytes often forget.

Contemporary

For some people, there's still a problem of labels here. Atheism is not a label which actually conveys something affirmative and specific. In many circumstances, it is really not helpful to say that you don't believe in God, i.e., which implies the God of the Abrahamic religions, though which one of those continues to be a problem. Labels sometimes get in the way. This is generally not a problem for self-identified atheists, though it is for some others. Michael Shermer offers the following about this:

> Atheism is typically used to mean 'disbelief in, or denial of, the existence of a God' (not to mention its pejorative permutations). But 'denial of a God' is an untenable position. It is no more possible to prove God's nonexistence than it is to prove His existence. 'There is no God' is no more defensible than 'there is a God.' The problem with the term *agnostic*, however, is that most people take it to mean that you are unsure or have yet to make up your mind, so the term *nontheist* might be more descriptive. (466).

I do take issue with this. We've dealt with the matter of "proof" already, including the "false equivalency" issue and the fact that one does not need to prove the nonexistence of a human fantasy. Further, identifying as a "nontheist" doesn't escape the problem. It only creates tangents in conversation about the issue.

Philosopher Eric Steinhart suggests we should look at two different types of atheism. (504) There is skeptical atheism, and then there is one that holds a certain type of "faith" that science can or will provide "final" answers. That would be absurd, of course. Science, by definition, continues to evolve as more is learned, but it provides no absolute answers. That would make it a religion, something theists regularly and nonsensically accuse atheists of being. He also suggests a kind of "spiritual atheism" in which an individual reveres the natural world, feels grateful for it (though to no being, it seems), has "mystical" experiences, and takes pleasure from art, beauty, etc. Some might see parallels to humanism or naturalism here. One can laud such a perspective, but it rather confuses the meaning of the word "spiritual," which sometimes has a different connotation, and misses the point of whether or not it is reasonable to believe in some sort of god. The idea of "spiritual atheism" misses the point and can likely lead to confusion.

Attempts to re-label atheism haven't really caught on. For example, philosopher Dennett proposed the term "Brights" for atheists, indicating an inquisitive nature (as opposed to demonstrating a high IQ). (122) Even though he suggests it as a noun, there's just too much baggage in this term. Many atheists refer to themselves as "freethinkers." This is a bit closer to a reasonable description of an identity, but it really doesn't convey the perspective of disbelief. "Atheism" still seems to have the broadest appeal.

Some describe themselves as "militant atheists," Richard Dawkins, for example. It's an unfortunate label since to most observers it connotes a more belligerent promulgation of a philosophy, which is not really the intent. There may be some linguistic hostility toward religion and its incursion into secular life, which is quite understandable; but there's no evidence of anything but writings, lectures, and polemic dialogue in the atheist effort. Of course, this is quite unlike some organized faiths which actually employ violence—Islam, Christianity—to promulgate their views. Some have called for a moderate atheism, what's been labeled "soft atheism" or a-theism. Philosopher Philip Kitcher, like many others, argues for respecting religion's traditions, reinforcing its social goods, and viewing scriptural myths as metaphors enabling moral progress (103), which would presumably weaken organized religion. I have my doubts.

Conclusion

Atheism is not a religion. Identifying as an atheist is still problematic in various circumstances, since it differentiates you from all "persons of faith," which can result in social ostracism, public discrimination, or worse. It was only a century ago that you could be prosecuted for it in America. In 1928, atheist Charles Lee Smith was prosecuted in Arkansas for the crime of blasphemy—for arguing against the state's anti-evolution laws and publicly promoting atheism. The charges were eventually dropped, but only after a long, drawn-out series of court proceedings and a hunger strike. (168) However, more recently atheism has become a somewhat more popular position in Western cultures (see the data below). People who endorse this idea are aware that it is not a religion, and they know it is not a major solution to the problems that theistic religion has created.

It would be naïve to assume that atheism is the cure for the tragedies of organized religion. (Then again, how many people have been killed in the name of atheism?) Atheism is a flexible approach to life, not a set of dogmatic edicts. It is a kind of humanism—a respect and empathy for other humans. I could not put it better than journalist and author Susan Jacoby did.

> The atheist is free to concentrate on the fate of this world — whether that means visiting a friend in a hospital or advocating for tougher gun control laws — without trying to square things with an unseen overlord in the next. Atheists do not want to deny religious believers the comfort of their faith. We do want our fellow citizens to respect our deeply held conviction that

> the absence of an afterlife lends a greater, not a lesser, moral importance to our actions on earth. (243)

Change?

I'd argue that the weakening of organized religion and the growth of secular positions on life is beneficial for our culture in general. Such is certainly consistent with the philosophies of some modern thinkers about morality and of the Enlightenment utilitarian thinkers, i.e., the greatest security and happiness for the greatest number. However, I can safely say that a major abandoning of theistic religion is unlikely to happen in the near future. People rarely change their minds about such things, as they are naturally invested in their views and behaviors developed over a lifetime. "Minds" do not change easily. If a person does change her mind about such things, how does that happen? We'll look at a few things that bear on that process and attitudes about the God question.

Science

One of the most potent forces against the problems caused by organized religion is advancing secular education, which is a countervailing force against misinformation and superstitious beliefs. Coming back again to Thomas Jefferson, who was an advocate for freedom *from* religion, "If a nation expects to be ignorant & free, in a state of civilisation, it expects what never was & never will be." (246) We've seen that the more educated you are, the less likely you are to believe the tenets of theistic religion. Whether you believe in a god or not, more education is certainly likely associated with an attitude consistent with secular culture and government. And in many cases, of course, also tolerance for people of other views.

And a respect for science is a critical piece of this. Enhanced knowledge and respect for science can render theism and religious ritual increasingly impotent in people's behavior. Of course, I do not mean "science" to the neglect of other, equally important disciplines and practices. One cannot overstate the importance to a civilized, critical-thinking culture of music and other fine arts, literature, sport, and other uniquely human endeavors. In point of fact, such activities tend to reinforce the value of being human in communities of humans—without having to rely on the fantasy of a god. Yet, adults don't easily change their views about science and what it tells us about our behavior. Some psychologists suggest that better acceptance of science can be achieved through appealing to underlying attitudes and group identities rather than confronting and debating the issues. (225) We might appeal, for instance, to the economic benefits of "green jobs" or the moral responsibility of caring for the Earth. Anti-vaccination attitudes might be reduced by sharing of personal information among parents, the overwhelming majority of whom do inoculate their children—against a number of diseases.

Still, such kinds of efforts are less effective if one hasn't received much substantive science education, obviously beginning in childhood. Unfortunately, there's very inconsistent teaching of science in K-12 schools. Without belaboring the details, there is no reason that good science cannot be taught very early in a child's education. Children are naturally curious, and exploration is natural and often very reinforcing, which enables reasonable science education early in childhood. Looking through a microscope at a leaf or watching an ant carry it around can be fascinating. Children are our future, and there is no good reason they shouldn't grow up with a respect for reason, critical thinking, and science, including certainly the whole subject of evolution. Even very young children can appreciate how evolution gave us the long-necked giraffes. Attitudes that grow out of science education help to guide children in realistic directions and toward a minimizing of dogmatic myths.[1]

Argument

The God question and various religious matters have been debated since time immemorial, and they continue to this day—in colleges, seminaries, public forums, and living rooms. Direct arguments rarely change anything. People hold their beliefs, including their often-biased understandings of the world, with incredible tenacity. And it often occurs because of attitudes and preconceptions about the world. It was the late Senator Daniel Patrick Moynihan who famously said, "Everyone is entitled to his own opinion, but not his own facts." This is clearly a wise perspective. However, when it comes to arguments or debates on religious matters, facts seldom make a difference in the real world. Adults generally don't respond well to new information on controversial topics, and they tend to often reject it out of hand. We all do have a rather reflexive tendency—as anything new could be a threat—to treat new information with at least a mild sense of skepticism, what's generally referred to as "epistemic vigilance." (499) It's a natural human proclivity. Even courteous debates can seem like unwanted personal confrontations. In sum, people generally do not change their views on controversial topics based on information. In fact, existing opinions are often strengthened when other, possibly contrary information is presented, what we know as the "backfire effect."

Then there's the significant factor of "cognitive dissonance" I've alluded to previously. Tension between beliefs is generally reduced by enhancing regard for one's existing belief and reducing it for any other one. It takes more energy, and usually a bit more time, and often some discomfort or stress, to acknowledge any new and possibly contrary information. Thus, old ideas retain power and new ones have a hard time breaking through. In fact, when facts in opposition to one's beliefs are presented, there is a tendency to discount them and seek out or use information that supports your position, what we know as "confirmation bias." We saw in Chapter 1 how Bertrand Russell eloquently laid this out a century ago. A perfect contemporary illustration of this principle is succinctly summarized in a song by Paul Simon.

> *Such are promises, all lies, and jest,*
> *Still a man hears what he wants to hear*
> *And disregards the rest*
> (The Boxer, 1968)

Intellectual debates and arguments about religious matters seldom go anywhere. As we saw earlier with Daniel Kahneman's work, we are not fully irrational or emotional creatures, but we do tend to become irrational when presented with challenges to our biases or when strong emotions are stimulated. (259) We especially rely on emotions and lean toward our preexisting biases when information is ambiguous. Discussions, interactions, or disagreements about God are probably the pinnacle of this phenomenon. What we believe is highly influenced by emotion. In general, the less emotional attachment to an idea that a person holds, the easier it is to change a belief. Without belaboring the details, this has even been demonstrated in neurological studies of the brain. (263) Where discussions minimize an emotional tone, some people will begin to change, or at least question, their own beliefs. Yet, we remember that such beliefs evolve because of a lifetime of conditioning. It is almost impossible for a "true believer" to converse rationally about any possibility that God does not exist. It's a bit fanciful to think many people will change their minds because of the prevailing arguments.

Respect

Changing minds begins with respect. We must respect individuals whose beliefs do not agree with ours. It is not reasonable to denigrate, suppress, or punish extant religious beliefs or those who hold them. Very little is achieved with punishment and aggressive criticism. The history of organized religion itself bears this out quite tragically.

In the end, the questions of reality, knowledge, and the idea of God can only be addressed if we respect each other and can communicate with each other on all sorts of topics. And in fact, some people seem to have a great need to do this, including many religious adherents. When people get to know each other on a personal basis, tension in ideas is often reduced. Such comes about when we consider commonalities. Do we share similar values—about children, social institutions, economics, foods, etc.? Do we agree on what is moral? Less tension in discussions about religion sometimes develops in such contexts.

And should we "respect" each other's beliefs? Not necessarily. What we should respect is the person who holds them. It is nearly impossible in most circumstances to convince people who hold religious beliefs that they are wrong or that the tenets of their religions are unfounded and fantastical. Yet some people are "on the fence" or leaning toward a more secular view. It is difficult to say, in most social situations, that one is disappointed in his religion or is thinking about agnosticism or atheism, but respect for the thought process and communications goes a long way. Shermer well-summarizes a reasonable strategy in this regard:

> If corrective facts only make matters worse, what can we do to convince people of the error of their beliefs? From my experience, 1 keep emotions out of the exchange, 2 discuss, don't attack (no ad hominem), 3 listen carefully and try to articulate the other position accurately, 4 show respect, 5 acknowledge that you understand why someone might hold that opinion, and 6 try to show how changing facts does not necessarily mean changing worldviews. These strategies may not always work to change people's minds, but now that the nation has just been put through a political fact-check wringer, they may help reduce unnecessary divisiveness. (471)

I agree. However, I might take a bit of exception to number five. If I say, "I know why you believe that," it may be difficult for a listener not to perceive a dismissive attitude about this. It can be seen as patronizing, demeaning another's views as "merely" opinion. Nevertheless, direct and respectful dialogue with all these ideas in mind does tend to make a difference. And it can often bring people closer together, which is what most of us want anyway.

Models

The other factor which influences beliefs is models. That is, we tend to lean our behavior toward the views of people who are trusted and respected—beginning with parents, of course, but also with many models throughout life. People do look to models to guide their beliefs, no matter their "expertise," whether it's for blind religious faith or, on the other hand, for atheistic or secular philosophies, or laundry detergent.

In religious contexts, this influence is provided by various authority figures, particularly clergy. The words, and sometimes the actions of such individuals, are potent models for others, particularly children. In this context, we should again acknowledge that organized religion is big business, and its leaders are the people who earn their living by proselytizing it. Unfortunately, people often forget or minimize this as they model religious behavior. What is modeled is what is being paid for, whether there is a so-called true belief associated with it or not. What has become a classic adage brilliantly summarizes the point. Truth is not enhanced when it's tied to money, and any clerical doubts about God don't survive well in that context. As the great American writer, Upton Sinclair, famously said, “It is difficult to get a man to understand something, when his salary depends upon his not understanding it.” One is hard-pressed to find a pastor, priest, rabbi, or imam who doesn’t openly endorse the idea of a supernatural god, as their livelihoods depend on it. In fact, there are many clergy who seriously doubt a god, but their overt behavior—controlled by financial contingencies—must support such things. And this economic reality ends up enabling a powerful model for the flock.

On the other side of the coin, just as religious authorities are models for God beliefs, other people may act as models for atheism. In America, certain individuals endorse ideas or model behavior for others because they are famous—from various fields like politics, sports, or entertainment. Obviously, people who are famous for one reason or another have no more knowledge or insight about this subject than any other human. However, they act as powerful models for the behavior of others—whether it’s social issues, product consumerism, charitable behavior, or many other things.[2] When it comes the God question, today there are few prominent individuals who openly profess that they are atheists. Only a brief list—entertainer and magician Penn Jillette, singer Mick Jagger, filmmakers James Cameron, Woody Allen, and Steven Soderbergh, comedians Bill Maher, Ricky Gervais, and the late George Carlin, and athletes like martial artist Bruce Lee and swimmer Diana Nyad. You’d be correct to infer that there aren’t that many people of fame who openly profess their atheism publicly.

In the political sphere, very few individuals openly acknowledge atheism. There are former Massachusetts Congressman Barney Frank and Jesse Ventura, former governor of Minnesota, who said he follows the teachings of George Carlin. Political leaders are, by their very nature, extremely shy about this, even though it is likely that many of them are atheists at heart. In the United States at least, it is almost impossible to get elected with that tag. There is still no one currently serving in U.S. national office who openly identifies as “atheist.”

Of course, leadership and models are often critical in social movements. We can certainly heap great credit for important social developments on such people as Susan B. Anthony, Mahatma Gandhi, Martin Luther King, and Nelson Mandela. It’s certain that the above atheists will not be a tipping point for a movement toward atheism, but change is slow, and history is long.

Trends

Attitudes about religion are changing. Despite the dominance of theistic religion, the trends do seem to be moving away from religiousness in many parts of the world. Theistic religion around the developed world, especially in the West and especially with younger people, is fading—albeit very slowly. I’ll quote a few statistics here, but we should remember that such are only approximations, and the data depend heavily on various methodologies, survey dates, how questions are posed, and other factors. With that caveat, here are some interesting findings.

United States

In surveys of religious attitudes, the United States is still one of the most religious of the large nations in the world. As of 2018, if we don't consider the Middle East, sub-Saharan Africa, or South America (where it is more difficult to obtain reliable data), the percentage of people who say religion is "very important in their lives" varies from China at 3% at the low end to the United States at 53% at the high end. (196)

Although the U.S. is a very religious country, in 2018, data show that there were similar percentages of "non-affiliated" Americans, those not identifying with any particular religion (23.1%) to both Christians (22.5%) and Catholics (23%). (364) And identification and participation are declining. In 2020, less than half of Americans (47%) said they actually belonged to some type of house of worship (church, mosque, synagogue); this was a significant decrease from 70% as of 1999. (253)

As of 2021 the percentage of the religiously "unaffiliated" in the U.S. had risen to about 29%. (490) In 2024, the percentage of all of those categorized as "nones" (identifying as atheist, agnostic, or "nothing in particular") in the United States had risen to 28%, up significantly from 16% in 2007. (491) In young people, the data are even more provocative. As of 2019, almost 40% of Americans under 30 described themselves as "non-religious." (7)

The data on "belief" and atheism in particular are provocative. Most recent surveys have about 80% of U.S. adults saying they believe in God, but belief in God among young adults has dropped to around 68% (254). It is difficult to get reliable data on the endorsement of atheism in particular, and it may be that the frequency of atheism is well underestimated. When most polls are taken, stigma can provide a reluctance to admit to an atheist view. Even with anonymous polling, it can't be ruled out that many Americans are unwilling to endorse atheism, as opposed to endorsement or denial of any particular religious affiliation. In a recent study with more rigorous methodology than typical polls on this issue, it was found that the frequency of atheism among U.S. adults is probably around 26%. (179)

The World

Religion in Europe and around the world does appear to be waning somewhat. In Europe, about a quarter of adults are "religiously unaffiliated." But there is an interestingly wide range of these "unaffiliateds," from highly Catholic Ireland at 15% to the Netherlands, where nearly about half have no religious identification. (441) The trends also show that most Christians in Europe are "non-practicing." (441) In recent years, in fact, many Christian churches have become increasingly empty; many have become gyms, supermarkets, and even bars. (44) Some other countries outside Europe are manifesting similar trends. In New Zealand, for example, as of 2018 nearly half of the population had no religious affiliation. (503)

The trends in religious behavior are clear. Younger people around the world are becoming less religious, and that trend is more acute in so-called Christian countries. In 41 countries around the world (other than almost all those in Africa, the Middle East, and southern and eastern Asia), younger adults are less "affiliated" with any religion than older adults. (197)

Conclusion

The overall trend in most of the world does appear to be skewed in the direction of less religious identification, and this is more pronounced with young people. Both "non-affiliation" and disbelief in God are slowly increasing, of course faster in some cultures and countries than others.

We have no way of predicting what will happen, of course. Trends do reverse at times. Nevertheless, it is possible that continuing movement away from religion is likely. We live in an information age, with rapid global communications and social media accessible around the world. Such platforms tend to have a homogenizing effect on social issues. (28) That is, they tend to facilitate people coming to believe in things in similar ways and developing shared views, and many of the algorithms they employ build upon themselves to create even stronger opinions and beliefs among some users, whether they're about current events, conspiracy theories, or religion. It also does appear that "social conventions" can be "tipped," in effect. "Tipping points" can be reached, where even when a minority viewpoint reaches a critical mass, resulting in an acceleration of people "coming on board." Such tipping points are often quoted as somewhere between 10% and 40%, a concept that has been studied mathematically by many, and some well-controlled experiments verify this estimate. (83) How this will relate to religious beliefs, of course, is still an open question. However, quite interestingly, we do see that about 17% of Americans are now adjusting their social media settings (e.g., blocking or unfriending) because of religious content. (133)

Nevertheless, the trend is clear. Belief and religiosity are declining in America and around the developed world, a problem for this business in some places. Recently, I noticed a large sign outside a church in my neighborhood. "Doubt is welcome here." I'm sure similar advertisements have popped up around the country. Despite a presumed motive to "spread the word" of a loving God, that's a clear indication of what organized religions really want—more paying customers.

A Fragile Species

We humans are not as special as we think, in all sorts of ways, as we've reviewed. Returning to the Copernican Principle, humans have existed but a microcosm of time compared to other life on this planet. Moreover, we are a fragile species, quite vulnerable to extinction. Although evolution tends to produce more and more complex life forms, there is a potential problem for us here. A fundamental principle of biological evolution is that the more complex the organism, the more vulnerable it is to extinction (other things being equal, of course). We currently *are* the most complex creature on the planet. Size is also a factor. As a species get larger, it is also more vulnerable to extinction. It often does a poorer job of adapting over time, since it reproduces more slowly and has a harder time enabling favorable genetic mutations in response to selection pressures. Contrary to what most people believe, we humans actually do not have a lot of genetic diversity and are therefore vulnerable to external factors that may hinder our response to external pressures and therefore our survival as a species.

Other species adapt and reproduce and evolve more quickly. Bacteria demonstrate much more adaptability than we do and existed for billions of years before we came along. The classic allusion to the lowly cockroach may not be just a cliché. There's some logic to this. Cockroaches are likely to be around long after we're gone. Ants are also likely to survive us. There are perhaps more than 3 million times as many ants on Earth as humans; they have complex and cooperative societies (if only we were so well-organized and cooperative); they're critical to many types of life on Earth; and they've been around for at least a 100 million years. The great biologist E. O. Wilson describes them as "The little things that run the world." (551) If I had to bet on whether humans or ants would survive the longest, it's obvious to me who'd win.

The tendency here is to say, "So what?" I think it's a matter of perspective and perhaps a bit of humility. There have already been five mass extinctions in history, with a significant percentage of life forms disappearing from the planet; remember, over 99.9% of the species that have ever existed are now extinct. If this was "God's plan," I guess He's learning by trial-and-error, with humans being one of the most fragile species in the current experiment.

We've Almost Been Extinct Before

We humans have almost been extinct before. Many experts believe that our population was reduced to less than a thousand individuals during a very dry ice age around 195,000 years ago. Then, about 140,000 years ago, while we all still lived in Africa, some dramatic climate change event reduced our population to a few hundred, a number of animals that would currently classify them as an "endangered species." (350) Then, a few millennia later, we were again on the brink. In the Stone Age, about 70,000 years ago, the so-called Toba super-volcano in Indonesia erupted. This, along with an encroaching ice age, produced climatic changes which resulted in the number of *Homo sapiens* falling to only a few thousand. We were again lucky to survive and multiply.

We might say that we're actually fortunate to be here at all. Like all evolution, we are an accident of history. If the mass extinction event of 65–66 million years ago that killed off the dinosaurs had not occurred, we probably wouldn't even be here. The dinosaur extinction was at least a major factor that enabled mammals to develop and evolve. Then, there is some initial evidence that, even before *Homo sapiens* existed, individuals in the evolutionary "stem" that led to us were nearly wiped out during the Pleistocene epoch about 800–900 thousand years ago.[3] Climate change may have been a culprit, though this new research is still a bit controversial. Nevertheless, the breeding populations may have been little more than a thousand individuals during that period, a reproductive "bottleneck" which lasted for over 100,000 years. (226) Our lineage might never have evolved into *Homo sapiens* in the first place.

Some Risks

It is certainly unknown how long our species will last. We are at risk from any number of possible cataclysmic events—nuclear holocaust, major asteroid strike, or some other unknown catastrophe. Yes, the above near-extinctions have likely been related to climate change scenarios, and these types of devastation might be considered "natural." But of course, nature doesn't care whether climatic effects on the environment and human and other populations are of natural origin or caused by humans themselves.

Human-produced climate change and increased probability of viral pandemics do threaten our population. They may not wipe us out, but they can seriously reduce our numbers, leading to all sorts of unknown changes. There have been many such events in the past. The Black Plague of the 14th century killed probably at least 25 million (an even larger percentage of humans than the recent COVID-19 pandemic). The worldwide flu pandemic of 1918 killed about 50 million people. At least a million Americans were wiped out by the COVID-19 pandemic in the early 2020s, and over 7 million had been lost worldwide as of 2024. Viruses are cleverer than we are, and one can't rule out that some future virus may significantly diminish human populations.

We're unlikely to be fully "wiped out" by some disease or calamitous event. However, we may be changed as a species because of it. Humans may also no longer be the dominant species on Earth. Some would argue this is already the case (remember those clever ants). Whether any differentiations are induced in our species by radiation, famine, disease, or other unknown environmental factors, the evolution of *Homo sapiens* will continue to follow its selfish genes. Adaptation to a changing environment is the pivotal factor in evolution and persistence of species. *No species has ever overcome that.* It would indeed be the height of arrogance to think we are so special as to be exempt from that biologic principle.

In the End

A long time ago, some primate brains evolved to employ a type of thinking and behavior that enabled a survival advantage for some hairless apes, with the ability to imagine, solve complex problems, develop stable social groups, and modify their own environments. Beavers, termites, and many other creatures also modify their own environments, though only to the degree they need to for their individual survival.[4] Our brains have also led to marvelous creations, including incredible technology, art, and culture. And they also created fantasies of gods—which facilitated both cultural development and extreme horror and destructiveness among populations of humans around the world.

I've covered a lot of ground. If you've gotten this far, you'll not be surprised at my conclusions. At this point, I think it's safe to say that there are a few things we know about the god problem. (1) the existence of gods cannot be supported by any logical or rational argument or subjective or personal belief, (2) there is no demonstrable need for a god to exist to explain our reality, (3) organized religion has been a major factor in the death of countless millions of people, (4) there is no practical reason to believe in a god and innumerable reasons not to, (5) we humans were not "created" and are unique and special only to ourselves and in one place and one very brief period of history, (6) belief in God and other religious memes are merely well-conditioned human behaviors arising in a determined universe, and (7) invoking god memes is not necessary for humans to live in some sort of harmony.

Imagine . . . No Religion

It's clearly unrealistic at this point in history, but take another gander at John Lennon's lyrics above. Theistic religion has created some of the greatest misunderstandings, hostility, and violence the world has ever seen. Is religion the "root of all evil?" Maybe not. But it's right up there. Can we imagine humanity moving away from worshiping superstitions and killing each other in the names of gods?

What factors may contribute to the waning of religion? Repression does not work to weaken religious behavior. Such attempts are inappropriate, ineffective, and unimaginably cruel, as history has well demonstrated. The Crusades didn't abolish Muslims or Jews; many Jews survived the German Holocaust. Both Protestants and Catholics have survived the "Troubles" in Northern Ireland. Muslim populations are still thriving despite efforts to suppress them, e.g., the Crusades, or the Chinese suppression of the Uyghurs, or the large efforts of one Muslim sect to kill the other. The Hindu population is thriving despite the genocidal massacres by sultans in the Middle Ages. There aren't enough pages to provide a full accounting. You can't "wipe out" a religion, though relentless attempts continue, as they always have.

Despite a determined universe, the future is unknown to us. As the great philosopher Yogi Berra once said, "It's tough to make predictions, especially about the future." (There are similar aphorisms attributed to physicist Niels Bohr and others, but I give most credit to Yogi). Nevertheless, we can certainly *imagine*, and surely, there are a number of positive factors which tend to weaken organized religion. Taking an overly optimistic venture into the matter, here's my admittedly rather unoriginal list—moving religions out of governments, more K-12 science education, movement toward racial, gender, and economic equality, and the global empowerment of women, particularly regarding access to education, economic opportunity, and contraception. In fact, it's abundantly clear that the suppression of women has been one of the major factors in the maintenance of theistic religion to this point, and reversing that trend can only weaken such faiths.

One can reasonably argue that extinction of religious behavior tends to be associated with increased economic and cultural cooperation. This is not a new idea, of course, and we do see examples of greater economic equality and growth and cultural cooperation in areas where religion empirically has become less important, for instance in Western Europe, as opposed to some other parts of the world. We also see religion's impact lessening with increasing education and all the other factors I've previously noted. Further, the "gaps" in our knowledge continue to be filled in—for life, for who we are, and for our world and a still expanding cosmos, making invocation of a personal god less and less palatable for some. This trend may be intensified by our rapid global communications and access to information. Perhaps this is too hopeful, but the accelerating access to information may render education and secular values more impactful. Of course, the current spate of misinformation will make this more difficult, actually a problem that can't be realistically addressed without more widespread education.

People will not give up their religions purely because of such social and cultural developments, and, as we saw, they'll not be argued out of their superstitions. The one thing we know about human behavior is that if it changes, it changes very slowly. Time will tell. As Thomas Paine said in *Common Sense* in 1776, "Time makes more converts than reason." But we can imagine that without organized religion the world would undoubtedly be a much better place—and for most everyone. Unless, of course, we wipe ourselves out first. The faithful might then get their apocalypse, though it will be of our own creation.

The Meaning of Life

Without a god fantasy, is there any meaning to life, as we asked in Chapter 9? One of my favorite answers comes from Douglas Adams' *The Hitchhiker's Guide to the Galaxy* (2). When the supercomputer, "Deep Thought," is asked the question about the meaning of "Life, the universe, and everything," it gives the answer "42."[5] Is there any better answer? No. There is no better answer to the "meaning" of "everything." I'll reiterate that humans do have the ability to ask questions to which there are no possible answers. If you believe that the "meaning" of your own life is validated only by a supernatural being, that is indeed sad. We do have a natural instinct to avoid death and come to grips with the notion that, in fact, death *is* the end, but, as the old saying goes, "Life is not a dress rehearsal." Gilgamesh expended great time and effort to find that wisdom. Orienting one's behavior toward an afterlife now seems more irrational than ever.

Where does this leave us? In the end, without religion, we're "left" with the opportunity to live in some sort of harmony and find our own "meaning" in each of our lives. As Camus noted, the only true philosophical question is whether life is worth living. If it is, you don't need a god to appreciate it. The only real answers are what individuals discover for themselves. Meaning is unique to each of us. For any of us, meaning and purpose may be in loving, appreciating beauty, building something, pushing boulders up a hill, reading poetry, exercising, or any other of the panoply of activities you find reinforcing. One does not need to believe in superstitions to enjoy the beauty and meaning of life. As best summarized by the poet John Keats:

> *Beauty is truth, truth beauty,—that is all*
> *Ye know on earth, and all ye need to know.*
> (Ode on a Grecian Urn)

A Final Word

In the end, it matters not what humans "believe." Arguments about religions and gods may persist forever, of course, and undoubtedly longer that we have available to us humans. Physicist Kaku sensibly predicted, "Robots owned by devout, churchgoing fundamentalists may get into arguments with robots owned by atheists." (261) I might go further. Given the accelerating growth of artificial intelligence, it is not difficult to envision that robots who are atheists themselves will argue with devout robots, whose god may be the supercomputer creating our reality.

We know that it is unnecessary to have any sort of divine inspiration for people to get along with each other, to behave in ways that enhance survival, harmony, and peace. To respect and care for each other doesn't depend on a god for moral guidance. Most enlightened humans implicitly endorse this way of thinking (even if not always acting that way), including many who still cling to a rather theistic view of reality. Even if they "believe" in a god, they more often depend on secular institutions for information, guidance, and support in their lives. They implicitly endorse what I think is the best closure here, the wisdom of Benjamin Franklin, *"Lighthouses are more useful than churches."*

We humans may consider ourselves unique or special in this universe, but as we've seen, there is no reason to believe that. We're here for an immeasurably small bit of time in our eternal universe. If we take a spatial perspective, it's just as tiny. Taking a view from very high above the Earth, we see a mostly blue planet with a bit of greenery and little organisms moving around trying to survive. There are tiny organisms, like bacteria and flatworms, and there are larger ones, like dogs, humans, and blue whales. But even from the window of an airliner, you can't even pick out one from the other. They disappear. In 1990, the Voyager 1 probe took a snapshot of the Earth from 3.7 billion miles away. Carl Sagan described the picture as a "pale blue dot," barely visible in the image. We humans are "merely" some of the inhabitants on that dot hurtling through space around and around the Sun. There is no choice but to enjoy the journey. No god will help us.

NOTES

Chapter 1

1. William James was a physician and is generally credited as being the founder of the science of psychology. (241)

Chapter 2

1. No, there are only five, ESP enthusiasts' claims notwithstanding. We do know that in humans, and in most animals, there are a couple of other sensory organs in the body, e.g., proprioceptors, which give us a sense of movement and position, and nociceptors, certain nerve endings which respond to potentially noxious or painful stimuli. But that's about it. Some other animals possess sensory skills that we don't have. Snakes, for example, can sense infrared radiation; and elephants can use sounds that we cannot hear. For us humans, our ability to sense the world is somewhat limited.
2. Contrary to what you might think, many experiments we hear about have not been repeated; and replicating results is very far from certain for many types of studies. This is an especially acute problem in biomedical (92) and behavioral sciences (381), though the problem does often appear in the physical sciences as well.
3. It is beyond our purpose here to opine on the enduring question of whether mathematics describing the natural world is a tool invented by humans or whether it is an essential feature of nature itself. Suffice it to say that the equations which describe the natural world are synonymous with the idea of the laws of nature. Einstein once summarized the problem as, "How can it be that mathematics, being after all a product of human thought which is independent of experience, is so admirably adapted to the objects of reality?" Einstein erred, of course, in assuming that human thought is independent of experience, but the dilemma remains. We should also acknowledge that we may not be the only species with mathematical ability. Many species have quantitative abilities. Ducks and lions, for example, "know" when a duckling or cub is missing from a large brood or litter; and some species of crows can actually count. (310)
4. Formal proof actually does not exist outside the discipline of mathematics. Scientists use mathematics all the time, but only in an applied manner. However, the "truth" of any mathematical proposition, and mathematical proofs themselves, are only exercises in numerical logic, though sometimes quite elaborate, i.e., how one can show that a numerical statement is true.
5. MMR, for measles, mumps, and rubella.
6. For current information, see https://www.cdc.gov/vaccine-safety/about/autism.html
7. A wonderful book for children about critical scientific issues of our time—from evolution to the origins of our universe to various god fantasies—I recommend Richard Dawkins' *The Magic of Reality: How We Know What's Really True.* (115) Its intended audience is primarily older children (and their parents), but certainly anyone can get a rich experience about science and reality from this excellent book.
8. In a letter to Jost Winteler, father in a family with whom he had boarded during some teenage years while attending secondary school in Aarau, Switzerland. Translations vary somewhat, but you get the picture.
9. Variation in such surveys, and the ones cited in subsequent chapters, are often likely secondary to slightly different survey years, how surveys are conducted, how questions are asked, and some other variations in methodology.

Chapter 3

1. For the interested lay reader, I recommend Tyson & Trefil's *Cosmic queries* (521), or Seife's *Alpha & Omega: The Search for the Beginning and End of the Universe* (458); or a bit more challenging—Greene's *The Fabric of the Cosmos* (192) and *The Hidden Reality: Parallel Universes and the Deep Laws of the Cosmos.* (193)
2. We also cannot "see" some of this "light" with our eyes. Other wavelengths we cannot "see"—x-rays, gamma rays, microwaves, ultraviolet light, etc.—are sensed with our instruments.
3. The age of 13.8 billion years is obviously a rough estimate, if you can say such things about cosmology. We'll ignore here the fact that different methods of "measuring" the age of the universe come up with slightly different ages, though the difference has been an important and problematic headache for cosmologists.
4. It is interesting to note Kepler's implied reference to an essentially Copernican Principle in his science fiction work, *The Dream*, written in about 1608. He envisioned the existence of moon people who believed that the Earth revolved around the moon.
5. We should note, however, that there is still no consensus on what space "is." Does it have some tangible quality or not? And if so, what is it? It certainly has energy, but the end of this story is still unwritten.
6. For a reasonably lay description, there are some good sources I'd recommend—an excellent article by Edward Frenkel (172) and for more substantial writings, Brian Greene's *The Elegant Universe* (191) or Adam Becker's *What Is Real? The Unfinished Quest for the Meaning of Quantum Physics.* (40)
7. There are numerous modern examples from the twentieth century, but a version was apparently first conducted by British physician and polymath Thomas Young back in the early 19th century to show that light coming through a double-slit experimental design showed interference effects, indicating that light, in fact, behaved as a wave. There is actually some indication that Isaac Newton may have conducted something similar, though he thought light was actually "corpuscles" of tiny mass.
8. There are somewhat more detailed lay descriptions of this. (172, 191) This is sometimes difficult to picture. There are many visual illustrations of the double slit experiment which can be found online.
9. This has been subsequently confirmed experimentally in various ways, though the search for any "missing pieces" goes on. There are some good, short, lay descriptions of this ongoing problem. (260, 178); but the matter is far from settled at this point. Many physicists reject the idea of nonlocality, maintaining that at its most basic level, the universe must be local and described with classical laws.
10. The specific language is actually, "Der Alte wurfelt nicht." A rough translation is "The Old Man [that is, God] doesn't roll the dice."
11. It should be noted that other physicists, principally physicist George Gamow, also contributed to this theory, along with Russian mathematician Alexander Friedmann, who also apparently had proposed the idea of an expanding universe. English astronomer Fred Hoyle had proposed the idea of an infinite steady-state universe, disagreeing with Gamow. In jest, he called Gamow's theory a "big bang." Obviously, Lemaître and Gamow were proved right, but the name has stuck ever since.
12. "Infinitely small," of course, is not mathematically possible. The initial condition is often described as smaller than some subatomic particle or being actually "dimensionless." The best answer so far is that we just don't know.
13. You may have heard of black holes, the center of which are also considered singularities. When a star of a certain size dies, it collapses under its own gravity to an incredibly dense area of space, so dense that it cannot be appreciated by Einstein's laws of general relativity. In fact, it is believed that the density at the

center of a black hole cannot be mathematically described. This "singularity" may or may not be analogous to the idea of that at the Big Bang.

14. We should acknowledge that there is some dispute about that figure. It comes from a certain type of measurement. Other measurements have a slightly different age. The discrepancy is important to physicists, but the only importance here is that it happened a long time ago, with that order of magnitude.
15. I would point out that some scientists disagree with the Big Bang theory, but it appears to be the one with the current consensus.
16. I should point out that word "creation" here is actually considered by theists to mean "out of nothing." But the alternate meaning is the development of something novel, like Aristotle's notion of organization coming out of chaos. In point of fact, all events of "creation" are *merely* reorganizations of mass and energy, though often by humans and often of great beauty and value. But they are not creations "out of nothing." They are reorganizations of matter and energy by humans (and perhaps some birds and other creatures) that render something unique or innovative.
17. Relatedly, it took humans until at least 4,000 BCE in Mesopotamia to mathematically indicate the absence of a quantity; and the number zero/"0" probably did not appear until the third century BCE in ancient Babylon, though other cultures seem to have invented it separately centuries later.
18. The actual, physical nature of space or spacetime itself still remains to be described.
19. In point of fact, the generally accepted conception is that the energy density of the universe is exactly what it should be if there is a balance of positive and negative energy, that is, where the total energy of the system comes out to *zero*. And it has always been that way. Given some quantum uncertainty, it is fully conceivable that is how our universe took on the character it did. (212) Although we might consider energy as "something," the universe technically began from a state of *zero* energy.
20. The human mind is incapable of appreciating the concept of infinity in any ordinary sense. It is a mathematical, not an ordinary concept. But it certainly implies the notion of there being no numerical limits. It is also seen as a physical concept, one of no limits to space and time, for example. In any case, these are "ideas," which the human mind cannot grasp. They are "merely" concepts used to discuss other matters.
21. It's unclear if dark energy will exert its influence with the same type of acceleration it has had "recently"; or perhaps it will "run out of steam," or slow down, or change in some other way; but no one knows. Further, there are theories about how all the remaining atoms might break up.
22. To clarify again, the concept of "infinite" suggesting a number is a bit misleading. There is no such thing as an infinite "number." If number is a concept, the intention here is that there is no conceivable limit. And infinity itself is not a quantity; it's an idea. As will appear a number of times in this book, this reference is merely a grammatical convenience.
23. Nobel Prize winning physicist Roger Penrose has proposed a variant of this model, called the "Conformal Cyclic Cosmology Model," though it is very provisional and not widely accepted. When the universe eventually evolves to its maximum entropy state, at that point nothing will exist except energy, in the form of photons, i.e., pure quantum fluctuations. Of course, photons "know" nothing of size, or distance, or time. Thus, it would be unclear if a universe is big or small, beginning or ending. The universe would then go through another cycle, leading to the kind of universe we see, perhaps a bit different, infinitely.
24. Skip forward to current times, and the metaphor may be more than that. One reality considered by cosmologists, which I'll not devote any time to here, is embodied in a so-called "holographic principle," that what we *perceive* as our three-dimensional reality is only as we conceive it; but it is actually a hologram dictated by events occurring on a distant, two-dimensional outer edge of our universe. (193)
25. We refer to "cosmic horizon" as though we could actually measure it. In point of fact, it is merely the distance from which we could possibly retrieve information,

i.e., photons. We don't know how much bigger it could be. Space is continuing to expand and at an ever-increasing rate. At some point, distant galaxies that we currently "see" may well be too far away to record, since space will be expanding faster than any light can travel to us. Remember, the speed of light is fixed and an unattainable maximum for any entity. But that is for movement *through* space. There is, as yet, no known limit to how fast space itself can expand. At some point astronomers will again believe that our galaxy is the only one, that the entire universe is made up of the stars we now group as the Milky Way.

26. As I've said, it has sometimes been described as dimensionless or an infinitely small point. Of course, this is at least grammatically and mathematically impossible. To be "infinitely small" is to not exist.
27. It should be noted that the fundamental constants we identify in our universe are only the ones we've identified so far. There may be others yet to be discovered.
28. Though there is some reason to believe that slight variations of them may still enable life as we do know it somewhere. (250)
29. For great, understandable descriptions of how such "thoughts" led Einstein to the development of those theories, see Isaacson's *Einstein: His life and Universe.* (235)
30. Physicists estimate that Einstein's spacetime didn't appear until at least about 10^{-43} seconds "afterward."
31. An intriguing fantasy about time and our perceptions is physicist Alan Lightman's little novel *Einstein's Dreams* (313), which I strongly recommend.
32. This is essentially the same concept used in statistics. A variable is "free" to take on any value within a certain range. In this case, the "fundamental constants" are not constants in a statistical sense. They are potential variables. The language is just unfortunate here.

Chapter 4

1. This shouldn't be taken literally. Dawkins was intending to signify the overwhelming weight of evidence. However, evolution is not literally a "theorem." A theorem is a mathematical statement that can be formally proved, based on certain assumptions. Natural science, broadly defined, doesn't work that way. Although all theories continue to be updated, the basic facts of evolution by natural selection remain fully supported.
2. The controversies and contributions of the British biologist Alfred Russel Wallace (1823-1913) to the theory of evolution by natural selection are not important here. Darwin is generally credited with putting forth the major explication of this theory, though it is clear Wallace did formulate aspects of this independently.
3. By random here I mean only that they are very hard to predict; they may occur spontaneously during replication of gametes, be caused by ubiquitous types of radiation, or other potential physical influences; but by "random" I do not mean that they are not determined by physical laws or cannot be potentially understood.
4. There do seem to be some other factors which may affect offspring, however. Certain chemical factors outside the genes themselves may affect genetic expression, what's now called epigenetics. For example, epigenetic mechanisms seem to be responsible for some role behavior in some ants, for example, determining whether an ant will become, in overgeneralized terms, a "warrior" or "nurse." (478) A particular species of crayfish has used epigenetics to adapt to becoming a successful invasive species in various novel environments around the globe. (77) In humans, epigenetic mechanisms are likely involved in certain diseases, such as smoking in the development of lung cancer. There is no evidence that epigenetic factors have had any measurable influence on the evolution of human beings—yet. Finally, there are other potential mechanisms in which genes can be modified. For example, there's the phenomenon of "horizontal gene transfer," in which genes are directly "swapped" between two organisms. (156) However, there is no evidence that HGT has had any significant influence on the evolution of

human beings as yet, though there is speculation that HGT may play a role in a number of human diseases, e.g., possibly cancers, and the evolution of drug-resistant bacteria. For our purposes here, I'm considering only classic Darwinian mechanisms, which is how such changes in DNA are passed to future generations in any case.

5. Further, we should realize that the definition of "species" continues to undergo revision, and consensus seems elusive. The "species" concept may even be grossly inadequate for a taxonomy to describe the diversity of earthly life. For example, are we humans or Neanderthals or human-Neanderthals or something else, given that we carry up to about 2% Neanderthal DNA in some parts of the world.
6. This is only a fairly modern variant. Many organisms reproduce asexually, which has been going on for billions of years.
7. There are a few other species that can escape after being eaten by predators, e.g., some insects and snails. The skills in those species appear to have evolved separately.
8. There's the legend of the "crazy cat-lady syndrome," which has been attributed to the parasite, but there's actually no specific evidence for that. However, there is some evidence of an association with a diagnosis of schizophrenia and possibly other mental disorders. (508) However, we should point out that despite a great deal of literature on the subject, the specific or multivariate causes of such mental problems remains unknown.
9. There are strains of MRSA which arise in the community and others that arise in hospitals; but the principal variant of severe illnesses usually derives from the community-based strains. There also appears to be a variant which has invaded cattle.
10. Despite the lore on this dynamic, there is no good evidence that social animals actually withdraw and isolate themselves when dying so as to protect the group, e.g., from disease or predators.
11. Temporal estimates can vary somewhat; and of course, there is some reasonable error of measurement here. We also cannot rule out that further discoveries will push the date even further back. There is also reasonable data to suggest that viruses were already extant at that time as well.
12. Some other creatures have better sensitivities, being able to "see" in the infrared range of light, e.g., some snakes, bats, and insects.
13. The term *hominin* refers generally to us, the extinct *Homo* species, and all our immediate ancestors, i.e., primarily the species who walked upright. The related term we often see is *hominids*, which includes all the modern and extinct great apes, humans, chimpanzees, gorillas, orangutans, and all their immediate ancestors. We should note that such types of classification have changed over time and continue to change as new discoveries are made.
14. The dates I quote here are rough approximations, based on the anthropological literature. And these do keep changing with new discoveries. The numbers offered here are for general illustrative purposes.
15. Yes, the legend is correct. Johnson did name the creature Lucy after the Beatles' song "Lucy in the Sky with Diamonds."
16. Whether they engaged in some type of music is still notably controversial. (93)
17. Other social animals may have actual language as well, which evolved separately, driven by social factors. Though this may seem controversial, actual language—phrasing, pitch, phonemes, timing, etc.—have been observed, for example, in the humpback whale, a very social animal. (542) This was obviously convergent evolution, but the whales are also known to pass language on culturally and even learn from other pods, which suggests the possibility of actual language.
18. Among other factors, it is a specific gene (the FOXP2 gene) that may have enabled most modern language capabilities. In fact, Neanderthals did have the same FOXP2 gene variant that we have. (283) As far as we know, all mammals have some variant of the FOXP2 gene, which is one part of the basis for communication, mimicry, language, and related behaviors. It's just that the very tiny variant which arose in

some *Homo* species seems to be responsible for most of the sophisticated communicative behaviors we show, including verbal language.

19. Found in 2015, a 1951 letter from Einstein to students in a Jesuit school in Porto Alegre, Brazil. (56)
20. The EQ, the precise measure used by biologists, is the ratio of actual brain mass to the predicted brain mass for an animal of a given size. Humans presently show the highest EQ, but this also does not consider the additional factor of the recent folding of the neocortex in higher mammals, particularly hominins, which increases the available surface area of the cortex in the advanced brain.
21. A small decrease in the human EQ around the time of the agricultural revolution has been reported by some. (203) We do know that brains require a great deal of energy. However, the theory that smaller brain size in domesticated humans was due to cognitive and other information being "stored socially," i.e., with others—for support, problem solving, and protection—thus requiring or enabling smaller brains, has not been rigorously tested (if that's even possible). If true, we're possibly in for some significant shrinkage in the computer/information age. In fact, that idea is gaining some traction, not specifically for an actual EQ but at least for some types of intelligence among modern humans.
22. Many birds defy the logic that larger brains are associated with greater problem-solving c1apacities. Their brains are miniscule in comparison to primates. And it may not be size or even EQ that's the variable.
23. Whether a bee "consciously" solves problems or "reasons it out" must be left to inference of their subjective states as much as it is for us, which we'll leave to later chapters.
24. I should mention another minority view on this. It is possible that certain bands of hunter gatherers may have dabbled in some form of primitive agriculture even before it "took off," as people began to form into larger tribes. (189)
25. Some readers may interpret this as a political statement since it is close to some utterances in recent publications and commentary. Nevertheless, it is merely a clear anthropological fact that humans depended on the formation and maintenance of close-knit groups in order to raise the young and survive.
26. In the latter three, as with a small number of other species, it is a senior female who is the dominant member of a tribe. In most primates, including apes, social order is dominated by males. In only one extant ape species, the bonobos, are the hierarchies dominated by females (though they can be just as violent as males). It's interesting to speculate what type of human culture we might have had if females were the dominant sex in the evolution of *Homo sapiens*.
27. That the evidence clearly suggests a burial site has been legitimately questioned (416), though more data may be forthcoming. There are also a number of technical problems at the site, which have interfered with dating the remains, resolution of which would shed some light on this.
28. However, it's unlikely that the "grandmother effect" is operative in chimpanzees. (306)
29. We'll leave for the moment the nonsense of both the Christian view of a Garden of Eden or the Scientology notion of our souls being deposited into earthly volcanoes, equally absurd superstitions; more on this in Chapter 7.
30. It should be pointed out that this dynamic is in no way equivalent to the notion of "race," which is a social construct, not a biological one.

Chapter 5

1. James is considered the father of the modern science of psychology. (242)
2. Watson is considered the founder of modern behavioral psychology. (533) More on that in Chapter 10.
3. He was referring to the organisms in our bodies that contribute to our microbiome, the bacteria, fungi, viruses, and other microorganisms that inhabit us. We do know

that there are at least as many cells of other organisms inhabiting our body as our own body's cells. (459)

4. We might take this as metaphor. Notwithstanding the idea that photons travel *at* the speed of light, by definition, and time does not exist for them. There is no "time" for any "feeling" or smelling.
5. Some would disagree with this; but even the idea of subjective language or "thoughts" is unconvincing. The physical brain does not experience itself.
6. Though he may not endorse the subjective notion of consciousness, the citation of "chaos theory" is often misconstrued. Chaos theory is not a rebuke of determinism, which would take consciousness out of the causal order. It merely suggests that certain conditions can lead to "unpredictable" outcomes. Rather than "unpredictable," of course, the calculational impasse would suggest that chaos theory is essentially a reflection of incomplete knowledge, often evoking the idea that systems may appear random because small changes can lead exponentially to more complex events.
7. Both of these can be very complex; but for the interested reader, these theories, along with the current conflicts and controversies, are elaborated in a couple of excellent summaries. (277, 340)
8. In actuality, research suggests that this is likely closer on average to about 86 billion. (23, 218)
9. For example, the various programs sponsored by the NIH Human Connectome Project, https://neuroscienceblueprint.nih.gov/human-connectome/connectome-programs
10. There are probably more glia than neurons, though the often-touted ratio of 10:1 for glial to neuron cells hasn't been well confirmed. (23, 236)
11. Verbal reports during neuro-stimulation of an exposed brain, which is sometimes done for patients with intractable epilepsy, would not meet such criterion since stimulated verbal reports identifying experience would be, by definition, conscious.
12. In this regard, studies of hypnotic subjects have not offered any usable information; and it is widely construed that hypnosis (which itself defies a consensus definition) is a "conscious" state.
13. There are a number of technical problems here, including the fact that (1) the same neuron activation can do many things, (2) the same general "thought" can be associated with different groups of neurons, and (3) the problem of what has been called "reverse inference" in neuroscience, the logical problem of inferring certain experiences from an image, for example. (408, 409) Also, there can be strange and absurd "false" positive results of brain activity from fMRI findings where certain technical and statistical techniques are not employed. (46)
14. Some progress has been made with EEG patterns across subjects, but resultant accuracy of individually recorded patterns in relation to group reference codes is significantly lower than it is for individual subject trainings. (141) The notion that there could be coherence of neural activity for any specific higher cognition across human brains is likely a fantasy at this time.
15. Eukaryotic cells are those with their DNA bound into chromosomes, a true nucleus with a membrane, and various other structures/organelles. Prokaryotes do not have those characteristics and are principally some other bacteria and archaea, primitive life forms similar to bacteria, the names of which few would recognize.
16. We should acknowledge that some quantum effects are biologically acknowledged and do appear useful in some ways to organisms. That quantum effects assist our sense of smell has gained a lot of ground. There are also now some data to suggest that even many organisms may use "quantum effects" for magnetic location and orientation—in birds, whales, honeybees, and other creatures. (81) The resonance signature of vibrating molecules and the "spin" of electrons with such are not well understood and will probably remain so until quantum theory is more mature. But this is not related to the problem of whether consciousness can be defined with quantum mechanics.

17. Psychologists will recognize this as the famous "Spearman's g."
18. For a more detailed description of these phenomena, there are some good lay sources. (45, 299)
19. It is theoretically conceivable that the mechanisms and algorithms an AGI uses to generate behavior are discoverable. However, this would be more than identifying information being used. It would necessitate identifying how information flows through the system, which may be as difficult here *as it is for the human brain.* This may be as much a calculational impasse problem for AGI as it is for us.
20. In fact, there is some indication that advanced AI is already generating hallucinations, not merely errors, but fabricating objects, ideas, and realities that actually don't exist. You could say that a system is "wrong" or merely "makes things up"; but human brains do the same thing when they hallucinate, whether secondary to some labeled mental aberration, when they construct some fantasy, or when there arises an idea of a supernatural god.
21. The debatable notion of whether an AGI can achieve a human level of intelligence without parallel physical experience of the world is really not relevant here. It is not important, from a theoretical perspective, whether a machine can mimic a human type of intelligence. We solve problems differently from other creatures who are intelligent in ways that are useful specifically to them. This is not a limitation to AGI, which could solve problems appropriate to its own sphere, which could be considerable.
22. As Skinner notes, "The real problem is not whether machines think but whether men do." (487) See Chapter 10.
23. They were actually first proposed in his short story, "Runaround," in 1942. (15)
24. Note that such assessments in nonhuman animals are based on observation of behavior over some period of time. When such "personality" assessments are done on humans, the evaluation is generally done with psychological tests, which attempt to have subjects rate or otherwise retrospectively report their own behaviors.
25. By affective states, they mean what we think of as emotions. The use of "intentional" behaviors and "decision making" in the statement should also be clarified here. Such references do not imply a notion of free, voluntary action. Rather, they merely suggest only a set of behaviors which are goal-directed. Also, I'd point out that the notion of "bird brain" is now a bit antiquated. In point of fact, some birds do possess a small, homologous structure that might be considered a neocortex. (219)
26. As incredible as it seems, some experiments have begun with inserting human pluripotent stem cells into other species' embryos, e.g., monkeys, which might lead to development of neural capacities more similar to humans. (305) If the embryos were allowed to be born and the chimeras develop, would the animal have something closer to human consciousness, as has been speculated? Philosophical musings aside, it would be unimaginable in practice. In fact, the experiments have been halted, likely out of ethical concerns or, relatedly, perhaps out fear of what would, in fact, happen.
27. Richard Dawkins' *The God Delusion* (113) does not examine the notion of "delusion" from a mental illness perspective, though it cogently examines why religious beliefs are unfounded or delusional.
28. There are a number of very interesting legal and moral issues with this, which are described well elsewhere, and we won't get into here. (455) We do now provide at least retirement "sanctuaries" for chimpanzees who were former experimental animals; and more and more are being moved there.

Chapter 6

1. In an address to the Spinoza Society of America, September 22, 1932. (235)
2. I might point out that some sects, e.g., Calvinists, believe in the idea of free will but endorse some "type" of determinism, whatever that means. The theological

gymnastics required to resolve such an inconsistency are not worthy of comment here.

3. In an interview with poet and pro-Nazi propagandist George Viereck in 1929. (235)
4. Lyrics from the song "Me and Bobby McGee," written by Kris Kristofferson & Fred Foster, 1969. Janis Joplin recorded the most popular version, although it had been first recorded by Roger Miller in 1969.
5. Again, this is a statistical concept, meaning a variable is "free" to take on any value in a certain range, but it has an equivalent meaning here.
6. With James having denounced the notion of consciousness (Chapter 5), we should take his meaning here to suggest he was likely referring to a totality of human behavior and experience, so to speak, not a tangible consciousness, per se.
7. The actual experiment involved $500 to each organization if no choice of nonprofit was made, and $1,000 to a particular one if there was a choice. The subjects' perceptions were shaped by signed commitments of donation by the experimenters. It's unknown if money was actually donated.
8. This is a painless procedure. There are no sensory organs in the brain, so there is no pain experienced with the brain stimulation. It is unknown if any noxious experiences would be reported if stimulation was provided to more primitive areas of the brain, where higher functions are not mediated.
9. Of course, what about the things that happen in your gut, your heart, the pain of a sprained limb, seeing a beautiful sunset, or falling in love—the kinds of realities that do not impinge on AI? These kinds of things are all massive contributors to our experiences as humans; but as far as the brain is concerned, they are "only" sources of *information* used in any decision process or emission of behavior.
10. However, we know that the CNS can override some of this. Breathing can be somewhat controlled. Even controlling things like heart rate and blood pressure, have been accomplished through various psychological techniques, by various meditations, or with biofeedback. (I'll return to some of these issues in Chapter 10.) However, we generally do not consider the unattended behavior of our organs or reflexes to be voluntary actions or a function of free will.
11. I should point out that reward and reinforcement are not exactly the same, but that's not important for our discussion here.
12. We'll see in Chapter 10 that some learning/conditioning can occur even during gestation, but this is a far cry from the spontaneously emitted behavior at birth and wouldn't generally be described as "voluntary."
13. For an excellent lay summary of their work, see Kahneman's *Thinking Fast and Slow.* (259) Unfortunately, Tversky had died a few years before the prize was awarded; and the Prize is not awarded posthumously.
14. The notion that there are other parts of the central and peripheral nervous systems that contribute to the phantom perception doesn't weaken the point. The experience we call phantom pain is "created" in the absence of the anatomy that is subjectively perceived. Phantom pain remains something of a mystery, even though we're a bit closer to understanding, in general, some of the physiology of pain itself.
15. And of course, there was some variability in the brain damage and images of the subjects, but there was a reasonable coherence in the results to identify the likely neural networks which subserve the results.
16. The implication here is that free will, as we've said, is irrelevant to human behavior. As an epiphenomenon, much as with the notion of consciousness or other perceptions, it is only correlated with behavior. The *perception* of "free will" is merely an idea, not an agent in the causal order—in the common meaning of that idea.
17. Not all cultures share this model, of course. The iTaukei people of Fiji, for example, discourage focusing on this type of reasoning and tend to judge others by their actions and consequences, even if such occur "accidentally." (338)
18. The issue of mental disorders continues to be debated in the scientific and professional communities. There is some consensus with respect to some "disorders" and very much less so with others. Most of the disorders listed in organized psychiatry's *Diagnostic and Statistical Manual of Mental Disorders*

(now DSM-5-TR), for example, do not have any tangible referents, which is why the label of mental "disease" might have been replaced with "disorder," a more socially acceptable term, though everyone knows what we're talking about. Nevertheless, they are often used in forensic proceedings to establish or rule out some sort of personal responsibility.

19. For instance, one model maintains a free will perspective balanced against a person's ability to make decisions—based on an individual's specific capacity to appreciate and reason about behavior. (391) This is a somewhat more detailed analysis of criminal behavior, which is beyond the scope of what we're addressing here; but it also hasn't penetrated American judicial proceedings at this point.
20. Just more reasons, along with racial and socioeconomic prejudice, that a criminal justice system based on notions of responsibility often provides unequal and rather retributive justice as opposed to contingent, proportional punishment and rehabilitation.

Chapter 7

1. (435) Lucretius was a first-century BCE Roman poet and philosopher. From his only known work, *De Rerum Natura* (On the Nature of Things).
2. From his archives. (144)
3. A notable exception with primates is among bonobos, where female dominance is the rule. They don't have one alpha member but are dominated by a few high-ranking females whose power derives from their bonding together. Whether you call this a matriarchy or not, it is clearly an exception to the rule of alpha male dominance in the primate world.
4. This is not convergent evolution in the Darwinian sense, with organisms evolving similar traits as gene expressions, as cacti might do on different continents, since there is no evidence that sophisticated cultural behaviors are promoted or enabled genetically. We know group selection *does* occur in some organisms, e.g., in some species of ants and termites. A particular colony might have a well-developed method of cooperative behavior enabling defense, for example, which would favor its survival, even if it was of no particular advantage to an individual ant. If a different colony was less skilled in this area, one can envision better adaptability and survival for the first. But for the most part, humans don't share the level of cooperation and cohesion that ants and termites do.
5. We know that even some modern leaders were alleged to have god-like personas ascribed to them, e.g., Kim Il-sung, the first leader of North Korea.
6. The only extant culture we know about that does not seem to have an origin story is a tribe of hunter-gatherers, the Pirahã people of the Amazon in South America. (157) Though they do not tell stories in the sense that we understand them, they just do not have any overt conception of creation. They tend to believe only the things which come from direct personal experience. One can only hope that the all too frequent clearings and fires in the Amazon do not threaten this proud people.
7. For all the inconsistencies and problems with the putative "evidence" for an historical Jesus, this is extensively and convincingly laid out in a number of excellent books and essays. (167, 536, 565)
8. Gender inequity and subjugation were even more pronounced in those times than it is today, albeit still persistent, often to a horrific extent, in many parts of the world.
9. Again, I note here that I'm in no way endorsing any validity of Freud's well-defunct theories on human behavior, pathology, or treatment of any aberrations. I merely offer here and elsewhere a few of his often sagacious observations of culture, which are relevant to these topics.
10. The creatures we know about that seem to be able to stave off death are the hydra, and possibly a couple of others, as I described in Chapter 4. Absent predation or disease, they have mechanisms that can conceivably allow them to live on indefinitely. Most other creatures, including humans, are not so equipped.

11. We acknowledge that the fear is somewhat variable in humans and other creatures. Mothers of many species, for example, will overcome fear of their own death to protect offspring. And the subjective level of fear for any person is a function of both innate tendencies and cultural and individual learning.
12. As I previously noted, I'm not endorsing any validity of Freud's well-defunct theories on human behavior. I merely offer here and elsewhere a few of his observations of culture, which are particularly relevant to these topics.
13. Freud's delusional argument well presages Dawkins' *The God Delusion* (113), though he has taken Freud's hypothesis further.
14. It is actually similar to what has been previously used in measuring a "galvanic skin response," a measure of electrical resistance in the skin. It was previously used in biofeedback treatments, but the measurement is notably fickle and unreliable and has been largely abandoned in this area. It's not clear why it continues to be employed in some polygraph "lie detector" examinations.
15. Rabbi Wolpe's comments are noted in a 2001 *Los Angeles Times* story. (531)
16. You might consider western governments or nation states, so to speak, as being richer. However, despite assets, the net value of these is almost always below zero.
17. The United States, England, and a few other countries do not tax religious organizations. And, by the way, it's pretty easy to identify as a religion and get the exemption. Most countries in western Europe and certain other parts of the world do have some kind of levy.
18. The only one generally acknowledged without a history of violence is Jainism, a religion a couple of millennia old, with some similarities to Buddhism. Jainists do not believe in gods, per se; and it's considered a pacifist religion.
19. The evolution of the split in Islam, resulting in these sects, occurred after the death of the supposed prophet Muhammad, c. 632 CE. The split came over different opinions on who would succeed him. Sunnis, the vast majority of the Muslim world today, believe that the religious guidance should come from a leader who emerges in the Muslim community. The Shia, a minority sect of Muslims, believed that the succession should remain in the "family," and they would follow a man named Ali, who was Muhammad's cousin and son-in-law.
20. A great deal has been written about this. One of the better volumes on this is constitutional attorney Andrew Seidel's *The Founding Myth. (*457)

Chapter 8

1. An interesting sidelight to Galileo's imprisonment was that he was allowed visitors during his confinement, even heretics, as long as they weren't mathematicians. One of his visitors was the poet John Milton. It is variously reported that Milton had made a visit to Florence and befriended Vincenzo Galilei, the illegitimate son of Galileo, who introduced him to his father. Subsequently, Milton would pay tribute to Galileo in several passages of his masterpiece, *Paradise Lost*, including one where the Angel Raphael is granted a clear view through the heavens, "As when by night the glass of Galileo, less assured, observes imagined lands and regions in the moon."
2. The scientifically inclined reader will note that this is not the same idea as statistically rejecting a null hypothesis. In science, we test the null hypothesis—that there is no likely effect in an experiment. The only way to conclude that an effect is "real" or reliable is to find that the phenomenon in question is otherwise statistically unlikely. One can never confirm a null hypothesis, only reject it statistically. In the god debate, we have never seen an event or any evidence which causes us to reject the null hypothesis. No miracle has ever been confirmed.
3. Interestingly, some were likely similar to tryptamine and monoamine oxidase inhibitors, which are relevant to the action of some of today's antidepressant and other psychotropic medications.

Chapter 9

1. Steven Weinberg, theoretical physicist and Nobel laureate. (538)
2. There have been many instances of this general comment, but it is most generally, and most recently attributed to journalist H. L. Mencken.
3. Physicist Max Born, in his Nobel Prize acceptance speech in 1954.
4. This is true for the most part, i.e., some so-called moral codes also proscribe a variety of personal, private behavior as well, especially sexual. These are dogmatic rules, and their effect is mostly to exercise control of individuals by authority figures and those who share different values.
5. There were others well before that, but Hammurabi's was a significant advance in that it was very detailed, covered a large range of offenses, and itemized various punishments.
6. Draco of Athens himself, and his legal prescriptions for punishments, would not come along for another millennium. We just obviously use the term for the notion of often disproportionally severe punishment.
7. We should note that there are essentially various versions of the Ten Commandments in the Bible, at least four sets. They are different in their content as well as the wording. And of course, there are numerous instances of self-contradiction.
8. Harris correctly points out that this is not the original meaning of the term. Originally, it meant that our "goodness" cannot be equated to other states like "happiness." (203) But for our purposes, the fallacy of employing "naturalism" as the presumed basis for morality is reasonable here.
9. I should point out that such designation is not an "official" label of a geologic time period, having been recently rejected by the organization that classifies such things. (553)
10. The details of the consequences for each such individual require more specific assessment, though psychologists and others often venture general speculation on such things. The eventual punishment for this notwithstanding, such "immoral" acts do occur for reasons that can be potentially analyzed behaviorally.
11. The term sometimes connotes a "morality" unique to a particular individual. However, when this is the case and any value in question is not shared by another, we seldom currently refer to such as morality.
12. Though similar sagacious observations have resurfaced throughout history, King's here is likely borrowed, or perhaps slightly paraphrased, from the words of Theodore Parker, a 19th-century abolitionist and Unitarian minister.

Chapter 10

1. From Aristotle's *Nicomachean Ethics.* Translations differ somewhat. This is the interpretation from Will Durant's *The Story of Philosophy* in 1926.
2. Greek historian. From *The Histories,* 7.49; *c.* 425 BCE. Herodotus is often considered the "father of history"; but he's been criticized for his use of legend, fables, and some fabrications in his writings. Nevertheless, he was sagacious in the comment, as history well demonstrates.
3. English critic, essayist, and reformer (1819-1900).
4. There are some approaches and techniques used in some methods of psychotherapy which do utilize objective principles of behavioral science, though these are quite in the minority.
5. We again should be clear that this is not the "inheritance of acquired characteristics," as proposed by Lamark. Rather, it is possible that organisms' exposure may trigger "adaptive fitness" through epigenetic influence on organisms' "neural plasticity," enabling more effective learning, which is then selected for in Darwinian fashion. Then, subsequent generations are each increasingly able to acquire similar responses with less exposure or practice,

leading to the evolution of more easily emitted behavior.

6. One might suspect, based on statistical probability, that if that percentage were again attempted for replication, additional ones may drop out, some may prove reliable, and so forth; but additional research has continued to find otherwise.
7. I should point out again that replication failures are not limited to the behavioral sciences; they do occur in other fields of science, though often to a lesser degree. It may seem surprising; but this does happen in the physical and biomedical sciences more often than you'd think.
8. But aren't there psychological tests and other ways to "reveal" some sort of character or personality? Yes, but those methods are no more than limited samples of behavior, observed or reported by subjects or examiners, which may correlate with other real-world behavior. They do not "reveal" an empirically existent, tangible personality or character. In Chapter 5, we also noted the similarity of some animals to humans in terms of behavioral traits, described in terms of "personalities," but again such are merely labels for collections of behavioral observations. A personality, as such, doesn't exist as a tangible phenomenon.
9. Psychologists reliably bristle at such a view. There's also the notion that neuroscience will reveal some coherent correlates of these concepts, even as Freud had anticipated, or how they actually "emerge." To call this unlikely is an understatement, as I noted in Chapter 5.
10. Mlodinow's reference is certainly from Born's *Natural Philosophy of Cause and Chance.* (Oxford: Clarendon Press, 1948).
11. Some psychologists might argue that this, technically, reflected habituation rather than a strict operant or respondent conditioning process; but that distinction is not important for our illustration here. It is clear that there was some associative responding by the plants.
12. I noted that there are now what can only be described as "primitive" efforts to assess the neurological correlates of "reported" experience. Despite their value in advancing neuroscience, these do *not* constitute measurement of subjective events, only the gross physiology that seems to be associated with them.
13. The citation of Raymond Williams, a twentieth-century cultural scholar, is from an old edition, but the text is unchanged in the later one. (548)
14. We probably will never know, but one can't rule out that this dynamic underlies how language evolved in *Homo* species in the first place. Primitive humans, and earlier members of our genus, did not likely begin communicating with complex subjective ideas. These undoubtedly arose from using sounds, and later some combinations of phonemes, to communicate actions of the speaker, other individuals' behaviors, and other objects and events in the environment. As our brains developed and more language capabilities evolved, we likely began to condense complex communications summarizing histories of events into more efficient language, which later merely became "subjective thoughts."
15. We should point out that ethical criticisms of this experiment have been frequently raised; and they may have been justified, as the child was not deconditioned from this fear. However, circumstances actually prevented the deconditioning from occurring. Albert (not his real name) died at an early age; and contrary to speculations actually from congenital neurological problems, which had nothing to do with the experiment.
16. Wilhelm Wundt, the German physiologist, often considered by some, along with William James, as the founder of modern psychology.
17. For the interested reader, Skinner best describes the principles and implications for the lay audience in his books *Science and Human Behavior* (482) and *About Behaviorism*. (485) He further details the behavioral approach to the realms of education, government, social issues, and self-management with his various papers in *Upon Further Reflection.* (486)
18. It's consistent for all fields of science. We formulate hypotheses based on observations of events and then apply questions to future events to see if the

predictions hold up. That theories are built up from such does not suggest that they have palpable qualities themselves. The theory of gravity or general relativity doesn't exist as such. It is only the label we use in place of the equations which describe the phenomenon, based on past observations. That is not the same as what I noted as "psychological theories" above. Those attach some sort of reality or palpable construct to psychological phenomena, which has always been unjustified. Behaviorism deals only with what can be observed and empirically predicted.

19. There are countless instances in which reinforcement coincides with behavior, e.g., when verbal behavior and others' attention coincide, though such is enabled by the history of attention following behavior; but that is insignificant for our purposes here.
20. There is the erroneous common notion that reinforcement or reward is a "bribe." Reinforcers appear after the relevant behavior. A bribe is generally given before.
21. Details are available at https://clergyproject.org/.
22. You'll notice again the problem of using "theory" here. If there exists something which is not measurable or potentially operational in some way, i.e., "free will," such would exist outside the bounds of science and cannot be considered in any "theory" of human behavior.
23. I would point out that the only peer-reviewed studies on the experimental conditioning of superstitious behavior in humans include a few meaningless computer-driven simulations. To artificially condition any real-world superstitious behavior in humans may be of questionable ethics.
24. There are still controversial genetic findings and a great deal we don't know about early linguistic differences among cultures in different parts of the world. These differences won't be reviewed here, but they are considerable. (238) Various cultures do seem to have "invented" language separately in different parts of the world. (350)
25. I would also point out that Chomsky's criticism of Skinner's behavioral language learning has not held up to any reasonable scrutiny. (389)
26. There is evidence that some species of birds pass on specific guidance to their offspring; but parental guidance is mostly a mammalian trait.
27. As I've noted, play extends well into the animal kingdom. The young of other species, principally mammals, especially social mammals, even some bees, engage in play, which enables familiarity and mastery of their environments.
28. Though the term is widely used now, it is still a somewhat controversial concept. Such types of aberrations have been continually revised and relabeled by the group which writes a psychiatric nomenclature for these types of things, published in the *Diagnostic and Statistical Manual of Mental Disorders* (now DSM-5-TR). Contrary to some popular opinion, despite any behaviors which would characterize persons so identified, there is no requirement for impaired general intelligence to be so "diagnosed"; and many persons who carry that label do manifest average, and sometimes well above average, intelligence, as well as occasional outstanding abilities. In fact, people so categorized have often been more appropriately construed as merely exemplifying the idea of "neurodiversity," as there is no one, unified set of symptoms or behaviors which characterize individuals who have been clinically labeled with this "diagnosis."
29. Dr. Collins was recently diagnosed with an aggressive form of prostate cancer. Whether or not he's invoked any religious reasoning about this, he has apparently responded well to treatment, and we absolutely can only wish him well.

Chapter 11

1. Lest you think this is a purely American problem, I'll note that many theocracies around the world eschew much science education in favor of religious indoctrination. And it's not just in the Islamic world, as some might suspect. Even in Israel, a young democracy, a significant percentage of students—those in orthodox publicly funded schools—often receive no education in math or science. (85)

2. I am not in a position to assess anything about information from what are now considered as "influencers" on social media. Nor do I have an opinion on the veracity or worth of the information so dispensed. Nevertheless, that significant modeling occurs in this context would be hard to deny.
3. Genetic analyses have revealed that in all likelihood, there was a speciation event some 765 to 550 thousand years ago which led to the emergence of the probably "last common ancestor" of Neanderthals, Denisovans, and *H. Sapiens.* (226)
4. One might argue that some human cohorts have modified their environments to the benefit of much larger groups and cultures. However, those modifications are of no benefit to other species and are generally destructive for them. The environmental modifications effected by other creatures—dams, nests, termite mounds, etc.—are short-lived, are sometimes used by other creatures, and contribute to environmental stability.
5. Although Adams has offered that the choice of 42 was merely impulsive and arbitrary, the number 42 does have a number of provocative historical uses and unique mathematical properties (118); but there's no evidence that Deep Thought or Douglas Adams himself considered such.

REFERENCES

A

1. Abington School District v. Schempp, 374 U.S. 203 (1963). Retrieved September 30, 2024, from https://caselaw.findlaw.com/court/us-supreme-court/374/203.html
2. Adams, D. (1979). *The Hitchhiker's Guide to the Galaxy.* New York: Harmony Books.
3. Adams, J. (1787). *A Defense of the Constitutions of Government of the United States of America (1787-1788).* Vol. 1. London. Retrieved February 27, 2024, from https://books.googleusercontent.com/books/content
4. Adolphs, R., & Anderson, D. J. (2018). *The neuroscience of emotion: A new synthesis.* Princeton, NJ: Princeton University Press.
5. Ahmed, Q. A. (2019). Legal child abuse must end. *Scientific American, 321(6),* 10. doi: 10.1038/scientificamerican1219-10
6. Altschul, D. M., et al., (2018). Personality links with lifespan in chimpanzees. *eLife,* 7:e33781. doi: 10.7554/eLife.33781
7. American Atheists. (2019, August 27). More people are leaving religion. Retrieved April 5, 2021, from https://www.atheists.org/
8. American Atheists. (2022, January 17). What is atheism? Retrieved April 1, 2022, from https://www.atheists.org/activism/resources/about-atheism/
9. Amoateng, A. Y, & Bahr, S. J. (1986). Religion, family, and adolescent drug use. *Sociological Perspectives, 29(1),* 53–76. doi: 10.2307/1388942
10. Amsterdam, B. (1972). Mirror self-image reactions before age two. *Developmental Psychobiology, 5,* 297–305. doi: 10.1002/dev.420050403
11. Ananthaswamy, A. (2020). Do we live in a simulation? Chances are about 50–50. *Scientific American, 323(6).* Retrieved from https://www.scientificamerican.com/article/do-we-live-in-a-simulation-chances-are-about-50-50/
12. Andreatta, D. (2024, June14). Physicists confirm quantum entanglement persists between top quarks, the heaviest known fundamental particles. *Phys-Org.* Retrieved June 22, 2024, from https://phys.org/news/2024-06-physicists-quantum-entanglement-persists-quarks.html
13. Appenzeller, T. (2018). Europe's first artists were Neanderthals. *Science, 359(6378),* 852-863. doi: 10.1126/science.359.6378.852
14. Arndt, M., et al. (1999). Wave–particle duality of C_{60} molecules. *Nature, 401,* 680-682. doi:10.1038/44348
15. Asimov, I. (1950). *I robot.* New York: Gnome Press.
16. Asser, S. M., & Swan, R. (1998). Child fatalities from religion-motivated medical neglect. *Pediatrics, 101(4),* 625–629. doi: 10.1542/peds.101.4.625
17. Atkinson, Q. D., & Bourrat, P. (2010). Beliefs about God, the afterlife and morality support the role of supernatural policing in human cooperation. *Evolution and Human Behavior, 32(1),* 41–49. doi:10.1016/j.evolhumbehav.2010.07.008
18. Aristotle. (2002). *Aristotle's Nicomachean Ethics: A New Translation* (Translation by R. C. Bartlett & S. D. Collins,). Chicago: University of Chicago Press.
19. Atran, S. (2002). *In gods we trust: The evolutionary landscape of religion.* New York: Oxford University Press.
20. Avery-Natale, E. (2009). The god metaphor. *Journal of Social, Evolutionary, and Cultural Psychology, 3(4),* 402-415. doi: 10.1037/h0099307
21. Avalon Project (The). (2008). *The Barbary Treaties 1786-1816, Treaty of Peace and Friendship.* Signed at Tripoli November 4, 1796. Yale Law School. Retrieved March 22, 2020, from https://avalon.law.yale.edu/18th_century/bar1796t.asp
22. Aymer, M., & Nasrallah, L. (2018, June 15). What Jeff Sessions got wrong when quoting the Bible. *The Washington Post.* https://www.washingtonpost.com/news/acts-of-faith/wp/2018/06/15/what-jeff-sessions-got-wrong-when-quoting-the-bible/?noredirect=on&utm_term=.83e9a3aefa1b

23. Azevedo, F. A., et al. (2009). Equal numbers of neuronal and nonneuronal cells make the human brain an isometrically scaled-up primate brain. *Journal of Comparative Neurology, 513(5),* 532–541. doi: 10.1002/cne.21974

B

24. Baars, B. J., & Edelman, D. B. (2012). Consciousness, biology and quantum hypotheses. *Physics of Life Reviews, 9(3)*, 285-294. doi: 10.1016/j.plrev.2012.07.001
25. Bae, C. J., et al. (2017). On the origin of modern humans: Asian perspectives. *Science, 358,* eaai9067. doi: 10.1126/science.aai9067
26. Bahr, A. (2003-4). A physicist's critique of the existence of a god. *American Atheist, 42(1),* 54-56. https://www.scribd.com/document/257434551/American-Atheist-Magazine-Winter-2003-2004
27. Baldri, I. (2017, May 31). Could cold spot in the sky be a bruise from a collision with a parallel universe? *The Conversation.* Retrieved February 10, 2025, from https://theconversation.com/could-cold-spot-in-the-sky-be-a-bruise-from-a-collision-with-a-parallel-universe-78563
28. Balogun, S. K., & Aruoture, E. (2024). Cultural homogenization vs. cultural diversity: Social media's double-edged sword in the age of globalization. *African Journal of Social & Behavioral Sciences, 13(4)*, 1491-1512. Retrieved November 11, 2014, from https://journals.aphriapub.com/index.php/AJSBS/article/view/2722/2515
29. Barquera, R., et al. (2024, June 12). Ancient genomes reveal insights into ritual life at Chichén Itzá. *Nature, 630*, 912–919. doi: 10.1038/s41586-024-07509-7
30. Barr, W. P. (2009, October 11). Remarks to the Law School and the de Nicola Center for Ethics and Culture at the University of Notre Dame, South Bend, IN. *Justice News*, The United States Department of Justice. Retrieved January 10, 2020, from https://www.justice.gov/opa/speech/attorney-general-william-p-barr-delivers-remarks-law-school-and-de-nicola-center-ethics
31. Barash, D. P. (2007). A review of Jay D. Glass, *The power of faith: Mother Nature's gift. Evolutionary Psychology, 5(2),* 383-386. https://cyberleninka.org/article/n/618751/viewer
32. Barlow, P. (2015). The natural history of consciousness, and the question of whether plants are conscious, in relation to the Hameroff-Penrose quantum-physical 'Orch OR' theory of universal consciousness. *Communicative & Integrative Biology 8(4),* e1041696-1. doi: 10.1080/19420889.2015.1041696.
33. Barth, M. (2022, June 19). Neurotic AI becomes too self aware. *Hard Drive*. Retrieved August 2, 2022, from https://hard-drive.net/hd/technology/neurotic-ai-becomes-too-self-aware/
34. Bartkowski, J. P., et al. (2008). Religion and child development: Evidence from the Early Childhood Longitudinal Study. *Social Science Research, 37,* 18–36. doi: 10.1016/j.ssresearch.2007.02.001
35. Baumeister, R. F. (2008). Free will in scientific psychology. *Perspectives on Psychological Science, 3(1),* 14-19. doi: 10.1111/j.1745-6916.2008.00057.x
36. Baumeister, R. F., et al. (2009). Prosocial benefits of feeling free: Disbelief in free will increases aggression and reduces helpfulness. *Personality and Social Psychology Bulletin, 35(2)*, 260–268. doi: 10.1177/0146167208327217
37. Bear, A. (2016, April 28). What neuroscience says about free will. *Scientific American Blog*. Retrieved January 19, 2019, from http://blogs.scientificamerican.com/mind-guest-blog/what-neuroscience-says-about-free-will/
38. Bear, A, & Bloom, P. (2016). A simple task uncovers a postdictive illusion of choice. *Psychological Science, 27(6)*, 914-922. doi: 10.1177/0956797616641943
39. Beauregard, M., & Paquette, V. (2006). Neural correlates of a mystical experience in Carmelite nuns. *Neuroscience Letters, 405,* 186–190. doi:10.1016/j.neulet.2006.06.060
40. Becker, A. (2018). *What is real? The unfinished quest for the meaning of quantum physics*. New York: Basic Books.
41. Becker, S., et al. (2019). Unified prebiotically plausible synthesis of pyrimidine and

purine RNA ribonucleotides. *Science, 366(6461),* 76-82. doi: 10.1126/science.aax2747
42. Beliles, M. A., & Newcombe, J. (2014). *Doubting Thomas? The religious life and legacy of Thomas Jefferson.* New York: Morgan James Publishing.
43. Bellier, L., et al. (2023). Music can be reconstructed from human auditory cortex activity using nonlinear decoding models. *PLOS Biology, 21(8)*: e3002176. doi: 10.1371/journal.pbio.3002176
44. Bendavid, N. (2015, January 2). Europe's empty churches go on sale. *The Wall Street Journal.* https://www.wsj.com/articles/europes-empty-churches-go-on-sale-1420245359
45. Bengio, Y. (2016). Machines who learn. *Scientific American, 314(6),* 46-51. doi:10.1038/scientificamerican0616-46
46. Bennett, C. M., et al. (2010). Neural correlates of interspecies perspective taking in the post-mortem Atlantic salmon: An argument for proper multiple comparisons correction. *Journal of Serendipitous and Unexpected Results, 1(1)*, 1-5. Retrieved January 5, 2019, from https://teenspecies.github.io/pdfs/NeuralCorrelates.pdf
47. Benson, H., et al. (2006). Study of the therapeutic effects of intercessory prayer (STEP) in cardiac bypass patients: A multicenter randomized trial of uncertainty and certainty of receiving intercessory prayer. *American Heart Journal, 151(4),* 934-942. doi: 10.1016/j.ahj.2005.05.028
48. Berkessel, J. B., et al. (2021). National religiosity eases the psychological burden of poverty. *PNAS, 118(39),* e2103913118. doi: 10.1073/pnas.2103913118
49. Bergan, A., & McConatha, J. T. (2001). Religiosity and life satisfaction. *Activities, Adaptation & Aging, 24(3),* 23-34. doi: 10.1300/J016v24n03_02
50. Bjork, D. W. (1993). *B. F. Skinner: A life*. New York: Basic Books.
51. Bielecki, J., et al., (2023, October 9). Associative learning in the box jellyfish *Tripedalia cystophora. Current Biology, 33(19),* 4150-4159.e5. doi: 10.1016/j.cub.2023.08.056
52. Blundell, S. J. (2016). Emergence, causation and storytelling: Condensed matter physics and the limitations of the human mind. University of Oxford, Dept. of Physics. Retrieved August 1, 2018, from https://arxiv.org/ftp/arxiv/papers/1604/1604.06845.pdf.
53. Boehm, C. (2010). *Moral Origins: The evolution of virtue, altruism, and shame*. New York: Basic Books.
54. Bourgeon, L., et al. (2017, January 6). Earliest human presence in North America dated to the last glacial maximum: New radiocarbon dates from Bluefish Caves, Canada. *PLOS ONE.* doi: 10.1371/journal.pone.0169486.
55. Bourke, E. (2018). Catholic Church national wealth estimated to be $30 billion, investigation finds. *ABC News Online*. https://www.abc.net.au/news/2018-02-12/catholic-church-worth-$30-billion-investigation-finds/9422246
56. Bowater, D. (2015, June 6). Don't be a chicken: Be a lark. *Pressreader*, https://www.pressreader.com/canada/saskatoon-starphoenix/20150606/282789240054618
57. Bradshaw, M., & Ellison, C. G. (2008). Do genetic factors influence religious life? Findings from a behavior genetic analysis of twin siblings. *Journal for the Scientific Study of Religion, 7(4),* 529-544. doi.org/10.1111/j.1468-5906.2008.00425.x
58. Brainard, J. (2018) Scrolls revealed as forged. *Science, 362(6414),* 504. Retrieved December 1, 2020, from https://science.sciencemag.org/content/sci/362/6414/504.full.pdf
59. Brefczynski-Lewis, J. A., et al. (2007). Neural correlates of attentional expertise in long-term meditation practitioners. *Proceedings of the National Academy of Sciences, 104*(27), 11483-11488. doi: 10.1073_pnas.0606552104
60. Brembs, B. (2011). Towards a scientific concept of free will as a biological trait: Spontaneous actions and decision-making in invertebrates. *Proceedings of the Royal Society B, 278*, 930–939. doi:10.1098/rspb.2010.2325
61. Brenan, N. (2019, July 26). 40% of Americans believe in creationism. *Gallup Survey*. https://news.gallup.com/poll/261680/americans-believe-creationism.aspx

62. Brennand, E. (2011, May 17). Tardigrades: Water bears in space. *Nature Features.* Retrieved October 1, 2019, from https://www.bbc.co.uk/nature/12855775
63. Brent, L. J. N, et al. (2015), Ecological knowledge, leadership, and the evolution of menopause in killer whales. *Current Biology, 25,* 746–750. doi: 10.1016/j.cub.2015.01.037
64. Brunet, M., et al. (2002). A new hominid from the Upper Miocene of Chad, Central Africa. *Nature, 418*, 145 - 151. doi: 10.1038/nature00879
65. Buckley, C., & Meyers, S. L. (2019, Aug. 9). China said it closed Muslim detention camps. There's reason to doubt that. *The New York Times.* https://www.nytimes.com/2019/08/09/world/asia/china-xinjiang-muslim-detention.html
66. Bufford, R. K. (1981). *The human reflex: Behavioral psychology in Biblical perspective.* San Francisco: Harper & Row.
67. Burkert, W. (1996). *Creation of the sacred: Tracks of biology in early religions.* Cambridge, MA: Harvard University Press.
68. Butler, S. (1912). *The note-books of Samuel Butler* (H. F. Jones, Ed.). London: A. C. Fifield. https://www.gutenberg.org/files/6173/6173-h/6173-h.htm

C

70. Callaway, E. (2017, June 7-8). Oldest *Homo sapiens* fossil claim rewrites our species' history. *Nature News*. doi: 10.1038/nature.2017.22114
71. Callaway, E. (2022, August 24). Seven-million-year-old femur suggests ancient human relative walked upright. *Nature News.* doi: 10.1038/d41586-022-02313-7
72. Calvo, P. (2023). *Planta sapiens: The new science of plant intelligence*. New York: W. W. Norton.
73. Cammaerts, M., & Cammaerts, R. (2015), Are ants (hymenoptera, formicidae) capable of self recognition? *Journal of Science, 5(7),* 521-532. Retrieved February 10, 2020, from https://www.journalofscience.net/showpdf/MjY4a2FsYWkxNDc4NTIzNjk=
74. Camus, A. (1955). *The myth of Sisyphus, and other essays.* (Originally published 1942). New York: Alfred A. Knopf.
75. Caplan, L., et al. (2014). Religiosity after a diagnosis of cancer among older adults. *Journal of Religion, Spiritual & Aging, 26(4),* 357–369. doi: 10.1080/15528030.2014.928922
76. Carlshausen, J. (2018). *Religion delusion.* Calwell, Australia: Australia Self-Publishing Group. IBSN: 978-0-648-2934-5-3.
77. Carneiro, V. C., & Lyko, F. (2020). Rapid epigenetic adaptation in animals and its role in invasiveness. *Integrative and Comparative Biology*, 60(2), 267–274. doi: 10.1093/icb/icaa023
78. Carroll, S. (2016). *The big picture.* New York: Dutton.
79. Carroll, S. (2019). *Something deeply hidden.* New York: Penguin Random House.
80. Caspar, E. A., et al. (2017, January 17). The influence of (dis)belief in free will on immoral behavior. *Frontiers in Psychology, 8,* article 20. doi:10.3389/fpsyg.2017.00020
81. Castelvecchi, D. (2012). Magnetic sense shows many animals the way to go. *Scientific American, 306(1),* 49-53. https://www.scientificamerican.com/article/the-compass-within/#
82. Cellan-Jones, R. (2014, December). Stephen Hawking warns artificial intelligence could end mankind. *BBC News.* https://www.bbc.com/news/technology-30290540
83. Centola, D., et al. (2018). Experimental evidence for tipping points in social convention. *Science, 360,* 1116–1119. doi: 0.1126/science.aas8827
84. Chen, B., et al. (2022). Fully body visual self-modeling of robot morphologies. *Science Robotics, 7(68).* doi: 10.1126/scirobotics.abn1944
85. Chabin, M. (2023). Israel's science-free schools for ultra-Orthodox Jews draw fire. *Science, 381 (6653)*, 17. doi: 10.1126/science.adj5606

86. Chalmers, D. (2014, March). How do you explain consciousness? (Text of *Ted Talk*). Retrieved September 9, 2018, from https://dorshon.com/wp-content/uploads/2015/06/how-do-you-explain-consciousness.pdf.
87. Chijiiwa, H., et al. (2015). Dogs avoid people who behave negatively to their owner: Third-party affective evaluation. *Animal Behaviour*, *106*, 123-127. Retrieved April 18, 2024, from https://repository.kulib.kyoto-u.ac.jp/dspace/bitstream/2433/198493/1/j.anbehav.2015.05.018.pdf.
88. Chirchir, H., et al., (2014). Recent origin of low trabecular bone density in modern humans. *PNAS, 112(2)*, 366-371. doi: 10.1073/pnas.1411696112
89. Clark, C. J., et al. (2014). Free to punish: A motivated account of free will belief. *Journal of Personality and Social Psychology, 106*(4), 501-513. doi: 10.1037/a0035880.supp
90. Clarke, A. C. (1973). Hazards of prophecy: The failure of imagination. (Originally published in 1961). In *Profiles of the future: An enquiry into the limits of the possible*. New York: Macmillan.
91. Clegg, L. F. (2012). Protean free will. Unpublished. Pasadena, CA: California Institute of Technology. Retrieved July 12, 2020, from https://authors.library.caltech.edu/29887/
92. Cobey, K. D., et al. (2024). Biomedical researchers' perspectives on the reproducibility of research. *PLOS Biology, 22(11)*: e3002870. doi: 10.1371/journal.pbio.3002870
93. Conard, N. J., et al., (2009). New flutes document the earliest musical tradition in southwestern Germany. *Nature, 460,* 737–740. doi: 10.1038/nature08169
94. Cook, G. (2011, November 15). Neuroscience challenges old ideas about free will. *Scientific American Blog*. Retrieved December 10, 2020, from https://www.scientificamerican.com/article/free-will-and-the-brain-michael-gazzaniga-interview/
95. Cooperman, A., et al. (2013, December 30). Public's view on human evolution. *Pew Research Center*. Retrieved January 10, 2020, from http://www.pewresearch.org/wp-content/uploads/sites/7/2013/12/Evolution-12-30.pdf.
96. Cooperman, A., et al. (2015, November 3). *Pew Research Center.* U. S. public becoming less religious. Retrieved February 10, 2018, from https://www.pewforum.org/wp-content/uploads/sites/7/2015/11/201.11.03_RLS_II_full_report.pdf.
97. Cooperman, A., et al. (2021, May 11). Jewish Americans in 2020. *Pew Research Center.* Retrieved September 12, 2024, from https://www.pewresearch.org/religion/2021/05/11/jewish-americans-in-2020/
98. Costafreda, S. G., et al. (2008). Predictors of amygdala activation during the processing of emotional stimuli: A meta-analysis of 385 PET and fMRI studies. *Brain Research Reviews, 58(1),* 57-70. doi:10.1016/j.brainresrev.2007.10.012
99. Coyne, J. A. (2009). *Why evolution is true.* Oxford, UK: Oxford University Press.
100. Coyne, J. A. (2015). *Faith vs. fact: Why science and religion are incompatible*. New York: Penguin Books.
101. Crabtree, S. (2021, August 31). Religiosity highest in world's poorest nations. *Gallup Survey*. Retrieved May 3, 2024, from https://news.gallup.com/poll/142727/religiosity-highest-world-poorest-nations.aspx
102. Crawford, C., & Galdikas, B. M. F. (1986). Rape in non-human animals: An evolutionary perspective. *Canadian Psychology, 27*(3), 215–230. doi: 10.1037/h0084335
103. Cutting, G. (2014, May 15). The case for 'soft atheism.' *The New York Times*. https://archive.nytimes.com/opinionator.blogs.nytimes.com/2014/05/15/the-case-for-soft-atheism/.

D

104. Dahl, R. (2006). *The gremlins*. (Originally published in 1943). Milwaukee, OR: Dark Horse Books.

105. Darby, R., et al. (2018). Lesion network localization of free will. *Proceedings of the National Academy of Sciences, 115(42),*10792-10797. doi: 10.1073/pnas.1814117115
106. Darwin, C. (1874). *The descent of man and selection in relation to sex.* New York: Rand, McNally and Company.
107. Davidson, R. J., & Lutz, A. (2008). Buddha's brain: Neuroplasticity and meditation. *IEEE Signal Process Magazine, 25(1)*, 176-174. Retrieved May 14, 2020, from https://pmc.ncbi.nlm.nih.gov/articles/PMC2944261/
108. Davies, P. (1983). *God and the new physics*. New York: Simon & Schuster.
109. Davies, P. (2004). Undermining free will. *Foreign Policy, 144,* 36-38. doi: 10.2307/4152977
110. Davis, L. G., et al. (2019). Late upper Paleolithic occupation at Cooper's Ferry, Idaho, USA, ~ 16,000 years ago. *Science, 365(6456),* 891-897. doi: 10.1126/science.aax9830.
111. Dawkins, R. (1976). *The selfish gene*. Oxford: Oxford University Press.
112. Dawkins, R. (1996). *The blind watchmaker*. New York: W. W. Norton.
113. Dawkins, R. (2006). *The god delusion.* New York, Boston: Houghton Mifflin.
114. Dawkins, R. (2009). *The greatest show on earth.* New York: Free Press.
115. Dawkins, R. (2011). *The magic of reality: How we know what's really true.* New York: Free Press.
116. Dawkins, R. (2019). *Outgrowing God: A beginner's guide*. New York: Random House.
117. Dehaene, S., et al. (2017). What is consciousness, and could machines have it? *Science, 358,* 486–492. doi: 10.1126/science.aan8871
118. Delahaye, J.-P. (2020, September 11). For math fans: A hitchhiker's guide to the number 42. *Scientific American*. Retrieved March 13, 2024, from https://www.scientificamerican.com/article/for-math-fans-a-hitchhikers-guide-to-the-number-42/
119. d'Huy, J. (2016). The evolution of myths. *Scientific American, 315(6)*, 62-69. Retrieved May 20, 2020, from https://www.scientificamerican.com/article/scientists-trace-society-rsquo-s-myths-to-primordial-origins/
120. Dennett, D. (1991). *Consciousness explained.* Boston: Little, Brown and Co.
121. Dennett, D. (2003a). *Freedom evolves*. New York: Penguin.
122. Dennett, D. (2003b, July 12). The bright stuff. *The New York Times, Opinion Section.* http://www.nytimes.com/2003/07/12/opinion/the-bright-stuff.html.
123. Dennett, D. (2006). *Breaking the spell: Religion as a natural phenomenon.* New York: Viking.
124. de Roode, J. C., et al. (2013). Self-medication in animals. *Science, 340,* 150-151, doi: 10.1126/science.1235824
125. Dervic, K., et al. (2006). Protective factors against suicidal behavior in depressed adults reporting childhood abuse. *The Journal of Nervous and Mental Disease, 194(12),* 971–974. doi: 10.1097/01.nmd.0000243764.56192.9c
126. Desmurget, M., et al. (2009). Movement intention after parietal vortex stimulation in humans. *Science, 324 (5928),* 811-813. doi: 10.1126/science.1169896
127. DeScioli, P. (2023). On the origin of laws by natural selection. *Evolution and Human Behavior, 44(3),* 195-209. doi: 10.1016/j.evolhumbehav.2023.01.004
128. de Waal, F. B. M. (1996). *Good natured: The origins of right and wrong in humans and other animals.* Cambridge, MA: Harvard University Press.
129. de Waal, F. (2013). *The bonobo and the atheist: In search of humanism among the primates.* New York: W. W. Norton.
130. de Waal, F. (2014). One for all. *Scientific American, 311(3),* 69-71. doi:10.1038/scientificamerican0914-68
131. de Waal, F. (2016). *Are we smart enough to know how smart animals are?* London: Granta Books.
132. Diamant, J. (2021, May 13). Jews in U.S. are far less religious than Christians and Americans overall, at least by traditional measures. *Pew Research Center*. Retrieved September 27, 2024, from https://www.pewresearch.org/short-reads/2021/05/13/jews-in-u-s-are-far-less-religious-than-christians-and-americans-overall-at-least-by-traditional-measures/

133. Diamant, J. (2023, June 21). One-in-six Americans have taken steps to see less of someone on social media due to religious content. *Pew Research Center*. Retrieved October 26, 2024, from https://www.pewresearch.org/short-reads/2023/06/21/one-in-six-americans-have-taken-steps-to-see-less-of-someone-on-social-media-due-to-religious-content/
134. Dick, P. K. (2002). Adjustment team. In *Selected stories of Philip K. Dick* (pp. 133-156). New York: Houghton Mifflin Harcourt.
135. D'Onofrio, B. M., et al. (1999). Adolescent religiousness and its influence on substance use: Preliminary findings from the Mid-Atlantic School Age Twin Study. *Twin Research and Human Genetics, 2(2),* 156-168. doi:10.1375/twin.2.2.156
136. Dirks, P., et al. (2015). Geological and taphonomic context for the new hominin species *Homo naledi* from the Dinaledi Chamber, South Africa. *eLife, 4*:e09561. doi: 10.7554/eLife.09561
137. Dona., H. S. G., et al. (2022). Do bumble bees play? *Animal Behavior*, *194*, 239-251. doi: 10.1016/j.anbehav.2022.08.013
138. Dorkenwald, S., et al. (2024). Neuronal wiring diagram of an adult brain. *Nature,* 634, 124–138. doi: 10.1038/s41586-024-07558-y
139. Drews, C., & Han, W. (2010) Dynamics of wind setdown at Suez and the eastern Nile delta. *PLOS ONE 5(8):* e12481. doi: 10.1371/journal.pone.0012481
140. Drummond, H. (2007). *The ascent of man.* (Originally published 1894). New York: Cosimo, Inc.
141. Duan, Y., et al. (2023). *DeWave: Discrete EEG waves encoding for brain dynamics to text translation. arXiv*:2309.14030. doi: 10.48550/arXiv.2309.14030
142. Dufour, et al., (2018). Life history and origin of the camelids provisioning a mass killing sacrifice during the Chimú period: Insight from stable isotopes. *Environmental Archaeology,* doi: 10.1080/14614103.2018.1498165
143. Dugatkin, L.A. (2018). The silver fox domestication experiment. *Evolution: Education and Outreach, 11(*16) doi: 10.1186/s12052-018-0090-x
144. Dukas, H., & Hoffman, B. (Eds.). (1979). *Albert Einstein: The human side, glimpses from his archives.* Princeton, NJ: Princeton University Press.

E

145. Eaves, L., et al. (1999). Transmission of religion and attitudes. *Twin Research, 2*(2), 59-61. doi:10.1375/twin.2.2.59
146. Edwards v. Aguillard, 482 U.S. 578 (1987). https://supreme.justia.com/cases/federal/us/482/578/
147. Ehrman, B. (2013). Heaven and Hell: When was Heaven and Hell invented? *The Bart Ehrman Blog*. Retrieved January 15, 2024, from https://ehrmanblog.org/the-invention-of-heaven-and-hell-for-members/
148. Ehrman, B. (2022). Why date the Gospels after 70 CE? *The Bart Ehrman Blog*. Retrieved February 5, 2024, from https://ehrmanblog.org/why-date-the-gospels-after-70-ce/
149. Einstein, A. (1901). Letter written July 8 to Jost Winteler. Princeton, NJ: Princeton University Press. Retrieved March 29, 2023, from https://einsteinpapers.press.princeton.edu/vol1-trans/198
150. Einstein, A. (1950). *Out of my later years*. (Essays originally published in *The world as I see it,* 1935). New York: The Philosophical Library, Inc.
151. Einstein, A. (1954). Letter to Eric Gutkind, March 1 (The "God Letter"). Retrieved August 1, 2020, from https://www.christies.com/lotfinder/Lot/einstein-albert-1879-1955-autograph-letter-signed-a-6167096-details.aspx.
152. Einstein, A. (1982). *Ideas and opinions* (Originally published 1954). New York: Crown Publishing Group.
153. Elgammal, A., et al. (2017). CAN: Creative adversarial networks generating "art" by learning about styles and deviating from style norms. Cornell University Library. Retrieved August 10, 2019, from https://arxiv.org/abs/1706.07068.

154. El-Islam, M. F. (2015). Religion and mental health. *The Arab Journal of Psychiatry, 25(1),* 1-3. doi: 10.12816/0010500
155. Elison, S., & Chou, H. G. (2014). Impact of birth order on religious behaviors among college students raised by highly religious Mormon parents, *Archive for the Psychology of Religion, 36(1)*, 105-117. doi: 10.1163/15736121-12341275
156. Emamalipour, et al. (2020, May 19). Horizontal gene transfer: From evolutionary flexibility to disease progression. *Frontiers in Cell and Developmental Biology, 8,* article 229. doi: 10.3389/fcell.2020.00229
157. Everett, D. (2010). The Pirahã: people who define happiness without God. *Freethought Today, 27(3).* Retrieved October 12, 2023, from https://ffrf.org/publications/freethought-today/item/13492-the-pirahae-people-who-define-happiness-without-god%3E

F

158. Feldman, G. et al. (2014). Free will is about choosing: The link between choice and the belief in free will. *Journal of Experimental Social Psychology, 55,* 239-245. doi: 10.1016/j.jesp.2014.07.012
159. Fernald, R. D. (2006). Casting a genetic light on the evolution of eyes. *Science,* 313, 1914-1918. doi: 10.1126/science.1127889
160. Ferris, J. (2012, October 2). The connectome debate: Is mapping the mind of a worm worth it? *Scientific American.* https://www.scientificamerican.com/article/c-elegans-connectome/
161. Festinger, L. (1962). Cognitive dissonance. *Scientific American, 207(4), 93–102.* doi:10.1038/scientificamerican1062-93.
162. Feynman, R. P. (1965). *The character of physical law*. Cambridge, MA: The MIT Press.
163. Feynman, R. P. (1985). Cargo cult science. In E. Hutchings (Ed.), *Surely you're joking, Mr. Feynman* (pp. 338-346). New York: W. W. Norton.
164. Feynman, R. P. (1988). *QED: The strange theory of light and matter*. Princeton: Princeton University Press.
165. Finkel, E. (2023). Researchers propose test for AI sentience, *Science*, *381(6660)*, 822-823. doi: 10.1126/science.adk4479
166. Fischer, L. (2021). How slime molds remember where they ate. *Scientific American, 324(6)*, 19. doi:10.1038/scientificamerican0621-19
167. Fitzgerald, D. (2017). *Jesus, mything in action: The complete heretics guide to western religion.* Amazon: CreateSpace Independent Publishing Platform. ISBN: 1542858887, 9781542858885
168. Flynn, T. (Ed.) (2007). *The new encyclopedia of unbelief.* New York: Prometheus Books.
169. Fox, D. (2018). The brain, reimagined. *Scientific American, 318(4),* 60-67. doi:10.1038/scientificamerican0418-60
170. Franklin, B. (1780, October 9). Letter to Richard Price. *Founders Online, The National Archives.* Retrieved September 1, 2022, from https://founders.archives.gov/documents/Franklin/01-33-02-0330
171. Frégnac, Y. (2017). Big data and the industrialization of neuroscience: A safe roadmap for understanding the brain? *Science, 358,* 470–477. doi: 10.1126/science.aan8866
172. Frenkel, E. (2015, February 23). The reality of quantum weirdness. *The New York Times, Sunday Review*. https://www.nytimes.com/2015/02/22/opinion/sunday/the-reality-of-quantum-weir.
173. Freud, S. (1913). *Totem and taboo: Resemblances between the psychic lives of savages and neurotics.* Australia: Emereo Publishing.
174. Freud, S. (1927). *The future of an illusion.* Toronto, Canada: Broadview Editions, 2012.
175. Fried, I., et al. (2011). Internally generated preactivation of single neurons in human medial frontal cortex predicts volition. *Neuron 69(3),* 548–562. doi: 10.1016/j.neuron.2010.11.045

G

176. Gagliano, M., et al. (2014). Experience teaches plants to learn faster and forget slower in environments where it matters. *Oecologia, 175,* 63–72. doi: 10.1007/s00442-013-2873-7
177. Gagliano, M., et al. (2016, December 2). Learning by association in plants. *Scientific Reports, 6:38427*. doi: 10.1038/srep38427
178. Garisto, D. (2023). The universe is not locally real, and the physics Nobel Prize winners proved it. *Scientific American, 328(1),* 48-53. doi: 10.1038/scientificamerican0123-48
179. Gervais, W. M., & Najle, M. B. (2018). How many atheists are there? *Social Psychological and Personality Science, 9(1),* 3-10. doi: 10.1177/1948550617707015
180. Gibbons, A. (2004). How we tamed ourselves—and became modern. *Science, 346(6208)*, 405-406. doi: 10.1126/science.346.6208.405
181. Gibbons, A. (2012). The ultimate sacrifice. *Science, 336,* 834-837. doi:10.1126/science.336.6083.834
182. Gibbons, A. (2017). Oldest members of our species discovered in Morocco. *Science, 356(6342)*, 993-994. doi: 10.1126/science.356.6342.993
183. Gibbons, A. (2024). Neanderthals and moderns mingled early and often. *Science, 385(6705)*, 132-133. Retrieved July 18, 2024, from https://www.science.org/doi/epdf/10.1126/science.adr6187
184. Glass, J. D. (2007). *The power of faith: Mother Nature's gift*. Corona del Mar, CA: Donington Press.
185. Goodall, J. (1971). *In the shadow of man.* London: William Collins.
186. Goodall, J. (2010). *Through a window: My thirty years with the chimpanzees of Gombe*. New York: Houghton Mifflin Harcourt.
187. Gould, S. J. (1997). Nonoverlapping magisteria. *Natural History, 106*, 16-22. https://bazhum.muzhp.pl/media/texts/filozoficzne-aspekty-genezy-philosophical-aspects-of-origin/2014-tom-11/filozoficzne_aspekty_genezy_philosophical_aspects_of_origin-r2014-t11-s7-21.pdf
188. Gould, S. J. (1999). *Rocks of ages: Science and religion in the fullness of life*. New York: The Ballantine Publishing Group.
189. Graeber, D., & Wengrow, D. (2021). *The dawn of everything: A new history of humanity*. New York: Farrar, Straus and Giroux.
190. Grayling, A. C. (2013). *The God argument: The case against religion and for humanism.* New York: Bloomsbury.
191. Greene, B. (1999). *The Elegant Universe*. New York: Vintage Books, Random House.
192. Greene, B. (2004). *The fabric of the cosmos*. New York: Vintage Books, Random House.
193. Greene, B. (2011). *The hidden reality: Parallel universes and the deep laws of the cosmos*. New York: Knopf.
194. Grimm, D. (2022). NIH can't deny chimps sanctuary retirement. *Science, 378(6626)*, 1265. doi: 10.1126/science.adg3415
195. Grimm, D. (2015). Dawn of the dog. *Science, 348 (6232),* 274-272. doi: 10.1126/science.348.6232.274

H

196. Hackett, C., et al. (2018a, June 13). How religious commitment varies by country among people of all ages. *Pew Research Center*. Retrieved October 25, 2024, from https://www.pewresearch.org/religion/2018/06/13/how-religious-commitment-varies-by-country-among-people-of-all-ages/).
197. Hackett, C., et al. (2018b, June 13). The age gap in religion around the world. *Pew Research Center*. Retrieved October 14, 2024, from https://www.pewresearch.org/wp-ontent/uploads/sites/20/2018/06/ReligiousCommitment-FULL-WEB.pdf
198. Haggard, P. (2011). Decision time for free will. *Neuron, 69(3),* 404-407. doi: 10.1016/j.neuron.2011.01.028
199. Hammer, M. F., et al. (2011). Genetic evidence for archaic admixture in Africa.

Proceedings of the National Academy of Sciences, 108(37), 15123-15128. doi: 10.1073/pnas.1109300108

200. Hameroff, S. (2012). How quantum brain biology can rescue conscious free will. *Frontiers in Integrative Neuroscience, 6*, 1-17. doi: 10.3389/fnint.2012.00093
201. Hameroff, S., & Chopra, D. (2012). Chapt. 5, The "Quantum soul": A scientific hypothesis. In Moreira-Almeida, A., & Santana Santos, F. (Eds), *Exploring frontiers of the mind-brain relationship: Mindfulness in behavioral health* (pp. 79-93). New York: Springer. doi: 10.1007/978-1-4614-0647-1_5
202. Hantula, D. A. (1991). A simple BASIC program to generate values for variable-interval schedules of reinforcement. *Journal of Applied Behavior Analysis, 24(4)*, 799-801. doi: 10.1901/jaba.1991.24-799
203. Harari, Y. N. (2015). *Sapiens: A brief history of humankind*. London: Harvill Secker.
204. Hare, B. (2017). Survival of the friendliest: *Homo sapiens* evolved via selection for prosociality. *Annual Review of Psychology, 68,* 155–186. doi: 10.1146/annurev-psych-010416-044201
205. Harris, H.M.B., & Hill, C. (2021, January 14). A place for viruses on the tree of life. *Frontiers in Microbiology, 11*. doi: 10.3389/fmicb.2020.604048
206. Harris, S. (2004). *The end of faith: Religion, terror, and the future of reason.* New York, Norton.
207. Harris, S. (2010). *The moral landscape: How science can determine human values.* New York: Free Press.
208. Harris, S. (2012). *Free will*. New York: Free Press.
209. Harris S, et al. (2009). The neural correlates of religious and nonreligious belief. *PLOS ONE, 4(10),* e0007272. doi:10.1371/journal.pone.0007272.
210. Hawking, S. W. (1988). *A brief history of time*. Bantam Books: New York.
211. Hawking, S. (2001). *The universe in a nutshell*. New York: Bantam Books.
212. Hawking, S. (2005). *The theory of everything: The origin and fate of the universe.* Beverly Hills, CA: Phoenix Books.
213. Hawking, S. (2018). *Brief answers to the big questions*. New York: Bantam Books. (Published posthumously).
214. Hawking, S., & Mlodinow, L. (2010). *The grand design*. New York: Bantam Books.
215. Haynes, S. D., et al. (2003). Free will, determinism, and punishment. *Psychological Reports, 93*, 1013-1021. doi: 10.2466/pr0.2003.93.3f.1013
216. Heinrich, B. & Bugnyar, T. (2007), Just how smart are ravens? *Scientific American, 296(4),* 64-71. doi: 10.1038/scientificamerican042007-4HVff2MXWv9yJGkrl5DWIY
217. Heraclitus. (2003). (Translated by B. Haxton). *Fragments.* New York, Penguin.
218. Herculano-Houzel, S. (2009). The human brain in numbers: A linearly scaled-up primate brain. *Frontiers in Human Neuroscience*, 3, article 31. doi: 10.3389/neuro.09.031.2009
219. Herculano-Houzel, S. (2020). Birds do have a brain cortex—and think. *Science, 369(6511),* 1567-1568. doi: 10.1126/science.abe0536
220. Hermann, H. R. (2017). Chapter 12, Dominance in religion. *Dominance and aggression in humans and other animals: The great game of life* (pp. 183-200). London: Academic Press. doi: 10.1016/B978-0-12-805372-0.00012-2
221. Hitchens, C. (2007). *God is not great: How religion poisons everything.* New York: Hachett Book Group.
222. Hoffman, D. L., et al. (2018). Symbolic use of marine shells and mineral pigments by Iberian Neandertals 115,000 years ago. *Science Advances,* 4: eaar5255, 1-6. doi: 10.1126/sciadv.aar5255
223. Holland, C. M., et al. (2023). On-demand entanglement of molecules in a reconfigurable optical tweezer array. *Science, 382*, 1143–1147. doi: 10.1126/science.adf4272
224. Horgan, J. (2023, June). A 25-year-old bet about consciousness has finally been settled. *Scientific American.* Retrieved August 1, 2023, from https://www.scientificamerican.com/article/a-25-year-old-bet-about-consciousness-has-finally-been-settled/

225. Hornsey, M. J., & Fielding, K. S. (2017). Attitude roots and Jiu Jitsu persuasion: Understanding and overcoming the motivated rejection of science. *American Psychologist, 72(5)*, 459–473. doi: 10.1037/a0040437
226. Hu, W. J. et al. (2023). Genomic inference of a severe human bottleneck during the Early to Middle Pleistocene transition. *Science, 381(6661)*, 979-984. doi: 10.1126/science.abq7487
227. Huenemann, C. (2014, March 17). David Hume on the mystery of promises, and falling into a bog. *3 Quarks Daily*. Retrieved January 3, 2020, from https://www.3quarksdaily.com/3quarksdaily/2014/03/david-hume-on-the-mystery-of-promises-and-falling-into-a-bog.html
228. Hume, D. (1988). *Dialogues concerning natural religion* (2nd ed.). (Originally published 1779, posthumously). Cambridge, MA: Hackett Publishing Co.
229. Hume, D. (2004). *An enquiry concerning human understanding.* (Originally published c. 1748). Mineola, NY: Dover Publication.
230. Hume, D. (2006). *A treatise of human nature*. Part II. Of the ideas of space and time, Sect. iii. Of the other qualities of our idea of space and time. (Written c. 1738-40). Adelaide, South Australia: The University of Adelaide Library (eBooks@Adelaide). Retrieved February 12, 2020, from https://philo-labo.fr/fichiers/Hume%20-%20A%20Treatise%20on%20Human%20Nature.pdf
231. Huxley, T. H. (1898). On the hypothesis that animals are automata, and its history. *Collected essays,* Vol. 1. London: Macmillan. Retrieved January 3, 2020, from http://www.sophia-project.org/uploads/1/3/9/5/13955288/huxley_animals.pdf

I

232. Ibbotson, P., & Tomasello, M. (2016). Language in a new key. *Scientific American, 315(5)*, 70-75. doi: 10.1038/scientificamerican1116-70
233. Iglesias, et al. (2019). Cyclic nature of the REM sleep-like state in the cuttlefish *Sepia officinalis*. *Journal of Experimental Biology, 222*, jeb174862. doi:10.1242/jeb.174862
234. Ilardo, et al., (2018). Physiological and genetic adaptations to diving in sea nomads. *Cell, 173*, 569–580. doi: 10.1016/j.cell.2018.03.054
235. Isaacson, W. (2007). *Einstein: His life and universe.* New York: Simon & Schuster.

J

236. Jabr, F. (2012). Know your neurons: What is the ratio of glia to neurons in the brain? *Scientific American Blog.* Retrieved January 5, 2019, from https://blogs.scientificamerican.com/brainwaves/know-your-neurons-what-is-the-ratio-of-glia-to-neurons-in-the-brain/
237. Jackson, J. C., et al. (2018). The faces of God in America: Revealing religious diversity across people and politics. *PLOS ONE, 13(6)*, e0198745. doi: 10.1371/journal.pone.0198745
238. Jacobson, R. (2018). Mother tongue: Genetic evidence fuels debate over a root language's origins. *Scientific American, 318(3),* 13-14. doi:10.1038/scientificamerican0318-12
239. James, W. (1890). *The principles of psychology. New York: Henry Holt & Co.*
240. James, W. (1892). *Psychology: Briefer course. Cambridge, MA: Harvard University Press.*
241. James, W. (1901-2). *The varieties of religious experience: A study in human nature.* (The Gifford Lectures on Natural Religion delivered at Edinburgh in 1901-1902, edited and annotated for the World Wide Web by LeRoy L. Miller). Retrieved January 2, 2019, from https://www.coursehero.com/file/25235135/william-james-varpdf/.
242. James, W. (1904). Does 'consciousness' exist? *The Journal of Philosophy, Psychology and Scientific Methods, 1(18),* 477-491. Retrieved March 20, 2020, from https://www.jstor.org/stable/pdf/2011942.pdf

243. Jacoby, S. (2013a, January 6). The blessings of atheism. *The New York Times, Sunday Review*. https://www.nytimes.com/2013/01/06/opinion/sunday/the-blessings-of-atheism.html.
244. Jacoby, S. (2013b). *The great agnostic: Robert Ingersoll and American freethought.* New Haven: Yale University Press.
245. Jefferson, T. (1784). *The Founders' Constitution,* Vol. 1, Chapter 18, Epilogue: Securing the republic. Document 16, Notes on the State of Virginia, Queries 14 and 19, 146-49, 164–65. Retrieved August 1, 2020, from http://press-pubs.uchicago.edu/founders/documents/v1ch18s16.html
246. Jefferson, T. (1816, January 6). Letter to Charles Yancey. *Founders Online, National Archives.* Retrieved September 30, 2024, from https://founders.archives.gov/documents/Jefferson/03-09-02-0209.
247. Jefferson, T. (1820a, April 13). Letter to William Short. *Library of Congress*. Retrieved June 16, 2020, from https://www.loc.gov/resource/mtj1.051_1224_1227/
248. Jefferson, T. (1820b, April 11). Letter to José Corrêa da Serra, *Founders Online, National Archives*. Retrieved August 1, 2020, from https://founders.archives.gov/documents/Jefferson/98-01-02-1213
249. Jefferson, T. (1823, April 11). Letter to John Adams, *American History.* Retrieved February 27, 2024, from https://www.let.rug.nl/usa/presidents/thomas-jefferson/letters-of-thomas-jefferson/jefl271.php
250. Jenkins, R., & Perez, G. (2010). Looking for life in the multiverse. *Scientific American, 302(1),* 42-49. doi: 10.1038/scientificamerican012010-vIeZ1Npt1VYIMfUjGr5Ww
251. Jernegan, M. W. (1916, April). Slavery and conversion in the American colonies. *The American Historical Review* (American Historical Association), *21(3)*, 504-527. Oxford, UK: Oxford University Press.
252. Johnson, D., & Bering, J. (2006). Hand of God, mind of man: Punishment and cognition in the evolution of cooperation. *Evolutionary Psychology, 4,* 219-233. doi: 10.1177/147470490600400119
253. Jones, J. M. (2021, March 29). U.S. church membership falls below majority for first time. Retrieved August 5, 2024, from https://news.gallup.com/poll/341963/church-membership-falls-below-majority-first-time.aspx
254. Jones, J. M. (2022, June 17). Belief in God in U.S. dips to 81%, a new low. *Gallup Survey.* Retrieved August 31, 2022, from https://news.gallup.com/poll/393737/belief-god-dips-new-low.aspx
255. Jones, N. (2001, July 11). Babies' musical memories formed in womb. *New Scientist.* Retrieved September 20, 2019, from https://www.newscientist.com/article/dn994-babies-musical-memories-formed-in-womb/
256. Joshi, P., et al., (2009). *Role of religiosity in the lives of the low-income population: A comprehensive review of the evidence* (RTI Project Number: 0211824.000). U.S. Department of Health & Human Services, Office of the Assistant Secretary for Planning and Evaluation. https://aspe.hhs.gov/sites/default/files/migrated_legacy_files//43171/report.pdf
257. Juergensmeyer, M. (1998). Christian violence in America. In *The Annals of the American Academy of Political and Social Science*, *558*, 88–100. doi: 10.1177/0002716298558001008
258. Jung, J. H. (2018). Childhood adversity, religion, and change in adult mental health. *Research on Aging, 40(2),* 155–179. doi: 10.1177/0164027516686662

K

259. Kahneman, D. (2011). *Thinking fast and slow*. New York: Farrar, Straus and Girouox.
260. Kaiser, D. (2014, November 16). Is quantum entanglement real? *The New York Times, Sunday Review*. https://www.nytimes.com/2014/11/16/opinion/sunday/is-quantum-entanglement-real.html.
261. Kaku, M. (2014). *The future of the mind.* New York: Doubleday.
262. Kandler, C., & Riemann, R. (2013). Genetic and environmental sources of individual

religiousness: The roles of individual personality traits and perceived environmental religiousness. *Behavior Genetics, 43(4),* 297–313. doi: 10.1007/s10519-013-9596-8

263. Kaplan, J. T., et al. (2016). Neural correlates of maintaining one's political beliefs in the face of counterevidence. *Scientific Reports, 6(39589).* doi: 10.1038/srep39589
264. Katz, B. (2019, January 4). Heavily abridged 'Slave Bible' removed passages that might encourage uprisings. *Smithsonian Magazine*. Retrieved February 5, 2024, from https://www.smithsonianmag.com/smart-news/heavily-abridged-slave-bible-removed-passages-might-encourage-uprisings-180970989/
265. Kelly, C. P., et al. (2015). Climate change in the Fertile Crescent and implications of the recent Syrian drought. *Proceedings of the National Academy of Sciences, 112(11),* 3241-3246. doi: 10.1073/pnas.1421533112
266. Kelley, J., & De Graaf, N. D. (1997). National context, parental socialization, and religious belief: Results from 15 nations." *American Sociological Review 62 (4),* 639–659. doi:10.2307/2657431.
267. Kelley, L. H. (1996). *War before civilization: The myth of the peaceful savage.* New York: Oxford University Press.
268. Kendler, K. S., et al. (1999). Clarifying the relationship between religiosity and psychiatric illness: The impact of covariates and the specificity of buffering effects. *Twin Research and Human Genetics, 2(2),* 137-144. doi: 10.1375/twin.2.2.137
269. Kertzer, D. L. (2022). *The pope at war: The secret history of Pius XII, Mussolini, and Hitler*. New York: Random House.
270. Khait, I., et al. (2018). *Plants emit informative airborne sounds under stress. bioRxiv.* doi: 10.1101/507590.
271. Kiehl, K. A., et al. (2006). Brain potentials implicate temporal lobe abnormalities in criminal psychopaths. *Journal of Abnormal Psychology, 115(3),* 443–453. doi: 10.1037/0021-843X.115.3.443
272. King, B. J. (2013). *How animals grieve*. Chicago: University of Chicago Press.
273. Kitzmiller, et al. v. Dover Area School District, et al. 400 F. Supp. 2d 707 (M.D. Pa. 2005). https://www.aclupa.org/sites/default/files/legacy/8813/1404/6697/Dec20opinion.pdf
274. Klemm, W. R. (2010). Free will debates: Simple experiments are not so simple. *Advances in Cognitive Psychology, 6,* 47-55. doi: 10.2478/v10053-008-0076-2
275. Kobayashi, T., et al. (2024). Cleaner fish with mirror self-recognition capacity precisely realize their body size based on their mental image. *Scientific Reports, 14(1)*, 20202. doi: 10.1038/s41598-024-70138-7
276. Koch, C. (2008). Consciousness redux: What is it like to be a bee? *Scientific American Mind, 19(6), 18-19.* doi:10.1038/scientificamericanmind1208-18
277. Koch, C. (2018). What is consciousness? *Scientific American, 318(6)*, 61-64. doi:10.1038/scientificamerican0618-60
278. Koenig, H. G. (2018). Chapter 3 - Religion and coping. In *Religion and Mental Health: Research and Clinical Applications* (pp. 49-77). London: Academic Press. doi: 10.1016/j.paid.2009.06.003
279. Koenig, L. B., et al. (2009). Rearing environmental influences on religiousness: An investigation of adolescent adoptees, *Personality and Individual Differences, 47(6),* 652-656. doi: 10.1016/j.paid.2009.06.003
280. Kolodny, O., & Feldman, M. W. (2017). A parsimonious neutral model suggests Neanderthal replacement was determined by migration and random species drift. *Nature Communications,* doi: 10.1038/s41467-017-01043-z
281. Kollipara (2014, January 22). Earth won't die as soon as thought. *Science,* News. doi: 10.1126/article.23634
282. Krachun, C. (2002). Are apes conscious? An overview of inconclusive evidence. *Carleton University Cognitive Science Technical Report 2002-10.* Retrieved February 10, 2019, from https://carleton.ca/ics/wp-content/uploads/2002-10.pdf
283. Krause, J., et al. (2007). The derived FOXP2 variant of modern humans was shared with Neandertals. *Current Biology, 17(21),* 1908–1912. doi: 10.1016/j.cub.2007.10.008

284. Krauss, L. M. (2012). *A universe from nothing: Why is there something rather than nothing*. New York: Simon & Schuster.
285. Krauss, L. M. (2017). *The greatest story ever told - so far: Why are we here?* New York: Atria Books.
286. Krebs, D. L. (2011). *The origins of morality: An evolutionary account.* New York: Oxford University Press.
287. Krebs, D. L., & Denton, K. K. (2013). How did morality originate? A review of Christopher Boehm, *Moral Origins: The Evolution of Virtue, Altruism, and Shame. Evolutionary Psychology, 11(1),* 9-17. doi: 10.1177/147470491301100102
288. Kristof, N. (2020, December 21). A conversation about the virgin birth that maybe wasn't. *The New York Times.* https://www.nytimes.com/2024/12/21/opinion/virgin-birth-jesus.html
289. Kühl, H. S. (2016). Chimpanzee accumulative stone throwing. *Scientific Reports, 6*:22219. doi: 10.1038/srep22219
290. Kushner, D., et al. (2018). Trepanation procedures/outcomes: Comparison of prehistoric Peru with other ancient, medieval, and American Civil War cranial surgery. *World Neurosurgery, 114,* 245–251. doi: 10.1016/j.wneu.2018.03.143

L

291. Laland, K. N. (2017). *Darwin's unfinished symphony: How culture made the human mind.* Princeton, NJ: Princeton University Press.
292. Lamb, T. D., et al. (2008). The origin of the vertebrate eye. *Evolution: Education and Outreach, 1(Supp. 4)*, 415–426. doi: 10.1007/s12052-008-0091-2
293. Langford, D. J., et al. (2006). Social modulation of pain as evidence for empathy in mice. *Science, 312(5782),* 1967-1970. doi: 10.1126/science.1128322
294. Langdon, S. (2011, March 15) *The Sumerian Law Code compared with the Code of Hammurabi.* (Originally published 1920). United Kingdom: Cambridge University Press. doi: 10.1017/S0035869X00148543
295. Larøi, F. et al. (2019). An epidemiological study on the prevalence of hallucinations in a general-population sample: Effects of age and sensory modality. *Psychiatry Research, 272,* 707-714. doi: 10.1016/j.psychres.2019.01.003
296. Larson, E. J., & Witham, L. (1998). Leading scientists still reject God. *Nature, 394*, 313. doi: 10.1038/28478
297. Lasure, L. C., & Mikulas, W. L. (1996). Biblical behavior modification. *Behaviour Research and Therapy, 34(7)*, 563-566. doi: 10.1016/0005-7967(96)00013-7
298. Leblanc, L. A., & Register, K. E. (2013). *Constant battles: The myth of the peaceful, noble savage.* New York: St. Martin's Publishing Group.
299. LeCun, Y., et al. (2015). Deep learning. *Nature, 521*, 436-444. doi: 10.1038/nature14539
300. Leeming, D. A. (2010). *Creation myths of the world: An encyclopedia* (2nd ed.). Santa Barbara, CA: ABC-CLIO.
301. Leibniz, G. (1908). On the ultimate origin of things. (Originally published 1697). In *Philosophical works* (pp. 106-113). New Haven, CT: The Tuttle, Morehouse & Taylor Company. Text retrieved March 14, 2024, from https://ia600205.us.archive.org/35/items/worksofleibniz00leibuoft/worksofleibniz00leibuoft.pdf
302. Leimgruber, K. L., et al. (2016). Capuchin monkeys punish those who have more. *Evolution and Human Behavior, 37(3),* 236–244. doi: 10.1016/j.evolhumbehav.2015.12.002
303. Lenharo, M. (2023, June 24). Decades-long bet on consciousness ends — and it's philosopher 1, neuroscientist 0. *Nature News*. doi: 10.1038/d41586-023-02120-8
304. Lenski, R.E. (2023). Revisiting the design of the long-term evolution experiment with *Escherichia coli. Journal of Molecular Evolution, 91*, 241–253 doi: 10.1007/s00239-023-10095-3
305. Leslie, M. (2021). Lab-grown embryos mix human and monkey cells. *Science, 372(6539)*, 223. doi: 10.1126/science.372.6539.223
306. Leslie, M. (2023, October 23). Menopause may be widespread among mammals, challenging famed hypothesis. Retrieved December 2, 2023 from https://

www.science.org/content/article/menopause-may-be-widespread-among-mammals-challenging-famed-hypothesis

307. Le Texier, T. (2019). Debunking the Stanford Prison Experiment. *American Psychologist, 74(7)*, 823–839. doi: 10.1037/amp0000401
308. Lewis, G. J., et al. (2011). The relationship between intelligence and multiple domains of religious belief: Evidence from a large adult US sample. *Intelligence, 39(6),* 468-472. doi: 10.1016/j.intell.2011.08.002
309. Lewis, K. (2000). Programmed death in bacteria *Microbiology and Molecular Biology Reviews, 64* (3), 503-514. doi: 10.1128/MMBR.64.3.503-514.2000
310. Liao, D. A., et al. (2024). Crows "count" the number of self-generated vocalizations. *Science 384,* 874–877. doi: 10.1126/science.adl0984
311. Libet, B., et al. (1983). Time of conscious intention to act in relation to onset of cerebral activities (readiness-potential): The unconscious initiation of a freely voluntary act. *Brain 106(Pt. 3)*, 623-642. doi: 10.1093/brain/106.3.623
312. Libet, B. (1985). Unconscious cerebral initiative and the role of conscious will in voluntary action. *The Behavioral Brain Sciences, 8*, 529-566. doi: 10.1017/S0140525X00044903
313. Lightman, A. (1993). *Einstein's dreams.* New York: Warner Books.
314. Lincoln, T. A., & Joyce, G. F. (2009). Self-sustained replication of an RNA enzyme. *Science, 323(5918)*, 1229-1232. doi: 10.1126/science.1167856
315. Lipka, M. (2017, August 9). Muslims and Islam: Key findings in the U.S. and around the world. *Pew Research Center.* Retrieved August 11, 2018, from http://www.pewresearch.org/fact-tank/2017/08/09/muslims-and-islam-key-findings-in-the-u-s-and-around-the-world/.
316. List, C. (2014). Free will, determinism, and the possibility doing otherwise. *Noûs, 8(1),* 156-178. doi: 10.1111/nous.12019
317. Liu, L., et al. (2018, October). Fermented beverage and food storage in 13,000 y-old stone mortars at Raqefet Cave, Israel: Investigating Natufian ritual feasting. *Journal of Archaeological Science: Reports, 21*, 783-793. doi: 10.1016/j.jasrep.2018.08.008
318. Longrich, N. (2020). War in the time of Neanderthals: how our species battled for supremacy for over 100,000 years. *The Conversation*. Retrieved February 15, 2024, from https://theconversation.com/war-in-the-time-of-neanderthals-how-our-species-battled-for-supremacy-for-over-100-000-years-148205
319. Loukola, O. J., et al. (2017). Bumblebees show cognitive flexibility by improving on an observed complex behavior. *Science, 355(6237),* 833-836. doi: 10.1126/science.aag2360
320. Low, P., et al. (2012, July 7). The Cambridge declaration on consciousness. Unpublished: Cambridge, UK. Retrieved August 11, 2018, from https://fcmconference.org/img/CambridgeDeclarationOnConsciousness.pdf
321. Ludes, et al. (2024, April 10). A ritual murder shaped the Early and Middle Neolithic across Central and Southern Europe. *Science Advances, 10,* eadl3374. Retrieved April 26, 2024, from https://www.science.org/doi/pdf/10.1126/sciadv.adl3374
322. Lynn, R., et al. (2009). Average intelligence predicts atheism rates across 137 nations. *Intelligence, 37(1),* 11-15. doi: 10.1016/j.intell.2008.03.004

M

323. Makin, S. (2023). Newfound hybrid brain cells send signals like neurons do. *Scientific American, 329(5),* 10-11. doi: 10.1038/scientificamerican1223-10
324. Maoz, U., & Yaffe, G. (2015). What does recent neuroscience tell us about criminal responsibility? *Journal of Law and the Biosciences,* 120–139. doi:10.1093/jlb/lsv051
325. Maoz, U., et al. (2017). Neural precursors of deliberate and arbitrary decisions in the study of voluntary action. *bioRxiv,* 097626. doi: 10.1101/097626
326. Masci, D. (2019, February 11). For Darwin Day, 6 facts about the evolution debate. *Pew Research Center.* Retrieved September 12, 2020, from http://www.pewresearch.org/fact-tank/2017/02/10/darwin-day/.

327. Marcus, G. (2017). Am I human? *Scientific American, 316(3)* 59-63. doi:10.1038/scientificamerican0317-58
328. Marx, K. (1844). A contribution to the ccritique of Hegel's philosophy of right. *Deutsch-Französische Jahrbücher* (*German–French Annals*), 10 February, only issue published. Retrieved September 10, 2017, from https://www.marxists.org/archive/marx/works/1843/critique-hpr/intro.htm
329. Mascaro, A., et al. (2022). Application of insects to wounds of self and others by chimpanzees in the wild. *Current Biology, 32(3),* R112-R113. doi: 10.1016/j.cub.2021.12.045
330. Mashour, G. A., et al. (2020). Conscious processing and the Global Neuronal Workspace hypothesis. *Neuron, 105(5),* 776-798. doi: 10.1016/j.neuron.2020.01.026
331. Mather, J. A. (2008). Cephalopod consciousness: Behavioural evidence. *Consciousness and Cognition, 17(1)*, 37-48. doi: 10.1016/j.concog.2006.11.006
332. Mather, J. (2019). What is in an octopus's mind? *Animal Sentience*, *4*(26), 1-29. doi: 10.51291/2377-7478.1370
333. Mcleod, S, (2023, November 17). Stanford Prison Experiment: Zimbardo's famous study. *Simply Psychology*. Retrieved April 27, 2024, from https://www.simplypsychology.org/zimbardo.html
334. McCollom, T. M. (2013). Miller-Urey and beyond: What have we learned about prebiotic organic synthesis reactions in the past 60 years? *Annual Review of Earth and Planetary Sciences, (41),* 207-229. doi: 10.1146/annurev-earth-040610-133457
335. McConnell, J. V., et al. (1959). The effects of regeneration upon retention of a conditioned response in the planarian. *Journal of Comparative and Physiological Psychology, 52(1)*, 1-5. doi: 10.1037/h0048028
336. McCullough, M. E., & Larson, D. B. (1999). Religion and depression: A review of the literature. *Twin Research and Human Genetics, 2(2),* 126-136. doi:10.1375/twin.2.2.126
337. McGovern, P., et al. (2017). Early Neolithic wine of Georgia in the South Caucasus. *Proceedings of the National Academy of Sciences, 114 (48)*, E10309-E10318. doi: 10.1073/pnas.1714728114
338. McNamara, R. A., et al. (2019). Weighing outcome vs. intent across societies: How cultural models of mind shape moral reasoning. *Cognition, 182,* 95–108. doi: 10.1016/j.cognition.2018.09.008
339. Meacham, J. (2012). *Thomas Jefferson: The art of power*. New York: Random House.
340. Melloni, L., et al. (2021). Making the hard problem of consciousness easier. *Science, 372(6545)*, 911-912. doi: 10.1126/science.abj3259.
341. Merlin, M. D. (2003). Archaeological evidence for the tradition of psychoactive plant use in the old world. *Economic Botany, 57(3,* 295–323. Retrieved March 25, 2024, from https://www.researchgate.net/profile/Mark-Merlin/publication/225632469_Archaeological_Evidence_for_the_Tradition_of_Psychoactive_Plant_Use_in_the_Old_World/links/0deec5313bdf5bf775000000/Archaeological-Evidence-for-the-Tradition-of-Psychoactive-Plant-Use-in-the-Old-World.pdf
342. Mervis, J. (2024). West Virginia OKs intelligent design in schools. *Science, 383(6690),* 1399. Retrieved September 10, 2021, from https://www.science.org/doi/pdf/10.1126/science.adp4707
343. Meslier, J. (2009), *Testament: Memoir of the thoughts and sentiments of Jean Meslier*. (Translation by M. Shreve). Amherst, NY: Prometheus Books.
344. Mikulas, W. L. (1981). Buddhism and behavior modification. *The Psychological Record, 31,* 331-342. doi: 10.1007/BF03394746
345. Miller J. D., et al. (2006). Public acceptance of evolution. *Science, 313,* 765–766. doi: 10.1126/science.1126746
346. Miller, J. D., et al. (2021, August 16). Public acceptance of evolution in the United States, 1985–2020. *Public Understanding of Science*. doi: 10.1177/09636625211035919
347. Miller, R. J. (2019). Religion and faith – the forgotten factor of the Civil War. *Civil War Book Review, 21(2)*. doi: 10.31390/cwbr.21.2.03

348. Miller, S. L., & Urey, H. C. (1959). Organic compound synthesis on the primitive Earth. *Science 130 (3370)*, 245-251. doi: 10.1126/science.130.3370.245.
349. Mlodinow, L. (2008). *The drunkard's walk: How randomness rules our lives*. New York: Pantheon Books.
350. Mlodinow, L. (2015). *The upright thinkers: The human journey from living in trees to understanding the cosmos.* New York: Pantheon Books.
351. Mohamed, B. et al. (2017, July 26). U.S. Muslims concerned about their place in society, but continue to believe in the American dream. *Pew Research Center*. Retrieved March 1, 2024, from https://www.pewresearch.org/religion/2017/07/26/political-and-social-views/
352. Monroe, A. E., & Malle, B. F. (2010). From uncaused will to conscious choice: The need to study, not speculate about people's folk concept of free will. *The Review of Philosophy and Psychology, 1(2)*, 211–224. doi: 10.1007/s13164-009-0010-7
353. Moody, E.R.R., et al. (2024, July 12). The nature of the last universal common ancestor and its impact on the early Earth system. *Nature Ecology & Evolution,* Published online. doi: 10.1038/s41559-024-02461-1
354. Morrogh-Bernard, H. C., et al. (2017). Self-medication by orang-utans (*Pongo pygmaeus*) using bioactive properties of *Dracaena cantleyi. Scientific Reports, 7:16653.* doi: 10.1038/s41598-017-16621-w.
355. Moskowitz, C. (2016, August 3). Did the universe boot up with a "Big Bounce?" *Scientific American Blog*. Retrieved August 10, 2018, from https://www.scientificamerican.com/article/did-the-universe-boot-up-with-a-big-bounce/.
356. Mote, F. W. (1971). *Intellectual foundations of China*. New York: Knopf.
357. Mueller, C. T. (2002, August 1). The use of classical conditioning in planaria to investigate a non-neuronal memory mechanism. Unpublished. Research Science Institute. Retrieved June 1, 2017, from http://www.drmichaellevin.org/caitlin.pdf_
358. Musser, G. (2015). Is the cosmos random? *Scientific American, 313(3),* 88-93. doi:10.2307/26046246

N

359. Nahmias, E. (2015). Why we have free will. *Scientific American, 312(1),* 77-79. doi:10.2307/26046069
360. NASA. (2022). About life detection. *Astrobiology at NASA: Life in the universe*. Retrieved August 31, 2022, from https://astrobiology.nasa.gov/research/life-detection/about/
361. National Academy of Sciences. (1998). *Teaching about evolution and the nature of science*. Washington, DC: The National Academies Press. doi: 10.17226/5787
362. National Academy of Sciences. (2008). *Science, evolution, and creationism.* Washington, DC: The National Academies Press. doi: 10.17226/11876.
363. National Academy of Sciences. (2024). Membership. Retrieved September 27, 2024, from https://www.nasonline.org/membership/membership-overview/
364. National Opinion Research Center (NORC). (2018). *The general social survey.* University of Chicago. Retrieved July 1, 2019, from http://gss.norc.org/
365. Navidad, A. E. (2023). Stanford marshmallow test experiment. Simply Psychology, updated online September 7. Retrieved January 22, 2025, from https://www.simplypsychology.org/marshmallow-test.html
366. Nesse, R., & Williams, G. C. (1996). *Evolution and healing: The new science of Darwinian medicine*. London: Phoenix (The Orion Publishing Group Ltd).
367. Nevo, Y. D., & Koren, J. (2003). *Crossroads to Islam: The origins of the Arab religion and the Arab state*, Amherst, NY: Prometheus Books.
368. Newberg, A., et al. (2001). *Why God won't go away: Brain science and the biology of belief.* New York: Ballantine.
369. Newberg, A. B., & Lee, B. Y. (2005). The neuroscientific study of religious and spiritual phenomena: Or why God doesn't use biostatistics. *Zygon: Journal of Religion and Science, 40(2)*, 469-490. doi: 10.1111/j.1467-9744.2005.00675.x

370. Newman, S. (2014, February 14). 1 In 4 Americans thinks the sun goes around the earth, survey says. *NPR, The Two Way*. Retrieved January 14, 2025, from https://www.npr.org/sections/thetwo-way/2014/02/14/277058739/1-in-4-americans-think-the-sun-goes-around-the-earth-survey-says.
371. Newitz, A. (2021). The origin of home. *Scientific American, 324(3),* 68-71. doi: 10.1038/scientificamerican0321-66
372. New York University. (2024, April 24). The New York declaration on animal consciousness. Retrieved May 3, 2024, from https://sites.google.com/nyu.edu/nydeclaration/declaration
373. Nietzsche, F. (1954). *Twilight of the idols* (W. Kaufmann, Translator). (Originally published 1889). New York: Penguin Books.
374. Nietzsche, F. (2016). *The joyful wisdom.* Section 125, The madman. (Originally published 1882). Retrieved July 10, 2019, from https://www.gutenberg.org/files/52881/52881-h/52881-h.htm
375. Nichols, S. (2011). Experimental philosophy and the problem of free will. *Science, 331,* 1401-1403. doi: 10.1126/science.1192931
376. Nichols, S. (2011). Is free will an illusion? *Scientific American Mind, 22(5),18-19.* Retrieved March 10, 2020, from https://www.scientificamerican.com/article/is-free-will-an-illusion/
377. Nieder, A. (2018). Honey bees zero in on the empty set. *Science, 360(6393),* 1069-1070. doi: 10.1126/science.aat8958
378. Nomura, Y. (2017). The quantum multiverse. *Scientific American, 316(6),* 29-35. doi:10.1038/scientificamerican0617-28
379. Norenzayan, A., et al. (2012). Mentalizing deficits constrain belief in a personal god. *PLOS ONE, 7(5),* e36880. doi:10.1371/journal.pone.0036880.
380. Northover, S. B., et al., (2024). Religious signaling and prosociality: A review of the literature. *Evolution and Human Behavior, 45(5)*, 106593. Published online. doi: 10.1016/j.evolhumbehav.2024.06.002
381. Nosek, B., et al. (Open Science Collaboration). (2015). Estimating the reproducibility of psychological science. *Science, 349(625)*. doi: 10.1126/science.aac4716
382. NPR. (2014, February 14). 1 In 4 Americans thinks the sun goes around the earth, survey says. Retrieved June 14, 2020, from https://www.npr.org/sections/thetwo-way/2014/02/14/277058739/1-in-4-americans-think-the-sun-goes-around-the-earth-survey-says.
383. Nyoff, M. A., & Johnson, C. N. (2017). Is God just a big person? Children's conceptions of God across cultures and religious traditions. *British Journal of Developmental Psychology, 35,* 60–75. doi: 10.1111/bjdp.12173

O

384. Oizumi, M., et al. (2014). From the phenomenology to the mechanisms of consciousness: Integrated information theory 3.0. *PLOS Computational Biology, 10(5),* e1003588. doi: 10.1371/journal.pcbi.1003588
385. Orwell, G. (1946). *The prevention of literature*. United Kingdom: Renard Press Ltd.
386. Overbye, D. (2023, January 17). Are we living in a computer simulation, and can we hack it? - A popular cosmological theory holds that the cosmos runs on quantum codes. So how hard could it be to tweak the supreme algorithm? *The New York Times*. https://www.nytimes.com/2023/01/17/science/cosmology-universe-programming.html

P

387. Pagel, J. F., & Kirshtein, P. (2017). Forms of machine dreaming. In *Machine dreaming and consciousness* (pp. 187-200). London: Academic Press.
388. Paley, W. (1802). *Natural Theology: Or, evidences of the existence and attributes of the deity.* United Kingdom: J. Faulder.
389. Palmer, D. C. (2006). On Chomsky's appraisal of Skinner's *Verbal Behavior*: A half

century of misunderstanding. *The Behavior Analyst, 29(2),* 253–267. doi: 10.1007/BF03392134
390. Pansani, T. R., et al. (2023). Evidence of artefacts made of giant sloth bones in central Brazil around the last glacial maximum. *Proceedings of the Royal Society, B/290:* 20230316. doi: 10.1098/rspb.2023.0316
391. Parmigiani, G., et al. (2017). Free will, neuroscience, and choice: Towards a decisional capacity model for insanity defense evaluations. *Rivista di Psichiatria (Journal of Psychiatry, Ital.), 52(1)*, 9-15. doi: 10.1708/2631.27049
392. Partanen, E., et al. (2013). Learning-induced neural plasticity of speech processing before birth. *Proceedings of the National Academy of Sciences, 110 (37)*, 15145-15150. doi:10.1073/pnas.1302159110
393. Paul, G. (2009). The chronic dependence of popular religiosity upon dysfunctional psychosociological conditions. *Evolutionary Psychology, 7(3),* 398-441. doi: 10.1177/147470490900700305
394. Penrose, R. (1994). *Shadows of the mind: A search for the missing science of consciousness*. New York, Oxford: Oxford University Press.
395. Perlman, L. M. (2005). Memories of the Child Development Center study of adopted monozygotic twins reared apart: An unfulfilled promise. *Twin Research and Human Genetics, 8(3),* 271–281. doi: 10.1375/twin.8.3.271
396. Persaud, R. (2012, September 8). Was it a psychotic episode in L. Ron Hubbard that led him to found the Church of Scientology? *Huffington Post Blog.* Retrieved April 1, 2020, from https://www.huffingtonpost.co.uk/dr-raj-persaud/scientology-l-ron-hubbard_b_1656883.html
397. Pershin, Y. V., et al. (2009, 21 August). Memristive model of amoeba's learning. *Physical Review E, 80(2)*. doi: 10.1103/PhysRevE.80.021926
398. Perry, S., & Smolla, M. (2020, February 24). *Capuchin monkey rituals: An interdisciplinary study of form and function. bioRxiv*. doi: 10.1101/2020.02.21.958223
399. Pesic, P. (2007). *Sky in a bottle*. Cambridge, MA: MIT Press.
400. Persinger, M. A. (1990). Experimental simulation of the God experience: Implications for religious beliefs and the future of the human species. In R. Joseph (Ed.), *Neurotheology* (pp. 267-284). San Jose, CA: University Press. Retrieved August 10, 2018, from http://www.neurosciarchive.byethost12.com/2003-Persinger-Neurotheology-Brain-Science-Spirituality-Religious-Experience-Experimental-simulation-of-the-God-experience.pdf
401. Pereira, T., & de Souza, T. (2018). Caenorhabditis elegans tolerates hyperaccelerations up to 400,000 x g. *Astrobiology, 18(7)*, 1-9. doi: 10.1089/ast.2017.1802
402. *Pew Research Center.* (2009, November 5). Scientists and belief. Retrieved September 27, 2024, from https://www.pewresearch.org/religion/2009/11/05/scientists-and-belief/.
403. Pika, S., et al. (2020). Ravens parallel great apes in physical and social cognitive skills. *Scientific Reports, 10(20617)*. doi: 10.1038/s41598-020-77060-8
404. Pinker, S. (1997). *How the mind works*. New York: W. W. Norton & Company.
405. Pinker, S. (2018). *Enlightenment now: The case for reason, science, humanism, and progress*. New York: Penguin Books.
406. Piore, A. (2018). Bringing Darwin back. *Scientific American 319(2),* 56-63. doi:10.1038/scientificamerican0818-56
407. Pogue, D. (2018). The robotic artist problem. *Scientific American, 318(2),* 23. doi:10.1038/scientificamerican0218-23
408. Poldrack, R. A. (2006). Can cognitive processes be inferred from neuroimaging data? *Trends in Cognitive Sciences, 10 (2),* 59-63. doi: 10.1016/j.tics.2005.12.004
409. Poldrack, R. A. (2018). *The new mind readers: What neuroimaging can and cannot reveal about our thoughts*. Princeton, NJ: University Press.
410. Pomeroy, E., et al. (2020). New Neanderthal remains associated with the 'flower burial' at Shanidar Cave. *Antiquity, 94 (373)*, 11–26. doi: 10.15184/aqy.2019.207
411. Powell, K. (2018, November 8). Biology from scratch. *Nature,* 563, 172-175. doi: 10.1038/d41586-018-07289-x

412. Power, E. A. (2016). Discerning devotion: Testing the signaling theory of religion. *Evolution and Human Behavior, 38(1),* 82–91. doi: 10.1016/j.evolhumbehav.2016.07.003
413. Powner, M. W., et. al. (2009). Synthesis of activated pyrimidine ribonucleotides in prebiotically plausible conditions. *Nature, 459,* 239–242. doi: 10.1038/nature08013
414. Premack, D. (1965). Reinforcement theory. In D. Levine (Ed.), *Nebraska symposium on motivation,* Vol. 13. Lincoln, NE: University of Nebraska Press.
415. Premack, D., & Woodruff, G. (1978). Does the chimpanzee have a theory of mind? *The Behavioral and Brain Sciences, 4,* 515-526. Retrieved March 25, 2024, from https://www.deempathischeorganisatie.nl/wp-content/uploads/2020/02/Premack-Woodruff-1978-Theory-of-Mind.pdf
416. Price, R. (2024). Explosive claim about ancient burials challenged. *Science, 385(6709),* 585. Retrieved August 16, 2024, from https://www.science.org/doi/epdf/10.1126/science.ads2942
417. Proffitt, T., et al. (2016). Wild monkeys flake stone tools. *Nature, 539,* 85–88. doi: 10.1038/nature20112
418. Protzko, J., et al. (2024). High replicability of newly discovered social-behavioural findings is achievable. *Nature Human Behaviour, 8,* 311–319. doi: 10.1038/s41562-023-01749-9
419. Purtill, C. (2023, October 19). Scientist, after decades of study, concludes: We don't have free will. *Phys.org*. Retrieved October 23, 2023, from https://phys.org/news/2023-10-scientist-decades-dont-free.html

R

420. Rachlin, H. (1978). A molar theory of reinforcement schedules. *Journal of the Experimental Analysis of Behavior, 30(3),* 345-360. doi: 10.1901/jeab.1978.30-345
421. Raia, et al. (2020). Past extinctions of *Homo* Species coincided with increased vulnerability to climatic change. *One Earth 3,* 480–490. doi: 10.1016/j.oneear.2020.09.007
422. Ramzy, A., & Buckley, C. (2019, November 17). 'Absolutely no mercy': Leaked files expose how China organized mass detentions of Muslims. *The New York Times.* https://www.nytimes.com/interactive/2019/11/16/world/asia/china-xinjiang-documents.html
423. Rees, T. J. (2009). Is personal insecurity a cause of cross-national differences in the intensity of religious belief? *Journal of Religion and Society, 11*. Retrieved May 3, 2024, from https://cdr.creighton.edu/server/api/core/bitstreams/c9036171-0138-4f23-b708-67a9b31bcb47/content
424. Reese, T. (2018, April 3). Pope Francis and Hell. *National Catholic Reporter*. Retrieved January 10, 2019, from https://www.ncronline.org/news/opinion/signs-times/pope-francis-and-hell
425. Richerson, P. J., & Boyd, R. T. (2024). Culture in humans and other animals. *Science, 386(6274),* 846-847.
426. Richerson, P., et al. (2016). Cultural group selection plays an essential role in explaining human cooperation: A sketch of the evidence. *Behavioral and Brain Sciences, 39,* 1-68. doi: 10.1017/S0140525X1400106X
427. Roberts, L., et al., (2009). Intercessory prayer for the alleviation of ill health. *Cochrane Database of Systematic Reviews*. Issue 2. Art. No.: CD000368. doi: 10.1002/14651858.CD000368.pub3.
428. Robinson, G. E., & Barron, A. B. (2017). Epigenetics and the evolution of instincts. *Science, 356(6333)*, 26-27. doi: 10.1126/science.aam6142
429. Robinson, D. N. (2012). Determinism: Did Libet make the case? *Philosophy, 87*(3), 395-401. doi:10.1017/S0031819112000253
430. Roes, F. L., & Raymond, R. (2003). Belief in moralizing gods. *Evolution and Human Behavior, 24(2),* 126–135. doi: 10.1016/S1090-5138(02)00134-4

431. Root-Bernstein, M., et al. (2019). Context-specific tool use by *Sus cebifrons. Mammalian Biology, 98,* 102-110. doi: 10.1016/j.mambio.2019.08.003
432. Rossi, L., & Chittka, N. (2023). Bees learn to dance: Experience yields precision in the waggle dance of honey bees. *Science, 379(6636),* 985-986. doi: 10.1126/science.adg6020
433. Rousseau, J. (1762). *On the social contract: Or, the principles of political rights.* Republished 1893, London: G. P. Putnam's Sons.
434. Russell, B. (1919). *Proposed roads to freedom: Socialism, anarchism and syndicalism*. New York: Henry Holt & Co.
435. Russell, B. (1927). Why I am not a Christian. (1927) (Lecture to the National Secular Society of Great Britain). Retrieved January 11, 2018, from http://www.users.drew.edu/~jlenz/whynot.html.
436. Russell, B. (1952). Is there a god? In J. G. Slater, (Ed.), *The collected papers of Bertrand Russell, Vol. II: Last philosophical testament, 1943–68* (pp. 542–548). New York: Routledge.
437. Russell, B. (2013) *History of western philosophy*. (Originally published 1947). London: Taylor & Francis.
438. Rutherford, M. (2007). The evolution of morality. *Groundings 1*. University of Glasgow. Retrieved April 10, 2017, from http://eprints.gla.ac.uk/3723/1/rutherford3723.pdf.

S

439. Sagan, C. (1996). *The demon-haunted world: Science as a candle in the dark.* New York: Random House.
440. Sagan, C. (2006). *The varieties of scientific experience: A personal view of the search for God.* (1985, Gifford Lectures). New York: The Penguin Press.
441. Sahgal, N., et al., (2018, May 29). *Pew Research Center.* Being Christian in western Europe. Retrieved April 20, 2019, from http://www.pewforum.org/wp-content/uploads/sites/7/2018/05/Being-Christian-in-Western-Europe-FOR-WEB1.pdf.
442. Saigusa, T., et al. (2008, 3 January). Amoebae anticipate periodic events. *Physical Review Letters, 100,* 018101. doi: 10.1103/PhysRevLett.100.018101
443. Sapir-Hen, L. & Ben-Yosef, E. (2013). The introduction of domestic camels to the Southern Levant: Evidence from the Aravah Valley. *TEL AVIV, 40*, 277–285. doi: 10.1179/033443513X13753505864089.
444. Sapolsky, R. M. (2017) *Behave: The biology of humans at our best and worst*. New York: Penguin Press.
445. Sapolsky, R. M. (2023). *Determined: A science of life without free will.* New York: Penguin Press.
446. Sayin, H. U. (2014). The consumption of psychoactive plants during religious rituals: The roots of common symbols and figures in religions and myths. *NeuroQuantology, 12(2),* 276-296. doi: 10.14704/nq.2014.12.2.753
447. Scalia, A. (2002, May). God's justice and ours. *First Things*. Retrieved January 3, 2021, from https://www.firstthings.com/article/2002/05/gods-justice-and-ours
448. Schaffner, S. (2021). Adam, Eve, and the evolution of humankind. *Science, 374(6564)*. 162. doi: 10.1126/science.abl8547
449. Scerri, E. M. L.et al. (2018). Did our species evolve in subdivided populations across Africa, and why does it matter? *Trends in Ecology & Evolution, 33(8)*, 582-594. doi: 10.1016/j.tree.2018.05.005
450. Schnell, A. K., et al. (2021). Cuttlefish exert self-control in a delay of gratification task. Proceedings of the Royal Society B, no. 1946, 20203161. doi: 10.1098/rspb.2020.3161
451. Schrödinger, E. (1944; reprint 1992). *What is life? The physical aspect of the living cell with mind and matter & autobiographical sketches*. Cambridge, UK: Cambridge University Press.
452. Schulting, R. J., et al. (2024). 'The darker angels of our nature': Early Bronze Age

butchered human remains from Charterhouse Warren, Somerset, UK. *Antiquity,* Online December 16, 1-17. doi: 10.15184/aqy.2024.180

453. Schurger, A., et al. (2012, August 12). An accumulator model for spontaneous neural activity prior to self-initiated movement. *Proceedings of the National Academy of Sciences*. doi: 10.1073/pnas.1210467109
454. Searle, J. (1980). Minds, brains, and programs. *Behavioral and Brain Sciences, 3(3),* 417-424. doi: 10.1017/S0140525X00005756
455. Sebo, J. (2018, April 7). Should chimpanzees be considered persons? *The New York Times, Opinion.* https://www.nytimes.com/2018/04/07/opinion/sunday/chimps-legal-personhood.html
456. Segal, N. L. (2005). Commentary: More thoughts on the Child Development Center Twin Study. *Twin Research and Human Genetics, 8(3),* 276-281. doi: 10.3316/informit.945604065129896
457. Seidel, A. L. (2019). *The founding myth: Why Christian nationalism is un-American.* New York: Sterling.
458. Seife, C. (2003). *Alpha & omega: The search for the beginning and end of the universe*. New York: Viking.
459. Sender, et al. (2016). *Revised estimates for the number of human and bacteria cells in the body. bioRxiv*. doi: 10.1101/036103
460. Seth, A. K., & Bayne, T. (2022). Theories of consciousness. *Nature Reviews Neuroscience, 23(7),* 439-452. doi: 10.1038/s41583-022-00587-4
461. Shariff, A. F., & Vohs, K. D. (2014). The world without free will. *Scientific American, 310(6),* 76-79. doi:10.1038/scientificamerican0614-76
462. Shariff, A. F., et al. (2014). Free will and punishment: A mechanistic view of human nature reduces retribution. *Psychological Science, 25,* 1563–1570. doi: 10.1177/0956797614534693
463. Sharot, T., & Sunstein, C. R. (2020). How people decide what they want to know. *Nature Human Behaviour*, *4(1)*, 14-19. doi:10.1038/s41562-019-0793-1
464. Sharp, D. (2002). Science, faith, and gods. *The Lancet, 359,* 806-807. doi: 10.1016/S0140-6736(02)07891-1
465. Shenhav, A., et al. (2012). Divine intuition: Cognitive style influences belief in God. *Journal of Experimental Psychology: General, 141(3),* 423-428. doi: 10.1037/a0025391
466. Shermer, M. (2000). *How we believe: Science, skepticism, and the search for God.* (2nd ed.). New York: Henry Holt and Co.
467. Shermer, M. (2002). Shermer's last law. Any sufficiently advanced extraterrestrial intelligence is indistinguishable from God. *Scientific American, 286(1)*, 33. Retrieved January 10, 2024, from https://www.scientificamerican.com/issue/sa/2002/01-01/
468. Shermer, M. (2012). Free won't. *Scientific American, 307(2)*, 86. doi:10.1038/scientificamerican0812-86
469. Shermer, M. (2015). *Scientia humanitatas*. *Scientific American, 312(6)*, 80. doi:10.1038/scientificamerican0615-80
470. Shermer, M. (2017a). Imagine no universe. *Scientific American, 316(2),* 73. doi:10.1038/scientificamerican0217-73
471. Shermer, M. (2017b). When facts backfire. *Scientific American, 316(1)*, 69. doi:10.1038/scientificamerican0117-69
472. Shermer, M. (2018). *Heavens on earth.* New York: Macmillan.
473. Shmakova, L., et al. (2021). A living bdelloid rotifer from 24,000-year-old Arctic permafrost. *Current Biology, 31(11),* R712 - R713. doi: 10.1016/j.cub.2021.04.077
474. Shonkoff, J. P., & Phillips, D. A. (Eds.). (2000). *From neurons to neighborhoods: The science of early childhood development.* Washington, DC: National Research Council, Institute of Medicine. doi: 10.17226/9824
475. Siegfried, T. (2000). *The bit and the pendulum: From quantum computing to M Theory—the new physics of information.* New York: John Wiley & Sons.
476. Siegel, R. K. (2005). *Intoxication: The universal drive for mind-altering substances.* Rochester, VT: Park Street Press.

477. Sillick, W. J., & Cathcart, S. (2014). The relationship between religious orientation and happiness: The mediating role of purpose in life. *Mental Health, Religion & Culture, 17(5)*, 494-507. doi: 10.1080/13674676.2013.852165
478. Simola, D. F., et al. (2013). A chromatin link to caste identity in the carpenter ant *Camponotus floridanus*. *Genome Research*, *23*(3), 486-496. doi:10.1101/gr.148361.112
479. Singh, M. et al. (2020). Small gods, rituals, and cooperation: The Mentawai water spirit Sikameinan. *Evolution and Human Behavior, 42(1)*, 61-72. doi: 10.1016/j.evolhumbehav.2020.07.008
480. Skinner, B. F. (1938). *The behavior of organisms: An experimental analysis*. Cambridge, MA: B. F. Skinner Foundation (reprinted, 1991).
481. Skinner, B. F. (1948). 'Superstition' in the pigeon. *Journal of Experimental Psychology*, *38*, 168-172. doi: 10.1037/h0055873
482. Skinner, B. F. (1953). *Science and human behavior.* New York: Macmillan.
483. Skinner B.F. (1957). *Verbal behavior.* New York: Appleton-Century-Crofts.
484. Skinner, B. F. (1971). *Beyond freedom and dignity.* New York: Knopf.
485. Skinner, B. F. (1974). *About behaviorism.* New York: Knopf.
486. Skinner, B. F. (1987). *Upon further reflection.* Englewood Cliffs, NJ: Prentice-Hall.
487. Skinner, B. F. (1989). The origins of cognitive thought. *American Psychologist, 44(1)*, 13-18. doi: 10.1037/0003-066X.44.1.13
488. Smith, G. A. (2017, October 16). A growing share of Americans say it's not necessary to believe in God to be moral. *Pew Research Center.* Retrieved February 10, 2018, from http://www.pewforum.org/2017/04/26/in-america-does-more-education-equal-less-religion/
489. Smith, G. A., et al. (2017, April 26). In America, does more education equal less religion? *Pew Research Center.* Retrieved January 10, 2019, from http://www.pewforum.org/2017/04/26/in-america-does-more-education-equal-less-religion/
490. Smith, G. A., et al. (2021, December 14). About three-in-ten U.S. adults are now religiously unaffiliated. *Pew Research Center.* Retrieved January 10, 2022, from https://www.pewresearch.org/religion/2021/12/14/about-three-in-ten-u-s-adults-are-now-religiously-unaffiliated/
491. Smith, G. A., et al. (2024, January 24). Religious 'nones' in America: Who they are and what they believe. *Pew Research Center.* Retrieved February 3, 2024, from https://www.pewresearch.org/religion/2024/01/24/religious-nones-in-america-who-they-are-and-what-they-believe/
492. Smithsonian, National Museum of Natural History. (2022). *Homo neanderthalensis*. Retrieved August 1, 202 from https://humanorigins.si.edu/evidence/human-fossils/species/homo-neanderthalensis.
493. Somers, J. (2021, December 6). The science of mind reading. *The New Yorker.* https://www.newyorker.com/magazine/2021/12/06/the-science-of-mind-reading
494. Soon, C. S., et al. (2008). Unconscious determinants of free decisions in the human brain. *Nature Neuroscience, 11*, 543–545. doi:10.1038/nn.2112
495. Sosis, R., & Alcorta, C. (2003). Signaling, solidarity, and the sacred: The evolution of religious behavior. *Evolutionary Anthropology*, *12*, 264–274. doi: 10.1002/evan.10120
496. Speed, D. (2016). What is spirituality, anyway? *Skeptic Magazine, 21(4)*, 34-36. Retrieved April 1, 2022, from https://aaagnostica.org/wp-content/uploads/2020/03/What-Is-Spirituality-Anyway.pdf
497. Speed, D. (2022). Throw BABE Out with the bathwater? Canadian atheists are no less healthy than the religious. *Journal of Religion and Health,* 61, 4608–4634. doi: 10.1007/s10943-022-01558-w
498. Speed, D., & Brewster, M. (2021). Love thy neighbour … or not: Christians, but not atheists, show high in-group favoritism. *Secularism and Nonreligion, 10(7)*, 1–15. doi: 10.5334/ snr.136
499. Sperber, D., et al. (2010). Epistemic vigilance. *Mind & Language, 25(4)*, 359–393. doi: 10.1111/j.1468-0017.2010.01394.x

500. Sripada, C. (2016). Free will and the construction of options. *Philosophical Studies, 173*, 2913–2933 doi: 10.1007/s11098-016-0643-1
501. Stairs, S., et al. (2017). Divergent prebiotic synthesis of pyrimidine and 8-oxo-purine ribonucleotides. *Nature Communications, 8*, 15270. doi: 10.1038/ncomms15270
502. Starr, M. (2019, November 25). Crows can pass the marshmallow test as well as human children, new study shows. Science Alert. Retrieved January 24, 2025, from https://www.sciencealert.com/once-again-crows-have-passed-the-marshmallow-test-and-demonstrated-self-control
503. Stats NZ. (2019, September 23). 2018 census population and dwelling counts. Retrieved September 20, 2020, from https://www.stats.govt.nz/information-releases/2018-census-population-and-dwelling-counts
504. Steinhart, E. (2013, September). Skeptical and spiritual atheisms (Conference presentation). *British Society for the Philosophy of Religion*, Oriel College, Oxford University. Retrieved September 20, 2018, from http://www.ericsteinhart.com/articles/atheisms.pdf
505. Steinhardt, P. J., & Turok, N. (2002). A cyclic model of the universe. *Science, 296*, 1436-1439. doi: 10.1126/science.1070462
506. Stenger, V. (2007) *God: The failed hypothesis. How science shows that God does not exist.* Amherst, NY: Prometheus Books.
507. Stillman, T. F., et al. (2011). Free will in everyday life: Autobiographical accounts of free and unfree actions. *Philosophical Psychology, 24(3)*, 381–394. doi: 10.1080/09515089.2011.556607
508. Sutterland, A. L., et al. (2015). Beyond the association. *Toxoplasma gondii* in schizophrenia, bipolar disorder, and addiction: Systematic review and meta-analysis. *Acta Psychiatrica Scandinavica, 132*, 161-179. doi: 10.1111/acps.12423
509. Swift, J. (1710). Untitled. *The Examiner*, No. 15 (November 2 to 9). In *The Whig Examiner* [by J. Addison]. Retrieved July 10, 2019, from https://play.google.com/books/reader?id=KigTAAAAQAAJ&pg=GBS.PA23-IA1&hl=en
510. Symons, S, & Tasker, E. (2015). Stars of the dead. *Scientific American, 313(4)*, 70-75. doi:10.2307/26046357
511. Szep, J. (2013, April 8). Buddhist monks incite Muslim killings in Myanmar. *Reuters*. Retrieved February 8, 2024, from https://www.pulitzer.org/files/2014/international-reporting/reuters/04reuters2014.pdf
512. Szewczyk, et al. (2005). *Caenorhabditis elegans* survives atmospheric breakup of STS-107, space shuttle Columbia. *Astrobiology, 5(6)*, 690-705. doi: 10.1089/ast.2005.5.690

T

513. Tang, J., et al. (2023, May 1). Semantic reconstruction of continuous language from non-invasive brain recordings. *Nature Neuroscience*. doi: 10.1038/s41593-023-01304-9
514. Taylor, R. (1967). Determinism. In P. Edwards (Ed.), *The encyclopedia of philosophy* (pp. 359-373). New York: Macmillan & Free Press.
515. Teehan, J. (2010). *In the name of God: The evolutionary origins of religious ethics and violence.* London: Wiley-Blackwell.
516. Teehan, J., & diCarlo, C. (2004). On the naturalistic fallacy: A conceptual basis for evolutionary ethics. *Evolutionary Psychology, 2*, 32-46. doi: 10.1177/147470490400200108
517. Tegmark, M. (2003). Parallel universes. *Scientific American, 288(5)*, 40-51. doi: 10.1038/scientificamerican0503-40
518. Tharwani, Z. H., et al., (2022). Polio amid a humanitarian crisis in Afghanistan: challenges and recommendations. *The Lancet, Infectious Diseases, 22(2)*, 168-169. doi: 10.1016/S1473-3099(22)00004-4
519. Torrey, E. F. (2017). *Evolving brains, emerging gods: Early humans and the origins of religion.* New York: Columbia University Press.
520. Turing, A. (1950). Computing machinery and intelligence. In B. J. Copeland (Ed.), (2004). *The essential Turing seminal writings in computing, logic, philosophy,*

artificial intelligence, and artificial life, plus the secrets of enigma (Chapter 11). New York: Oxford University Press.

521. Tyson, N. D., & Trefil, J. (2021). *Cosmic queries*. Washington, DC: National Geographic Partners.

U

522. Ubeda, Y., et al., (2018). Personality in captive killer whales (*Orcinus orca*): A rating approach based on the five-factor model. *Journal of Comparative Psychology, 133(2),* 252-261. doi: 10.1037/com0000146
523. Uecker, J. E., et al. (2007). Losing my religion: The social sources of religious decline in early adulthood. *Social Forces, 85(4)I,* 1667–1692. doi:10.1353/sof.2007.0083.
524. United States v. Grayson, 438 U.S. 41, (1978). Retrieved July 11, 2019, from https://caselaw.findlaw.com/us-supreme-court/438/41.html

V

525. Velt, W. (2023, Nov. 28). An interview with Daniel Dennett. *Psychology Today,* Retrieved November 30, 2023, from https://www.psychologytoday.com/us/blog/science-and-philosophy/202311/an-interview-with-daniel-dennett.
526. Vilenkin, A., & Tegmark, A. (2011, July 19). The case for parallel universes. *Scientific American Blog*. Retrieved July 11, 2019, from https://www.scientificamerican.com/article/multiverse-the-case-for-parallel-universe/
527. Vohs, K. D., & Schooler, J. W. (2008). The value of believing in free will: Encouraging a belief in determinism increases cheating. *Psychological Science, 19(1)*, 49-54. doi: 10.1111/j.1467-9280.2008.02045.x

W

528. Wade, L. (2018). Feeding the gods. *Science, 360(6395)*, 1288-1292. doi: 10.1126/science.360.6395.1288
529. Wade, L. (2021). Footprints support claim of early arrival in the Americas. *Science, 373(6562)*, 1426. doi: 10.1126/science.acx9159
530. Wade, N. (2007, March 20). Scientist finds the beginnings of morality in primate behavior. *The New York Times, Science Section*. https://www.nytimes.com/2007/03/20/science/20moral.html
531. Watanabe, T. (2001, April 13). Doubting the story of Exodus. *Los Angeles Times.* https://www.latimes.com/archives/la-xpm-2001-apr-13-mn-50481-story.html
532. Wascher, C. A., & Bugnyar, T. (2013). Behavioral responses to inequity in reward distribution and working effort in crows and ravens. *PLOS ONE, 8(2),* e56885. doi:10.1371/journal.pone.0056885
533. Watson, J. B. (1924). *Behaviorism*. New York: W. W. Norton & Co.
534. WBGH Radio, Anonymous. (2004). An interview with Francis Collins. Boston, MA: WBGH Public Radio. Retrieved August 11, 2019, from https://www.pbs.org/wgbh/questionofgod/voices/collins.html
535. Wells, H. G. (2016). *Men like gods*. (Originally published 1923). Mineola, NY: Dover Publications.
536. Wells, G. A. (2010). *Did Jesus exist?* (2nd ed.: Revised, corrected, and expanded). Amherst, NY: Prometheus Books.
537. Wegner, D. M., & Wheatley, T. (1999). Apparent mental causation: Sources of the experience of will. *American Psychologist, 54(7)*, 480-492. doi: 10.1037/0003-066X.54.7.480
538. Weinberg, S. (1999, April). Address to the Conference on Cosmic Design, Washington, D.C.: *American Association for the Advancement of Science*. Edited version in "A designer universe?" *Skeptical Inquirer,* September/October, 2001. Retrieved April 1, 2024, from https://cdn.centerforinquiry.org/wp-content/uploads/sites/29/2001/09/22164801/p64.pdf

539. Whang, O. (2023, January 6). 'Consciousness' in robots was once taboo. Now it's the last word. *The New York Times.* https://www.nytimes.com/2023/01/06/science/robots-artificial-intelligence-consciousness.html
540. Wheeler, J. A. (1992). Toward 'It from bit.' In J. S. Anandan & J. L. Safko (Eds.), *Quantum coherence and reality: International Conference on Fundamental Aspects of Quantum Theory,* December 10-12 (pp. 281-297). University of South Carolina, Columbia: World Scientific Publishing Co. ISBN: 978-981-4549-81-3
541. White, T. D. (2001). Once were cannibals. *Scientific American, 285(2)*, 58-65. doi: 10.1038/scientificamerican0801-58
542. Whiten, A., & Youngblood, M. (2025). Convergent evolution in whale and human vocal cultures. *Science, 387(6734),* 581-582. doi: 10.1126/science.adv2318
543. Whiten, A., et al. (2001). Charting cultural variation in chimpanzees. *Behaviour, 13(11),* 1481–1516. doi: 10.1163/156853901317367717
544. Whitmarsh, T. (2015). *Battling the gods: Atheism in the ancient world.* New York: Knopf.
545. Wilford, J. N. (2015, September 10). *Homo naledi*, new species in human lineage, is found in South African cave. *The New York Times.* https://www.nytimes.com/2015/09/11/science/south-africa-fossils-new-species-human-ancestor-homo-naledi.html
546. Willard, A. K. et al. (2020). Rewarding the good and punishing the bad: The role of karma and afterlife beliefs in shaping moral norms. *Evolution and Human Behavior, 41*, 385–396. doi: 10.1016/j.evolhumbehav.2020.07.001
547. Williams, P. S. (2012). A universe from someone: Against Lawrence M. Krauss' *A Universe from Nothing*: *Why there is something rather than nothing.* Gimlekollen School of Journalism and Communication, Norway. Retrieved August 11, 2020, from http://www.bethinking.org/is-there-a-creator/a-universe-from-someone-against-lawrence-krauss
548. Williams, R. (2014). *Keywords: A vocabulary of culture and society. New* York: Oxford University Press.
549. Wilson, D. S. (2002). *Darwin's cathedral: Evolution, religion, and the nature of society.* Chicago: University of Chicago Press.
550. Wilson, D. S., & Sober, E. (1994). Reintroducing group selection to the human behavioral sciences. *Behavioral and Brain Sciences, 17,* 585-608. doi: 10.1017/S0140525X00036104
551. Wilson, E. O. (1987). The little things that run the world (The importance and conservation of invertebrates). *Conservation Biology, 1(4)*, 344-346. Retrieved November 18, 2022, from https://faculty.washington.edu/timbillo/Readings%20and%20documents/ABRIDGED%20READINGS%20for%20PERU/Wilson_1987_Little_things_that_run.pdf
552. Wine, S. T. (2012). *A provocative people: A secular history of the Jews*. Farmington Hills, MI: International Institute for Secular Humanistic Judaism. (Published posthumously)
553. Witze, A, (2024, March 20). It's final: the Anthropocene is not an epoch, despite protest over vote. *Nature News.* doi: 10.1038/d41586-024-00868-1
554. Witzel, J. G., et al. (2016). Increased frequency of brain pathology in inmates of a high-security forensic institution: A qualitative CT and MRI scan study. *European Archives of Psychiatry and Clinical Neuroscience, 266(6),* 533-541. doi: 10.1007/s00406-015-0620-2
555. Wofford, T. (2014, October 30). Pope Francis's remarks on evolution are not that controversial among Roman Catholics. *Newsweek.* https://www.newsweek.com/pope-franciss-remarks-evolution-are-not-controversial-among-roman-catholics-281115
556. Wong, K. (2016). Mystery human. *Scientific American, 314(3)*, 29-37. doi:10.1038/scientificamerican0316-28
557. Wong, K. (2017). New evidence of mysterious *Homo naledi* raises questions about how humans evolved. *Scientific American,316(5), 28-35.* doi:10.1038/

scientificamerican0517-28
558. Wong, E., & Buckley, C. (2021, January 19). The U.S. calls China's repression of the Uighurs 'genocide,' echoing Biden's earlier comments. *The New York Times.* https://www.nytimes.com/2021/01/19/us/politics/the-us-calls-chinas-repression-of-the-uighurs-genocide-echoing-bidens-earlier-comments.html
559. Wong, S. (2016, January 20). Grisly massacre site in Africa reveals one of the earliest wars. *New Scientist.* Retrieved February 20, 2024, from https://www.newscientist.com/article/2074066-grisly-massacre-site-in-africa-reveals-one-of-the-earliest-wars/
560. Worsley, P. M. (2009). 50 years ago: Cargo cults of Melanesia. *Scientific American, 300(5),* 12. doi: 10.1038/scientificamerican052009-4Pq1QBTMdZOY9QIaZ0T9ig
561. Wright, W. (2024). Shortcuts to God. *Scientific American, 330(4),* 37-39. doi: 10.1038/scientificamerican0424-37
562. Wunn, I. (2000). Beginning of religion. *Numen, 47(4)*, 417-452. doi: 10.1163/156852700511612

Y

563. Yam, M. J. Z., et al. (2013, September). The role of attachment to God on reduction of emotional-behavioral disorders among male adolescents in Tabriz [Conference presentation]. *6th International Congress on Child and Adolescent Psychiatry*, Tabriz, Iran. Tabtiz University of Medical Sciences.

Z

564. Zheng, J., & Meister, M. (2025). The unbearable slowness of being: Why do we live at 10 bits/s? *Neuron, 113(2),* 192-204. doi: 10.1016/j.neuron.2024.11.008
565. Zindler, F. (1998, summer). Did Jesus exist? *American Atheist Magazine*. Retrieved May 15, 2024, from https://www.atheists.org/activism/resources/did-jesus-exist/
566. Zuckerman, M., et al. (2013). The relation between intelligence and religiosity: A meta-analysis and some proposed explanations. *Personality and Social Psychology Review, 17(4)*, 325-354. doi: 10.1177/1088868313497266.

ABOUT THE AUTHOR

Dr. Corey D. Fox is a psychologist, specializing in human learning and conditioning, behavior modification, and human health and neuropsychology. He has served on the faculty of several medical and graduate schools, with numerous scientific and professional publications and extensive teaching history, throughout the United States. Dr. Fox has served as a consultant for business, professional, government, and healthcare organizations. As a board-certified health psychologist, he has helped to develop treatment guidelines for people with chronic intractable pain, behavioral health problems, and disabilities. He directed a number of successful clinical programs for people suffering from chronic pain and related disorders. Dr. Fox has observed how religion affects people, and he has been able to help them better manage their suffering and recover their lives without reliance on religion.

www.ingramcontent.com/pod-product-compliance
Lightning Source LLC
Chambersburg PA
CBHW071407230926
51846CB00001B/28
* 9 7 8 1 9 6 3 1 1 7 9 2 9 *